Ergebnisse der Mathematik und ihrer Grenzgebiete

Volume 52

3. Folge

A Series of Modern Surveys in Mathematics

Chris A. M. Peters · Joseph H. M. Steenbrink

Mixed Hodge Structures

Springer

Prof. Dr. Chris A. M. Peters
Université de Grenoble I
Institut Fourier
UFR de Mathématiques
100 rue de Maths
38402 Saint-Martin d'Hères
France
chris.peters@ujf-grenoble.fr

Prof. Dr. Joseph H. M. Steenbrink
Radboud University Nijmegen
Department of Mathematics
Toernooiveld 1
6525 ED Nijmegen
Netherlands
J.Steenbrink@math.ru.nl

ISBN 978-3-540-77015-2 e-ISBN 978-3-540-77017-6

DOI 10.1007/978-3-540-77017-6

Ergebnisse der Mathematik und ihrer Grenzgebiete. 3. Folge / A Series of Modern Surveys in Mathematics ISSN 0071-1136

Library of Congress Control Number: 2007942592

Mathematics Subject Classification (2000): 14C30, 14D07, 32C38, 32G20, 32S35, 58A14

Typesetting: by the author using a Springer TEX macro package
Production: LE-TEX Jelonek, Schmidt & Vöckler GbR, Leipzig
Cover design: WMX Design GmbH, Heidelberg

Printed on acid-free paper

9 8 7 6 5 4 3 2 1

springer.com

Preface

The text of this book has its origins more than twenty-five years ago. In the seminar of the Dutch Singularity Theory project in 1982 and 1983, the second-named author gave a series of lectures on Mixed Hodge Structures and Singularities, accompanied by a set of hand-written notes. The publication of these notes was prevented by a revolution in the subject due to Morihiko Saito: the introduction of the theory of Mixed Hodge Modules around 1985. Understanding this theory was at the same time of great importance and very hard, due to the fact that it unifies many different theories which are quite complicated themselves: algebraic D-modules and perverse sheaves.

The present book intends to provide a comprehensive text about Mixed Hodge Theory with a view towards Mixed Hodge Modules. The approach to Hodge theory for singular spaces is due to Navarro and his collaborators, whose results provide stronger vanishing results than Deligne's original theory. Navarro and Guillén also filled a gap in the proof that the weight filtration on the nearby cohomology is the right one. In that sense the present book corrects and completes the second-named author's thesis.

Many suggestions and corrections to this manuscript were made by several colleagues: Benoît Audoubert, Alex Dimca, Alan Durfee, Alexey Gorinov, Dick Hain, Theo de Jong, Rainer Kaenders, Morihiko Saito, Vasudevan Srinivas, Duco van Straten, to mention a few. Thanks to all of you!

During the preparation of the manuscript the authors received hospitality and support from the universities of Grenoble and Nijmegen. Moreover, we thank Annie for providing us excellent working conditions at Veldhoven.

Grenoble/Nijmegen, August 2007 — Chris Peters, Joseph Steenbrink

Contents

Part IV Hodge Structures and Local Systems

Part V Appendices

Introduction

Brief History of the Subject

One can roughly divide the history of mixed Hodge theory in four periods; the period up to 1967, the period 1967–1977, the period 1977–1987, the period after 1987.

The **first period** could be named *classical.* The "prehistory" consists of work by Abel, Jacobi, Gauss, Legendre and Weierstrass on the periods of integrals of rational one-forms. It culminates in Poincaré's and Lefschetz's work, reported on in Lefschetz's classic monograph [Lef]. The second landmark in the classical era proper is Hodge's decomposition theorem for the cohomology of a compact Kähler manifold [Ho47]. To explain the statement, we begin by noting that a complex manifold always admits a hermitian metric. As in differential geometry one wants to normalise it by choosing holomorphic coordinates in which the metric osculates to second order to the constant hermitian metric. This turns out not be always possible and one reserves for such a special metric the name *Kähler metric.* The existence of such a metric implies that the decomposition of complex-valued differential forms into type persists on the level of cohomology classes. We recall here that a complex form α has type (p,q), if in any local system of holomorphic coordinates $(z_1,\ldots,z_n)$, the form α is a linear combinations of forms of the form (differentiable function)$\cdot(dz_{i_1}\wedge\cdots\wedge dz_{i_p}\wedge\overline{dz}_{j_1}\wedge\cdots\wedge\overline{dz}_{j_q})$. Indeed, Hodge's theorem (See Theorem 1.8) states that this induces a decomposition

$$H^m(X;\mathbb{C}) = \bigoplus_{p+q=m} H^{p,q}(X), \qquad \text{(HD)}$$

where the term on the right denotes cohomology classes representable by closed forms of type (p,q). The space $H^{p,q}(X)$ is the complex conjugate of $H^{q,p}$, where the complex conjugation is taken with respect to the real structure given by $H^m(X;\mathbb{C}) = H^m(X;\mathbb{R})\otimes_{\mathbb{R}}\mathbb{C}$. A decomposition (HD) with this reality constraint by definition is the prototype of *a weight m Hodge structure.*

The Hodge decomposition fails in general, as demonstration by the Hopf manifolds, complex m-dimensional manifolds homeomorphic to $S^1 \times S^{2m-1}$. Indeed H^1 being one-dimensional for these manifolds, one can never have a splitting $H^1 = H^{1,0} \oplus H^{0,1}$ with the second subspace the complex conjugate of the first. It follows that complex manifolds do not always admit Kähler metrics. A complex manifold which does admit such a metric is called a *Kähler manifold*. Important examples are the complex projective manifolds: the Fubini-Study metric (Examples 1.5.2) on projective space is Kähler and restricts to a Kähler metric on every submanifold.

It is not hard to see that the fundamental class of a complex submanifold of a Kähler manifold is of pure type (c, c), where c is the codimension (Prop. 1.14). This applies in particular to submanifolds of complex projective manifolds. By the GAGA-principle these are precisely the algebraic submanifolds. Also singular codimension c subvarieties can be shown to have a fundamental class of type (c, c), and by linearity, so do cycles: finite formal linear combinations of subvarieties with integral or rational coefficients. Hodge's famous conjecture states that, conversely, any rational class of type (c, c) is the fundamental class of a rational cycle of codimension c. This conjecture, stated in [Ho50], is one of the millennium one-million dollar conjectures of the Clay-foundation and is still largely open.

The **second period** starts in the late 1960's with the work of Griffiths [Grif68, Grif69] which can be considered as neo-classical in that this work goes back to Poincaré and Lefschetz. In the monograph [Lef], only weight one Hodge structures depending on parameters are studied. In Griffiths's terminology these are weight one *variations of Hodge structure*. Indeed, in the cited work of Griffiths this notion is developed for any weight and it is shown that there are remarkable differences with the classical weight one case. For instance, although the ordinary Jacobian is a polarized abelian variety, their higher weights equivalents, the intermediate Jacobians, need not be polarized. Abel-Jacobi maps generalize in this set-up (see § 7.1.2) and Griffiths uses these in [Grif69] to explain that higher codimension cycles behave fundamentally different than divisors.

All these developments concern smooth projective varieties and cycles on them. For a not necessarily smooth and/or compact complex algebraic variety the cohomology groups cannot be expected to have a Hodge decomposition. For instance H^1 can have odd rank. Deligne realized that one could generalize the notion of a Hodge structure to that of a mixed Hodge structure. There should be an increasing filtration, the *weight filtration*, so that m-th graded quotient has a pure Hodge structure of weight m. This fundamental insight has been worked out in [Del71, Del74].

Instead of looking at the cohomology of a fixed variety, one can look at a family of varieties. If the family is smooth and projective all fibres are complex projective and the cohomology groups of a fixed rank m assemble to give the prototype of a *variation of weight m Hodge structure*. An important observation at this point is that giving a Hodge decomposition (HD) is equivalent to giving a *Hodge filtration*

$$F^p H^m(X;\mathbb{C}) := \bigoplus_{r\geq p} H^{r,s}(X), \quad F^p \oplus \overline{F}^{m-p+1} = H^m(X;\mathbb{C}), \qquad \text{(HF)}$$

where the last equality is the defining property of a Hodge filtration. The point here is that the Hodge filtration varies holomorphically with X while the subbundles $H^{p,q}(X)$ in general don't.

If the family acquires singularities, one may try to see how the Hodge structure near a singular fibre degenerates. So one is led to a *one-parameter degeneration* $X \to \Delta$ over the disk Δ, where the family is smooth over the punctured disk $\Delta^* = \Delta - \{0\}$. So for $t \in \Delta^*$ cohomology group $H^m(X_t;\mathbb{C})$ has a classical weight m Hodge structure. In order to capture the degeneration Hodge theoretically this classical structure has to be replaced by a *mixed* Hodge structure, the so-called *limit mixed Hodge structure*. Griffiths conjectured in [Grif70] that the monodromy action defines a weight filtration which together with a certain limiting Hodge filtration should give the correct mixed Hodge structure. Moreover, this mixed Hodge structure should reveal restrictions on the monodromy action, and notably should imply a local invariant cycle theorem: all cohomology classes in a fibre which are invariant under monodromy are restriction from classes on the total space. In the algebraic setting this was indeed proved by Steenbrink in [Ste76]. Clemens [Clem77] treated the Kähler setting, while Schmid [Sch73] considered abstract variations of Hodge structure over the punctured disk. We should also mention Varchenko's approach [Var80] using asymptotic expansions of period integrals, and which goes back to Malgrange [Malg74].

The **third period**, is a period of on the one hand consolidation, and on the other hand widening the scope of application of Hodge theory. We mention for instance the extension of Schmid's work to the several variables [C-K-S86] which led to an important application to the Hodge conjecture [C-D-K]. In another direction, instead of varying Hodge structures one could try to enlarge the definition of a variation of Hodge structure by postulating a second filtration, the weight filtration which together with the Hodge filtration (HF) on every stalk induces a mixed Hodge structure. Indeed, this leads to what is called a *variation of mixed Hodge structure*. On the geometric side, the fibre cohomology of families of possible singular algebraic varieties should give such a variation, which for obvious reasons is called "geometric". These last variations enjoy strong extra properties, subsumed in the adjective *admissible*. Their study has been started by Steenbrink and Zucker [St-Z, Zuc85], and pursued by Kashiwara [Kash86].

On the abstract side we have Carlson's theory [Car79, Car85b, Car87] of the *extension classes* in mixed Hodge theory, and the related work by Beilinson on *absolute Hodge cohomology* [Beil86]. Important are also the *Deligne-Beilinson cohomology groups*; these can be considered as extensions in the category of pure Hodge complexes and play a central role in unifying the classical class map and the Abel-Jacobi map. For a nice overview see [Es-V88].

Continuing our discussion of the foundational aspects, we mention the alternative approach [G-N-P-P] to mixed Hodge theory on the cohomology of a singular algebraic variety. It is based on cubical varieties instead of simplicial varieties used in [Del74]. See also [Car85a].

In this period a start has been made to put mixed Hodge structures on other geometric objects, in the first place on *homotopy groups* for which Morgan found the first foundational results [Mor]. He not only put a mixed Hodge structure on the higher homotopy groups of complex algebraic manifolds, but showed that the minimal model of the Sullivan algebra for each stage of the rational Postnikov tower has a mixed Hodge structure. The fundamental group being non-abelian a priori presents a difficulty and has to be replaced by a suitably abelianized object, the De Rham fundamental group. Morgan relates it to the 1-minimal model of the Sullivan algebra which also is shown to have a mixed Hodge structure. In [Del-G-M-S] one finds a striking application to the formality of the cohomology algebra of Kähler manifolds. For a further geometric application see [C-C-M]. Navarro Aznar extended Morgan's result to possibly singular complex algebraic varieties [Nav87]. Alternatively, there is Hain's approach [Hain87, Hain87b] based on Chen's iterated integrals. At this point we should mention that the Hurewicz maps, which are natural maps from homotopy to homology, turn out to be morphisms of mixed Hodge structure.

A second important development concerns *intersection homology and cohomology* which is a Poincaré-duality homology theory for singular varieties. The result is that for any compact algebraic variety X the intersection cohomology group $IH^k(X;\mathbb{Q})$ carries a weight k pure Hodge structure compatible with the pure Hodge structure on $H^k(\tilde{X};\mathbb{Q})$ for any desingularization $\pi : \tilde{X} \to X$ in the sense that π^* makes $IH^k(X;\mathbb{Q})$ a direct factor of $IH^k(\tilde{X};\mathbb{Q}) = H^k(\tilde{X};\mathbb{Q})$.

There are two approaches. The first, which still belongs to this period uses L_2-cohomology and degenerating Hodge structures is employed in [C-K-S87] and [Kash-Ka87b]. The drawback of this method is that the Hodge filtration is not explicitly realized on the level of sheaves as in the classical and Deligne's approach. The second method remedies this, but belongs to the next period, since it uses D-modules.

We now come to this last period, the **post D-modules period**. Let us explain how D-modules enter the subject. A variation of Hodge structure with base a smooth complex manifold X in particular consists of an underlying local system $\mathbb{V}$ over X. The associated vector bundle $\mathcal{V} = \mathbb{V} \otimes \mathcal{O}_X$ thus has a canonical flat connection. So one has directional derivatives and hence an action of the sheaf $\mathcal{D}_X$ of germs of holomorphic differential operators on X. In other words, $\mathcal{V}$ is a a $\mathcal{D}_X$-module.

At this point we have a pair $(\mathcal{V}, \mathbb{V})$ consisting of a $\mathcal{D}_X$-module and a local system which correspond to each other. A Hodge module as defined by Saito incorporates a third ingredient, a so called "good" filtration on the $\mathcal{D}_X$-module. In our case this is the Hodge filtration $\mathcal{F}^\bullet$ which for historical reasons is written as as increasing filtration, i.e. one puts $\mathcal{F}_k = \mathcal{F}^{-k}$. The axiom of Griffiths tranversality just means that this filtration is good in the technical sense. The resulting triple $(\mathcal{V}, \mathcal{F}_\bullet, \mathbb{V})$ indeed gives an example of a Hodge module of weight n. It is called a smooth Hodge module. [1]

Saito has developed the basic theory of Hodge modules in [Sa87, Sa88, Sa90]. The actual definition of a Hodge module is complicated, since it is by induction on the dimension of the support. To have a good functorial theory of Hodge modules, one should restrict to polarized variations of Hodge structure and their generalizations the polarized Hodge modules. If we are "going mixed", any polarized admissible variation of mixed Hodge structure over a smooth algebraic base is the prototype of a mixed Hodge module. But, again, the definition of a mixed Hodge module is complex and hard to grasp. Among the successes of this theory we mention the existence of a natural pure Hodge structure on intersection cohomology groups, the unification of the proofs of vanishing theorems, and a nice coherent theory of fundamental classes.

A second important development that took place in this period is the emergence of non-abelian Hodge theory. Classical Hodge theory treats harmonic theory for maps to the abelian group $\mathbb{C}^*$ which governs line bundles: in contrast, non-abelian Hodge theory deals with harmonic maps to non-abelian groups like $\mathrm{GL}(n)$, $n \geq 2$. This point of view leads to so-called *Higgs bundles* which are weaker versions of variations of Hodge structure that come up when one deforms variations of Hodge structure. It has been developed mainly by Simpson, [Si92, Si94, Si95], with contributions of Corlette [Cor]. This work leads to striking limitations on the kind of fundamental group a compact Kähler manifold can have. A similar approach for the mixed situation is still largely missing.

There are many other important developments of which we only mention two. The first concerns the relation of Hodge theory to the logarithmic structures invented by Fontaine, Kato and Illusie, which was studied in [Ste95]. A second topic is mixed Hodge structures on Lawson homology, a subject whose study started in [F-M], but which has not yet been properly pursued afterwards.

Contents of the Book

The book is divided in four parts which we now discuss briefly. The first part, entitled *basic Hodge theory* comprises the first three chapters.

In Chapter 1 in order to motivate the concept of a Hodge structure we give the statement of the Hodge decomposition theorem. Likewise, polarizations are motivated by the Lefschetz decomposition theorem. It has a surprising

[1] If you want such a triple to behave well under various duality operators it turns out to be better to replace $\mathbb{V}$ by a complex placed in degree $-n = -\dim X$ so that it becomes a perverse sheaf. See Chapter 13 for details.

topological consequence: the Leray spectral sequence for smooth projective families degenerates at the E_2-term. In particular, a theorem alluded to in the Historical Part holds in this particular situation: the invariant cycle theorem (cycles invariant under monodromy are restrictions of global cycles).

Chapter 2 explains the basics about pure Hodge theory. In particular the crucial notions of a Hodge complex of weight m and a Hodge complex of sheaves of weight m are introduced. The latter makes Hodge theory local in the sense that if a cohomology group can be written as the hypercohomology groups of a Hodge complex of sheaves, such a group inherit a Hodge structure. This is what happens in the classical situation, but it requires some work to explain it. In the course of this Chapter we are led to make an explicit choice for a Hodge complex of sheaves on a given compact Kähler manifold, the *Hodge-De Rham complex of sheaves* $\underline{\mathbb{Z}}_X^{\mathrm{Hdg}}$. Incorporated in this structure are the *Godement resolutions* which we favour since they behave well with respect to filtrations and with respect to direct images. The definition and fundamental properties are explained in Appendix B.

These abstract considerations enable us to show that the cohomology groups of X can have pure Hodge structure even if X itself is *not* a compact Kähler manifold, but only bimeromorphic to such a manifold. In another direction, we show that the cohomology of a possibly singular V-manifold posses a pure Hodge structure.

The foundations for mixed Hodge theory are laid down in Chapter 3. The notions of Hodge complexes and Hodge complexes of sheaves are widened to mixed Hodge complexes and mixed Hodge complexes of sheaves. The idea is as in the pure case: the construction of a mixed Hodge structure on cohomological objects can be reduced to a local study. Crucial here is the technique of spectral sequences which works well because the axioms imply that the Hodge filtration induces only one filtration on the successive steps in the spectral sequence (Deligne's comparison of three filtrations). Next, the important construction of the cone in the category of mixed Hodge complexes of sheaves is explained. Since relative cohomology can be viewed as a cone this paves the way for mixed Hodge structures on relative cohomology, on cohomology with compact support, and on local cohomology. The chapter concludes with Carlson's theory of extensions of mixed Hodge structures and Beilinson's theory of absolute Hodge cohomology.

The second part of the book deals with *mixed Hodge structures on cohomology groups* and starts with Chapter 4 on smooth algebraic varieties. The classical treatment of the weight filtration due to Deligne is complemented by a more modern approach using logarithmic structures. This is needed in Chapter 11 which deals with variations of Hodge structure.

Chapter 5 treats the cohomology of singular varieties. Instead of Deligne's simplicial approach we explain the cubical treatment proposed by Guillén, Navarro Aznar, Pascual-Gainza and F. Puerta.

The results from Chapter 5 are further extended in Chapter 6 where Arapura's work on the Leray spectral sequence is explained, followed by a treatment of cup and cap products and duality. This chapter ends with an application to the cohomology of two geometric objects, halfway between an algebraic and a purely topological structure: deleted neighbourhoods and links of closed subvarieties of a complex algebraic variety.

In Chapter 7 we give applications of the theory which we developed so far. First we explain the Hodge conjecture as generalized by Grothendieck, secondly we briefly discuss Deligne cohomology and the relation to algebraic cycles. Finally we introduce Du Bois's filtered de Rham complex and give applications to singularities.

The third part is entitled *mixed Hodge structures on homotopy groups.* We first give the basics from homotopy theory enabling to make the transition from homotopy groups to Hopf algebras. Next, we explain Chen's homotopy de Rham theorem and Hain's bar construction on Hopf algebras. These two ingredients are necessary to understand Hain's approach to mixed Hodge theory on homotopy which we give in Chapter 8. The older approach, due to Sullivan and Morgan is explained in Chapter 9.

The fourth and last part is about *local systems in relation to Hodge theory* and starts with the foundational Chapter 10. In Chapter 11 Steenbrink's approach to the limit mixed Hodge structure is explained from a more modern point of view which incorporates Deligne's vanishing and nearby cycle sheaves. The starting point is that the cohomology of any smooth fibre in a one-parameter degeneration can be reconstructed as the cohomology of a particular sheaf on the singular fibre, the nearby cycle sheaf. So a mixed Hodge structure can be put on cohomology by extending the nearby cycle sheaf to a mixed Hodge complex of sheaves on the singular fibre. This is exactly what we do in Chapter 11. Important applications are given next: the monodromy theorem, the local invariant cycle theorem and the Clemens-Schmid exact sequence.

Follows Chapter 12 with applications to singularities (the cohomology of the Milnor fibre and the spectrum), and to cycles (Grothendieck's induction principle).

The fourth part is leading up to Saito's theory which, as we explained in the historical part, incorporates D-modules into Hodge theory through the Riemann-Hilbert correspondence. This is explained in Chapter 13, where the reader can find some foundational material on D-modules and perverse sheaves. In the final Chapter 14 Saito's theory is sketched. In this chapter we axiomatize his theory and directly deduce the important applications we mentioned in the Historical Part. We proceed giving ample detail on how to construct Hodge modules as well as mixed Hodge modules, and briefly sketch how the axioms can be verified. Clearly, many technical details had to be omitted, but we hope to have clarified the overall structure. Many mathematicians consider Saito's formidable work to be rather impenetrable. The final chapter is meant as an introductory guide and hopefully motivates an interested researcher to penetrate deeper into the subject by reading the original articles.

The book ends with three appendices: Appendix A with basics about derived categories, spectral sequences and filtrations, Appendix B where several fundamental results about the algebraic topology of varieties is assembled, and Appendix C about stratifications and singularities.

Finally a word about what is *not* in this book. Due to incompetence on behalf of the authors, we have not treated mixed Hodge theory from the point of view of L_2-theory. Hence we don't say much on Zucker's fundamental work about L_2-cohomology. Neither do we elaborate on Schmid's work on one-parameter degenerations of abstract variations of Hodge structures, apart from the statement in Chapter 10 of some of his main results. In the same vein, the work of Cattani-Kaplan-Schmid on several variables degenerations is mostly absent. We only give the statement of the application of this theory to Hodge loci (Theorem 10.15), the result about the Hodge conjecture alluded to in the Historical Part.

The reader neither finds many applications to singularities. In our opinion Kulikov's monograph [Ku] fills in this gap rather adequately. For more recent applications we should mention Hertling's work, and the work of Douai-Sabbah on Frobenius manifolds and tt^*-structures [Hert03, D-S03, D-S04].

Mixed Hodge theory on Lawson homology is not treated because this falls too far beyond the scope of this book. For the same reason non-abelian Hodge theory is absent, as are characteristic p methods, especially motivic integration, although the motivic nearby and motivic vanishing cycles are introduced (Remark 11.27).

Part I

Basic Hodge Theory

1

Compact Kähler Manifolds

We summarize classical Hodge theory for compact Kähler manifolds and derive some important consequences. More precisely, in § 1.1.1 we recall Hodge's Isomorphism Theorem for compact oriented Riemannian manifolds, stating that in any De Rham-cohomology class one can find a unique representative which is a harmonic form. This powerful theorem makes it possible to check various identities among cohomology classes on the level of forms. By definition a Kähler manifold is a complex hermitian manifold such that the associated metric form is closed and hence defines a cohomology class. The existence of such metrics has deep consequences. In § 1.1.2 and § 1.2.2 we treat this in detail, the highlights being the Hodge Decomposition Theorem and the Hard Lefschetz theorem. Here some facts about representation theory of $\mathrm{SL}(2,\mathbb{R})$ are needed which, together with basic results needed in Chapt. 10, are gathered in § 1.2.1.

1.1 Classical Hodge Theory

1.1.1 Harmonic Theory

Let X be a compact n-dimensional Riemannian manifold equipped with a Riemannian metric g. This is equivalent to giving an inner product on the tangent bundle $T(X)$. So g induces inner products on the cotangent bundle and on its exterior product, the bundles of m-forms

$$\mathcal{E}_X^m := \Lambda^m T(X)^\vee.$$

We denote the induced metrics also by g. We normalize these metrics starting from an orthonormal frame $\{e_1, \dots, e_n\}$ for the cotangent bundle. We then declare that the vectors $\{e_{i_1} \wedge e_{i_2} \wedge \cdots \wedge e_{i_m}\}$ form an orthonormal frame for $\mathcal{E}_X^m$ where the indices range over all strictly increasing m-tuples $\{i_1, i_2, \cdots, i_m\}$ with $i_k \in \{1, \dots, m\}$, $k = 1, \dots, n$.

Assume that X can be oriented, i.e. that there is a global n-form which nowhere vanishes. If we choose the local frames for $\mathcal{E}_X^1$ such that they are compatible with the orientation, a canonical choice for the Riemannian volume form is given by

$$\mathrm{vol}_g = e_1 \wedge e_2 \cdots \wedge e_n.$$

The **Hodge $*$-operators** $\Lambda^m T_x^\vee X \xrightarrow{*} \Lambda^{n-m} T_x^\vee X$ defined by

$$\alpha \wedge *\beta = g(\alpha, \beta)[\mathrm{vol}_g]_x \, \forall \alpha, \beta \in \Lambda^m T_x^\vee X \tag{I–1}$$

induce linear operators on $\mathcal{E}_X^m$. The spaces of global differential forms on X, the **De Rham spaces**

$$E^m_{\mathrm{DR}}(X) := \Gamma(X, \mathcal{E}_X^m)$$

also carry (global) inner products given by

$$(\alpha, \beta) := \int_X g(\alpha, \beta) \,\mathrm{vol}_g = \int_X \alpha \wedge *\beta, \quad \alpha, \beta \in E^m_{\mathrm{DR}}(X).$$

The de **Rham groups** are defined by

$$H^k_{\mathrm{DR}}(X) := H^k\left(E^\bullet_{\mathrm{DR}}(X), d\right).$$

The operator $d^* = (-1)^{nm+1} * d *$ can be shown to be an adjoint of the operator d with respect to this inner product, i.e.,

$$(d\alpha, \beta) = (\alpha, d^*\beta), \quad \alpha, \beta \in E^m_{\mathrm{DR}}(X).$$

Its associated Laplacian is $\triangle_d = dd^* + d^*d$. The m-forms that satisfy the Laplace equation $\triangle_d = 0$ are called **d-harmonic** and denoted

$$\mathrm{Har}^m(X) = \{\alpha \in \Gamma(X, \mathcal{E}_X^m) \mid \triangle_d \alpha = 0\}.$$

The next result, originally proven by Hodge (for a modern proof see e.g. [Dem, § 4] states that any De Rham group, which in fact is a real vector space of *equivalence classes of forms*, can be replaced by the corresponding vector space of harmonic *forms*:

Theorem 1.1 (HODGE'S ISOMORPHISM THEOREM). *Let X be a compact differentiable manifold equipped with a Riemannian metric. Then we have:*

1) $\dim \mathrm{Har}^m(X) < \infty$.
2) *Let*

$$H : E^m_{\mathrm{DR}}(X) \to \mathrm{Har}^m(X)$$

be the orthogonal projection onto the harmonic forms. There is an orthogonal direct sum decomposition

$$E^m_{\mathrm{DR}}(X) = \mathrm{Har}^m(X) \oplus dE^{m-1}_{\mathrm{DR}}(X) \oplus d^* E^{m+1}_{\mathrm{DR}}(X)$$

and H induces an isomorphism

$$H^m_{\mathrm{DR}}(X) \xrightarrow{\cong} \mathrm{Har}^m(X).$$

There is a useful additional statement concerning holonomy groups. To explain what these are we start from the Levi-Civita connection, the unique metric connection without torsion. It defines parallel displacement along curves, and for a closed curve based at $x \in X$ it defines an isometry of the tangent space T_xX. These isometries by definition generate the **holonomy group** $G_x \subset \mathrm{O}(T_xX)$. For a connected X the holonomy groups G_x are abstractly isomorphic, say to $G \subset \mathrm{O}(T)$, for some vector space T isomorphic to T_x. The basic result we need is [Ch]:

Theorem 1.2 (Chern's theorem). *Let (X, g) be a compact connected Riemannian manifold of dimension n. Let $A \in \mathrm{End}(\Lambda T^\vee)$ an operator which on each fiber commutes with the holonomy representation. Then A through its action on $E_{\mathrm{DR}}X$ commutes with the Laplacian Δ and hence preserves the subspace of harmonic forms.*

Next, we assume that X is a complex manifold equipped with a hermitian metric h. Identifying $T(X)$ with the underlying real bundle $T(X)_{\mathrm{hol}}$, the real part $\mathrm{Re}(h)$ of h is a Riemannian metric, while

$$\omega_h := \mathrm{Im}(h)$$

is a real valued skew-form. The almost complex structure J on $T(X)$ preserves this form, which means that it is of type $(1,1)$. To fix the normalization, if in local coordinates h is given by $h = \sum_{j,k} h_{jk} dz_j \otimes d\bar{z}_k$, the associated form is given by

$$\omega_h = \frac{\mathrm{i}}{2} \sum_{j,k} h_{jk} dz_j \wedge d\bar{z}_k.$$

As before, the metric h induces point-wise metrics on the bundles of complex-valued smooth differential forms as well as on each of the bundles of complex-valued (p,q)-forms $\mathcal{E}_X^{p,q}$. The differential $d : \mathcal{E}_X^m(\mathbb{C}) \to \mathcal{E}_X^{m+1}(\mathbb{C})$ splits as $d = \partial + \bar{\partial}$ with $\partial : \mathcal{E}_X^{p,q} \to \mathcal{E}_X^{p+1,q}$, $\bar{\partial} : \mathcal{E}_X^{p,q} \to \mathcal{E}_X^{p,q+1}$.

The volume form associated to h defines then global inner products, the **Hodge inner products** on the spaces of complex valued smooth forms as well as on the spaces of smooth (p,q)-forms. With respect to these metrics we have an orthogonal splitting ([Wells, Chapt. V, Prop. 2.2])

$$\Gamma(\mathcal{E}_X^m(\mathbb{C})) = \bigoplus_{p+q=m} \Gamma(\mathcal{E}_X^{p,q}).$$

The fibre-wise conjugate-linear operator $\bar{*} : \mathcal{E}_X^{p,q} \xrightarrow{\sim} \mathcal{E}_X^{n-q,n-p}$, defined by $\bar{*}(\alpha) = *\bar{\alpha}$ extends to global (p,q)-forms. We also may consider forms with coefficients in a holomorphic vector bundle E equipped with a hermitian metric h_E. The bundle of differentiable E-valued forms of type (p,q) by definition is the bundle

$$\mathcal{E}_X^{p,q}(E) = \mathcal{E}_X^{p,q} \otimes E.$$

The operator $\overline{\partial}_E : \Gamma(\mathcal{E}_X^{p,q}(E)) \to \Gamma(\mathcal{E}_X^{p,q+1}(E))$ given by $\overline{\partial}_E(\alpha \otimes s) = \overline{\partial}\alpha \otimes s$ is well-defined and since $\overline{\partial}_E \circ \overline{\partial}_E = 0$ we obtain a complex $\Gamma(\mathcal{E}_X^{p,\bullet}(E))$.

The Hodge metric on the space of E-valued m-forms is obtained as follows. First choose a conjugate linear isomorphism $\tau : E \to E^\vee$ and define

$$\bar{*}_E : \mathcal{E}_X^{p,q}(E) \to \mathcal{E}_X^{n-q,n-p}(E^\vee)$$

by $\bar{*}_E(\alpha \otimes e) = \bar{*}\alpha \otimes \tau(e)$. Then the **global Hodge inner product** on $\Gamma(\mathcal{E}_X^{p,q}(E))$ is given by

$$(\alpha, \beta) = \int_X \alpha \wedge \bar{*}_E \beta, \quad \alpha, \beta \in \Gamma(\mathcal{E}_X^{p,q}(E)). \tag{I–2}$$

With respect to this metric, one defines the (formal) adjoint $\overline{\partial}_E^*$ of $\overline{\partial}_E$ and the Laplacian $\Delta_{\overline{\partial}_E} := \overline{\partial}_E \overline{\partial}_E^* + \overline{\partial}_E^* \overline{\partial}_E$ with respect to which one computes the harmonic forms $\mathrm{Har}^{p,q}(E)$. We can now state:

Theorem 1.3 (Hodge's isomorphism theorem, second version). *Let X be a compact complex manifold and E be a holomorphic vector bundle. Suppose that both T_X and E are equipped with a hermitian metric. We have:*

1) $\dim \mathrm{Har}^{p,q}(E) < \infty$.
2) *Let*

$$H : \Gamma(X, \mathcal{E}_X^{p,q}(E)) \to \mathrm{Har}^{p,q}(E)$$

be the orthogonal projection onto the harmonic forms. There is a direct sum decomposition

$$\Gamma(\mathcal{E}_X^{p,q}(E)) = \mathrm{Har}^{p,q}(E) \oplus \overline{\partial}_E \Gamma(\mathcal{E}_X^{p,q-1}(E)) \oplus \overline{\partial}_E^* \Gamma(\mathcal{E}_X^{p,q+1}(E))$$

and H induces an isomorphism

$$H_{\overline{\partial}}^{p,q}(E) \xrightarrow{\cong} \mathrm{Har}^{p,q}(E)$$

where

$$H_{\overline{\partial}}^{p,q}(E) := \frac{\overline{\partial}\text{-closed } (p,q)\text{-forms with values in } E}{\overline{\partial}\mathcal{E}^{p,q-1}(E)}.$$

The operator $\bar{*}_E$ commutes with the Laplacian $\Delta_{\overline{\partial}}$ as acting on $\mathcal{E}_X^{p,q}(E)$ and hence harmonic (p,q)-forms with values in E go to harmonic $(n-p, n-q)$-forms with values in $E^\vee$. In particular $\mathrm{Har}^{p,q}(E)$ and $\mathrm{Har}^{n-q,n-p}(E^\vee)$ are conjugate-linearly isomorphic. For reference we state the following classical consequence.

Corollary 1.4 (Serre duality). *The operator $\bar{*}_E$ defines an isomorphism*

$$H^q(X, \Omega_X^p(E)) \xrightarrow{\cong} H^{n-q}(X, \Omega_X^{n-p}(E^\vee))^\vee.$$

1.1.2 The Hodge Decomposition

A hermitian metric h on a complex manifold is called **Kähler** if the associated $(1,1)$-form ω_h is closed. Such a form is called a **Kähler form**. Any manifold equipped with a Kähler metric is called a **Kähler manifold**. It is well known that h is Kähler if and only if there exist local coordinates in which h is the standard euclidean metric up to second order.

Examples 1.5. 1) Any hermitian metric on a Riemann surface is Kähler.

2) The Fubini-Study metric on $\mathbb{P}^n$ is Kähler. We recall the definition. Introduce coordinates $(Z_0, \ldots, Z_n)$ on $\mathbb{C}^{n+1}$ and for $Z \in \mathbb{C}^{n+1}$ put $||Z|| = \sum_{i=0}^{n} |Z_i|^2$. This defines an $\mathbb{C}^*$-invariant $(1,1)$ form on $\mathbb{C}^{n+1}$, the **Fubini-Study form**

$$\omega_{\mathrm{FS}} := \frac{\mathrm{i}}{2\pi} \partial\bar{\partial} \log ||Z||^2$$

and one considers it as a form on $\mathbb{P}^n$. The constant is chosen in such a way that the class of ω_{FS} is the fundamental cohomology class of a hyperplane. So the Kähler class is *integral* in this case.

3) Any submanifold of a Kähler manifold is Kähler. Indeed, the restriction of the Kähler form restricted to the submanifold is a Kähler form on this submanifold. An important special case is formed by the projective manifolds. For these, the restriction of the Fubini-Study form defines an integral Kähler class. More generally, we say that we have a **Hodge metric** whenever the Kähler class is *rational*. Kodaira showed (see e.g. [Wells, Chapter VI]) that a Hodge metric exists if and only if the manifold is projective.

For a complex manifold X the (real) tangent space T_xX, $x \in X$ has a complex structure $J \in \mathrm{End}(T_xX)$. If in addition X is hermitian, it can be shown [Helg, Ch. VIII. §2] that the metric is Kähler if and only if J is parallel with respect to the Levi-Civita connexion. The almost complex structure extends $\mathbb{C}$-linearly to the complex tangent bundle and since J is parallel the splitting $T_xX^\vee \otimes \mathbb{C} = (T_x^{1,0}X)^\vee \oplus (T_x^{0,1}X)^\vee$ into $\pm\mathrm{i}$-eigenspaces is preserved by holonomy. We let

$$\Lambda^{p,q}(T_xX)^\vee := \Lambda^p(T_x^{1,0}X)^\vee \otimes \Lambda^q(T_x^{0,1}X)^\vee$$

be the vector space of (p,q)-covectors at $x \in X$. On these spaces J acts as multiplication by i^{p-q}; they are likewise preserved by holonomy. Another consequence of J being parallel is that the hermitian structure on the tangent space T_xX is preserved by the holonomy group G_x:

Lemma 1.6. *Let (X,h) be a Kähler manifold. Then the holonomy group G_x is contained in $\mathrm{U}(T_xX) \simeq \mathrm{U}(n)$, $n = \dim_{\mathbb{C}} X$.*

At this point we recall that the hermitian metric on the tangent space induces hermitian metrics on the associated vector spaces of covectors; the spaces of covectors of different type are mutually orthogonal. The value at x of the Kähler form is an invariant covector. Hence:

Lemma 1.7. *The group* $\mathrm{U}(n)$ *acts on each of the spaces* $\Lambda^{p,q}(T_xX)^\vee$. *The operator "multiplication with the Kähler form" commutes with this action.*

A central result is the following theorem.

Theorem 1.8 (HODGE DECOMPOSITION THEOREM). *Let* X *be a compact Kähler manifold. Let* $H^{p,q}(X)$ *be the space of cohomology classes whose harmonic representative is of type* (p,q). *There is a direct sum decomposition*

$$H^m_{\mathrm{DR}}(X) \otimes \mathbb{C} = \bigoplus_{p+q=m} H^{p,q}(X).$$

Moreover $H^{p,q}(X) = \overline{H^{q,p}(X)}$.

Proof. Since the holonomy group preserves the (p,q)-covectors, the projections $\Lambda^{p+q}(T_xX \otimes \mathbb{C})^\vee \to \Lambda^{p,q}(T_xX)^\vee$ commute with holonomy and hence, by Chern's theorem 1.2, the (p,q)-components of a complex harmonic form remain harmonic. This proves the theorem. □

The Hodge decomposition is in fact independent of the chosen Kähler metric. This is seen by comparing it with the **Bott-Chern** cohomology groups

$$H^{p,q}_{\mathrm{BC}}(X) := \frac{d\text{-closed forms of type } (p,q)}{\partial\bar\partial \Gamma \mathcal{E}_X^{p-1,q-1}}.$$

We have ([Dem, Lemma 8.6])

Lemma 1.9 ($\partial\bar\partial$-LEMMA). *For a d-closed* (p,q)*-form* α *the following statements are equivalent:*

a) $\alpha = d\beta$ *for some* $p+q-1$*-form* β;
b) $\alpha = \bar\partial\beta''$ *for some* $(p,q-1)$*-form* β'';
c) $\alpha = \partial\bar\partial\gamma$ *for some* $(p-1,q-1)$*-form* γ;
d) α *is orthogonal to the harmonic* (p,q)*-forms.*

Corollary 1.10. *The natural morphism*

$$H^{p,q}_{\mathrm{BC}}(X) \to H^{p+q}_{\mathrm{DR}}(X) \otimes \mathbb{C}$$

which sends the class of a d-closed (p,q)*-form to its De Rham class is injective with image* $H^{p,q}(X)$. *Therefore the latter subspace is independent of the Kähler metric: it consists precisely of the De Rham classes representable by a closed form of type* (p,q).

Proof. The injectivity follows from the equivalence b)⇔c). Any class whose harmonic representative α is of type (p,q) is the image of the Bott-Chern class of α. □

Corollary 1.11. *Let* $f : X \to Y$ *be a holomorphic map between compact Kähler manifolds. Then* $f^* : H^m_{\mathrm{DR}}(Y) \otimes \mathbb{C} \to H^m_{\mathrm{DR}}(X) \otimes \mathbb{C}$ *maps* $H^{p,q}(Y)$ *to* $H^{p,q}(X)$ $(p+q=m)$.

1.1.3 Hodge Structures in Cohomology and Homology

We now introduce the following fundamental concepts.

Definition 1.12. A **Hodge structure of weight** k on a $\mathbb{Z}$-module $V_{\mathbb{Z}}$ of finite rank is a direct sum decomposition

$$V_{\mathbb{C}} := \mathbb{V}_{\mathbb{Z}} \otimes_{\mathbb{Z}} \mathbb{C} = \bigoplus_{p+q=k} V^{p,q} \quad \text{with } V^{p,q} = \overline{V^{q,p}}.$$

The numbers

$$h^{p,q}(V) := \dim V^{p,q}$$

are the **Hodge numbers** of the Hodge structure.

Let $V_{\mathbb{Z}}$ and $W_{\mathbb{Z}}$ be two Hodge structures of weight k. A **morphism of Hodge structures** $f : V_{\mathbb{Z}} \to W_{\mathbb{Z}}$ is a homomorphism of $\mathbb{Z}$-modules such that its complexification $f_{\mathbb{C}}$ preserves types: $f_{\mathbb{C}} : V^{p,q} \to W^{p,q}$.

With these definitions at hand, here is a concise reformulation of the preceding results:

Corollary 1.13. *Let X be a compact Kähler manifold. Then the integral cohomology group $H^k(X)$ carries a weight k Hodge structure. If $f : X \to Y$ is a holomorphic map between compact Kähler manifolds, then $f^* : H^k(Y) \to H^k(X)$ is a morphism of Hodge structures.*

Proof. We use the De Rham isomorphism (B–17) to identify $H^k(X; \mathbb{C}) = H^k(X) \otimes \mathbb{C}$ with $H^m_{\mathrm{DR}}(X) \otimes \mathbb{C}$ and then invoke Theorem 1.8. The last assertion follows by functoriality and Cor. 1.11 . □

Other basic examples of Hodge structures are the **Hodge structures of Tate** $\mathbb{Z}(m)$:

$$\mathbb{Z}(m) = [2\pi \mathrm{i}]^m \mathbb{Z} \subset \mathbb{C}, \quad \mathbb{Z}(m) \otimes \mathbb{C} = [\mathbb{Z}(m) \otimes \mathbb{C}]^{-m,-m}. \tag{I–3}$$

If we have any Hodge structure $V_{\mathbb{Z}}$ of weight k, the **Tate twist** $V(m)$ is a Hodge structure of weight $k - 2m$. It has $V \otimes (2\pi \mathrm{i})^m$ as underlying $\mathbb{Z}$-module, while $V(m)^{p,q} = V^{p-m.q-m}$.

Fundamental examples are provided by the fundamental classes of subvarieties as we now explain. Consider a subvariety Y of a compact complex manifold X of codimension c. Integration of smooth forms of degree $2(\dim X - c)$ over Y defines the integration current. The class of this current in cohomology with complex coefficients in fact comes from the topologically defined fundamental cohomology class $\mathrm{cl}(Y) \in H^{2c}(X)$ (see Remark B.31). Let $y \in Y$ be a smooth point and choose coordinates such that Y is locally given by the vanishing of c of the coordinates. The restriction to Y of a (p,q)-form with $p + q = 2n - 2c$ necessarily vanishes around y if either $p > n - c$ or $q > n - c$, since such a form involves differentials of more than $n - c$ coordinates (or conjugates thereof). We conclude

Proposition 1.14. *The fundamental cohomology class* $\mathrm{cl}(Y) \in H^{2c}(X)$ *of a codimension* c *subvariety* Y *of a compact complex manifold* X *has pure type* (c, c). *In particular, if* X *is connected, the twisted trace map (see* (B–38) *in Appendix B.2.8 for the untwisted version) is an isomorphism*

$$\mathrm{tr}_X : H^{2n}(X) \xrightarrow{\simeq} \mathbb{Z}(-n)$$
$$[\alpha] \mapsto \left[\frac{1}{2\pi \mathrm{i}}\right]^n \int_X \alpha.$$

An element $\sum n_i Y_i$ of the free group $Z_k(X)$ on k-dimensional subvarieties of X is called an **algebraic** k**-cycle**. Conventionally, the codimension is used as an upper-index so that $Z^c(X) = Z_{n-c}(X)$ The assignment $\sum n_i Y_i \mapsto \sum_i n_i \,\mathrm{cl}(Y_i)$ defines the *cycle class map*

$$\mathrm{cl} : Z^c(X) \to H^{2c}(X).$$

An element in the image is called an *algebraic class*. If X is projective and $c = 1$, the algebraic classes are exactly the integral $(1, 1)$-classes (Lefschetz' theorem on $(1, 1)$-classes).

Remark 1.15. It is not true in general that all integral classes of pure type (c, c) are in the image of the class map. In fact, for $c > 1$, using a construction of Serre, Atiyah and Hirzebruch [At-Hir] have given examples of classes of finite order in $H^{2c}(X; \mathbb{Z})$ which are not algebraic. Going over to rational classes, we put

$$H^{2c}_{\mathrm{Hdg}}(X) := H^{c,c}(X) \cap \mathrm{Im}\{H^{2c}(X; \mathbb{Q}) \to H^{2c}(X; \mathbb{C})\}. \tag{I–4}$$

The following celebrated conjecture still is open.

Conjecture 1.16 (Hodge conjecture). *Let* X *be a smooth projective variety. Every* (c, c)*-class with rational coefficients is algebraic, i.e. every class in* $H^{2c}_{\mathrm{Hdg}}(X)$ *is a rational combination of fundamental cohomology classes of subvarieties of* X.

Remark. For a compact Kähler manifold X the Hodge conjecture fails, even if we replace the definition of an algebraic class by a rational combination of Chern classes of holomorphic vector bundles. See [Vois02].

Using the isomorphism

$$H_k(X, \mathbb{C}) \xrightarrow{\cong} \mathrm{Hom}(H^k(X), \mathbb{C})$$

we define a Hodge decomposition on $H_k(X, \mathbb{C})$:

Definition 1.17.

$$H_k(X, \mathbb{C})^{-p,-q} = \{\phi : H^k(X, \mathbb{C}) \to \mathbb{C} \mid \phi(H^{r,s}\,(X)) = 0 \text{ whenever } (r, s) = (p, q)\}.$$

This endows $H_k(X)$ with a Hodge structure of weight $-k$. If $f : X \to Y$ is a holomorphic map between compact Kähler manifolds, the induced maps in homology preserve this Hodge structure.

Note that we get a Hodge structure of weight $k+\ell$ on $H^k(X)\otimes H^\ell(X)$ by declaring that $H^{p,q}(X)\otimes H^{r,s}(X)$ has type $(p+r,q+s)$. It follows that the cup product map

$$H^k(X)\otimes H^\ell(X)\to H^{k+\ell}(X)$$

is a morphism of Hodge structures. Similarly, we get a Hodge structure of weight $k-\ell$ on $H^k(X)\otimes H_\ell(X)$ and the cap product

$$H^k(X)\otimes H_\ell(X)\to H^{\ell-k}(X)$$

is a morphism of Hodge structures.

Proposition 1.18. *The Poincaré isomorphism (B.24)*

$$D_X : H^{2n-k}(X)(n)\xrightarrow{\cong} H_k(X)$$

is an isomorphism of Hodge structures.

Proof. The map D_X is cap-product with the fundamental class in homology, which has type $(-n,-n)$, i.e. it factors as

$$H^{2n-k}(X)\otimes\mathbb{Z}(n)\xrightarrow{\cong} H^{2n-k}(X)\otimes H_{2n}(X)\xrightarrow{\cap} H_k(X)$$

which is the composition of two morphisms of Hodge structures. □

Remark. One could have used Poincaré duality to put a Hodge structure on homology. One the one hand this seems more natural, since one can work directly with integral structures. On the other hand, one needs a Tate twist, and, more seriously, for singular varieties there is no Poincaré-duality whereas the approach we have chosen remains valid. See also § 6.3.1.

Recall (B–41) that Gysin maps are induced by the maps in homology after applying the Poincaré duality isomorphisms so that the foregoing implies:

Lemma 1.19. *Let $f : X\to Y$ be a holomorphic map between compact Kähler manifolds. Let $n=\dim X$ and $m=\dim Y$. The twisted Gysin map $f_! : H^{-k+2n}(X)(n)\to H^{-k+2m}(Y)(m)$, $k=0,\dots,2n$ is of pure type* $(0,0)$.

As a side remark, the Gysin map in real cohomology can also be defined using the formulas (B–40). This yields:

Addendum 1.20. *The formula*

$$\left(\frac{1}{2\pi \mathrm{i}}\right)^n\int_X a\cup f^*b=\left(\frac{1}{2\pi \mathrm{i}}\right)^m\int_Y f_!a\cup b, \qquad \text{(I–5)}$$
$$a\in H_{\mathrm{DR}}^{-k+2n}(X)(n),\quad b\in H^k_{\mathrm{DR}}(Y)$$

determines the twisted Gysin map uniquely (up to torsion) in terms of cohomology.

1.2 The Lefschetz Decomposition

1.2.1 Representation Theory of SL(2, ℝ)

A basis for the Lie algebra $\mathfrak{sl}(2,\mathbb{R})$ is [1]

$$\ell = \begin{pmatrix} 0 & 0 \\ 1 & 0 \end{pmatrix}, \quad \lambda = \begin{pmatrix} 0 & 1 \\ 0 & 0 \end{pmatrix}, \quad b = \begin{pmatrix} -1 & 0 \\ 0 & 1 \end{pmatrix}.$$

The commutators are given by

$$[\ell, \lambda] = b, \quad [\ell, b] = -2\ell, \quad [\lambda, b] = 2\lambda.$$

To give a representation of $\mathfrak{sl}(2,\mathbb{R})$ in some real vector space V we need to have three linear maps L, Λ, B of V which satisfy the same commutator relations. This can be elegantly be rephrased as follows.

Lemma 1.21. *Let $V = \bigoplus_{k\in\mathbb{Z}} V^k$ be a finite dimensional graded real vector space. Let L be a degree 2 endomorphism, Λ a degree -2 endomorphism such that V^k is an eigenspace for $B := [L, \Lambda]$ with eigenvalue k. Then there is a unique Lie-algebra morphism $\rho : \mathfrak{sl}(2,\mathbb{R}) \to \operatorname{End} V$ for which $\rho(\ell) = L$, $\rho(\lambda) = \Lambda$, $\rho(b) = [L, \Lambda]$.*

We need a few more facts about representations of $\mathrm{SL}(2,\mathbb{R})$ and its Lie-algebra $\mathfrak{sl}(2,\mathbb{R})$. The standard representation ρ_m of $\mathrm{SL}(2,\mathbb{R})$ is defined to be that of the vector space P_m of homogeneous polynomials of degree m with g acting as $P \mapsto P \circ g^{-1}$. It is irreducible. If $X \in \mathfrak{sl}(2,\mathbb{R})$, we have $d\rho_m(X)P = \frac{d}{dt}(P \circ \exp(-tX))\Big|_{t=0}$. This gives an irreducible Lie-algebra representation. Explicitly:

$$\left\{ \begin{array}{lll} d\rho_m(\ell) & = & -y\partial_x, \qquad d\rho_m(\lambda) = -x\partial_y, \\ d\rho_m(b)x^a y^{m-a} & = & (m-2a)x^a y^{m-a}. \end{array} \right\} \tag{I–6}$$

Since $\mathrm{SL}(2,\mathbb{R})$ is a connected Lie group the standard representation is the unique representation of $\mathrm{SL}(2,\mathbb{R})$ which gives the Lie-algebra representation (I–6).

Let $\rho : \mathfrak{s}l(2,\mathbb{R}) \to \operatorname{End} V$ be a real representation and *define*

$$L := \rho(\ell), \quad \Lambda := \rho(\lambda), \quad B := [L, \Lambda]. \tag{I–7}$$

The action of Λ gives rise to the **primitive subspace**:

$$V_{\mathrm{prim}} := \operatorname{Ker} \Lambda.$$

For the standard representation $V = P_m$, the primitive space is the 1-dimensional space $V^{-m} = \mathbb{R}x^m$, the eigenspace for B with eigenvalue $-m$. Applying successive powers of L we get all of P_m. In fact

---

[1] In most reference books, e.g. [Wells] the ordered basis $\{\lambda, \ell, [\lambda, \ell]\}$ is used.

$$L^a(x^m) = (-1)^a \frac{m!}{a!} x^{m-a} y^a \tag{I–8}$$

is an eigenvector for B with eigenvalue $2a - m$. Also, for all $r \geq 0$ the map L^r sends V^{-r} isomorphically onto V^r.

Any representation of $\mathrm{SL}(2,\mathbb{C})$ is completely reducible. This is well known. See for instance [Wells, Chap. V, Coroll. 3.3]. This remains true for real representations of $\mathrm{SL}(2,\mathbb{R})$. We show this below (Corollary 1.24) and this explains why the properties we just discussed for the standard representation of $\mathfrak{sl}(2,\mathbb{R})$ hold in general. We turn this around by first proving them:

Lemma 1.22. *Let V be a finite dimensional representation of $\mathfrak{sl}(2,\mathbb{R})$. We use the notation* (I–7)*. Moreover we let V^μ be the generalized eigenspace for B with eigenvalue μ. The following assertions hold:*

a) *L and Λ are nilpotent and B is diagonalisable; V splits as a direct sum $V = \bigoplus_\mu V^\mu$ of eigenspaces for B and $LV^\mu \subset V^{\mu+2}$ and $\Lambda V^\mu \subset V^{\mu-2}$.*

b) *If $V \neq 0$, at least one eigenspace space V^μ for B contains primitive vectors.*

c) *Let $w \in V_{\text{prim}}$ a* primitive eigenvector *for B with eigenvalue μ. Then μ is a non-negative integer. We have*

$$\Lambda(L^k w) = k(-\mu - k + 1) L^{k-1} w. \tag{I–9}$$

If m is chosen such that $L^m w \neq 0$, but $L^{m+1} = 0$, then $\mu = -m$ and the subspace of V spanned by the $L^k w$ is an irreducible $\mathfrak{sl}(2,\mathbb{R})$-module isomorphic to P_m.

d) *Setting*

$$V^\mu_{\text{prim}} = V_{\text{prim}} \cap V^\mu,$$

there is a direct sum decomposition $V_{\text{prim}} = \bigoplus_{m \in \mathbb{N}} V^{-m}_{\text{prim}}$. We have

$$\begin{cases} L^r : V^{-m}_{\text{prim}} \hookrightarrow V^{-m+2r}, & r \leq m \\ L^r \, | V^{-m}_{\text{prim}} = 0, & r > m. \end{cases}$$

e) *We have* ***primitive decompositions***

$$V = \bigoplus_{r \in \mathbb{N}} L^r V_{\text{prim}},$$
$$V^\mu = \bigoplus_{r \in \mathbb{N}, r \geq \mu} L^r \left(V^{\mu-2r}_{\text{prim}} \right)$$

and L^r maps $V^{\mu-2r}_{\text{prim}}$ bijectively onto its image. In fact,

$$V = \bigoplus_{m \in \mathbb{N}} \left(V^{-m}_{\text{prim}} \oplus \cdots \oplus L^m V^{-m}_{\text{prim}} \right)$$
$$= \bigoplus_{m \in \mathbb{N}} W_m \otimes V^{-m}_{\text{prim}},$$

with W_m isomorphic to the standard representation P_m and

$$L^m : V^{-m} \xrightarrow{\sim} V^m.$$

Proof. a) The Lie-algebra homomorphisms $\mathrm{ad}(\ell)$ and $\mathrm{ad}(\lambda)$ are obviously nilpotent, while $\mathrm{ad}(b)$ is semi-simple. It then follows from general theory of representations of semi-simple Lie-algebras (such as $\mathfrak{sl}(2, \mathbb{R})$) that L and Λ are nilpotent, while B is semi-simple. See for instance [Se65, Theorem 5.7]. It follows that V^μ is an eigenspace for B so that the direct sum decomposition follows. Since

$$\begin{aligned} BLv &= LBv + [B, L]v = L(\mu v) + 2Lv = (\mu + 2)Lv, \\ B\Lambda v &= \Lambda Bv + [B, \Lambda]v = \Lambda(\mu v) - 2\Lambda v = (\mu - 2)\Lambda v, \end{aligned}$$

the map L sends V^μ to $V^{\mu+2}$ and Λ sends V^μ to $V^{\mu-2}$.

b) Suppose that $v \in V^\mu$ is a non-zero vector. Since there cannot be an infinity of eigenvectors with different eigenvalues for B, there must be a finite string of vectors $v, \Lambda v, \ldots, \Lambda^k v \neq 0, \Lambda^{k+1} v = 0$ so that $\Lambda^k v$ is primitive.

c) Set $w_k = L^k w$; then by a) we have $Bw_k = (\mu + 2k)w_k$ while by definition $Lw_k = w_{k+1}$. Hence

$$\begin{aligned} \Lambda w_k = \Lambda L^k w &= L^k \Lambda w - \sum_{0 \leq j \leq k-1} L^{k-j-1}[L, \Lambda] L^j w \\ &= 0 - \sum L^{k-j-1} B L^j w = - \sum_{0 \leq j \leq k-1} (\mu + 2j) L^{k-1} w \\ &= k(-\mu - k + 1) w_{k-1}. \end{aligned}$$

Applying this for $k = m + 1$ (so that $w_{m+1} = 0$), one sees that $\mu = -m \leq 0$. Finally, lemma 1.21 implies that the real vector space W spanned by the w_k defines a real representation isomorphic to P_m.

d) The relation $[B, \Lambda] = -2\Lambda$ shows that V_{prim} is preserved by B; the asserted direct sum decomposition follows from this and assertion a). Using c) we easily calculate that $\Lambda^s \circ L^r$ acts as multiplication by $(s!)r(r-1)\cdots(r-s+1)$ on V^{-r}_{prim}. The assertions bout L^r then follow.

e) Let $x \in V^r$ and suppose that $\Lambda^{s+1} x = 0$. Then $y = \Lambda^s x \in V_{\mathrm{prim}}^{-r-2s}$ and from $\Lambda^s \circ L^s y = (s!)y$ we deduce that $x' = x - (1/s!)L^s y$ belongs to the kernel of $\Lambda^s | V^{-r}$. Continuing the argument with x' we inductively find an expression

$$x = x_0 + Lx_1 + Lx_2 + \cdots + L^s x_s, \quad x_j \in V_{\mathrm{prim}}^{r-2j}.$$

This expression is unique: if $x = 0$ and j is the largest integer for which $L^j x_j \neq 0$, then, by d) $j \geq r$ and if we apply L^{j-r} to both sides, d) also implies $0 = L^{2j-r} x_j$ and hence $x_j = 0$ contrary to our assumption. This shows that we have a primitive direct sum decomposition of V^μ. The assertion of the weights follow since primitive weight vectors have negative weights and then the assertion about L is a consequence of d). The last assertion of c) implies the last assertion of e). □

Corollary 1.23. *Let $0 \neq x \in V^\mu$. Then x is primitive if and only if both $\mu \leq 0$ and $L^{\mu+1}x = 0$.*

Corollary 1.24. *Every finite dimensional real representation V of $\mathfrak{sl}(2,\mathbb{R})$ can be written as a direct sum $\bigoplus_{m\in\mathbb{N}} V_{-m} \otimes V_{\text{prim}}^{-m}$ with V_{-m} isomorphic to P_m. In particular, V is completely reducible.*

Proof. Everything follows from the above lemma, except for complete reducibility. This is however a direct consequence since any real sub representation U gives a real subspace $U_{\text{prim}}^{-m} \subset V_{\text{prim}}^{-m}$ and we can choose any complement W' so that, setting $W = W' \oplus \cdots \oplus L^m W'$ we get an $\mathfrak{sl}(2,\mathbb{R})$ invariant splitting $V = U \oplus W$. □

Corollary 1.25. *Let $V = \bigoplus_{j\in\mathbb{Z}} V^j$ be a graded finite dimensional real vector space and let $L : V \to V$ be an endomorphism of degree 2 such that*

$$L^j : V^{-j} \xrightarrow{\cong} V^j$$

for all $j \in \mathbb{Z}$. Then there exists a unique representation ρ of $SL(2,\mathbb{R})$ on V such that $d\rho(\ell) = L$ and $d\rho(b)(v) = jv$ for all $v \in V^j$.

Proof. Define $B \in \operatorname{End}(V)$ as multiplication by j on V^j. First extend $\ell \mapsto L, b \mapsto B$ to a representation of $\mathfrak{sl}(2,\mathbb{R})$ on V. This can be done in a unique way. Indeed, let $V_0^{-j} = \operatorname{Ker}\left[L^{j+1} : V^{-j} \to V^{j+2}\right]$. If $w \in V^{-j}$, there is a unique $u \in V^{-j-2}$ such that $L^{j+2}u = L^{j+1}w$. Then $Lu - w \in V_0^{-j}$ showing that there is a direct sum decomposition $V^{-j} = V_0^{-j} \oplus \operatorname{Im}(L : V_{-j-2} \to V_{-j})$. By induction it follows that there is a decomposition

$$V^t = \bigoplus_{r\geq 0} L^r V_0^{t-2r}.$$

It suffices to define $\Lambda = d\rho(\lambda)$ on each of the factors. The idea is that $v \in V_0^{-j}$ is primitive so that Λ is zero on this space. Copying (I–9) for $v \in V_0^{-j}$ and $k = 0, 1, \ldots, j$ we set

$$\Lambda(L^k v) = k(j - k + 1)L^{k-1}v.$$

Since this formula is dictated by the commutation rules, the result must be a representation of $\mathfrak{sl}(2,\mathbb{R})$. Then we use again that any finite dimensional irreducible representation of $\mathfrak{sl}(2,\mathbb{R})$ can be lifted to $SL(2,\mathbb{R})$ in a unique way. □

For later applications (§ 11.3.2) we need to know the action of the involutions

$$w = \begin{pmatrix} 0 & 1 \\ -1 & 0 \end{pmatrix} \quad W = \rho(w)$$

on primitive cohomology. We compute this using Lemma 1.22:

Corollary 1.26. *For a primitive vector $v \in V^{-r}$ we have*

$$W(L^j v) = (-1)^j \frac{j!}{(r-j)!} L^{r-j} v, \quad j = 0, 1, \ldots, r.$$

Proof. We may assume that V is irreducible and hence isomorphic to P_r. We saw that x^r generates the space of primitive vectors and the action of L is given by formula (I–8). Hence, since then $W(x^{r-j}y^j) = (-1)^j y^{r-j} x^j$ we have

$$\begin{aligned} W(L^j x^r) &= (-1)^j W \left[\frac{r!}{j!} x^{r-j} y^j \right] \\ &= (-1)^r \frac{r!}{(r-j)!} x^j y^{r-j} \\ &= (-1)^j \frac{j!}{(r-j)!} L^{r-j} x^r . \square \end{aligned}$$

1.2.2 Primitive Cohomology

Let X be an n-dimensional complex manifold equipped with a hermitian metric h. Let L denote the operator defined by multiplication against the Kähler form

$$L(\alpha) = \omega_h \wedge \alpha.$$

and put

$$\Lambda = L^* = \bar{*}^{-1} L \bar{*}.$$

These two operators are real and act pointwise on covectors and we say that $u \in \Lambda^*(T_x X)^\vee$ is **primitive** if $\Lambda u = 0$. Since L commutes with the unitary action on covectors (Lemma 1.7), also Λ commutes with this action. Together they define a representation of $\mathfrak{sl}(2, \mathbb{R})$. This follows from Lemma 1.22 and the formula

$$[L, \Lambda] u = (m - n) u, \quad u \in \Lambda^m T_x X^\vee \tag{I–10}$$

proven for instance in [Wells, Chap. V §1]. This entire represention thus commutes with the unitary action. Since the holonomy group is contained in the unitary group (Lemma 1.6), Chern's theorem 1.2 then implies:

Lemma 1.27. *The action of L and Λ on $E_{\mathrm{DR}} X$ induces an $\mathfrak{sl}(2, \mathbb{R})$-representation on harmonic forms. The first Hodge isomorphism allows to transport this representation to cohomology.*

Now note that the space of *complex* k-covectors forms an $\mathfrak{sl}(2,\mathbb{C})$-representation space which by (I–10) has weight $k-n$. So, using the notation

$$n_+ = \max(n,0) \tag{I–11}$$

Lemma 1.22 and Corollary 1.23 translate as:

Lemma 1.28. 1) *There are no non-zero primitive k-covectors when $k>n$ and for $k\le n$ a primitive k-covector is annihilated by L^r whenever $r > n-k$.*

2) *Any k-covector u can be written uniquely as a sum $u=\sum L^r u_{k-2r}$, where u_{k-2r} is a primitive $k-2r$-covector and we sum over non-negative $r\ge k-n$. Note that this is compatible with the decomposition into types. Hence we have direct sum decompositions*

$$\Lambda^k(T_xX\otimes\mathbb{C})^\vee = \bigoplus_{r\ge(k-n)_+} L^r(\Lambda^{2n-k}T_xX\otimes\mathbb{C})^\vee_{\mathrm{prim}}$$

$$\Lambda^{p,q}(T_xX)^\vee = \bigoplus_{r\ge(p+q-n)_+} L^r(\Lambda^{p,q}T_xX)^\vee_{\mathrm{prim}}$$

3) *Suppose that $k\le n$. Then we have*

$$\text{A k-covector u is primitive} \iff L^{n-k+1}u=0.$$

For all integers $k\le n$ and (p,q) with $p+q\le n$ we have ***Lefschetz-isomorphisms***

$$L^{n-k} : \Lambda^k(T_xX\otimes\mathbb{C})^\vee \xrightarrow{\cong} \Lambda^{2n-k}(T_xX\otimes\mathbb{C})^\vee$$
$$L^{n-p-q} : \Lambda^{p,q}(T_xX)^\vee \xrightarrow{\cong} \Lambda^{n-p,n-q}(T_xX)^\vee.$$

In particular L^r is injective on k-forms as long as $r\le n-k$.

Using the characterization (3) of primitive covectors, one can easily give examples of such covectors:

Example 1.29. The (p,q)-covector $dz_1\wedge\cdots\wedge dz_p\wedge d\bar z_{p+1}\wedge\cdots\wedge d\bar z_{p+q}$ is primitive.

We next define a **primitive form** as a form which at each point is a primitive covector. By Lemma 1.27 there is a primitive decomposition for harmonic forms as well:

$$\mathrm{Har}^{p,q}(X) = \bigoplus_{r\ge(p+q-n)_+} L^r\left(\mathrm{Har}^{p-r,q-r}\right)_{\mathrm{prim}}(X).$$

Via the complexification of the first Hodge isomorphism this transports to cohomology. To make this explicit, define **primitive cohomology** as

$$\begin{aligned} H^m_{\mathrm{prim}}(X) &= \mathrm{Ker}\left(\Lambda : H^m(X;\mathbb{C}) \to H^{n-2}(X;\mathbb{C})\right) \\ &= \mathrm{Ker}\left(L^{n-m+1} : H^m(X;\mathbb{C}) \to H^{2n-m+2}(X;\mathbb{C})\right), \quad \text{if } m \le n. \end{aligned}$$

Here we define the operator Λ on cohomology through the action on the harmonic forms, i.e. we use the harmonic projection to identify cohomology and harmonic forms. The last equality follows formally from Corollary 1.28, part 3) and shows that this action does not depend on the metric. Likewise, define the primitive (p,q)-spaces by

$$\begin{aligned} H^{p,q}_{\mathrm{prim}}(X) &= \mathrm{Ker}\left(\Lambda : H^{p,q}(X) \to H^{p-1,q-1}(X)\right) \\ &= \mathrm{Ker}\left(L^{n-p-q+1} : H^{p,q}(X;\mathbb{C}) \to H^{n-q+1,n-p+1}(X;\mathbb{C})\right) \\ &\quad (\text{for } p+q \le n). \end{aligned}$$

Then there is an induced Hodge decomposition on primitive cohomology

$$H^m_{\mathrm{prim}}(X) = \bigoplus_{p+q=m} H^{p,q}_{\mathrm{prim}}(X)$$

and, using the notation (I–11), we have **Lefschetz-decompositions**

$$\begin{aligned} H^m(X;\mathbb{C}) &= \bigoplus_{r \ge (m-n)_+} L^r H^{m-2r}_{\mathrm{prim}}(X;\mathbb{C}) \\ H^{p,q}(X) &= \bigoplus_{r \ge (p+q-n)_+} L^r H^{p-r,q-r}_{\mathrm{prim}}(X). \end{aligned} \tag{I–12}$$

Now we can apply Corollary 1.28 either directly or first to harmonic forms, proving the

Theorem 1.30 (Hard Lefschetz theorem). *For any Kähler manifold (X,ω), cup product with the Kähler class $[\omega]$ induces isomorphisms*

$$\begin{aligned} L^{n-k} : H^k(X;\mathbb{C}) &\xrightarrow{\cong} H^{2n-k}(X;\mathbb{C}), \quad k \le n \\ L^{n-(p+q)} : H^{p,q}(X) &\xrightarrow{\cong} H^{n-q,n-p}(X), \quad p+q \le n. \end{aligned}$$

In particular L^r is injective on k-cohomology as long as $r \le n-k$.

Remark 1.31. It is quite formal that the first isomorphism (for all k) implies the Lefschetz decomposition for cohomology. Indeed, we may assume that the Lefschetz decomposition has been proven for all ranks $< m$ and we consider $\alpha \in H^m(X;\mathbb{C})$. There is a unique $\beta \in H^{m-2}(X;\mathbb{C})$ such that $L^{n-m+1}\alpha = L^{n-m+2}\beta$ and $\alpha - L\beta$ is primitive. By induction there is a unique decomposition $\beta = \beta_1 + L\beta_2 + \dots$ with $\beta_r \in H^{m-2r}_{\mathrm{prim}}(X)$. Now apply L to β, and the decomposition follows. Uniqueness follows from the injectivity of L^r on $H^k(X;\mathbb{C})$ for $r \le n-k$.

We have seen at the start of this section that the primitive covectors form a $\mathrm{U}(n)$-representation. They give in fact an *irreducible* representation:

Lemma 1.32. *The space of primitive covectors* $(\Lambda^{p,q}T_xX)^\vee_{\rm prim}$ *is an irreducible* $\mathrm{U}(n)$*-representation.*

Proof. Example 1.29 shows that the space $(\Lambda^{p,q}T_xX)^\vee_{\rm prim}$ is non-trivial. Hence, with $m = \min(p,q)$, there are at least $m+1$ irreducible components in

$$(\Lambda^{p,q}T_xX)^\vee = \bigoplus_{0\le r\le m} L^r(\Lambda^{p-r,q-r}T_xX^\vee)_{\rm prim}.$$

It suffices therefore to see that there are at most $m+1$ irreducible $\mathrm{U}(n)$-components in the latter $\mathrm{U}(n)$-module. It is well known that the irreducible $\mathrm{U}(n)$-modules are in bijection with the eigenvectors of the action of the diagonal matrices. But these all act differently on the $m+1$ covectors

$$L^r(dz_{n-p+r+1}\wedge\cdots\wedge dz_n\wedge d\bar z_1\wedge\cdots d\bar z_{q-r}).$$

Indeed, if $\epsilon_1,\ldots,\epsilon_n$ is the canonical basis for the characters of this diagonal subgroup, the preceding vector has weight $\epsilon_1\cdots+\epsilon_{q-r}-(\epsilon_{n-p+r+1}+\cdots+\epsilon_n)$.
□

The final major ingredients in classical Hodge theory are the Hodge-Riemann bilinear relations. to formulate these, we use action on $H^*_{\rm DR}(X;\mathbb{C})$ induced by the complex structure $J\in\operatorname{End}T_xX^\vee$. This is the **Weil-operator** C which explicitly is given by

$$C\big|_{H^{p,q}(X)} = \mathrm{i}^{p-q}. \tag{I–13}$$

The Hard Lefschetz theorem leads directly to the Hodge-Riemann bilinear relations:

Theorem 1.33 (Hodge Riemann bilinear relations). *For* $k\in\mathbb{Z}$ *put*

$$\varepsilon(k) := (-1)^{\frac12 k(k-1)}. \tag{I–14}$$

Let (X,ω) *be a Kähler manifold of dimension* n*. The* ***Hodge-Riemann form*** *is the bilinear form*

$$Q(\alpha,\beta) = \varepsilon(k)\int_X \alpha\wedge\beta\wedge\omega^{n-k} \quad [\alpha],[\beta]\in H^k_{\rm DR}(X;\mathbb{C}).$$

It is $(-1)^k$*-symmetric (symmetric for* k *even and skew-symmetric for* k *odd).*

Using the Weil-operator C (I–13)*, the two Hodge-Riemann relations can be written as*

$$\begin{cases} Q(H^{p,q},H^{r,s}) = 0 & \text{if } (r,s)\ne(q,p)\\ \text{For } u\in H^{p,q}_{\rm prim},\ \mathrm{i}^{p-q}Q(u,\bar u) = Q(Cu,\bar u) = (u,u) & \text{and hence} > 0 \text{ if } u\ne 0.\end{cases}$$

Here, the right hand side uses the global Hodge inner product (I–2)*.*
If the (real) Kähler class $[\omega]$ *belongs to* $H^2(X;R)$ *for a subring* R *of* $\mathbb{R}$*, the Hodge-Riemann form can be evaluated on* $H^k(X;R)$ *and takes values in* R*.*

Proof. The $(-1)^k$-symmetry and the first bilinear relation is clear. So it remains to prove the second relation which is clearly a consequence of

$$\varepsilon(k)\mathrm{i}^{p-q}\omega^{n-k} \wedge \bar{u} = (n-k)!(*\bar{u}), \tag{I–15}$$

valid for any primitive (p,q)-form u with $p+q=k$. To prove this formula, we use Lemma 1.32 so that we need to verify the equation only for a well-chosen u, for example for $u = dz_1 \wedge \cdots \wedge dz_p \wedge d\bar{z}_{p+1} \wedge \cdots \wedge d\bar{z}_{p+q}$. Assume, as we may, that the metric is the standard metric $\omega = \mathrm{i}\sum_{k=1}^{n} dz_k \wedge d\bar{z}_k$. We may further assume that, if we let I and J run over all possible strictly increasing multi-indices with $|I| = p$, $|J| = q$ we obtain a unitary basis $\{dz_I \wedge d\bar{z}_J\}$ for $\Lambda^{p,q}T_x X^\vee$. For any subset $R \subset \{1, \ldots, n\}$ with $|R| = r$ we have

$$\omega^r = (-2\mathrm{i})^r \sum_{|R|=r} dz_R \wedge d\bar{z}_R$$

and since $\mathrm{vol} = \varepsilon(n)(-\mathrm{i})^n dz_1 \wedge \cdots \wedge dz_n \wedge d\bar{z}_1 \wedge \cdots \wedge d\bar{z}_n$, using formula (I–1) we deduce (I–15). □

1.3 Applications

As a first application, following [Hart75, proof of Theorem 6.1], we shall show how the Hard Lefschetz Theorem implies Barth's theorem [Barth].

Theorem 1.34 (Barth's theorem). *Let $Y \subset \mathbb{P}^N$ be a smooth subvariety of dimension n. The inclusion induces isomorphisms*

$$H^k(\mathbb{P}^N;\mathbb{C}) \xrightarrow{\cong} H^k(Y;\mathbb{C}), \quad \forall k \le 2n - N.$$

Proof. By B.30 $i_! \circ i^*$ is multiplication with the fundamental class $\mathrm{cl}(X)$. Let c be the codimension of Y. We have $i_! \circ i^* = a \cdot L_X^c$, where $a \neq 0$ and L_X is the Kähler class on X. Similarly, $i^* \circ i_! = a \cdot L_Y^c$, where $L_Y = i^* L_X$ is the Lefschetz-operator on the cohomology of Y. So we get a commutative diagram

$$\begin{array}{ccc} H^k(\mathbb{P}^N;\mathbb{C}) & \xrightarrow{\;i^*\;} & H^k(Y;\mathbb{C}) \\ \downarrow{\scriptstyle a\cdot L^c} & \swarrow{\scriptstyle i_!} & \downarrow{\scriptstyle a\cdot L_Y^c} \\ H^{k+2c}(\mathbb{P}^N;\mathbb{C}) & \xrightarrow[\;i^*\;]{} & H^{k+2c}(Y;\mathbb{C}). \end{array}$$

Given that $c \le \dim Y - k$, the Hard Lefschetz theorem for Y implies that the second vertical arrow is injective. It follows that the oblique arrow $i_!$ is injective. The first vertical arrow is injective (it is an isomorphism) and hence the upper horizontal arrow is injective. To show that it is onto, one remarks that, since $i_!$ is injective, it suffices to remark that since L^c is an isomorphism, it is, in particular, surjective. □

Remark 1.35. 1) Since the Kähler class is integral, the preceding proof also works over $\mathbb{Q}$.

2) The theorem implies the Weak Lefschetz Theorem: this is the case of a hypersurface ($N = n+1$): the (complex) cohomology groups of a smooth n-dimensional hypersurface in $\mathbb{P}^{n+1}$ are the same as those of $\mathbb{P}^{n+1}$, except the middle one $H^n(X;\mathbb{C})$. This theorem is a special case of Lefschetz' Hyperplane Theorem (C.15) which is valid even with integral coefficients (take $Y = \mathbb{P}^{n+1}$ and consider a degree d-hypersurface as a hyperplane via the d-fold Veronese embedding).

As a second application of the Lefschetz decomposition we explain how to deduce the degeneration of the Leray spectral sequence for smooth projective morphisms.

Recall (see § A.3) that the Leray spectral sequence associated to a continuous map $f : X \to S$ between topological spaces converges to the cohomology of X (with coefficients in some ring R) and reads

$$E_2^{p,q} = H^p(S, R^q f_* \underline{R}_X) \Longrightarrow H^{p+q}(X;R).$$

Cup product with a class $h \in H^2(X;R)$ defines an action on the cohomology of X and, by restriction, on the cohomology of any subset of X, in a way which is compatible with inclusions. It follows that there are induced homomorphisms

$$(*)_k \qquad [\cup h]^k : R^{m-k} f_* \underline{R}_X \to R^{m+k} f_* \underline{R}_X, \quad k > 0$$

compatible with the action on the cohomology of the fibres of f.

Definition 1.36. Assume that X and S are differentiable manifolds and that the general fibre of f has dimension $2m$. If for this m and all $k \geq 0$ $(*)_k$ is an isomorphism, we say that the fibration f **has the hard Lefschetz property** (with respect to the class $h \in H^2(X;R)$).

Examples 1.37. 1) Let X and S be smooth complex manifolds. A holomorphic map $f : X \to S$ is **projective** if we can embed X into a locally trivial fibre bundle $P \to S$ with fibre a projective space, and such that the following self evident diagram is commutative

$$\begin{array}{ccc} X & \overset{j}{\hookrightarrow} & P \\ & {\scriptstyle f}\searrow \quad \swarrow & \\ & S & \end{array}$$

Let us see what the hard Lefschetz property amounts to in the case of a smooth projective family. This condition being local, we may assume that $P = \mathbb{P}^N \times S$ so that each fibre is naturally embedded in $\mathbb{P}^N$. The hyperplane class on $\mathbb{P}^N$ pull back first to a class on $\mathbb{P}^N \times S$ (via projection) and then to a class $h \in H^2(X;\mathbb{Q})$ giving a rational Kähler class on each fibre.

The stalk at s of the direct image sheaf $R^m f_* \underline{\mathbb{Q}}_X$ is the cohomology group $H^m(X_s;\mathbb{Q})$ of the fibre X_s of f at s. Since f is locally differentiably trivial (Theorem C.10), the direct image sheaves $R^m f_* \underline{\mathbb{Q}}_X$ are locally constant. The hard Lefschetz property can thus be verified fibre by fibre. On a fibre this assertion is exactly the hard Lefschetz theorem.

2) The same remarks can be applied in the Kähler setting for cohomology with real coefficients, provided we assume that there is a closed 2-form on X which restricts to a Kähler form on each fibre. In this case we say that f is a **smooth Kähler** family.

3) A fibration $X \to C$, with X Kähler and C a curve sometimes has the hard Lefschetz property. See Lemma C.13 for conditions which guarantee this. As an application, by Cor. C.22 most Lefschetz pencils have this property.

Proposition 1.38. *If f satisfies the hard Lefschetz property, the Leray spectral sequence for f degenerates at the E_2-term. In particular this holds for smooth projective morphisms. It follows also that the restriction maps*

$$H^m(X;R) \to E^{0,m} = H^0(S, R^m f_* R)$$

are surjective.

Proof. We have seen (see Remark 1.31) that the Lefschetz property implies the Lefschetz decomposition. This argument being formal, we introduce

$$(R^{m-k} f_* R)_{\mathrm{prim}} = \mathrm{Ker}\{R^{m-k} f_* R \to R^{m+k+2} f_* R\}$$

and then we have decompositions

$$R^m f_* R = \bigoplus_{r \geq (m-n)_+} h^r (R^{m-2r} f_* R)_{\mathrm{prim}}.$$

The class h acts also on the terms of the Leray spectral sequence, and we define

$${}_{\mathrm{prim}}E_r^{p,m-k} = \mathrm{Ker}\{h^{k+1} : E_r^{p,m-k} \to E_r^{p,m+k+2}\}.$$

Now by induction we assume that $d_2 = d_3 = \cdots = d_{r-1} = 0$ so that $E_2 = E_r$ and since $E_2^{p,m-k} = H^p(S, R^{m-k} f_* R)$, the preceding decompositions induce a direct decomposition of the E_2-terms into isomorphic images of primitive pieces. It suffices to show that $d_r = 0$ on each primitive piece. We have a commutative diagram

$$\begin{array}{ccc}
{}_{\mathrm{prim}}E_r^{p,m-k} & \xrightarrow{d_r} & E_r^{p+r,m-k-r+1} \\
\downarrow {\scriptstyle h^{k+1}=0} & & \downarrow {\scriptstyle h^{k+1}} \\
E_r^{p,m+k+2} & \xrightarrow{d_r} & E_r^{p+r,m+k-r+3}.
\end{array}$$

The vertical arrow on the right is an injection: the Lefschetz property implies that there is an isomorphism

$$h^{k+r-1}: E_r^{p+r,m-k-r+1} = H^{p+r}(S, R^{m-(k+r-1)} f_* R) \longrightarrow \\ H^{p+r}(S, R^{m+k+r-1} f_* R) = E_r^{p+r,m+k+r-1}$$

and hence h^{k+1} is injective, since $r \geq 2$. The commutativity of the diagram implies that $d_r = 0$.

The final assertion follows from the fact that the restriction map in this case is the natural surjective map

$$H^m(X; R) \twoheadrightarrow E_\infty^{0,m} \cong E_2^{0,m} \square$$

Remark 1.39. 1) This Proposition applies to smooth Kähler families over a base of arbitrary dimension. In [Del68] it is shown that, more generally, the Leray spectral sequence degenerates if $f : X \to S$ is a proper smooth morphism between smooth algebraic varieties.

2) Similar arguments (loc. cit) can be used to show that we have a decomposition

$$Rf_* \underline{\mathbb{Q}}_X \simeq \bigoplus_q R^q f_* \underline{\mathbb{Q}}_X[-q] \tag{I–16}$$

in the derived category of sheaves of $\mathbb{Q}$-vector spaces on S. This statement in fact can easily be seen to imply degeneracy of the Leray spectral sequence.

3) The Leray spectral sequence always degenerates at E_2 for morphisms from a complex projective manifold onto a smooth curve. See Theorem 4.24. For a vast generalization see Theorem 14.11.

To interpret the final assertion of Proposition 1.38 in case f is a smooth projective family, we look a bit more carefully at the local system $R^m f_* \underline{R}_S$ with stalks $H^m(X_s; R)$. The global sections of this local system are the invariants under the monodromy action. Thus we have:

Corollary 1.40 (Global invariant cycle theorem or locus of an invariant cycle theorem). *Let $f : X \to S$ be a smooth Kähler family (see Examples 1.37, 2). Then for all $s \in S$ the invariants in $H^m(X_s; \mathbb{Q})$ under the monodromy action come from restriction of global classes on X.*

Historical Remarks. The topics treated in § 1.1.1–1.2.2 are by now classical and for most details we refer to standard texts such as [Weil] or [Wells]. The idea to make use of the holonomy group in order reprove the Lefschetz decomposition theorem is due to Chern [Ch]. This geometric idea reduces in fact the existence of such a decomposition to linear algebra. Apparently Hecht refined the linear algebra part by making use of representation theory of SL(2)(see [Wells, p. 183]), an idea we have followed. In this context the complex theory suffices, but later, in Chap. 10 we need the theory of real representations. This is not so standard and we have preferred to give full details by adapting the already streamlined presentation one can find in [Dem, § 6C]. That the Lefschetz decomposition can be used to show degeneration of the Leray spectral sequence, as explained in § 1.3 was first observed by Deligne [Del68].

The terminology "locus of an invariant cycle" is due to Lefschetz and can be explained as follows. Suppose that the fibres of a family have dimension n. Consider a $(2n-m)$-cycle γ on a fixed fibre over a point s which is invariant under monodromy. It can be displaced along any path starting at s defining a cycle in the fibre over the end point fibre. Take the union of such cycles arising for all possible paths in the base. It is the locus inside the total space traced out by the invariant cycle in the fibre over t when t varies over the base S. This locus, according to Lefschetz [Lef], should be a $(2n + 2\dim S - m)$-cycle on the total space restricting to the invariant cycle we started with.

Lefschetz ultimate goal, constructing algebraic cycles in an inductive manner, has been pursued later by Hodge [Ho50] and Grothendieck [Groth69], resulting in the "Induction Principle". To explain the latter we need a lot more Hodge theory and we return to this later (§ 12.2).

2

Pure Hodge Structures

The Hodge decomposition of the n-th cohomology group of a Kähler manifold is the prototype of a Hodge structure of weight n. In this chapter we study these from a more abstract point of view. In § 2.1 and § 2.2 the foundations are laid. Hodge theoretic considerations for various sorts of fundamental classes associated to a subvariety are given in § 2.4.

In § 2.3 some important concepts are developed which play a central role in the remainder of this book, in particular the concept of a Hodge complex, which is introduced in § 2.3. The motivating example comes from the holomorphic De Rham complex on a compact Kähler manifold and is called the *Hodge-De Rham complex*. However, to show that this indeed gives an example of a Hodge complex follows only after a strong form of the Hodge decomposition is shown to hold. This also allows one to put a Hodge structure on the cohomology of any compact complex manifold which is bimeromorphic to a Kähler manifold, in particular algebraic manifolds that are not necessarily projective.

In Chapter 3 we shall extend the notion of a Hodge complex of sheaves to that of a mixed Hodge complex of sheaves.

We finally show in § 2.5 that the cohomology of varieties with quotient singularities also admits a pure rational Hodge structure.

2.1 Hodge Structures

2.1.1 Basic Definitions

We place the definition of a weight k Hodge structure (Def. 1.12) in a wider context. Let V be a finite dimensional real vector space and let $V_\mathbb{C} = V \otimes \mathbb{C}$ be its complexification.

Definition 2.1. A **real Hodge structure** on V is a direct sum decomposition

$$V_\mathbb{C} = \bigoplus_{p,q\in\mathbb{Z}} V^{p,q}, \text{with } V^{p,q} = \overline{V^{q,p}} \quad \textbf{(the Hodge decomposition.)}$$

The numbers

$$h^{p,q}(V) := \dim V^{p,q}$$

are **Hodge numbers** of the Hodge structure. The polynomial

$$P_{\text{hn}}(V) = \sum_{p,q \in \mathbb{Z}} h^{p,q}(V) u^p v^q \tag{II–1}$$

its associated **Hodge number polynomial**.

If the real Hodge structure V is of the form $V = V_R \otimes_R \mathbb{R}$ where R is a subring of $\mathbb{R}$ and V_R is an R-module of finite type we say that V_R carries carries an **R-Hodge structure**.

A **morphism of R-Hodge structures** is a morphism $f : V_R \to W_R$ of R-modules whose complexification maps $V^{p,q}$ to $W^{p,q}$.

If V is real Hodge structure, the **weight k part** $V^{(k)}$ is the real vector space underlying $\bigoplus_{p+q=k} V^{p,q}$. If $V = V^{(k)}$, we say that V is a weight k real Hodge structure and if $V = V_R \otimes_R \mathbb{R}$ we speak of a weight k R-Hodge structure. Usually, if $R = \mathbb{Z}$ we simply say that V or $V_{\mathbb{Z}}$ **carries a weight k Hodge structure**.

Examples 2.2. i) The De Rham group $H^k_{\text{DR}}(X)$ of a compact Kähler manifold has canonical real Hodge structure of weight k defined by the classical Hodge decomposition. We have seen (Corr. 1.13) that it is in fact an integral Hodge structure.

ii) The Hodge structure $\mathbb{Z}(1)$ of Tate (I–3) has variants over any subring R of $\mathbb{R}$: we put $R(k) := R \otimes_{\mathbb{Z}} \mathbb{Z}(k)$.

iii) The top cohomology of a compact complex manifold X of dimension say n, can be identified with a certain Tate structure. Indeed, the **trace map** is the isomorphism given by

$$\text{tr} : H^{2n}(X; \mathbb{R}) \xrightarrow{\sim} \mathbb{R}(-n), \quad \omega \mapsto \left(\frac{1}{2\pi \mathrm{i}}\right)^n \int_X \omega. \tag{II–2}$$

Let $V = V^{(k)}$ be a weight k Hodge structure. The **Hodge filtration** associated to this Hodge structure is given by

$$F^p(V) = \bigoplus_{r \geq p} V^{r,s}.$$

Conversely, a decreasing filtration

$$V_{\mathbb{C}} \supset \cdots \supset F^p(V) \supset F^{p+1}(V) \cdots$$

on the complexification $V_{\mathbb{C}}$ with the property that $F^p \cap \overline{F^q} = 0$ whenever $p + q = k + 1$ defines a weight k Hodge structure by putting

$$V^{p,q} = F^q \cap \overline{F^q}.$$

The condition that $F^p \cap \overline{F^q} = 0$ whenever $p + q = k + 1$ is equivalent to $F^p \oplus \overline{F^{k-p+1}} = V_{\mathbb{C}}$ and we say that the *filtration $F^\bullet$ is k-opposed to its complex conjugate filtration.*

Definition 2.3 (Multi-linear algebra constructions). Suppose that V, W are real vector spaces with a Hodge structure of weight k, respectively ℓ, the Hodge filtration on $V \otimes W$ is given by

$$F^p(V \otimes W)_{\mathbb{C}} = \sum_m F^m(V_{\mathbb{C}}) \otimes F^{p-m}(W_{\mathbb{C}}) \subset V_{\mathbb{C}} \otimes_{\mathbb{C}} W_{\mathbb{C}}.$$

This gives $V \otimes W$ a Hodge structure of weight $k + \ell$ with Hodge number polynomial given by

$$P_{\mathrm{hn}}(V \otimes W) = P_{\mathrm{hn}}(V)P_{\mathrm{hn}}(W). \tag{II–3}$$

Similarly, the multiplicative extension of the Hodge filtration to the tensor algebra

$$TV = \bigoplus_a T_a V \quad \text{with } T_a V := \bigotimes^a V$$

of V is defined by

$$F^p T_a V = \sum_{k_1 + \cdots k_a = p} F^{k_1} V_{\mathbb{C}} \otimes \ldots F^{k_a} V_{\mathbb{C}}$$

and gives a Hodge structure of weight ak on $T_a V$. It induces a Hodge structure of the same weight on the degree a-piece of the symmetric algebra SV of V and the exterior algebra ΛV of V. We can also put a Hodge structure on duals, or, more generally spaces of homomorphisms as follows:

$$F^p \operatorname{Hom}(V, W)_{\mathbb{C}} = \{f : V_{\mathbb{C}} \to W_{\mathbb{C}} \mid f F^n(V_{\mathbb{C}}) \subset F^{n+p}(W_{\mathbb{C}}) \quad \forall n\}$$

This defines a Hodge structure of weight $\ell - k$ on $\operatorname{Hom}(V, W)$ with Hodge number polynomial

$$P_{\mathrm{hn}}(\operatorname{Hom}(V, W)(u, v) = P_{\mathrm{hn}}(V)(u^{-1}, v^{-1})P_{\mathrm{hn}}(W)(u, v). \tag{II–4}$$

In particular, taking $W = \mathbb{R}$ with $W_{\mathbb{C}} = W^{0,0}$ we get a Hodge structure of weight $-k$ on the dual $V^\vee$ of V with Hodge number polynomial

$$P_{\mathrm{hn}}(V^\vee)(u, v) = P_{\mathrm{hn}}(V)(u^{-1}, v^{-1}). \tag{II–5}$$

Finally, we can define a Hodge structure of weight $ak - b\ell$ on $T_a V \otimes T_b V^\vee = V^{\otimes a} \otimes (V^\vee)^{\otimes b}$ using the multiplicative extension of F to the tensor algebra $TV \otimes TV^\vee$. The multiplication in each of the algebras TV, SV, ΛV, $TV \otimes TV^\vee$ is a morphism of Hodge structures.

Given any R-Hodge structure V, define its r**-th Tate twist** by

$$V(r) := V \otimes_R R(r).$$

If V has weight m, $V(r)$ has weight $m - 2r$ and

$$V(r)^{p,q} = V^{p+r,q+r}.$$

Note that one has:

$$P_{\text{hn}}(V(r)) = P_{\text{hn}}(V)(uv)^{-r}. \tag{II–6}$$

If W is another R-Hodge structure, giving A morphism $V(-r) \to W$ is also called a **morphism of Hodge structures** $V \to W$ **of type** (r, r). Morphisms of Hodge structures preserve the Hodge filtration. The converse is also true:

Proposition 2.4. *Let V, W be R-Hodge structures of weight k. Suppose that $f : V \to W$ is an $\mathbb{R}$-linear map preserving the R-structures and such that*

$$f_{\mathbb{C}}(F^pV) \hookrightarrow F^pW.$$

Then f is a morphism of R-Hodge structures.

Proof. One has $f_{\mathbb{C}}(\overline{F^qV}) \hookrightarrow \overline{F^qW}$, so, if $p + q = k$, we have

$$f_{\mathbb{C}}(V^{p,q}) = f_{\mathbb{C}}(F^pV) \cap \overline{F^qV} \hookrightarrow F^pW \cap \overline{F^qW} = W^{p,q}. \qquad \square$$

Clearly, the image of a morphism of Hodge structures is again a Hodge structure. By the above constructions the duality operation preserves Hodge structures, and so the kernel of a morphism of Hodge structures is a Hodge structure. Using the preceding multi-linear algebra constructions, it is not hard to see that we in fact have:

Corollary 2.5. *The category of R-Hodge structures is an abelian category which we denote $\mathfrak{hs}_R$. If $R = \mathbb{Z}$ we simply write $\mathfrak{hs}$.*

Hodge structures can also be defined through group representations and this is useful in the context of Mumford-Tate groups (see § 2.2). Introduce the algebraic group

$$\mathbb{S} := \{\text{the restriction of scalars from } \mathbb{C} \text{ to } \mathbb{R} \text{ à la Weil of the group } \mathbb{G}_m\}.$$

By definition, the complex points of $\mathbb{S}$ correspond to pairs of points $z, z' \in \mathbb{C}^*$. The point z corresponds to the standard embedding $\mathbb{C}^\times \hookrightarrow \mathbb{C}$ while z' corresponds to the complex conjugate embedding. Hence complex conjugation sends (z, z') to $(\bar{z}', \bar{z})$ and the real points $\mathbb{S}(\mathbb{R})$ consists of $\mathbb{C}^\times$ embedded into the group $\mathbb{S}(\mathbb{C}) = \mathbb{C}^\times \times \mathbb{C}^\times$ of complex points through $a \mapsto (a, \bar{a})$. So $\mathbb{S}$ is just the group $\mathbb{C}^\times$ considered as a *real* algebraic group.

Note that there is a natural embedding $w : \mathbb{G}_m \to \mathbb{S}$ of algebraic groups which on complex points is the diagonal embedding $a \mapsto (a, a)$ and on real points is just the embedding of $\mathbb{R}^\times \hookrightarrow \mathbb{C}^\times$. Note that $\mathbb{C}^\times = \mathbb{R}^\times \cdot S^1$ where S^1 are the real points of the unitary group $\mathrm{U}(1)$. We can extend the embedding $S^1 \hookrightarrow \mathbb{C}^\times$ to an embedding $\mathrm{U}(1) \hookrightarrow \mathbb{S}$ and then

$$\mathbb{S} = \mathrm{U}(1) \cdot w(\mathbb{G}_m).$$

Definition 2.6. A **complex Hodge structure on a complex vector space** W is a representation of $\mathbb{S}(\mathbb{C})$ on W. This amounts to a bigrading

$$W = \bigoplus_{p,q} W^{p,q}, \quad W^{p,q} = \{w \in W \mid (a,b)w = a^{-p}b^{-q}w,\ (a,b) \in \mathbb{S}(\mathbb{C})\}.$$

Now suppose that $W = V_{\mathbb{C}}$, where V is a real vector space. Then the above representation is a real representation if and only if the action of $\mathbb{S}(\mathbb{C})$ on the complex conjugate of any of the above summands is the summand on which the action is the conjugate action. This means precisely that the complex conjugate of $W^{p,q}$ is $W^{q,p}$. Looking at the action of the subgroup $\mathbb{G}_m(\mathbb{R}) = \mathbb{R}^\times$ we obtain the decomposition of V into weight spaces

$$V^{(k)} = \{v \in V \mid av = a^{-k}v,\ a \in \mathbb{R}^\times\},$$

i.e. $V^{(k)}$ is a real Hodge structure of weight k. If the representation is defined over a subring R of $\mathbb{R}$, these are weight k R-Hodge structures and conversely.

Suppose that we only have an U(1)-action on V. Then W splits into eigenspaces W^ℓ on which u acts via the character u^ℓ. Again W^ℓ is the conjugate of $W^{-\ell}$ and we would have a weight k Hodge structure if we declare its weight to be k: just put $W^{p,q} = W^{k-2q} = W^{-k+2p}$. Conversely, a real Hodge structure of weight k is an U(1)-action on W defined over $\mathbb{R}$ plus the specification of the number k. In fact, the argument shows:

Lemma 2.7. *let V_R be an R-module of finite rank. Then V_R admits the structure of an R-Hodge structure if and only if there is a homomorphism*

$$h : \mathbb{S} \to \mathrm{GL}(V \otimes_R \mathbb{R})$$

defined over $\mathbb{R}$, such that $h \circ w : \mathbb{G}_m \to \mathrm{GL}(V \otimes_R \mathbb{R})$ is defined over R.

Equivalently, an R-Hodge structure consists of an R-space V_R equipped with an action of U(1) *defined over R.*

As an example, consider the one-dimensional Hodge structures. These are exactly the Hodge structures of Tate. The group U(1) acts trivially on these. So the action of U(1) defined by a Hodge structure F on V is the same as the one given by $F(\ell)$ on $V(\ell)$. This illustrates the fact that $\mathbb{S} = \mathbb{G}_m \cdot$ U(1) where the action of the subgroup $\mathbb{G}_m$ registers the weight and this gives another interpretation of the preceding weight shift as the multiplication with a character of $\mathbb{G}_m$. In this setting we have the **Weil operator**

$$C|W^{p,q} = \mathrm{i}^{p-q}, \tag{II–7}$$

the image of $\mathrm{i} \in \mathbb{S}(\mathbb{R})$ under the representation (recall that i is identified with $(\mathrm{i}, -\mathrm{i}) \in \mathbb{S}(\mathbb{C})$).

Recall the construction of the Grothendieck group (Def. A.4) 3). It is defined for any abelian category such as the category $\mathfrak{hs}_R$ of R-Hodge structures:

it is the free group on the isomorphism classes $[V]$ of Hodge structures V modulo the subgroup generated by $[V]-[V']-[V'']$ where $0 \to V' \to V \to V'' \to 0$ is an exact sequence of R-Hodge structures. It carries a *ring structure* coming from the tensor product. Because the Hodge number polynomial (II–1) is clearly additive and by (II–3) behaves well on products, we have:

Lemma 2.8. *The Hodge number polynomial defines a ring homomorphism*

$$P_{\mathrm{hn}} : K_0(\mathfrak{hs}_R) \to \mathbb{Z}[u, v, u^{-1}, v^{-1}].$$

Inside $K_0(\mathfrak{hs}_R)$ Tate twisting r-times can be expressed as $[H] \mapsto [H] \cdot \mathbb{L}^{-r}$ where

$$\mathbb{L} = H^2(\mathbb{P}^1) \in K_0(\mathfrak{hs}_R). \tag{II–8}$$

2.1.2 Polarized Hodge Structures

The classical example of polarized Hodge structures is given by the primitive cohomology groups on a compact Kähler manifold (X, ω). If the Kähler class $[\omega]$ belongs to $H^2(X; R)$ for some subring R of $\mathbb{R}$, the Hodge-Riemann form Q restricts to an R-valued form on

$$H_R = H^k_{\mathrm{prim}}(X; R) := \mathrm{Im}\left[H^k(X; R) \to H^k(X; \mathbb{C})\right] \cap H^k_{\mathrm{prim}}(X)$$

where the homomorphism is the coefficient homomorphism. Recall the Hodge-Riemann bilinear relations with respect to the Hodge-Riemann form Q (see Definition 1.33). The first of these relations states that the primitive (p,q)-classes are Q-orthogonal to (r,s)-classes as long as $(p,q) \neq (s,r)$. This can be conveniently reformulated in terms of the Hodge filtration $F^m = \bigoplus_{p \geq m} H^{p,q}_{\mathrm{prim}}$ as follows. Note that F^m is Q-orthogonal to F^{k-m+1} since in the latter only (r,s)-forms occur with $r \geq k-m+1$ while in the first (p,q)-forms occur with $q \leq k-m$. The dimension of F^m being complementary to $\dim F^{k-m+1}$, we therefore have that the Q-orthogonal complement of F^m equals F^{k-m+1}.

The second Hodge-Riemann relation can be reformulated using the Weil operator C, which – as we saw before (II–7)– acts as multiplication by i^{p-q} on (p,q)-forms. We find that in writing $\mathrm{i}^{p-q} Q(u, \bar{u}) = Q(Cu, \bar{u})$, u a primitive (p,q)-form, the right hand side makes sense for any k-form. In this way we arrive at the following

Definition 2.9. A **polarization** of an R-Hodge structure V of weight k is an R-valued bilinear form

$$Q : V \otimes V \longrightarrow R$$

which is $(-1)^k$-symmetric and such that

1) The orthogonal complement of F^m is F^{k-m+1};

2) The hermitian form on $V \otimes \mathbb{C}$ given by

$$Q(Cu, \bar{v})$$

is positive-definite.

Any R-Hodge structure that admits a polarization is said to be **polarizable**.

Example 2.10. The m-th cohomology of a compact Kähler manifold is an integral Hodge structure of weight m. If R is a *field*, this Hodge structure is R-polarizable if there exists a Kähler class in $H^2(X;R)$. In fact, since R is a field, the Lefschetz decomposition (I–12) yields a direct splitting of Hodge structures

$$H^m(X;R) \simeq \bigoplus_{r \geq (k-n)_+} H^{m-2r}_{\text{prim}}(X;R)(-r)$$

and each of the summands carries a polarization. The Tate twist arises naturally: instead of the Kähler class we take $1/(2\pi\mathrm{i})$ times this class, which is represented by the curvature form (Def. B.39) of the Kähler metric. It belongs to $H^2(X;R)(-1)$ and cup product with it defines the modified Lefschetz-operator, say $\tilde{L} : H^k(X;R) \to H^{k+2}(X;R)(-1)$. To have a polarization on all of $H^m(X;R)$ we demand that the direct sum splitting be orthogonal and we *change signs on the summands* (see [Weil, p. 77]):

$$\begin{aligned} Q(\textstyle\sum_r L^r a_r, \sum_s L^s b_s) &:= \epsilon(k) \textstyle\sum_r (-1)^r \int_X \tilde{L}^{n-m+2r}(a_r \wedge b_r), \\ a_r, b_r &\in H^{m-2r}(X;R). \end{aligned}$$

Now there is a particularly concise reformulation of Definition 2.9 if we consider $S = (2\pi\mathrm{i})^{-k} Q$ as a *morphism* of Hodge structures $V \otimes V \to R(-k)$. Since $F^m(V \otimes V) = \sum_{r+s=m} F^r V \otimes F^s V$, this demand is equivalent to the first relation. For the second, note that it follows as soon as we know that the real-valued symmetric form $Q(Cu, v)$ is positive definite on the *real* primitive cohomology. This then leads to the following

Definition 2.9 (bis). A **polarization** of an R-Hodge structure V of weight k is a homomorphism of Hodge structures

$$S : V \otimes V \longrightarrow R(-k)$$

which is $(-1)^k$-symmetric and such that the real-valued symmetric bilinear form

$$Q(u, v) := (2\pi\mathrm{i})^k S(Cu, v) \tag{II–9}$$

is positive-definite on $V \otimes_R \mathbb{R}$.

Corollary 2.11. *Let V be an R-polarizable weight k Hodge structure. Any choice of a polarization on V induces an isomorphism R-Hodge structures $V \xrightarrow{\sim} V^\vee(-k)$ of weight k.*

We finish this section with an important principle:

Corollary 2.12 (Semi-simplicity). *Let (V, Q) be an R-polarized Hodge structure and let W be a Hodge substructure. Then the form Q restricts to an R-polarization on W. Its orthogonal complement $W^{\perp}$ likewise inherits the structure of an R-polarized Hodge structure and V decomposes into an orthogonal direct sum $V = W \oplus W^{\perp}$. Hence, the category of R-polarized Hodge structures is semi-simple.*

Proof. Since W is stable under the action of the Weil operator, the form S given by (II–9) restricts to a positive definite form on $W \otimes_R \mathbb{R}$ and so we have an orthogonal sum decomposition as stated.

2.2 Mumford-Tate Groups of Hodge Structures

In this section (V, F) denotes a finite dimensional $\mathbb{Q}$-Hodge structure of weight k. We have seen in § 2.1 that this means that we have a homomorphism

$$h_F : \mathbb{S} \to \mathrm{GL}(V)$$

of algebraic groups such that $t \in w(\mathbb{G}_m(\mathbb{R}))$ acts as $v \mapsto t^{-k}v$. Recall also that $\mathbb{S} = \mathrm{U}(1) \cdot w(\mathbb{G}_m)$. Restricting h_F to the subgroup $\mathrm{U}(1)$ gives the homomorphism of algebraic groups

$$h_F|\mathrm{U}(1) : \mathrm{U}(1) \to \mathrm{GL}(V).$$

The group $\mathbb{S}$ has two characters z and $\bar{z}$ which on complex points $\mathbb{S}(\mathbb{C}) = \mathbb{C}^* \times \mathbb{C}^*$ correspond to the two projections and hence on $\mathbb{S}(\mathbb{R})$ give the identity, respectively the complex conjugation, which explains the notation.

Definition 2.13. 1) The **Mumford-Tate group** $\mathrm{MT}(V, F)$ of the Hodge structure (V, F) is the Zariski-closure of the image of h_F in $\mathrm{GL}(V)$ over $\mathbb{Q}$, i.e. the smallest algebraic subgroup G of $\mathrm{GL}(V)$ defined over $\mathbb{Q}$ such that $G(\mathbb{C})$ contains $h_F(\mathbb{S}(\mathbb{C}))$.

2) The **extended Mumford-Tate group** $\widetilde{\mathrm{MT}}(V, F)$ is the Zariski-closure of the image of $[h_F \times z]$ in $\mathrm{GL}(V) \times \mathbb{G}_m$, i.e. the smallest subgroup $\tilde{G}$ of $\mathrm{GL}(V) \times \mathbb{G}_m$ defined over $\mathbb{Q}$ and such that $\tilde{G}(\mathbb{C})$ contains $(h_F \times z)\mathbb{S}(\mathbb{C})$.

3) The **Hodge group** or **special Mumford-Tate group** $\mathrm{HG}(V, F)$ is the Zariski-closure of the image of $h_F|\mathrm{U}(1)$.

Remark 2.14. Projection onto the first factor identifies $\widetilde{MT}(V, F)$ up to isogeny with $\mathrm{MT}(V, F)$, unless V has weight 0 and then it equals $\mathrm{MT}(V, F) \times \mathbb{G}_m$. As an illustration, consider $V = \mathbb{Q}(p)$. Then for $(u, v) \in \mathbb{C}^* \times \mathbb{C}^* = \mathbb{S}(\mathbb{C})$, $h_F(u, v)t = (uv)^{-p}t$ and the extended Mumford-Tate group equals $\mathbb{G}_m$ embedded in $\mathbb{G}_m \times \mathbb{G}_m$ via $u \mapsto (u^{-2p}, u)$ where the situation with respect to projection onto the first factor differs for the cases $p = 0$ and $p \neq 0$.

To have a more practical way of determining the Mumford-Tate group, we use as a motivation that all representations of $\mathrm{GL}(V)$ can be found from looking at the induced action on tensors

$$T^{m,n}V = V^{\otimes m} \otimes (V^{\vee})^{\otimes n}.$$

Indeed, this is a property of reductive algebraic groups as we shall see below. Together with the action of $\mathbb{G}_m$ on the Hodge structure of Tate $\mathbb{Q}(p)$ this defines a natural action of $\mathrm{GL}(V) \times \mathbb{G}_m$ on $T^{m,n}V(p)$ and hence an action of the Mumford-Tate group $\widetilde{\mathrm{MT}}(V, F)$ on $T^{m,n}V(p)$. The induced Hodge structure on $T^{m,n}V(p)$ has weight $(m-n)k - 2p$. Assume it is even, say $w = 2q$. Then $\mathrm{HG}(V, F)$ acts trivially on Hodge vectors (i.e. rational type (q, q)-vectors) inside $T^{m,n}V(p)$, while any $t \in w(\mathbb{G}_m(\mathbb{R}))$ multiplies an element in $T^{m,n}V(p)$ of pure type (q, q) by $|t|^{2q}$. Hence, if the weight of $T^{m,n}V(p)$ is zero, the Hodge vectors inside $T^{m,n}V(p)$ are fixed by the entire Mumford-Tate group. The content of the following theorem is the main result of this section.

Theorem 2.15. *The Mumford-Tate group $\widetilde{MT}(V, F)$ is exactly the (largest) algebraic subgroup of $\mathrm{GL}(V) \times \mathbb{G}_m$ which fixes all Hodge vectors inside $T^{m,n}V(p)$ for all (m, n, p) such that $(m-n)k - 2p = 0$. The Hodge group is the subgroup of $\mathrm{GL}(V)$ which fixes all Hodge vectors in all tensor representations $T^{m,n}V$.*

Before embarking on the proof let us recall that an algebraic group is **reductive** if it is the product of an algebraic torus and a (Zariski-connected) semi-simple group, both of which are normal subgroups. A group is **semi-simple** if it has no closed connected commutative normal subgroups except the identity. The groups $\mathrm{SL}(n), \mathrm{SO}(n), \mathrm{SU}(n), \mathrm{Sp}(n)$ are examples of semi-simple groups. The group $\mathrm{GL}(n)$ itself is reductive. In the sequel we use at several points (see [Sata80, I.3]:

Theorem 2.16. *An algebraic group over a field of characteristic zero is reductive if and only if all its finite-dimensional representations decompose into a direct product of irreducible ones.*

We need now a general result about the behaviour of tensor representations for reductive groups G with respect to algebraic subgroups H. For simplicity, assume that $G \subset \mathrm{GL}(V)$. and consider $T^{m,n}V$ as a G-representation. For any subgroup H of G, the set of vectors inside $T^{m,n}V$ fixed by H is as usual denoted by $(T^{m,n}V)^H$. We then put

$$\tilde{H} := \{g \in G \mid \text{there is some}\, (m, n) \text{ such that} \quad g|(T^{m,n}V)^H = \mathrm{id}\}.$$

If g fixes $(T^{m,n})^H$ and g' fixes $(T^{m',n'}V)^H$, then gg'^{-1} fixes $(T^{m-m',n-n'}V)^H$ so that $\tilde{H}$ is a subgroup of G. This group obviously contains H and we want to know when the two groups coincide. This is the criterion:

Lemma 2.17. *In the above notation $H = \tilde{H}$ if H is reductive or if every character of H lifts to a character of G.*

Proof. The crucial remark is that any representation of G is contained in a direct sum of representations of type $T^{m,n}V$ (see [DMOS, I, Prop 3.1]). Also, by Chevalley's theorem (loc. cit.) the subgroup H is the stabilizer of a line L in some finite dimensional representation V, which we may assume to be such a direct sum. If H is reductive, $V = V' \oplus L$ for some H-stable V' and $V^\vee = (V')^\vee \oplus L^\vee$ so that H is exactly the group fixing a generator of $L \otimes L^\vee$ in $V \otimes V^\vee$ and so $H = \tilde{H}$. If all characters of H extend to G, the one-dimensional representation of H given by L comes from a representation of G. Then H is the group fixing a generator of $L \otimes L^\vee$ inside $V \otimes V^\vee$, a tensor representation of the desired type. □

Proof (of the Theorem): We apply the preceding with $G = \mathrm{GL}(V) \times \mathbb{G}_m$ and H the extended Mumford-Tate group. By definition, the largest algebraic subgroup of $\mathrm{GL}(V) \times \mathbb{G}_m$ which fixes all Hodge vectors inside $T^{m,n}V(p)$, $(m-n)k - 2p = 0$ is the group $\tilde{H}$. We must show that $\tilde{H} = H$. To do this, we use the criterion that any rationally defined character $\chi : \mathrm{MT}(V) \to \mathbb{G}_m$ should extend to all of $\mathrm{GL}(V) \times \mathbb{G}_m$. Look at the restriction of this character to the diagonal matrices $\mathbb{G}_m \subset \mathrm{MT}(V, F)$. By Example 2.2 2), it defines a Hodge structure of Tate $\mathbb{Q}(k)$ and so, after twisting W by $\mathbb{Q}(-k)$ the character becomes trivial and so extends to $\mathrm{GL}(V) \times \mathbb{G}_m$ as the trivial character. Then also the original character extends to $\mathrm{GL}(V) \times \mathbb{G}_m$. □

The importance of the previous theorem stems from the following

Observation 2.18. *The rational Hodge substructures of $T^{m,n}V$ are exactly the rational sub-representations of the Mumford-Tate group acting on $T^{m,n}V$.*

Proof. Suppose that $W \subset T^{m,n}V$ is a rational sub-representation of the Mumford-Tate group. Then the composition $h : \mathbb{S} \hookrightarrow \mathrm{MT}(V, F) \to \mathrm{GL}(W)$ defines a rational Hodge structure on W. The converse can be seen in a similar fashion. □

Next, suppose that we have a *polarized* Hodge structure. Almost by definition of a polarization (Def. 2.9-bis) the Hodge group preserves the polarization: for all $t \in \mathrm{U}(1)$ and $u, v \in V$ one has $S(t \cdot u, t \cdot v) = S(u, v)$. Using this one shows:

Theorem 2.19. *The Mumford-Tate group of a Hodge structure which admits a polarization is a reductive algebraic group.*

Proof. It suffices to prove this for the Hodge group. The Weil element $C = h_F(\mathrm{i})$ is a real point of this group. The square acts as $(-1)^k$ on V and hence lies in the centre of $\mathrm{MT}(V, F)$. The inner automorphism $\sigma := \mathrm{ad}(C)$ of $\mathrm{HG}(V, F)$ defined by C is therefore an involution. Such an involution defines a real-form

G_σ of the special Mumford-Tate group. By definition this is the real algebraic group G_σ whose real points are

$$G_\sigma(\mathbb{R}) = \{g \in \mathrm{HG}(V,F)(\mathbb{C}) \mid \sigma(g) = g\}.$$

There is an isomorphism

$$G_\sigma(\mathbb{C}) \xrightarrow{\cong} \mathrm{HG}(V,F)(\mathbb{C})$$

such that complex conjugation on $G_\sigma(\mathbb{C})$ followed by σ corresponds to complex conjugation on $\mathrm{HG}(V,F)(\mathbb{C})$. This means that

$$\sigma(\bar{g}) = \mathrm{ad}(C)(\bar{g}) = g. \tag{II–10}$$

If the Hodge structure (V,F) admits a polarization Q, the following computation shows that G_σ admits a positive definite form and hence is compact. For $u,v \in V_\mathbb{C}$ and $g \in \mathrm{HG}(V,F)(\mathbb{C})$ we have, applying (II–10)

$$Q(Cu,\bar{v}) = (\bar{g}Cu, \bar{g}\bar{v}) = Q(CC^{-1}\bar{g}Cu) = Q(C\,\mathrm{ad}(C)(\bar{g})u, \overline{gv}) = Q(Cgu, \overline{gv}).$$

It follows that the positive definite form on $V_\mathbb{R}$ given by $Q(C-,-)$ is invariant under G_σ.

The compactness of G_σ implies that any finite dimensional representation of it completely decomposes into a direct product of irreducible ones and so, by the characterization of reductive groups, G_σ and also the special Mumford-Tate group is reductive. □

Since $\mathrm{MT}(V,F)$ is the product of the Hodge group and the diagonal matrices and since a group is semi-simple if and only if the identity is the only normal closed connected abelian subgroup, the previous theorem implies:

Corollary 2.20. *The Hodge group is semi-simple precisely when the centre of the Mumford Tate group consists of the scalar matrices.*

2.3 Hodge Filtration and Hodge Complexes

2.3.1 Hodge to De Rham Spectral Sequence

Recall (Theorem 1.8) that for a Kähler manifold X, we have a Hodge decomposition and an associated Hodge filtration

$$H^k(X,\mathbb{C}) = \bigoplus_{p+q=k} H^{r,s}(X), \quad F^pH^k(X,\mathbb{C}) = \bigoplus_{r\geq p} H^{r,k-r}(X).$$

Let us first explain how to define a putative Hodge filtration on De Rham cohomology of any compact complex manifold X in terms of a spectral sequence relating the holomorphic and differentiable aspects. First embed the holomorphic De Rham complex into the complexified De Rham complex

$$\Omega_X^\bullet \xrightarrow{j} \mathcal{E}_X^\bullet(\mathbb{C}).$$

The decomposition into types of the sheaf complex $\mathcal{E}^\bullet(\mathbb{C})$ gives the filtered complex

$$F^p(\mathcal{E}^\bullet(\mathbb{C})) = \bigoplus_{r \geq p} \mathcal{E}_X^{r,\bullet - r}$$

and the homomorphism j becomes a filtered homomorphism provided we put the trivial filtration

$$\sigma^{\geq p}\Omega_X^\bullet = \{0 \to \cdots \to 0 \to \Omega_X^p \to \Omega_X^{p+1} \cdots \to \Omega_X^n\} \quad (n = \dim X).$$

on the De Rham complex. Then $\mathrm{Gr}^p(j)$ gives the Dolbeault complex

$$0 \to \Omega_X^p \to \mathcal{E}_X^{p,0} \xrightarrow{\bar{\partial}} \to \mathcal{E}_X^{p,1} \xrightarrow{\bar{\partial}} \cdots$$

By Dolbeault's lemma this is exact and so j induces a quasi-isomorphism on the level of graded complexes. So the E_1-terms of the first spectral sequence, which computes the hypercohomology of the graded complex (see equation (A–29)) is just the De Rham-cohomology of the preceding complex, i.e. ${}'E_1^{p,q} = H^q(X, \Omega_X^p)$. The first spectral sequence of hypercohomology (viewed as coming from the trivial filtration) reads therefore

$${}'E_1^{p,q} = H^q(X, \Omega_X^p) \Longrightarrow \mathbb{H}^{p+q}(X, \Omega_X^\bullet) = H_{\mathrm{DR}}^{p+q}(X; \mathbb{C})$$

(**Hodge to De Rham spectral sequence**).

Consider now the filtration on the abutment:

Definition 2.21. The **putative Hodge filtration** on $H_{\mathrm{DR}}^k(X; \mathbb{C})$ is given by

$$F^p H_{\mathrm{DR}}^k(X; \mathbb{C}) = \mathrm{Im}\left(\mathbb{H}^k(X, \sigma^{\geq p}\Omega^\bullet) \xrightarrow{\alpha_p} \mathbb{H}^k(X, \Omega^\bullet)\right).$$

The **Hodge subspaces** are given by

$$H^{p,q}(X) = F^p H_{\mathrm{DR}}^{p+q}(X; \mathbb{C}) \cap \overline{F^q H_{\mathrm{DR}}^{p+q}(X; \mathbb{C})}.$$

The terminology is justified by considering a Kähler manifold.

Proposition 2.22. *Let X be a compact Kähler manifold. Then the Hodge to De Rham spectral sequence degenerates at E_1; the putative Hodge filtration coincides with the actual Hodge filtration, and the Hodge subspaces $H^{p,q}(X)$ coincide with the subspace of the De Rham classes having a harmonic representative of type (p, q).*

Proof. As seen before (see the discussion following Theorem B.18), we have a canonical isomorphism $H^{r,s}(X) \cong H^s(X, \Omega_X^r)$ (Dolbeault's theorem) and so

$$\sum_{p+q=k} \dim {}'E_1^{p,q} = \sum_{p+q=k} \dim H^{p,q} = \dim H^k(X; \mathbb{C}) = \sum_{p+q=k} \dim E_\infty^{p,q}$$

which implies that the spectral sequence degenerates at E_1 (since E_{r+1} is a subquotient of E_r). Hence the map α_p is injective and $h^{p,k-p}(X) = \dim \mathbb{H}^k(X, \sigma^{\geq p}\Omega^\bullet)) - \dim \mathbb{H}^k(X, \sigma^{\geq p+1}\Omega^\bullet))$ and so

$$\dim \mathbb{H}^k(X, \sigma^{\geq p}\Omega^\bullet)) = \sum_{r\geq p} \dim H^{r,k-r}(X) = \dim F^p H^k(X;\mathbb{C})$$

which means that the image of j_p^* is $F^pH^k(X;\mathbb{C})$. Also $\mathrm{Gr}^p(j)$ induces an isomorphism $H^q(\Omega_X^p) \to H^{p,q}(X)$ and so

$$F^pH^k(X;\mathbb{C}) = \bigoplus_{r\geq p} H^{r,k-r}(X).\square$$

Remark 2.23. The proof of the degeneration of the Hodge to De Rham spectral sequence hints at an algebraic approach to the Hodge decomposition. In fact Faltings [Falt] and Deligne-Illusie [Del-Ill] found a purely algebraic proof for the degeneracy of the Hodge to De Rham which works in any characteristic. The De Rham cohomology in this setting by definition is the hypercohomology of the algebraic De Rham complex, the algebraic variant of the holomorphic De Rham complex. The Hodge filtration is again induced by the trivial filtration on the De Rham complex. The proof then proceeds by first showing it first in characteristic p for smooth varieties of dimension $> p$ which can be lifted to the ring of Witt vectors of length 2. Since this can be arranged for if the variety is obtained from a variety in characteristic zero by reduction modulo p the result then follows in characteristic zero. In passing we note that there are many examples of surfaces in characteristic p for which the Hodge to De Rham spectral sequence does *not* degenerate. See [Del-Ill, 2.6 and 2.10] for a bibliography.

2.3.2 Strong Hodge Decompositions

Since by Corollary 1.10 the space $H^{p,q}(X)$ can be characterized as the subspace of $H^{p+q}_{\mathrm{DR}}(X;\mathbb{C})$ of classes representable by closed (p,q)-forms, the previous proposition motivates the following definition.

Definition 2.24. Let X be a compact complex manifold. We say that $H^k(X;\mathbb{C})$ admits a **Hodge decomposition in the strong sense** if

1) For all p and q with $p+q=k$ the Hodge (p,q)-subspace $H^{p,q}(X)$ as defined above can be identified with the subspace of $H^k(X;\mathbb{C})$ consisting of classes representable by closed forms of type (p,q). The resulting map

$$H^{p,q}(X) \to H^{p,q}_{\bar\partial}(X) \cong H^q(\Omega_X^p)$$

is required to be an isomorphism.

2) There is direct decomposition

$$H^k_{\mathrm{DR}}(X;\mathbb{C}) = \bigoplus_{p+q=k} H^{p,q}(X).$$

3) The natural morphism from Bott-Chern cohomology to De Rham cohomology

$$H^{p,q}_{\mathrm{BC}}(X) = \frac{d\text{-closed forms of type } (p,q)}{\partial\bar{\partial}\Gamma\mathcal{E}_X^{p-1,q-1}} \to H^{p+q}_{\mathrm{DR}}(X)\otimes\mathbb{C}$$

which sends the class of a d-closed (p,q)-form to its De Rham class is injective with image $H^{p,q}(X)$.

Example 2.25. For any compact Kähler manifold X the Hodge decomposition on $H^k(X;\mathbb{C})$ is a Hodge decomposition in the strong sense.

By definition, the graded pieces of the putative Hodge filtration sequence give the $'E_\infty$-terms of the spectral sequence. If the Hodge to De Rham spectral sequence degenerates at E_1 it follows therefore that these graded pieces are canonically isomorphic to the Dolbeault groups. It does *not* imply that the putative Hodge filtration defines a Hodge structure on the De Rham groups. It is for instance not true in general that the graded pieces are isomorphic to the Hodge subspaces, even when the spectral sequence degenerates at E_1.

Example 2.26. As is well known (see e.g. [B-H-P-V, Chapter IV]), for surfaces the Hodge to De Rham spectral sequence, also called the Fröhlicher spectral sequence, degenerates at E_1 whereas there is no Hodge decomposition on $H^1(X)$ for a non-Kähler surface X since $b_1(X)$ is odd for those. This is for example the case of a Hopf surface which is the quotient of $\mathbb{C}^2 - \{0\}$ by the cyclic group of dilatations $z \to 2^k z$, $k \in \mathbb{Z}$. Such a surface is indeed diffeomorphic to $S^1 \times S^3$ and its first Betti number is 1 and so H^1 can never admit a Hodge decomposition. In fact the two Hodge subspaces are equal and hence equal to $F^1 = F^0$, the Dolbeault group $H^1(\mathcal{O}_X)$ maps isomorphically onto these, whereas the other Dolbeault group $H^0(\Omega_X)$ is zero and maps to $F^0/F^1 = 0$.

The following proposition summarizes what one can say in general. We first introduce some terminology. We say that a filtration F on a the complexification of a real vector space V is *k-transverse* if $F^p \cap \overline{F^{q+1}} = \{0\}$ whenever $p+q=k$. Note that this is automatic when F defines a real Hodge structure of weight k on V and a k-transverse filtration is a Hodge filtration if $\dim F^p + \dim F^{q+1} = \dim V$ whenever $p+q=k$.

Proposition 2.27. *Suppose that the Hodge to De Rham spectral sequence degenerates. Then the Dolbeault group $H^q(X,\Omega^p_X)$ is canonically isomorphic to $\mathrm{Gr}^p_F H^{p+q}_{\mathrm{DR}}(X;\mathbb{C})$ and one has the equality*

$$b_k := \dim H^k(X;\mathbb{C}) = \sum_{p+q=k} \dim H^q(\Omega_X^p).$$

Suppose that the putative Hodge filtration on $H^k(X;\mathbb{C})$ is k-transverse, and that it is $(2n-k)$-transverse on $H^{2n-k}(X;\mathbb{C})$. Then the putative Hodge filtrations on $H^k(X;\mathbb{C})$ and $H^{2n-k}(X;\mathbb{C})$ are both Hodge filtrations. For $p+q=k$ and $p+q=2n-k$ the spaces $H^q(X,\Omega_X^p) \cong \mathrm{Gr}_F^p H_{\mathrm{DR}}^{p+q}(X;\mathbb{C})$ get canonically identified with $H^{p,q}(X)$.

Proof. We only need to prove the statements about the putative Hodge filtration. For this, we provisionally set $h^{p,q} = \dim H^q(\Omega_X^p)$ so that $b_k = \sum_{p+q=k} h^{p,q}$. Now for any t we have $\dim F^t = \dim \overline{F^t} = \sum_{r\geq t} h^{r,k-r}$. The assumption on the putative Hodge filtration then implies

$$\sum_{r\geq p} h^{r,k-r} + \sum_{r\geq k-p+1} h^{r,k-r} \leq b_k = \sum_r h^{r,k-r}$$

and hence

$$\sum_{r\geq p} h^{r,k-r} \leq \sum_{r\leq k-p} h^{r,k-r}.$$

This inequality for $2n-k$-cohomology with p replaced by $n-k-p$, together with Serre duality ($h^{p,q} = h^{n-p,n-q}$) yields the reverse inequality. So we have equality and hence the dimensions of F^p and $\overline{F^{q+1}}$ add up to $\dim H^{p+q}(X;\mathbb{C})$ when $p+q=k$ or $p+q=2n-k$. So we get Hodge structures and $F^pH^k(X;\mathbb{C}) = \bigoplus_{r\geq p} H^{r,s}(X)$. Since $H^q(\Omega_X^p)$ is canonically isomorphic to $\mathrm{Gr}_F^p H^{p+q}(X;\mathbb{C}) \cong H^{p,q}(X)$, the last assertion follows as well. □

In fact, we can even show that the assumptions of the preceding Proposition guarantee a Hodge decomposition in the strong sense on $H^k(X)$ and $H^{2m-k}(X)$. Indeed, we have the following statement which is an algebraic version of the $\partial\bar\partial$-Lemma (1.9). For a proof see [B-H-P-V, I, Lemma 13.6].

Corollary 2.28. 1) *Under the assumptions of Proposition 2.27, any cohomology class in degree k or in degree $(2n-k)$ can be represented by a form which is ∂- as well as $\bar\partial$-closed.*

2) *For a d-closed (p,q)-form α, $p+q=k$ or $p+q=2n-k$ the following statements are equivalent:*

a) *$\alpha = d\beta$ for some $p+q-1$-form β;*

b) *$\alpha = \bar\partial\beta''$ for some $(p,q-1)$-form β'';*

c) *$\alpha = \partial\bar\partial\gamma$ for some $(p-1,q-1)$-form γ;*

4) *The natural morphism*

$$H_{\mathrm{BC}}^{p,q}(X) = \frac{d\text{-closed forms of type } (p,q)}{\partial\bar\partial\Gamma\mathcal{E}_X^{p-1,q-1}} \to H_{\mathrm{DR}}^{p+q}(X)\otimes\mathbb{C}$$

which sends the class of a d-closed (p,q)-form to its De Rham class is injective with image $H^{p,q}(X)$. In particular, the latter space consists precisely of the De Rham classes representable by a closed form of type (p,q).

5) *For $p+q=k$ or $p+q=2n-k$ the natural map*

$$H^{p,q}(X) \to H^{p,q}_{\bar\partial}(X) \cong H^q(\Omega^p_X)$$

resulting from the identification of $H^{p,q}(X)$ as the space consisting of the De Rham classes representable by a closed form of type (p,q) is an isomorphism.

Despite the fact that holomorphic images of Kähler manifolds of the same dimension are not always Kähler [Hart70, p. 443] we can show:

Theorem 2.29. *Let X, Y be compact complex manifolds. Suppose that X is Kähler and that $f : X \to Y$ is a surjective holomorphic map. Then $H^k(Y)$ admits a Hodge decomposition in the strong sense. In fact $f^* : H^k(Y;\mathbb{R}) \to H^k(X;\mathbb{R})$ is injective and $f^*H^k(Y;\mathbb{R})$ is a real Hodge substructure of $H^k(X;\mathbb{R})$.*

Proof. We first show that f^* is injective. In fact, this holds for any surjective differentiable map $f : X \to Y$ between compact differentiable manifolds. To see this, first reduce to the equi-dimensional case by choosing a submanifold $Z \subset X$ to which f restricts as a generically finite map, say of degree d. With $f_!$ Poincaré dual to f_*, the composition $f_! \circ f^*$ is multiplication with d and so f^* is injective.

Next, we observe that for $m = \dim Y$, a generator of $H^{2m}(Y;\mathbb{C}) = H^{m,m}(Y) = H^m(\Omega^m_Y)$ is represented by the volume form vol_h with respect to some hermitian metric h on Y. If ω is the Kähler form on X, the form ω^c, $c = \dim X - \dim Y$ restricts to a volume form on the generic fibre F of f and hence

$$\int_X f^*(\mathrm{vol}_h) \wedge \omega^c = \int_Y \mathrm{vol}_h \int_F \omega^c \neq 0.$$

So $f^* : H^m(\Omega^m_Y) \to H^m(\Omega^m_X)$ is non-zero. Now one uses Serre duality to prove injectivity on $H^q(\Omega^p_Y)$ for all p and q. Indeed, given any non-zero class $\alpha \in H^q(\Omega^p_Y)$ choose $\beta \in H^{m-q}(\Omega^{m-p}_Y)$ such that $\alpha \wedge \beta \neq 0$. Then $f^*\alpha \wedge f^*\beta = f^*(\alpha \wedge \beta) \neq 0$ and hence $f^*\alpha \neq 0$.

Now compare the Hodge to De Rham spectral sequence for Y with that for X. What we just said shows that the E_1-term of the first injects into the E_1-term of the latter. For X the Hodge to De Rham spectral sequence degenerates and so $d_r = 0, r \geq 1$ and $E_1 = E_2 = \cdots$. It follows recursively that the same holds for the Hodge to De Rham spectral sequence for Y. In particular, it degenerates. But more is true. The map f^* on the level of spectral sequences induces an injection $F^pH^k(Y) \hookrightarrow F^pH^k(X)$ and since f^* commutes with complex conjugation, we conclude that $F^pH^k(Y)$ meets $\overline{F^{k-p+1}H^k(Y;\mathbb{C})}$ only in $\{0\}$ and so the hypothesis of Prop. 2.27 is satisfied and the result follows upon applying Corollary 2.28. □

By Hironaka's theorem [Hir64] the indeterminacy locus of a meromorphic map $X \dashrightarrow Y$ can be eliminated by blowing up. Since the blow up of a Kähler manifold is again Kähler (see [Kod54, Sect. 2, Lemma 1]) we can apply the previous theorem to a manifold bimeromorphic to a Kähler manifold.

Corollary 2.30. *Let X be a compact complex manifold bimeromorphic to a Kähler manifold. Then $H^k(X;\mathbb{C})$ admits a strong Hodge decomposition. This is in particular true for a (not necessarily projective) compact algebraic manifold. In particular, the previous theorem remains true when X is only bimeromorphic to a Kähler manifold.*

2.3.3 Hodge Complexes and Hodge Complexes of Sheaves

Comparison between complexes should take place in suitable derived categories. We prefer however to give explicit morphisms realizing these comparison morphisms. To fix ideas we introduce the following definitions.

Definition 2.31. Let $K^\bullet$, $L^\bullet$ two bounded below complexes in an abelian category. A **pseudo-morphism** between $K^\bullet$ and $L^\bullet$ is a chain of morphisms of complexes

$$K^\bullet \xrightarrow{f} K_1^\bullet \xleftarrow[\sim]{\text{qis}} K_2^\bullet \xrightarrow[\sim]{\text{qis}} \cdots \xrightarrow[\sim]{\text{qis}} K_{n+1}^\bullet = L^\bullet.$$

It induces a morphism in the derived category. We shall denote such a pseudo-morphism by

$$f : K^\bullet \dashrightarrow L^\bullet.$$

If also f is a quasi-isomorphism we speak of a **pseudo-isomorphism**. It becomes invertible in the derived category. We denote these by

$$f : K^\bullet \overset{\text{qis}}{\underset{\sim}{\dashrightarrow}} L^\bullet.$$

A **morphism** between two pseudo-morphisms $K^\bullet \xrightarrow{f} K_1^\bullet \xleftarrow[\sim]{\text{qis}} \cdots \xrightarrow[\sim]{\text{qis}} K_m^\bullet$ and $L^\bullet \xrightarrow{g} L_1^\bullet \xleftarrow[\sim]{\text{qis}} \cdots \xrightarrow[\sim]{\text{qis}} L_m^\bullet$ consists of a sequence of morphism $K^j \to L^j$, $j = 1, \ldots, m$ such that the obvious diagrams commute. Note that such morphisms are only possible between sequences of equal length.

Definition 2.32. 1) Let R a noetherian subring of $\mathbb{C}$ such that $R \otimes \mathbb{Q}$ is a field (mostly R will be $\mathbb{Z}$ or $\mathbb{Q}$). An **R-Hodge complex $K^\bullet$ of weight** m consist of

- A bounded below complex of R-modules $K_R^\bullet$ such that the cohomology groups $H^k(K_R^\bullet)$ are R-modules of finite type,
- A bounded below filtered complex $(K_\mathbb{C}^\bullet, F)$ of complex vector spaces with differential **strictly compatible** with F and a
- **comparison morphism** $\alpha : K_R^\bullet \dashrightarrow K_\mathbb{C}^\bullet$, which is a pseudo-morphism in the category of bounded below complexes of R-modules and becomes a pseudo-isomorphism after tensoring with $\mathbb{C}$.

$$\alpha \otimes \mathrm{id} : K_R^\bullet \otimes \mathbb{C} \overset{\text{qis}}{\underset{\sim}{\dashrightarrow}} K_\mathbb{C}^\bullet,$$

and such that the induced filtration on $H^k(K^\bullet_{\mathbb{C}})$ determines an R-Hodge structure of weight $m+k$ on $H^k(K^\bullet_R)$.

Its associated **Hodge-Grothendieck characteristic** is

$$\chi_{\mathrm{Hdg}}(K^\bullet) := \sum_{k\in\mathbb{Z}}(-1)^k \left[H^k(K^\bullet)\right] \in K_0(\mathfrak{hs}_R).$$

2) Let X be a topological space. An R-**Hodge complex of sheaves** of weight m on X consists of the following data

- A bounded below complex of sheaves of R-modules $\mathcal{K}^\bullet_R$ such that the hypercohomology groups $\mathbb{H}^k(X,\mathcal{K}^\bullet)$ are finitely generated as R-modules,
- A filtered complex of sheaves of complex vector spaces $\{\mathcal{K}^\bullet_{\mathbb{C}},F\}$ and a pseudo-morphism $\alpha : \mathcal{K}^\bullet_R \dashrightarrow \mathcal{K}^\bullet_{\mathbb{C}}$ in the category of sheaves of R-modules on X inducing a pseudo-isomorphism (of sheaves of $\mathbb{C}$-vector spaces)

$$\alpha\otimes\mathrm{id} : \mathcal{K}^\bullet_R\otimes\mathbb{C} \overset{\text{qis}}{\underset{\sim}{\dashrightarrow}} \mathcal{K}^\bullet_{\mathbb{C}},$$

and such that the R-structure on $\mathbb{H}^k(\mathcal{K}^\bullet_{\mathbb{C}})$ induced by α and the filtration induced by F determine an R-Hodge structure of weight $k+m$ for all k. Moreover, one requires that the spectral sequence for the derived complex $R\Gamma(X,\mathcal{K}^\bullet_{\mathbb{C}})$ (see (B–12) with the induced filtration

$$\mathbb{H}^{p+q}(X,\mathrm{Gr}^p_F\,\mathcal{K}^\bullet_{\mathbb{C}}) \Longrightarrow \mathbb{H}^{p+q}(X,\mathcal{K}^\bullet_{\mathbb{C}})$$

degenerates at E_1 (by Lemma A.42 this is equivalent to saying that the differentials of the derived complex are strict).

3) A **morphism** of Hodge complexes (of sheaves) of weight m, consists of a triple$(h_R,h_{\mathbb{C}},\kappa)$ where h_R is a morphism of (of sheaves of) R-modules, $h_{\mathbb{C}}$ a homomorphism of (sheaves of) $\mathbb{C}$-vector spaces and $\kappa : \alpha\to\beta$ is a morphism of pseudo-morphisms.

The notions of a Hodge complex and that of a Hodge complex of sheaves are related in the following way.

Proposition 2.33. *Given an R-Hodge complex of sheaves on X of weight m, say*

$$\mathcal{K}^\bullet = (\mathcal{K}^\bullet_R,(\mathcal{K}^\bullet_{\mathbb{C}},F),\alpha)\,,$$

any choice of representatives for the triple

$$R\Gamma\mathcal{K}^\bullet = (R\Gamma(\mathcal{K}^\bullet_R),(R\Gamma(\mathcal{K}^\bullet_{\mathbb{C}},F),R\Gamma(\alpha))$$

yields an R-Hodge complex. With $a_X : X\to \mathrm{pt}$ the constant map, we have

$$\chi_{\mathrm{Hdg}}(R\Gamma(\mathcal{K}^\bullet)) = [R(a_X)_*\mathcal{K}^\bullet] \in K_0(\mathfrak{hs}_R).$$

Here we view $R(a_X)_*\mathcal{K}^\bullet$, a complex of sheaves over the point pt, as a complex of R-modules whose (finite rank) cohomology groups $\mathbb{H}^k(X,\mathcal{K}^\bullet)$ are R-Hodge structures so that the right hand side makes sense in $K_0(\mathfrak{hs}_R)$.

Example 2.34. The existence of a strong Hodge decomposition for Kähler manifolds (Example 2.25) in fact tells us that for X a compact Kähler manifold, the constant sheaf $\underline{\mathbb{Z}}_X$, the holomorphic De Rham complex $\Omega_X^\bullet$ with the trivial filtration σ together with the inclusion $\underline{\mathbb{Z}}_X \hookrightarrow \Omega_X^\bullet$ (which gives the pseudo-isomorphism $\underline{\mathbb{C}}_X \to \Omega_X^\bullet$) is an integral Hodge complex of sheaves of weight 0. The same is true for any complex manifold bimeromorphic to a Kähler manifold. This complex will be called the **Hodge-De Rham complex of sheaves** on X and be denoted by

$$\mathcal{H}dg^\bullet(X) = (\underline{\mathbb{Z}}_X, (\Omega_X^\bullet, \sigma), \underline{\mathbb{Z}}_X \hookrightarrow \Omega_X^\bullet).$$

Taking global sections on the Godement resolution gives $R\Gamma\mathcal{H}dg^\bullet(X)$, the canonically associated **De Rham complex** of X with Hodge-Grothendieck characteristic

$$\chi_{\mathrm{Hdg}}(X) = \sum_{k\in\mathbb{Z}}(-1)^k \left[H^k(X)\right] = [R(a_X)_*\underline{\mathbb{Z}}_X^{\mathrm{Hdg}}] \in K_0(\mathfrak{hs}). \quad \text{(II–11)}$$

Lemma-Definition 2.35. 1) For an R-Hodge complex of sheaves $\mathcal{K}^\bullet = (\mathcal{K}_R^\bullet, (\mathcal{K}_\mathbb{C}^\bullet, F), \alpha)$ of weight m, and $k \in \mathbb{Z}$ we define the **k-th Tate-twist** by

$$\mathcal{K}^\bullet(k) := (\mathcal{K}_R^\bullet \otimes \mathbb{Z}(2\pi\mathrm{i})^k, (\mathcal{K}_\mathbb{C}^\bullet, F[k]), \alpha \cdot (2\pi\mathrm{i})^k).$$

It is an R-Hodge sheaf of weight $m - 2k$. This operation induces the Tate-twist in hypercohomology

$$\mathbb{H}^\ell(X, \mathcal{K}_\mathbb{C}^\bullet(k)) = \mathbb{H}^\ell(X, \mathcal{K}_\mathbb{C}^\bullet)(k).$$

A similar definition holds for $K^\bullet(k)$ where $K^\bullet$ is an R-Hodge complex.
2) We define the **shifted complex** by

$$\mathcal{K}^\bullet[r] := (\mathcal{K}_R^\bullet[r], (\mathcal{K}_\mathbb{C}^\bullet[r], F[r]), \alpha[r]).$$

It is a Hodge complex of sheaves of weight $m + r$ A similar definition holds for $K^\bullet[r]$ where $K^\bullet$ is an R-Hodge complex.

2.4 Refined Fundamental Classes

We recall (Proposition 1.14) that for any irreducible subvariety Y of codimension d in a compact algebraic manifold X the integral fundamental class $\mathrm{cl}(Y) \in H^{2d}(X)$ has pure type (d, d). This means that the fundamental class belongs to the d-th Hodge filtration level. So we can also define a **fundamental Hodge cohomology class**

$$\mathrm{cl}_{\mathrm{Hdg}}(Y) \in F^d H^{2d}(X; \mathbb{C}) = \mathbb{H}^{2d}(X, F^d \Omega_X^\bullet)$$

and the integral class maps to it under the inclusion $\mathbb{Z} \hookrightarrow \mathbb{C}$. To keep track of various powers of $2\pi\mathrm{i}$ introduced when integrating forms, it is better to replace this inclusion by

$$\epsilon^d : \mathbb{Z}(d) \hookrightarrow \mathbb{C} \tag{II–12}$$

and we consider the fundamental class as a class $\mathrm{cl}(Y) \in H^{2d}(X, \mathbb{Z}(d))$ which, under ϵ^d, maps to the image of the Hodge class in $H^{2d}(X;\mathbb{C})$. This is summarized in the following diagram

$$\begin{array}{ccccc}
\mathrm{cl}_{\mathrm{Hdg}}(Y) & & \longmapsto & & \mathrm{cl}_{\mathbb{C}}(Y) \\
\cap & & & & \cap \quad \uparrow \\
\mathbb{H}^{2d}(X, F^d\Omega_X^\bullet) & \hookrightarrow & \mathbb{H}^{2d}(X, \Omega_X^\bullet) & = & H^{2d}(X;\mathbb{C}) \quad \Big| \\
 & & & & \uparrow{\scriptstyle (\epsilon^d)_*} \quad \Big| \\
 & & & & H^{2d}(X;\mathbb{Z}(d)) \ni \mathrm{cl}(Y)
\end{array}$$

Remark. There is a much more intrinsic reason to consider $\mathrm{cl}(Y)$ as a class inside $H^{2d}(X, \mathbb{Z}(d))$ rather than as an integral class. The reason is that the only algebraically defined resolution of $\mathbb{C}$ is the holomorphic De Rham complex $\Omega_X^\bullet$ and the only algebraically defined fundamental class is coming from Grothendieck's theory of Chern classes. To *algebraically* relate the first Chern class which is naturally living in $H^1(\mathcal{O}_X^*) = \mathbb{H}^2(X, 0 \to \mathcal{O}_X^* \to 0)$ to a class in $H^2(X,\mathbb{C}) = \mathbb{H}^2(X, \Omega_X^\bullet)$ one uses $d\log : \mathcal{O}_X^* \to \Omega_X^1$ and zero else. This misses out the factor $2\pi\mathrm{i}$ which is inserted in the C^∞ De Rham theory. It follows that $\mathrm{cl}(Y)$ as defined in this way is no longer integral, but has values in $\mathbb{Z}(d)$. See [DMOS, I.1] where this is carefully explained. This remark becomes relevant when one wants to compare fundamental classes for algebraic varieties defined over fields $k \subset \mathbb{C}$ when one changes the embedding of k in $\mathbb{C}$.

Remark 2.36. Contininuing the preceding Remark, suppose that X is a non-singular algebraic variety defined over a field k of finite transcendence degree over $\mathbb{Q}$. Any embedding $\sigma : k \hookrightarrow \mathbb{C}$ defines a complex manifold $X^{(\sigma)}$ and a codimension d cycle Z on X defines a fundamental class $\mathrm{cl}^{(\sigma)}(Z) \in H^{2d}(X^{(\sigma)};\mathbb{C})$ which is rational in the sense that it belongs to $H^{2d}(X^{(\sigma)}; 2\pi\mathrm{i}\mathbb{Q})$. On the other hand, we have the algebraic De Rham groups $H^m_{\mathrm{DR}}(X/k)$ which are k-spaces, they are the hypercohomology groups of the algebraic De Rham complex $\Omega^\bullet_{X/k}$. These compare to complex cohomology through a canonical comparison isomorphism

$$\iota_\sigma : H^m_{\mathrm{DR}}(X/k) \otimes_{\sigma,k} \mathbb{C} \xrightarrow{\sim} H^m(X^{(\sigma)};\mathbb{C})$$

and under this isomorphism for $m = 2d$ the class $\mathrm{cl}(Z)$ on the right corresponds to a class

$$\mathrm{cl}_{\mathrm{B}}(Z) \in H^{2d}_{\mathrm{DR}}(X/k) \otimes (2\pi\mathrm{i})^d := H^{2d}(X)(d).$$

Then the class $\iota_\sigma\, \mathrm{cl}_{\mathrm{B}}(Z)$ is rational in the above sense. This motivates the definition of an absolute Hodge class:

Definition 2.37. Let X be a non-singular algebraic variety defined over a field k of finite transcendence degree over $\mathbb{Q}$. A class $\beta \in H^{2d}(X)(d)$ is **absolute Hodge** if for all embeddings $\sigma : k \hookrightarrow \mathbb{C}$ the image $\iota_\sigma(\beta) \in H^{2d}(X^{(\sigma)}; \mathbb{C})$ is rational.

If such a class β has the property that $\iota_\sigma(\beta)$ is rational for just one embedding we speak of a **Hodge class**. These come up in the Hodge conjecture 1.16 for a complex projective variety. To explain this, note that such a variety is of course defined over a given subfield k of $\mathbb{C}$ of finite transcendence degree over $\mathbb{Q}$ and there is a preferred embedding $k \hookrightarrow \mathbb{C}$.

Deligne's "hope" is that like the algebraic cycle classes, all such Hodge classes are absolute Hodge. This has been verified only for abelian varieties [DMOS].

We now continue our study of refined cycle classes in the setting of local cohomology, the main result being as follows.

Theorem 2.38. *Let X be a compact algebraic manifold and let $Y \subset X$ be an irreducible d-dimensional subvariety. Then the following variants of interrelated fundamental classes exist:*

1) *There is a* ***refined Thom class***

$$\tau_{\mathrm{Hdg}}(Y) \in \mathbb{H}^{2d}_Y(X, F^d\Omega^\bullet_X)$$

whose image under the map $\mathbb{H}^{2d}_Y(X, F^d\Omega^\bullet_X) \to \mathbb{H}^{2d}_Y(X, \Omega^\bullet_X) = H^{2d}_Y(X; \mathbb{C})$ coincides with the image under the map (II–12) of the Thom class $\tau(Y) \in H^{2d}_Y(X, \mathbb{Z}(d))$.

2) *There is a class $\tau^{d,d} \in H^d_Y(X, \Omega^d_X)$ which is the projection of the refined Thom class.*
3) *Forgetting supports, the class $\tau_{\mathrm{Hdg}}(Y)$ maps to $\mathrm{cl}_{\mathrm{Hdg}}(Y)$.*
4) *The various classes in this construction are related as follows*

$$\begin{array}{ccccc} \tau^{d,d}(Y)) & \longleftarrow & \tau_{\mathrm{Hdg}}(Y) & \longmapsto & \tau(Y) \\ \downarrow & & \downarrow & & \downarrow \\ \mathrm{cl}^{d,d}(Y) & \longleftarrow & \mathrm{cl}_{\mathrm{Hdg}}(Y) & \longmapsto & \mathrm{cl}_{\mathbb{C}}(Y) \end{array}$$

and where the elements come from the commutative diagram

$$\begin{array}{ccccc} H^d_Y(\Omega^d_Y) & \longleftarrow & \mathbb{H}^{2d}_Y(X, F^d\Omega^\bullet_X) & \longrightarrow & H^{2d}_Y(X; \mathbb{C}) \\ \downarrow & & \downarrow & & \downarrow \\ H^d(\Omega^d_X) & \longleftarrow & \mathbb{H}^{2d}(X, F^d\Omega^\bullet_X) = F^dH^{2d}(X; \mathbb{C}) & \longrightarrow & H^{2d}(X; \mathbb{C}) \end{array}$$

We start with a localizing tool. Let $\mathcal{F}$ be any sheaf on X. The assignment $U \mapsto H^k_Y(U, \mathcal{F})$ defines a presheaf on X whose associated sheaf is denoted by $\mathcal{H}^k_Y(\mathcal{F})$. These sheaves are related to the local cohomology groups through a spectral sequence

$$E_2^{r,s} = H^r(X, H_Y^s(\mathcal{F})) \Longrightarrow H_Y^{r+s}(X, \mathcal{F}) \qquad \text{(II–13)}$$

which is the second spectral sequence associated to the functor of taking sections with support in Y.

Lemma 2.39. *Let X be a complex manifold, $Y \subset X$ a codimension c subvariety and $\mathcal{E}$ a locally free sheaf on X. Then*

1) *the cohomology sheaf satisfies*

$$H_Y^q(\mathcal{E}) = 0, \quad q < c;$$

2) *there is an isomorphism*

$$H_Y^c(X, \mathcal{E}) \xrightarrow{\sim} H^0(X, H_Y^c(\mathcal{E})).$$

Proof. For a proof of the first assertion see [S-T, Prop. 1.12]. The second assertion then follows from the spectral sequence (II–13). □

We state a consequence for hypercohomology. We assume that we have a complex $\mathcal{K}^\bullet$ of locally free sheaves on X and we consider the first spectral sequence with respect to the trivial filtration $\sigma^{\geq p} = F^p$ for the functor of hypercohomology with supports in Y whose E_1-terms are

$$E_1^{q,r} = H_Y^q(X, F^s\mathcal{K}^r) \Longrightarrow \mathbb{H}_Y^{q+r}(F^s\mathcal{K}_X^\bullet), \quad F^s\mathcal{K}^r = \begin{cases} \mathcal{K}^r & \text{if } r \geq s \\ 0 & \text{if } r < s. \end{cases}$$

We find:

Corollary 2.40. *For a codimension c subvariety $Y \subset X$, we have*

$$\mathbb{H}_Y^m(X, F^s\mathcal{K}^\bullet) = 0, \quad m < s + c$$

and

$$\mathbb{H}_Y^{s+c}(X, F^s\mathcal{K}^\bullet) \cong H^0(X, H_Y^c(X, \mathcal{K}^s)).$$

Proof of Theorem 2.38. Step 1: Reduction to the case where Y is a smooth subvariety.

We let Y_{reg}, Y_{sing} be the regular locus, respectively the singular locus of Y and we put

$$X^0 := X - Y_{\text{sing}}$$

Let us combine the usual exact sequences for cohomology with support together with the excision exact sequences (B–36) to a commutative diagram

$$\begin{array}{ccccccc}
\mathbb{H}^{2d}_{Y_{\text{sing}}}(X, F^d\Omega_X^\bullet) & \to & \mathbb{H}^{2d}_Y(X, F^d\Omega_X^\bullet) & \to & \mathbb{H}^{2d}_{Y_{\text{reg}}}(X^0, F^d\Omega_X^\bullet) & \to & \mathbb{H}^{2d+1}_{Y_{\text{sing}}}(X, F^d\Omega_X^\bullet) \\
\| & & \downarrow r & & \downarrow r_{\text{reg}} & & \| \\
\mathbb{H}^{2d}_{Y_{\text{sing}}}(X, F^d\Omega_X^\bullet) & \to & \mathbb{H}^{2d}(X, F^d\Omega_X^\bullet) & \to & \mathbb{H}^{2d}(X^0, F^d\Omega_X^\bullet) & \to & \mathbb{H}^{2d+1}_{Y_{\text{sing}}}(X, F^d\Omega_X^\bullet).
\end{array}$$

In this diagram the first terms on the left vanish by Prop. 2.40. So one can define a unique Hodge class $\mathrm{cl}_{\mathrm{Hdg}}(Y_{\mathrm{reg}}) \in \mathbb{H}^{2d}(X^0, F^d\Omega^\bullet_X)$ which comes from the Hodge class of the pair (X, Y). A diagram chase then shows that one can reduce the construction of a Thom class to the smooth case (X^0, Y_{reg}).

In what follows we are going to construct a refined Thom class for (X^0, Y_{reg}) which maps to the usual Thom class for this pair. This suffices to complete the proof, in view of the commutative diagram

$$\begin{array}{ccc} \mathbb{H}^{2d}_{Y_{\mathrm{reg}}}(X^0, F^d\Omega^\bullet_X) & \longrightarrow & \mathbb{H}^{2d}_{Y_{\mathrm{reg}}}(X^0, \Omega^\bullet_{X^0}) = H^{2d}_{Y_{\mathrm{reg}}}(X^0; \mathbb{C}) \\ \downarrow & & \downarrow \\ \mathbb{H}^{2d}(X^0, F^d\Omega^\bullet_{X^0}) & \longrightarrow & \mathbb{H}^{2d}(X^0, \Omega^\bullet_{X^0}) = H^{2d}(X^0; \mathbb{C}) \end{array}$$

Step 2: Construction of $\tau^{d,d}(Y) \in H^d_Y(X, \Omega^d_X)$ for Y a complete intersection in a smooth (not necessarily compact) algebraic manifold X.

Let us cover X by Stein open sets $\{U_\alpha\}, \alpha \in I$. Suppose that $U_\alpha \cap Y$ is given by $f^{(k)}_\alpha = 0$, $k = 1, \ldots, d$. The open sets $U^k_\alpha := U_\alpha - \{f^{(k)}_\alpha = 0\}$, $k = 1, \ldots, d$ form an acyclic covering of $U_\alpha - Y \cap U_\alpha$. Consider the Čech $(d-1)$-cocycle

$$(U^1_\alpha \cap \cdots \cap U^d_\alpha) \longmapsto \eta_\alpha := \left[d \log f^{(1)}_\alpha \wedge \cdots \wedge d \log f^{(d)}_\alpha\right].$$

If we take other equations it is easy to write down a $(d-2)$ co-chain whose coboundary gives the difference. Under the isomorphism

$$H^{d-1}(U_\alpha - (Y \cap U_\alpha), \Omega^d_X) \xrightarrow{\sim} H^d_Y(U_\alpha, \Omega^d_X)$$

its class maps to a class $c_\alpha \in H^d_Y(U_\alpha, \Omega^d_X)$ which is therefore independent of the choice of equations for Y. Hence the c_α patch together to a section of the sheaf $H^d_Y(\Omega^d_Y)$. We then apply Lemma 2.39.

Step 3: Lifting of the class $\tau^{d,d}(Y)$ to a class $\tau_{\mathrm{Hodge}}(Y) \in \mathbb{H}^{2d}_Y(X, F^d\Omega^\bullet_X)$.

To do this, we consider the long exact sequence in hypercohomology with supports in Y associated to the exact sequence of complexes

$$0 \to F^{d+1}\Omega^\bullet_X \to F^d\Omega^\bullet_X \to \Omega^d_X[-d] \to 0.$$

It reads

$$\begin{array}{ccccccc} \mathbb{H}^{2d}_Y(X, F^{d+1}\Omega^\bullet_X) & \to & \mathbb{H}^{2d}_Y(X, F^d\Omega^\bullet_X) & \to & H^d_Y(X, \Omega^d_X) & \xrightarrow{\partial} & \mathbb{H}^{2d+1}_Y(X, F^{d+1}\Omega^\bullet_X) \\ \| & & & & & & \|\wr \\ 0 & & & & & & H^0(X, H^d_Y(X, \Omega^{d+1}_X)). \end{array}$$

Here we use Cor. 2.40. It follows that to calculate $\partial(\tau^{d,d}(Y))$, it suffices to do this locally. We use the same notation as in the previous step. So $\partial\tau^{d,d}(Y)|_{U_\alpha}$ is represented by the co-cycle

$$(U^1_\alpha \cap \cdots \cap U^d_\alpha) \mapsto d\eta_\alpha = d\left[d \log f^{(1)}_\alpha \wedge \cdots \wedge d \log f^{(d)}_\alpha\right] = 0.$$

So $\partial(\tau^{d,d}(Y)) = 0$ and there is a unique lift of this class to $\tau_{\mathrm{Hodge}}(Y) \in \mathbb{H}^{2d}_Y(X, F^d\Omega^\bullet_X)$.

Step 4: Proof that the class $\tau_{\mathrm{Hodge}}(Y) \in \mathbb{H}^{2d}_Y(X, F^d\Omega^\bullet_X)$ maps to the Thom class $\tau_{\mathbb{C}}(Y) \in H^{2d}_Y(X;\mathbb{C})$.

Recall (B.2.9) that Poincaré-duality implies that $\tau_{\mathbb{C}}(Y)$ generates local cohomology. Suppose that $\tau_{\mathrm{Hodge}}(Y)$ maps to $m\tau_{\mathbb{C}}(Y)$. To show that $m = 1$ a local computation suffices. Hence, by functoriality, we can reduce to the case of the origin in $\mathbb{C}^d$. Again, by functoriality we can further restrict down to a complex line passing through the origin. Next, we look at the closed 1-form dz/z on $\mathbb{C} - \{0\}$. It defines a De Rham class in $H^1(\mathbb{C} - \{0\})$ which generates the first integral cohomology of $H^1(\mathbb{C} - \{0\})$ under the embedding $\epsilon : \mathbb{Z}(1) \to \mathbb{C}$. This is simply the residue formula. The corresponding image $\partial(dz/z) \in H^2_0(\mathbb{C})$ generates integral cohomology with support in 0. It follows that $m = 1$. □

Remark 2.41. This construction also provides us with refined Thom classes for cycles $Y = \sum n_i Y_i$ of codimension d with support in $|Y| = \bigcup_i Y_i$. Indeed, one merely uses the isomorphism

$$H^{2d}_{|Y|}(X; F^d\Omega^\bullet_X) \cong \bigoplus_i H^{2d}_{Y_i}(X; F^d\Omega^\bullet_X)$$

coming from restriction and puts

$$\tau_{\mathrm{Hdg}}(Y) = \sum_i n_i \tau_{\mathrm{Hdg}}(Y_i).$$

To verify that restriction induces an isomorphism, one first remarks that this is obvious if the Y_i are disjoint, while the general case can be reduced to this case by comparing cohomology with support in $|Y|$ with cohomology with support in $\bigcup_i Y_i - (Y_i \bigcap \bigcup_{j\neq i} Y_j)$ using the excision exact sequence and the previous vanishing results.

2.5 Almost Kähler V-Manifolds

In this section we shall see that the Hodge decomposition is valid for the cohomology groups of a class of varieties that are possibly singular.

A **V-manifold** of dimension n is a complex space which can be covered by charts of the form U_i/G_i, $i \in I$, with $U_i \subset \mathbb{C}^n$ open and G_i a finite group of holomorphic automorphisms of U_i.

An **almost Kähler V-manifold** is a V-manifold X for which there exists a manifold Y bimeromorphic to a Kähler manifold and a proper modification $f : Y \to X$ onto X. Here we recall that a **proper modification** is a proper holomorphic map which induces a biholomorphic map over the complement of a nowhere dense analytic subset.

Examples 2.42. 1) A global quotient of a complex manifold by a finite group of holomorphic automorphisms. An important example is the case of a **weighted projective space** $\mathbb{P}(q_0, \ldots, q_n)$, where the q_j are non-negative integers, the weights. It is defined as the quotient of $\mathbb{P}^n$ by the coordinate-wise action of the product $\mu_{q_0} \times \cdots \times \mu_{q_n}$ of the q_j-th roots of unity μ_j, $j = 0, \ldots, n$. It can also be described as the quotient of $\mathbb{C}^{n+1} - \{0\}$ by the action of $\mathbb{C}^\times$ given by $t \cdot (z_0, \ldots, z_n) = (t^{q_0} z_0, \ldots, t^{q_n} z_n)$. The natural quotient map is denoted

$$p : \mathbb{C}^{n+1} - \{0\} \to \mathbb{P}(q_0, \ldots, q_n).$$

The subgroup $\mu(q_j) \subset \mathbb{C}^\times$ stabilizes $V_j = \{z_j = 1\}$ and p identifies $p(V_j)$ with the quotient $V_j = U_j/\mu(q_j)$. These together form the standard open affine covering of $\mathbb{P}(q_0, \ldots, q_n)$. Without loss of generality one may assume that the q_j have no factor in common and we may even assume that this is true for any $(n-2)$-tuple of weights.

A subvariety X of $\mathbb{P}(q_0, \ldots, q_n)$ is called **quasi-smooth** if the cone $p^{-1}X \subset \mathbb{C}^{n+1} - \{0\}$ is smooth. In other words, the only singularity of the corresponding affine cone is the vertex. It is not hard to see that a quasi-smooth subvariety of weighted projective space is a V-manifold.

2) The quotient of any torus by the cyclic group of order two generated by the involution $x \mapsto -x$, a **Kummer variety**.

3) A complete complex algebraic V-manifold admits a resolution of singularities Y and by Chow's lemma, Y is bimeromorphic to a smooth projective variety. It follows that a complete complex algebraic V-manifold is an almost Kähler V-manifold

4) Let us refer to [Oda] for the subject of **toric varieties**. We only say that to each convex polytope Π with integral vertices spanning $\mathbb{R}^n$ as a vector space there corresponds an n-dimensional toric variety X_Π and vice-versa. Each vertex v determines the cone $\bigcup_{n \geq 1} n\Pi_v$, where Π_v is the polytope Π translated over $-v$. If this cone has exactly n 1-dimensional faces it is called simplicial and Π is simplicial if all Π_v are simplicial. The singularities are in general rather bad, but if Π is simplicial, X_Π is a V-manifold.

The main result is

Theorem 2.43. *Let X be an almost Kähler V-manifold. Then $H^k(X;\mathbb{Q})$ admits a Hodge structure of weight k.*

Before we can prove this theorem, we need some preparations. First we note that locally a V-manifold is obtained as the quotient of a ball B by a finite group G of linear unitary automorphisms (see [Cart57, proof of Theorem 4]). The quotient B/G is smooth if and only if G is generated by generalized reflections (elements whose fixed locus is a hyperplane). In general, if we let G_{big} the subgroup of G generated by the generalized reflections and $G_{\text{small}} = G/G_{\text{big}}$, the smooth quotient $B' = B/G_{\text{big}}$ is acted upon by G_{small} with quotient B/G.

This description also shows that X is a rational homology manifold and hence Poincaré-duality holds with respect to rational coefficients.

Next, we need to digress on singularities. Recall that a module M over a local noetherian local ring $(R, \mathfrak{m})$ of Krull dimension n is called **Cohen-Macaulay** if it has a regular sequence of maximal length n (an ordered sequence $(t_1, \ldots, t_m)$ of elements $t_j \in \mathfrak{m}$ is called an **M-regular sequence** if each of the t_j is not a zero-divisor in $M/(t_1, \ldots, t_{j-1})M$). A local ring is called Cohen-Macaulay if it is Cohen-Macaulay as a module over itself.

A (germ of a) singularity (X, x) is called Cohen-Macaulay if $\mathcal{O}_{X,x}$ is a Cohen-Macaulay ring.

Examples 2.44. 1) Smooth points are of course Cohen-Macaulay.
2) Reduced curve singularities are Cohen-Macaulay.
3) Quotient singularities are quotients of a germ of smooth manifold (Y, y) by the action of a finite group G of holomorphic automorphisms. These are Cohen-Macaulay, since the local ring at the point $x \in X = Y/G$ corresponding to y is the ring of G-invariants $\mathcal{O}_{Y,y}^G$ of $\mathcal{O}_{Y,y}$ and hence a direct factor of the Cohen-Macaulay ring $\mathcal{O}_{Y,y}$ which itself is Cohen-Macaulay over $\mathcal{O}_{Y,y}^G$.

By [R-R-V], every equi-dimensional complex analytic space X of dimension n has a **dualizing complex** $\omega_X^\bullet$ which actually is an object in the derived category of bounded below complexes of $\mathcal{O}_X$-modules. It can be defined locally as follows. Suppose $U \subset X$ is an open subset embeddable into an open set $V \subset \mathbb{C}^N$, say $i : U \hookrightarrow V$. Then the complex

$$\omega_U^\bullet := R\,\mathrm{Hom}_{\mathcal{O}_V}(\mathcal{O}_U, \Omega_V^N[N])[-n]$$

is supported on U and is actually independent of the choice of V.

The dualizing complex intervenes in a duality statement of which we only need some special cases:

Theorem 2.45. 1) ***Serre-Grothendieck duality:*** *Let X be a compact complex space. For any $\mathcal{O}_X$-coherent sheaf $\mathcal{F}$ we have*

$$H^q(X, \mathcal{F})^\vee = \mathrm{Ext}^{n-q}(\mathcal{F}, \omega_X^\bullet).$$

2) *Let $f : Z \to X$ be a finite morphism between complex spaces. For any $\mathcal{O}_Z$-coherent sheaf $\mathcal{F}$ we have*

$$f_* \mathcal{E}xt^i_{\mathcal{O}_Z}(\mathcal{F}, \omega_Z^\bullet) = \mathcal{E}xt^i_{\mathcal{O}_X}(f_*\mathcal{F}, \omega_X^\bullet).$$

It can be shown that for a normal Cohen-Macaulay space X with singular locus X_{sing} and inclusion $i : X_{\mathrm{reg}} = X - X_{\mathrm{sing}} \hookrightarrow X$ of the smooth locus, the dualizing complex is actually a *sheaf*

$$\omega_X := i_* \Omega^n_{X_{\mathrm{reg}}}$$

viewed as a complex placed in degree 0. In the special case of a V-manifold X, this sheaf, or more precisely, the complex $i_* \Omega^\bullet_{X_{\mathrm{reg}}}$ can be described in terms of the local geometry of X:

Lemma 2.46. *Let $B \subset \mathbb{C}^n$ be an open ball and let G be a finite unitary subgroup acting on B. Let $p : B \to X = B/G$ be the quotient map. Then we have an equality of complexes*

$$\tilde{\Omega}^{\bullet}_X := i_* \Omega^{\bullet}_{X_{\text{reg}}} = (p_* \Omega^{\bullet}_B)^G.$$

In particular, $\tilde{\Omega}^{\bullet}_X$ is a resolution of the constant sheaf $\underline{\mathbb{C}}_X$.

Proof. If $G = G_{\text{small}}$ the subvariety $p^{-1}X_{\text{sing}}$ has codimension ≥ 2 in B and p induces the finite unramified cover $q : B' = B - p^{-1}X_{\text{sing}} \to X_{\text{reg}}$. Then $\Omega^{\bullet}_{X_{\text{reg}}} = (q_* \Omega^{\bullet}_B)^G$. Let $j : B' \hookrightarrow B$ be the inclusion. The assertion follows from

$$i_* \Omega^{\bullet}_{X_{\text{reg}}} = (i_* q_* \Omega^{\bullet}_B)^G = (p_* j_* \Omega^{\bullet}_{B'})^G = (p_* \Omega^{\bullet}_B)^G,$$

where the last equality follows since $q^{-1}X_{\text{sing}}$ has codimension ≥ 2 in B.

If $G = G_{\text{big}}$ the map p is ramified along hypersurfaces and locally on B, the map is given by $(z_1, z_2, \ldots, z_n) \mapsto (z_1^e, z_2, \ldots, z_n)$. Remembering that $X = X_{\text{reg}}$, as before we have $\Omega^{\bullet}_X = (p_* \Omega^{\bullet}_B)^G$ and the result follows in this case as well.

In the general case, we factor the map p into $B \xrightarrow{p'} B/G_{\text{big}} \xrightarrow{p''} B/G$ and we use that

$$(p_* \Omega^{\bullet}_B)^G = \left(p''_* (p'_* \Omega^{\bullet}_B)^{G_{\text{big}}}\right)^{G_{\text{small}}}.$$

The last assertion follows from the corresponding assertion on B upon taking G-invariants. □

If we apply the relative duality statement above to the quotient map p, we find

Corollary 2.47. *Let X be an n-dimensional V-manifold. Then*

1) $\mathcal{H}om_{\mathcal{O}_X}(\tilde{\Omega}^p_X, \omega_X) = \tilde{\Omega}^{n-p}_X$ *for all p;*
2) $\mathcal{E}xt^i_{\mathcal{O}_X}(\tilde{\Omega}^p_X, \omega_X) = 0$ *for all p and all $i > 0$.*

Using the local to global spectral sequence for Ext we conclude from this that

$$\operatorname{Ext}^p_{\mathcal{O}_X}(\tilde{\Omega}^q_X, \omega_X) = H^p(X, \tilde{\Omega}^{n-p}_X).$$

Combining this with Serre-Grothendieck duality this shows

Corollary 2.48. *$H^q(X, \tilde{\Omega}^p_X)$ is dual to $H^{n-q}(X, \tilde{\Omega}^{n-p}_X)$.*

Proof of Theorem 2.43:. Since $\tilde{\Omega}^{\bullet}_X$ is a resolution of the constant sheaf $\underline{\mathbb{C}}_X$, the spectral sequence in hypercohomology now reads

$$E_1^{pq} = H^q(X, \tilde{\Omega}^q_X) \Longrightarrow H^{p+q}(X; \mathbb{C}).$$

Let $f : Y \to X$ be a proper modification with Y bimeromorphic to a Kähler manifold. There is a natural morphism of sheaf complexes

$$\tilde{\Omega}_X^\bullet \to f_* \Omega_Y^\bullet$$

which can be seen to be an isomorphism. The local calculation showing this can be found in [Ste77a, Lemma 1.11]. It follows that there is a morphism f^* between the above spectral sequence and the Hodge-to De Rham spectral sequence for Y. We claim that f^* is already injective on the level of the E_1-terms. To see this, we use the previous Corollary: for every non-zero $\alpha \in E_1^{p,q}$, there exists a $\beta \in E_1^{n-p,n-q}$ with $\alpha \wedge \beta \neq 0$. Then $f^*\alpha \wedge f^*\beta = f^*(\alpha \wedge \beta) \neq 0$, since f^* is an isomorphism in the top cohomology. It follows that α is non-zero and so f^* is injective. But then the spectral sequence we started with degenerates at E_1 as well and f^* induces an isomorphism

$$H^q(X, \tilde{\Omega}_X^p) \xrightarrow{\sim} H^{p,q}(Y) \cap f^* H^{p+q}(X; \mathbb{C}).$$

We thus obtain a Hodge decomposition on $H^k(X;\mathbb{C})$ making f^* a morphism of Hodge structures.

Historical Remarks. The group theoretic point of view of the notion of Hodge structure is due to Mumford and has been exploited by Deligne in his study of absolute Hodge cycles (see the monograph [DMOS]). It has been used as a tool in approaching the Hodge conjecture on abelian varieties. See also the Appendix by Brent Gordon in [Lewis].

The Hodge complexes of sheaves are one of the basic building blocks for later constructions of mixed Hodge structures in geometric situations. This notion is inspired by Deligne [Del71], [Del74] but is different from his in that we prefer working with (filtered) complexes of sheaves instead of classes of these up to quasi-isomorphism. The algebraic version of the $\partial\overline{\partial}$-Lemma is a variation of an argument due to Deligne [Del71, Prop. 4.3.1]. The Hodge theoretic study of V-manifolds has been carried out in [Ste77b]. The notion of V-manifold is due to Satake [Sata56].

The Hodge theoretic aspects of the fundamental class have been extensively studied by El Zein in [ElZ].

3

Abstract Aspects of Mixed Hodge Structures

We continue to study the more formal aspects of Hodge theory, this time for the case of mixed Hodge structures. In § 3.1 the basic definitions are given; the Deligne splittings are introduced which make it possible to prove strictness of morphisms of mixed Hodge structures and to show that the category of mixed Hodge structures is abelian.

The complexes which come up in constructions for mixed Hodge structures have two filtrations and any one of these defines *a priori* different natural filtrations on the terms of the spectral sequence for the other filtration. We compare these in § 3.2. This study reveals (§ 3.3) that certain abstract properties built in the definition of a mixed Hodge complex of sheaves guarantee that their hypercohomology groups carry a mixed Hodge structure. If one can interpret a geometric object as a hypercohomology group of a complex underlying a mixed Hodge complex of sheaves, this object carries a mixed Hodge structure. This is the technique which will be employed in subsequent chapters.

Given a morphism of mixed Hodge structures, there is no canonical way to put the structure on the cone of the morphism. However, as we show in § 3.4, for a morphism of mixed Hodge complexes of sheaves the mixed cone is a canonical mixed Hodge structure on the cone of the underlying morphism of complexes of sheaves. It depends explicitly on the comparison morphisms, but this is built in in the definitions. The mixed cone construction will often be used later. As an example of its geometric significance we explain how to put a mixed Hodge structure on relative cohomology of a pair of compact smooth Kähler manifolds.

In § 3.5 we return to the categorical study of mixed Hodge structures. We first study extensions of two mixed Hodge structures and after that the higher Ext-groups. The category of mixed Hodge structures is abelian, but it does not have enough injectives; we use Verdier's direct approach (§ A.2.2) to the derived category. The higher Ext-groups turn out to be zero if $R = \mathbb{Z}$ or if R is a field. This is related to Beilinson's construction of absolute Hodge cohomology as we shall briefly indicate.

3.1 Introduction to Mixed Hodge Structures: Formal Aspects

We let R be a noetherian subring of $\mathbb{C}$ such that $R \otimes \mathbb{Q}$ is a field and we let V_R be a finite type R-module.

Definition 3.1. An R-**mixed Hodge structure** on V_R consists of two filtrations, an increasing filtration on $V_R \otimes_R (R \otimes \mathbb{Q})$, the **weight filtration** $W_\bullet$ and a decreasing filtration $F^\bullet$ on $V_\mathbb{C} = V \otimes_R \mathbb{C}$, the **Hodge filtration** which has the additional property that it induces a pure $(R \otimes \mathbb{Q})$-Hodge structure of weight k on each graded piece

$$\mathrm{Gr}_k^W(V_R \otimes_\mathbb{Z} \mathbb{Q}) = W_k/W_{k-1}.$$

We say that the R-mixed Hodge structure is **graded-polarizable** if the $\mathrm{Gr}_k^W(V_R \otimes_\mathbb{Z} \mathbb{Q})$ are pure, polarizable $(R \otimes \mathbb{Q})$-Hodge structures.

The mixed Hodge structure on V defines a class in the Grothendieck group (see Def. A.4.3) of pure R-Hodge structures

$$[V] := \sum_{k \in \mathbb{Z}} \left[\mathrm{Gr}_k^W\right] \in K_0(\mathfrak{hs}_R). \tag{III–1}$$

The Hodge numbers of these pure Hodge structures

$$h^{p,q}(V) := \dim_\mathbb{C} \mathrm{Gr}_F^p \mathrm{Gr}_{p+q}^W(V_\mathbb{C})$$

are the **Hodge numbers** of the mixed Hodge structure. These are the coefficients of the **Hodge-Euler polynomial**

$$e_{\mathrm{Hdg}}(V) := P_{\mathrm{hn}}([V]) = \sum_{p,q \in \mathbb{Z}} h^{p,q}(V) u^p v^q \in \mathbb{Z}[u, v, u^{-1}, v^{-1}]. \tag{III–2}$$

A **morphism** $f : V_R \to V'_R$ **of mixed Hodge structures** is an R-linear map which is compatible with the two filtrations W and F. In view of Prop. 2.4 the morphism f induces for all $m \in \mathbb{Z}$ morphisms $\mathrm{Gr}_m^W(f) : \mathrm{Gr}_W^m V \to \mathrm{Gr}_m^W V'$ of Hodge structures.

Examples 3.2. 1) A Hodge structure as defined in § 2.1.1 is a direct sum of Hodge structures of various weights and as such it is also a mixed Hodge structure. By definition, this is a mixed Hodge structure **split over** $\mathbb{R}$. By § 2.1 Hodge structures split over $\mathbb{R}$ are precisely the finite-dimensional representation of the group $\mathbb{S}$.

2) Let H, H' be two R-mixed Hodge structures. Then $\mathrm{Hom}(H, H')$ and $H \otimes H'$ are R-mixed Hodge structures. To see this, we put

$$\mathrm{Hom}(H, H')_R = \mathrm{Hom}_R(H_R, H'_R)$$
$$(H \otimes H')_R = H_R \otimes_R H'_R.$$

Let R stand for $\mathbb{Q}$ or $\mathbb{C}$. As we have seen in (see Def. (2.3) in Chap. 2) for any two filtrations T, respectively T' on H_R, respectively H'_R we have their multiplicative extensions to $\mathrm{Hom}_R(H_R, H'_R)$ and $H_R \otimes_R H'_R$. Explicitly,

$$T^p \mathrm{Hom}_R(H_R, H'_R) = \{f : H_R \to H'_R \mid fT^n H_R \subset (T')^{n+p} H'_R \quad \forall n\}$$
$$T^p(H_R \otimes_R H'_R) = \sum_m T^m H_R \otimes (T')^{p-m} H'_R \subset H_R \otimes_R H'_R$$

This procedure enables us to put Hodge, respectively weight filtrations on $\mathrm{Hom}_{\mathbb{C}}(H_{\mathbb{C}}, H'_{\mathbb{C}})$, $H_C \otimes H'_{\mathbb{C}}$, respectively $\mathrm{Hom}_{\mathbb{Q}}(H_{\mathbb{Q}}, H'_{\mathbb{Q}}), H_{\mathbb{Q}} \otimes H'_{\mathbb{Q}}$. It is straightforward to check that this indeed defines mixed Hodge structures and that the same formulas (II–3) and (II–4) for the Hodge Euler polynomials in the pure setting are valid. In a similar fashion, one can put mixed Hodge structures on the tensor algebras TH, $TH \otimes TH^\vee$, the symmetric algebra SH and the exterior algebra ΛH. The multiplication in these algebras is easily seen to be a morphism of mixed Hodge structures.

3) (Tate twists) Let $H = (H_{\mathbb{Z}}, F, W)$ be a mixed Hodge structure. We define

$$\left.\begin{array}{ll} H(m)_{\mathbb{Z}} & := (2\pi \mathrm{i})^{2m} \cdot H_{\mathbb{Z}} \\ W_k(H(m)) & := W_{k+2m} H \\ F^p(H(m)) & := F^{p+m} H. \end{array}\right\} \qquad \text{(III–3)}$$

Then $H(m)$ is also a mixed Hodge structure, the m-th **Tate twist of** H.

A morphism $f : (V, F) \to (V', F')$ of filtered vector spaces induces linear maps $\mathrm{Gr}^p_F(f) : \mathrm{Gr}^p_F V \to \mathrm{Gr}^p_F V'$ on the gradeds, but even if f is injective, the maps $\mathrm{Gr}^p_F(f)$ need not be injective. However, if f is *strict*, this is the case. Recall (A–26) that strictness of f means

$$f(V) \cap F^p V' = f(F^p V) \quad \text{for all } p. \qquad \text{(III–4)}$$

Example 3.3. Every morphism of Hodge structures, using the associated Hodge filtrations, gives a morphism of filtered complex vector spaces. The existence of the Hodge decomposition implies that such a morphism is strict (with respect to the Hodge filtration).

To prove strictness of morphisms of mixed Hodge structures with respect to both the Hodge and the weight filtrations we would like to have a Hodge decomposition in the mixed case. In general there is no such "Hodge decomposition" of $V_{\mathbb{C}}$, but we may introduce the (p, q)-component of the pure Hodge structure $\mathrm{Gr}^W_{p+q} V_{\mathbb{C}}$:

$$V^{p,q} := [\mathrm{Gr}^W_{p+q} V_{\mathbb{C}}]^{p,q}, \quad \dim_{\mathbb{C}} V^{p,q} = h^{p,q}(V), \qquad \text{(III–5)}$$

so that $V_{\mathbb{C}} \cong \bigoplus V^{p,q}$ as complex vector spaces. The components on the right hand side are of course only subquotients and we would like to find subspaces of $\mathbb{V}_C$ mapping isomorphically to these subquotients. Formally, we look for a bigrading $V = \bigoplus J^{p,q}$ of V such that

$$\left.\begin{array}{l} W_k^{\mathbb{C}} := W_k \otimes \mathbb{C} = \bigoplus_{p+q \leq k} J^{p,q} \\ F^p = \bigoplus_{r \geq p} J^{r,s} \end{array}\right\}.$$

Suppose that moreover $J^{p,q} = \overline{J^{q,p}}$ holds for all $p, q \in \mathbb{Z}$. Then the direct sum $V_k = \bigoplus_{p+q=k} J^{p,q}$ has a real structure; it is a weigh k real Hodge structure and $V = \bigoplus V_k$ is a splitting of the mixed Hodge structure over $\mathbb{R}$. In general $J^{p,q}$ need not be not the complex conjugate of $J^{q,p}$, but only modulo $W_{p+q-1}^{\mathbb{C}}$. We call such a bigrading a **weak splitting of the real mixed Hodge structure** (V, W, F). These do exist:

Lemma-Definition 3.4. The bigrading of $V_{\mathbb{C}}$ given by the subspaces

$$I^{p,q} := F^p \cap W_{p+q} \cap \left(\overline{F^q} \cap W_{p+q} + \sum_{j \geq 2} \overline{F^{q-j+1}} \cap W_{p+q-j} \right)$$

defines a weak a splitting of the real mixed Hodge structure, the **Deligne splitting**.

Proof. Observe that $I^{p,q} = F^p \cap \overline{F^q} \cap W_{p+q}^{\mathbb{C}} \bmod W_{p+q-2}^{\mathbb{C}}$ and so writing $x \in I^{p,q}$ accordingly as $x = y + w$, $y \in F^p \cap \overline{F^q} \cap W_{p+q}^{\mathbb{C}}$, $w \in W_{p+q-2}^{\mathbb{C}}$ we see that $x \in W_{p+q-1}$ precisely when $y \in W_{p+q-1}$, But since $F^p \cap \overline{F^q} \cap \mathrm{Gr}_W^{p+q-1} = \{0\}$ because Gr_W^{p+q-1} is a pure weight $(p+q-1)$ Hodge structure, it follows that $y \in W_{p+q-2}^{\mathbb{C}}$ and thus can be pulled into w. But then we apply the same argument to $x = w \in F^p \cap \overline{F^{q-1}} \cap W_{p+q-2}^{\mathbb{C}} \bmod W_{p+q+3}^{\mathbb{C}}$ to show $w \in W_{p+q+3}^{\mathbb{C}}$. Since the W-filtration is bounded below, we see that eventually $x \in W_{p+q-1}$, implying $x = 0$. So $I^{p,q}$ projects injectively to $V^{p,q}$.

Next, we show that this projection is a surjection. Let $[v] = [\bar{u}] \in V^{p,q}$ where $v \in F^p \cap W_{p+q}^{\mathbb{C}}$ and $u \in F^q \cap W_{p+q}^{\mathbb{C}}$. The equality $[v] = [\bar{u}]$ in Gr_{p+q}^W means $v = \bar{u} + w$ with $w \in W_{p+q-1}$. We are going to modify the image of w in Gr_{p+q-1}^W. The fact that F induces a pure Hodge structure of weight $p+q-1$ on this space yields a splitting $\mathrm{Gr}_{p+q-1}^W = F^p \cap W_{p+q-1}^{\mathbb{C}} + \overline{F^q} \cap W_{p+q-1}^{\mathbb{C}}$ so we can write $w = v' + \overline{u'} + w_1$, $v' \in F^p \cap W_{p+q-1}^{\mathbb{C}}$, $u' \in F^q \cap W_{p+q-1}^{\mathbb{C}}$ and $w_1 \in W_{p+q-2}^{\mathbb{C}}$. Now set

$$\begin{aligned} v_1 &:= v - v' = \bar{u} + \bar{u}' + w_1 \\ u_1 &:= u + u' \in F^q \cap W_{p+q}^{\mathbb{C}}. \end{aligned}$$

So $[v_1] = [v] = [\bar{u}] = [\bar{u}_1]$ and $v_1 = \overline{u_1} + w_1$. Next, we do the same thing with $w_1 \in \mathrm{Gr}_{p+q-2}^W$ and we find v_2 with $[v] = [v_1] = [v_2] = [\bar{u}_2]$, $v_2 = \bar{u}_2 + w_2$, $u_2 \in F^q \cap W_{p+q}^{\mathbb{C}} + F^{q-1} \cap W_{p+q-2}^{\mathbb{C}}$ and $w_2 \in W_{p+q-3}^{\mathbb{C}}$. Since $W_n = 0$ for n sufficiently large, this process terminates when we have $[v] = [\overline{u_n}]$ with $u_n \in F^q \cap W_{p+q}^{\mathbb{C}} + F^{q-1} \cap W_{p+q-2}^{\mathbb{C}} + F^{q-2} \cap W_{p+q-3}^{\mathbb{C}} + \cdots$, $v_n \in F^p \cap W_{p+q}^{\mathbb{C}}$ and $\bar{u}_n = v_n$ so that this last element (which still projects to the original class $[v]$) belongs to $I^{p,q}$ as desired. □

Remark 3.5. It can be shown (see [C-K-S86]) that the Deligne splitting is uniquely characterized by the following congruence:

$$I^{p,q} \equiv \overline{I^{q,p}} \bmod \bigoplus_{r<p,s<q} I^{r,s},$$

Corollary 3.6. *Any morphism $f : (V, F, W) \to (V', F', W')$ of mixed Hodge structures is strict* (III–4), *that is, any element of F'^p in the image of f comes from F^p and similarly for the weight filtration.*

The proof is the same as in the pure case, using the Deligne splitting instead of the Hodge decomposition. Indeed, any morphism of mixed Hodge structures preserves the Deligne splitting by its very definition. As an immediate consequence we have

Corollary 3.7. *Any morphism of mixed Hodge structures which is an isomorphism on the level of R-modules is an isomorphism of mixed Hodge structures.*

The following assertion is another immediate consequence of strictness.

Corollary 3.8. *Let*

$$H' \to H \to H''$$

be an exact sequence of mixed Hodge structures. Then for all k, p the sequences

$$\begin{gathered}
\mathrm{Gr}_k^W H'_{\mathbb{Q}} \to \mathrm{Gr}_k^W H_{\mathbb{Q}} \to \mathrm{Gr}_k^W H''_{\mathbb{Q}} \\
\mathrm{Gr}_F^p H'_{\mathbb{C}} \to \mathrm{Gr}_F^p H_{\mathbb{C}} \to \mathrm{Gr}_F^p H''_{\mathbb{C}} \\
\mathrm{Gr}_F^p \mathrm{Gr}_k^W H'_{\mathbb{C}} \to \mathrm{Gr}_F^p \mathrm{Gr}_k^W H_{\mathbb{C}} \to \mathrm{Gr}_F^p \mathrm{Gr}_k^W H''_{\mathbb{C}}
\end{gathered}$$

are also exact. If, moreover, the exact sequence extends to an exact sequence

$$0 \to H' \to H \to H'' \to 0,$$

we have

$$[H] = [H'] + [H''] \quad \text{in } K_0(\mathfrak{hs}). \tag{III–6}$$

If, in the preceding Corollary, $H' \to H$ is an injective morphism, we say that H' is a **mixed Hodge substructure** of H. In this case there is a unique mixed Hodge structure on the quotient H/H' making the quotient map $H \to H/H'$ a morphism of mixed Hodge structures; it is called the **quotient mixed Hodge structure**. If $\varphi : H_1 \to H_2$ is just any morphism of mixed Hodge structures, the kernel $\mathrm{Ker}(\phi)$ is a mixed Hodge substructure of H_1, the image $\mathrm{Im}(\phi)$ is a mixed Hodge substructure of H_2 and the natural map

$$H/\,\mathrm{Ker}(\phi) \to \mathrm{Im}(\phi)$$

is an isomorphism of mixed Hodge structures. All these facts follow from the compatibility of the respective morphisms with the Deligne splitting. We conclude:

Corollary 3.9. *The category of R-mixed Hodge structures is abelian. Its Grothendieck group is the same as the Grothendieck group* $K_0(\mathfrak{hs}_R)$ *for pure R-Hodge structures.*

For later reference we need the following criterion.

Criterion 3.10. *Suppose that*

$$0 \to H' \xrightarrow{f} H \xrightarrow{g} H'' \to 0$$

is an exact sequence of $\mathbb{Q}$*-vector spaces each endowed with an increasing 'weight filtration'* W *and a decreasing 'Hodge filtration'* F*. Suppose* f *and* g *preserve both filtrations and that* F *and* W *induce mixed Hodge structures on* H' *and* H''*. Then* F *and* W *induce a mixed Hodge structure on* H *if and only if* f *and* g *are strict with respect to both* F *and* W*.*

Proof. Necessity follows from the preceding. Let us prove that the condition is sufficient. Strict compatibility with the 'weight' filtration implies that we get induced exact sequences for the graded parts. By assumption, the extremes carry a pure Hodge structure of the same degree and so we may suppose that H', H'' are pure of weight n and that f and g strictly preserve the Hodge filtration. It suffices to prove that F induces a pure Hodge structure on H of weight n. Strictness implies that there is an induced exact sequence for the 'Hodge components'

$$0 \to (H')^{p,q} \xrightarrow{f} H^{p,q} \xrightarrow{g} (H'')^{p,q} \to 0.$$

Since $\bigoplus_{p+q=n}(H')^{p,q} = H'$ and $\bigoplus_{p+q=n}(H'')^{p,q} = H''$, we must also have $H = \bigoplus_{p+q=n} H^{p,q}$ and hence F induces a pure Hodge structure of weight n on the middle term. □

3.2 Comparison of Filtrations

Here we consider two filtrations and we compare various spectral sequences. The conventions we use for spectral sequences as well as certain basic facts concerning those can be found in § A.3.

We let $K^\bullet$ be a complex in an abelian category endowed with an increasing filtration W and a decreasing filtration F. There are three filtrations on the spectral sequence $E_r(K^\bullet, W)$ induced by F.

- The **first direct filtration** F_{dir} obtained by considering $E_r^{p,q}$ as a subquotient of K^{p+q} (recall that $E_r^{p,q} = Z_r^{p,q}/B_r^{p,q} \cap Z_r^{p,q}$ by definition). It is the same filtration as the one given by

$$F^p_{\text{dir}} E_r(K^\bullet, W) = \operatorname{Im}\left(E_r(F^p(K^\bullet, W)) \to E_r(K^\bullet, W)\right).$$

- The **second direct filtration** F^*_{dir} obtained by writing the term $E_r^{p,q} = Z_r^{p,q}/B_r^{p,q} \cap Z_r^{p,q}$ dually as a sub-object of a quotient of K^{p+q}, i.e. $E_r^{p,q} = \mathrm{Ker}\,(K^{p+q}/B_r^{p,q} \to K^{p+q}/(Z_r^{p,q} + B_r^{p,q}))$. It is the same as the filtration

$$[F^*_{\mathrm{dir}}]^p E_r(K^\bullet, W) = \mathrm{Ker}\,(E_r(K^\bullet, W) \to E_r((K^\bullet/F^pK^\bullet, W)))\,.$$

- The **inductive filtration** F_{ind} defined by induction on r. On the term $E_0^{p,q} = \mathrm{Gr}^W_{-p} K^{p+q}$ we take the filtration $F_{\mathrm{dir}} = F^*_{\mathrm{dir}}$ (it is straightforward to see this equality) and we define F_{ind} on E_{r+1} by considering $E_{r+1}^{p,q}$ as a sub-quotient of $E_r^{p,q}$.

Lemma 3.11 (Comparison of the three filtrations).

1) *On E_0 and E_1 the three filtrations $F_{\mathrm{dir}}, F^*_{\mathrm{dir}}, F_{\mathrm{ind}}$ coincide;*
2) *One has the inclusions*

$$F_{\mathrm{dir}} \subset F_{\mathrm{ind}} \subset F^*_{\mathrm{dir}};$$

3) *Suppose that W is a biregular filtration. The filtration F on $E_\infty^{p,q}$ induced by the isomorphism $E_\infty^{p,q} \cong \mathrm{Gr}^W_{-p} H^{p+q}(K^\bullet)$ is related to the first and second direct filtrations on E_∞ (obtained on E_r by taking r big enough) by means of the inclusions*

$$F_{\mathrm{dir}}(E_\infty) \subset F(E_\infty) \subset F^*_{\mathrm{dir}}(E_\infty).$$

Proof. 1) This is a direct consequence of the definitions and we omit the proof. 2) Using the second description of the filtration F_{dir}, we see immediately that the differentials d_r are compatible with F_{dir} and so there is an induced filtration G on E_{r+1}. We have

$$\begin{aligned}\mathrm{Ker}\,\big[E_r(F^pK^\bullet, W) \xrightarrow{d_r} \ & E_r(F^pK^\bullet, W)\big] \\ & \subset F^p_{\mathrm{dir}} \bigcap \mathrm{Ker}\,\big[E_r(K^\bullet, W) \xrightarrow{d_r} E_r(K^\bullet, W)\big],\end{aligned}$$

and $G^p(E_{r+1})$ is a quotient of the right hand space. The left hand space gives $F^p_{\mathrm{dir}} E_{r+1}(K^\bullet, W) \subset E_{r+1}(K^\bullet, W)$. So on the E_{r+1}-terms we have $F_{\mathrm{dir}} \subset G$. Dually, the filtration F^*_{dir} on E_r induces G^* on E_{r+1} and $G^* \subset F^*_{\mathrm{dir}}$. The assertion now follows by induction: if it is true on E_r, we have on E_{r+1} that F_{ind} interpolates G and G^* so that

$$F_{\mathrm{dir}}(E_{r+1}) \subset G(E_{r+1}) \subset F_{\mathrm{ind}}(E_{r+1}) \subset G^*(E_{r+1}) \subset F^*_{\mathrm{dir}}(E_{r+1}).$$

3) Consider the commutative diagram

$$\begin{array}{ccc} W_{-p} \cap F^iK^\bullet & \xrightarrow{j_p^i} & F^iK^\bullet \\ \big\downarrow k_i & & \big\downarrow j_i \\ W_{-p}K^\bullet & \xrightarrow{j_p} & K^\bullet. \end{array}$$

Let $H(\alpha)$ be the map induced by $\alpha = j_i, j_p, k_i, j_p^i$ in cohomology. The first inclusion is a direct consequence of the fact that $\mathrm{Im}\, H(j_p \circ k_i) \subset \mathrm{Im}(H(j_i)) \cap \mathrm{Im}(H(j_p))$ and the second is dual to it. □

Next we give Deligne's criterion from [Del71, Theorem 1.3.16] , [Del73, Proposition 7.2.8]:

Theorem 3.12. 1) *Suppose that $K^\bullet$ is equipped with two filtrations W and F, the first one being biregular. Suppose that for $r = 0, \dots, r_0$ the differentials d_r are strictly compatible with the inductive filtration. Then for $r \leq r_0 + 1$ the sequence of complexes*

$$0 \to E_r(F^pK^\bullet, W) \to E_r(K^\bullet, W) \to E_r(K^\bullet/F^pK^\bullet, W) \to 0 \qquad \text{(III–7)}$$

is exact. In particular the three filtrations $F_{\text{dir}}, F^*_{\text{dir}}$ and F_{ind} coincide on $E_0, \dots, E_{r_0+1}$.
2) *If for every $r \geq 0$ the differentials d_r are strictly compatible with the inductive filtration on E_r, then the three filtrations $F_{\text{dir}}, F^*_{\text{dir}}, F_{\text{ind}}$ coincide on E_∞ and coincide with the filtration induced by F on the sub-quotients $\text{Gr}^W_j H^\bullet(K^\bullet)$ of $H^\bullet(K^\bullet)$.*
3) *Under the assumption of 2), the spectral sequence for F degenerates at E_1, and one has an isomorphism of spectral sequences*

$$\text{Gr}^p_F(E_r(K^\bullet, W)) \cong E_r(\text{Gr}^p_F K^\bullet, W).$$

Proof. 1) We prove by induction on r that $E_r(F^pK^\bullet, W)$ injects into $E_r(K^\bullet, W)$ and that its image is $F^p_{\text{ind}}E_r(K, W)$. By definition this image is also $F^p_{\text{dir}}E_r(K, W)$. If we have shown this, the first map in (III–7) is into and the dual assertion asserts surjectivity of the second map and identifies the kernel with $F^p_{\text{ind}}E_r(K, W)$ as well. So the sequence is exact and the three filtrations coincide on E_r.

So assume that we have shown the above assertion for some $r < r_0$ and we want to prove it for $r + 1$. We have

$$\begin{aligned} F^p_{\text{ind}}E_{r+1}(K^\bullet, W) &= \text{Im}[\text{Ker}(F^p_{\text{ind}}E_r(K^\bullet, W) \xrightarrow{d_r} E_r(K^\bullet, W) \longrightarrow E_{r+1}(K^\bullet, W)] \\ &= \text{Im}[\text{Ker}\, E_r(F^p_{\text{ind}}K^\bullet, W) \xrightarrow{d_r} E_r(K^\bullet, W) \longrightarrow E_{r+1}(K^\bullet, W)] \\ &= \text{Im}[(E_{r+1}(F^p_{\text{ind}}K^\bullet, W) \longrightarrow E_{r+1}(K^\bullet, W)]. \end{aligned}$$

This shows the assertion about the F-filtration. As to injectivity, we use that d_r is strictly compatible with the inductive filtration (by the induction hypothesis) and so

$$d_rE_r(K^\bullet, W) \cap E_r(F^pK^\bullet, W) = d_rE_r(F^pK^\bullet, W)$$

and since E_{r+1} is obtained from $\text{Ker}(E_r \xrightarrow{d_r} E_r)$ upon taking the quotient by $\text{Im}(d_r)$ it follows that $E_{r+1}(F^pK^\bullet, W)$ injects into $E_{r+1}(K^\bullet, W)$.
2) Note that by 1) the sequences (III–7) are exact for $r \leq r_0$. If the W-filtration is biregular this also holds for $r = \infty$ so that, using Lemma 3.11.2), the three filtrations coincide on $E_\infty(K^\bullet, W)$.

3) The preceding exact sequences for r and $r+1$ fit into a commutative diagram whose rows are exact and in which the oblique arrows form exact 3-term complexes

$$\begin{array}{ccccc}
0 \leftarrow E_r(\operatorname{Gr}_F^p K^\bullet, W) \longleftarrow & E_r(F^p K^\bullet, W) & \longleftarrow & E_r(F^{p+1} K^\bullet, W) \leftarrow 0 \\
\| & & \searrow^{\beta} \quad {}^{\alpha}\swarrow & \\
\| & & E_r(K^\bullet, W) & \\
\| & & {}_{\alpha'}\swarrow \quad \searrow_{\beta'} & \\
0 \rightarrow E_r(\operatorname{Gr}_F^p K^\bullet, W) \longrightarrow & E_r(K^\bullet/F^{p+1} K^\bullet, W) & \longrightarrow & E_r(K^\bullet/F^p K^\bullet, W) \rightarrow 0
\end{array}$$

Here $F = F_{\mathrm{dir}} = F^*_{\mathrm{dir}} = F_{\mathrm{ind}}$. From the diagram we obtain an equality

$$\operatorname{Gr}_F^p(E_r(K^\bullet, W)) = \operatorname{Im}(\beta)/\operatorname{Im}(\alpha) = \operatorname{Ker}(\beta')/\operatorname{Ker}(\alpha').$$

Since $\operatorname{Im}(\beta)/\operatorname{Im}(\alpha)$ is a quotient of $E_r(\operatorname{Gr}_F^p K^\bullet, W)$, while $\operatorname{Ker}(\beta')/\operatorname{Ker}(\alpha')$ is a subspace, we obtain two dual isomorphisms $\operatorname{Gr}_F^p(E_r(K^\bullet, W)) \simeq E_r(\operatorname{Gr}_F^p K^\bullet, W)$ which are compatible with the differentials d_r (since F_{ind} is).

Finally, the sequence for $r = \infty$ says that

$$0 \rightarrow \operatorname{Gr}^W H(F^p K^\bullet) \rightarrow \operatorname{Gr}^W H(K^\bullet) \rightarrow \operatorname{Gr}^W H(K^\bullet/F^p K^\bullet) \rightarrow 0$$

is exact. This is the complex $\operatorname{Gr}^W L^\bullet$ with

$$L^\bullet = \{0 \rightarrow H(F^p K^\bullet) \rightarrow H(K^\bullet) \rightarrow H(K^\bullet/F^p K^\bullet) \rightarrow 0\}.$$

It is a general fact (see also Lemma A.42) that $\operatorname{Gr}^W(L^\bullet)$ is exact if and only if $L^\bullet$ is exact and the differentials are strictly compatible with W. Exactness means that the differentials of $K^\bullet$ are strictly compatible with F: if $a \in F^p K^{i+1} \cap d(K^i)$, the class of a is zero in $H^{i+1}(K^\bullet)$ and so $a \in d(F^p K^\bullet)$ and hence the spectral sequence $E(K^\bullet, F)$ degenerates at E_1 (Lemma A.42). □

3.3 Mixed Hodge Structures and Mixed Hodge Complexes

The following notion is a variant of Deligne's "Complexe de Hodge mixte cohomologique" from [Del73]. The notion of (mixed) Hodge complex of sheaves differs from his notion of "Complexe de Hodge (mixte) cohomologique" in that the morphisms between these objects are pseudo-morphisms (Def. 2.31) and hence are given by a chain of genuine morphism.

Definition 3.13. Let R a noetherian subring of $\mathbb{C}$ such that $R \otimes \mathbb{Q}$ is a field (mostly this will be $\mathbb{Z}$ or $\mathbb{Q}$).

1) A **(graded-polarizable) mixed R-Hodge complex**

$$K^\bullet = \left(K_R^\bullet, (K_{R\otimes\mathbb{Q}}^\bullet, W), \alpha, (K_\mathbb{C}^\bullet, W, F), \beta\right)$$

consists of the following data

- A bounded below complex $K_R^\bullet$ of R-modules such that $H^p(K_R^\bullet)$ is an R-module of finite type,
- A bounded below filtered complex $(K_{R\otimes\mathbb{Q}}, W)$ of $R \otimes \mathbb{Q}$-vector spaces, and a pseudo-morphism (Def. 2.31) $\alpha : K_R^\bullet \dashrightarrow K_{R\otimes\mathbb{Q}}^\bullet$ of bounded below complexes of R-modules (**the first comparison morphism**) inducing a pseudo-isomorphism

$$\alpha \otimes \mathrm{id} : K_R^\bullet \otimes \mathbb{Q} \overset{\text{qis}}{\underset{\sim}{\dashrightarrow}} K_{R\otimes\mathbb{Q}}^\bullet;$$

- A bi-filtered complex $(K_\mathbb{C}^\bullet, W, F)$ of complex vector spaces and a pseudo-morphism $\beta : (K_{R\otimes\mathbb{Q}}^\bullet, W) \dashrightarrow (K_\mathbb{C}^\bullet, W)$ in the category of bounded below filtered complexes of $R \otimes \mathbb{Q}$-modules (**the second comparison morphism**), inducing a pseudo-isomorphism

$$\beta \otimes \mathrm{id} : (K_{R\otimes\mathbb{Q}}^\bullet, W) \otimes \mathbb{C} \overset{\text{qis}}{\underset{\sim}{\dashrightarrow}} (K_\mathbb{C}^\bullet, W),$$

and such that the following axiom is satisfied (we refer to Def. 2.32.1) for the relevant notions):

for all $m \in \mathbb{Z}$ the triple $\mathrm{Gr}_m^W K^\bullet := \left(\mathrm{Gr}_m^W K_{R\otimes\mathbb{Q}}^\bullet, (\mathrm{Gr}_m^W K_\mathbb{C}^\bullet, F), \mathrm{Gr}_m^W(\beta)\right)$ *is a (polarizable) $R \otimes \mathbb{Q}$-Hodge complex of weight m.*

In the Grothendieck ring of pure Hodge structures (cf. Def. A.4), the associated **Hodge-Grothendieck characteristic** is given by

$$\chi_{\mathrm{Hdg}}(K^\bullet) := \sum_{k,m\in\mathbb{Z}} (-1)^k \left[H^k(\mathrm{Gr}_m^W K^\bullet)\right] \in K_0(\mathfrak{hs}).$$

2) Let X be a topological space. A **(graded-polarizable) mixed R-Hodge complex of sheaves** on X

$$\mathcal{K}^\bullet = \left(\mathcal{K}_R^\bullet, (\mathcal{K}_{R\otimes\mathbb{Q}}^\bullet, W), \alpha, (\mathcal{K}_\mathbb{C}^\bullet, W, F), \beta\right)$$

consists of the following data

- A bounded below complex of sheaves of R-modules $\mathcal{K}_R^\bullet$ such that the hypercohomology groups $\mathbb{H}^k(X, \mathcal{K}_R^\bullet)$ are finitely generated as R-modules,
- A complex of sheaves of $\mathbb{Q}$-vector spaces $\mathcal{K}_{R\otimes\mathbb{Q}}^\bullet$ equipped with an *increasing* filtration W and a pseudo-morphism in the category (of sheaves of R-modules on X) (**first comparison morphism**) $\alpha : \mathcal{K}_R^\bullet \dashrightarrow \mathcal{K}_{R\otimes\mathbb{Q}}^\bullet$ inducing a pseudo-isomorphism

$$\alpha \otimes \mathrm{id} : \mathcal{K}_R^\bullet \otimes \mathbb{Q} \overset{\text{qis}}{\underset{\sim}{\dashrightarrow}} \mathcal{K}_{R\otimes\mathbb{Q}}^\bullet$$

– A complex of sheaves of complex vector spaces $\mathcal{K}^\bullet_{\mathbb{C}}$ equipped with an increasing filtration W and a decreasing filtration F, together with a pseudo-morphism in the category (of sheaves of *filtered* $R \otimes \mathbb{Q}$-modules on X (**second comparison morphism**): $\beta : (\mathcal{K}^\bullet_{R\otimes\mathbb{Q}}, W) \dashrightarrow (\mathcal{K}^\bullet_{\mathbb{C}}, W)$ inducing a pseudo-isomorphism

$$\beta \otimes \mathrm{id} : \left(\mathcal{K}^\bullet_{R\otimes\mathbb{Q}} \otimes \mathbb{C}, W\right) \overset{\text{qis}}{\underset{\sim}{\dashrightarrow}} (\mathcal{K}^\bullet_{\mathbb{C}}, W)$$

such that the following axiom is satisfied (see Def. 2.32.2) for the relevant notions):

for all $m \in \mathbb{Z}$ the triple $\mathrm{Gr}^W_m \mathcal{K}^\bullet := \left(\mathrm{Gr}^W_m \mathcal{K}^\bullet_{R\otimes\mathbb{Q}}, (\mathrm{Gr}^W_m \mathcal{K}^\bullet_{\mathbb{C}}, F), \mathrm{Gr}^W_m(\beta)\right)$ *is a (polarizable) $R \otimes \mathbb{Q}$-Hodge complex of sheaves of weight m*

By Definition 2.32 this means
1. for all $k, m \in \mathbb{Z}$ the R-structure on $\mathbb{H}^k(\mathrm{Gr}^W_m \mathcal{K}^\bullet_{\mathbb{C}})$ induced by α and the filtration induced by F determine an R-Hodge structure of weight $k+m$;
2. the differentials of the derived complexes $R\Gamma(X, \mathrm{Gr}^W_m \mathcal{K}^\bullet_{\mathbb{C}})$ (see (B–12)) are strict with respect to the induced F-filtration.

We also need **Tate twists**:

Definition 3.14. Let $K^\bullet = \left(K^\bullet_R, (K^\bullet_{R\otimes\mathbb{Q}}, W), \alpha, (K^\bullet_{\mathbb{C}}, W, F), \beta\right)$ be an R-mixed Hodge complex. Its k-th Tate twist is defined by

$$K^\bullet(k) = \Big(K^\bullet_R \otimes_{\mathbb{Z}} \mathbb{Z}(2\pi \mathrm{i})^k, (K^\bullet_{R\otimes\mathbb{Q}} \otimes_{\mathbb{Q}} \mathbb{Q}(2\pi\mathrm{i})^k, W[2k]), \alpha, \\ (K^\bullet_{\mathbb{C}}, W[2k], F[k]), \beta(k)\Big)$$

where $\beta(k)$ is induced by β and multiplication with $(2\pi\mathrm{i})^k$. A similar definition holds for $\mathcal{K}^\bullet(k)$, where $\mathcal{K}^\bullet$ is an R-mixed Hodge complex of sheaves.

Remark 3.15. One can normalise the choice of the comparison morphisms as follows. Since comparison morphisms become morphisms in the derived category, they can be represented by a left fraction. A choice of such a left fraction for both comparison morphisms is called a **normalisation**. We speak of **normalized mixed Hodge complexes**. Concretely, we have a tent-like structure

$$\left.\mathcal{K}^\bullet_R \xrightarrow{\alpha_1} {}'\mathcal{K}^\bullet_{R\otimes\mathbb{Q}} \xleftarrow[\sim]{\alpha_2,\ \text{qis}} (\mathcal{K}^\bullet_{R\otimes\mathbb{Q}}, W) \xrightarrow{\beta_1} ({}'\mathcal{K}^\bullet_{\mathbb{C}}, W) \xleftarrow[\sim]{\beta_2,\ \text{qis}} (\mathcal{K}^\bullet_{\mathbb{C}}, W, F)\right\} \qquad \text{(III–8)}$$

where α_1 is a morphism of complexes of R-modules, α_2 is a quasi-isomorphism of $\mathbb{Q}$-vector spaces, β_1 is a filtered morphism of $R \otimes \mathbb{Q}$-modules and β_2 is a quasi-isomorphism of filtered $\mathbb{C}$-vector spaces. The morphisms α_1 and α_2 respectively become quasi-isomorphisms after tensoring with $\mathbb{Q}$, respectively

$\mathbb{C}$. A morphism between normalised mixed Hodge complexes of sheaves is a morphisms of the underlying mixed Hodge complexes of sheaves preserving the normalisations.

The definition of a morphism between mixed Hodge complexes or mixed Hodge complexes of sheaves is a bit subtle, since the comparison pseudomorphisms must be related by a morphisms between those, i.e. there should be morphisms between the constituents of the chains defining the pseudomorphisms verifying the obvious commutativity relations.

Definition 3.16. 1) Let $K^\bullet = \left(K^\bullet_R, (K^\bullet_{R\otimes\mathbb{Q}}, W), \alpha_K, (K^\bullet_\mathbb{C}, W, F), \beta_K\right)$ and $L^\bullet = \left(L^\bullet_R, (L^\bullet_{R\otimes\mathbb{Q}}, W), \alpha_K, (L^\bullet_\mathbb{C}, W, F), \beta_L\right)$ be two mixed Hodge complexes. A **morphism** $K^\bullet \to L^\bullet$ consists of

- a morphism of bounded below complexes of R-modules

$$\phi : K^\bullet_R \to L^\bullet_R;$$

- a morphism of bounded below filtered complexes of $R \otimes \mathbb{Q}$-modules

$$\phi_{R\otimes\mathbb{Q}} : (K^\bullet_{R\otimes\mathbb{Q}}, W) \to (L^\bullet_{R\otimes\mathbb{Q}}, W);$$

- a morphism of bounded below $\mathbb{C}$-vector spaces equipped with two filtrations

$$\phi_\mathbb{C} : (K^\bullet_\mathbb{C}, W, F) \to (L^\bullet_\mathbb{C}, W, F);$$

- morphisms of pseudo-morphisms $\alpha_K \to \alpha_L$ and $\beta_K \to \beta_L$.

2) Let $\mathcal{K}^\bullet = \left(\mathcal{K}^\bullet_R, (\mathcal{K}^\bullet_{R\otimes\mathbb{Q}}, W)\alpha_\mathcal{K}, (\mathcal{K}^\bullet_\mathbb{C}, W, F), \beta_\mathcal{K}\right)$ and $\mathcal{L}^\bullet = \left(\mathcal{L}^\bullet_R, (\mathcal{L}^\bullet_{R\otimes\mathbb{Q}}, W), \alpha_\mathcal{L}, (\mathcal{L}^\bullet_\mathbb{C}, W, F), \beta_\mathcal{L}\right)$ be two mixed Hodge complexes of sheaves. A **morphism** $\mathcal{K}^\bullet \to \mathcal{L}^\bullet$ consists of

- a morphism of bounded below complexes of sheaves of R-modules

$$\phi : \mathcal{K}^\bullet_R \to \mathcal{L}^\bullet_R;$$

- a morphism of bounded below filtered complexes of sheaves of $R \otimes \mathbb{Q}$-modules

$$\phi_{R\otimes\mathbb{Q}} : (\mathcal{K}^\bullet_{R\otimes\mathbb{Q}}, W) \to (\mathcal{L}^\bullet_{R\otimes\mathbb{Q}}, W);$$

- a morphism of bounded below sheaves of $\mathbb{C}$-vector spaces equipped with two filtrations

$$\phi_\mathbb{C} : (\mathcal{K}^\bullet_\mathbb{C}, W, F) \to (\mathcal{L}^\bullet_\mathbb{C}, W, F);$$

- morphisms of pseudomorphisms $\alpha_\mathcal{K} \to \alpha_\mathcal{L}$ and $\beta_\mathcal{K} \to \beta_\mathcal{L}$.

Remark 3.17. It should be clear what is meant by a short exact sequence of mixed Hodge complexes or of mixed Hodge complexes of sheaves. Given a short exact sequence $0 \to K^\bullet \to L^\bullet \to M^\bullet \to 0$ of mixed Hodge complexes, in view of (III–6), we have

$$\chi_{\mathrm{Hdg}}(L^\bullet) = \chi_{\mathrm{Hdg}} K^\bullet) + \chi_{\mathrm{Hdg}}(M^\bullet). \qquad \text{(III–9)}$$

The following result is the main tool to construct mixed Hodge structures.

Theorem 3.18. *Let*

$$\mathcal{K}^\bullet = \big(\mathcal{K}^\bullet_R, (\mathcal{K}^\bullet_{R\otimes\mathbb{Q}}, W), \alpha, (\mathcal{K}^\bullet_{\mathbb{C}}, W, F), \beta\big)$$

be a mixed R-Hodge complex of sheaves on X. Put

$$K^\bullet_R = R\Gamma(\mathcal{K}^\bullet_R),\ (K^\bullet_{R\otimes\mathbb{Q}}, W) = R\Gamma(\mathcal{K}^\bullet_{R\otimes\mathbb{Q}}, W),\ (K^\bullet_{\mathbb{C}}, W, F) = R\Gamma(\mathcal{K}^\bullet_{\mathbb{C}}, W, F).$$

I) *The triple*

$$R\Gamma\mathcal{K}^\bullet := \big(K^\bullet, (K^\bullet_{R\otimes\mathbb{Q}}, W)), (K^\bullet_{\mathbb{C}}, W, F)\big)$$

together with the induced comparison isomorphisms is an R-mixed Hodge complex. Or, more concretely, the global De Rham complexes of the Godement resolutions of the three (filtered) complexes $\mathcal{K}^\bullet_R$, $(\mathcal{K}^\bullet_{R\otimes\mathbb{Q}}, W)$, $(\mathcal{K}^\bullet_{\mathbb{C}}, W, F)$ together with the two induced comparison isomorphisms forms a mixed Hodge complex. This is compatible with Tate twists:

$$R\Gamma[\mathcal{K}^\bullet(k)] = [R\Gamma\mathcal{K}^\bullet](k).$$

II) *The filtrations $W[k]$ and F induce a mixed Hodge structure on the hypercohomology groups $\mathbb{H}^k(X, \mathcal{K}^\bullet_R)$. In fact, we have:*

i) *The filtration on $\mathbb{H}^k(X, \mathcal{K}^\bullet_{R\otimes\mathbb{Q}})$ defined by*

$$W_m\mathbb{H}^k(X, \mathcal{K}^\bullet_{R\otimes\mathbb{Q}}) = \operatorname{Im}\big(\mathbb{H}^k(X, W_{m-k}\mathcal{K}^\bullet_{R\otimes\mathbb{Q}}) \to \mathbb{H}^k(X, \mathcal{K}^\bullet_{R\otimes\mathbb{Q}})\big)$$

and the F-filtration induced on $\mathbb{H}^k(X, \mathcal{K}^\bullet_{\mathbb{C}})$ induce a mixed Hodge structure on $\mathbb{H}^k(X, \mathcal{K}^\bullet_R)$; this is compatible with Tate twists:

$$\mathbb{H}^k(X, \mathcal{K}^\bullet_R(m)) = \mathbb{H}^k(X, \mathcal{K}^\bullet_R)(m).$$

ii) *The differentials d_1 for $E_1(R\Gamma(X, \mathcal{K}^\bullet), W)$ are strictly compatible with the filtration induced by F.*

iii) *The spectral sequence for $(R\Gamma(X, \mathcal{K}^\bullet), W)$ whose E_1-term is given by*

$$E_1^{-m,k+m} = \mathbb{H}^k(X, \operatorname{Gr}^W_m \mathcal{K}^\bullet_{R\otimes\mathbb{Q}})$$

degenerates at E_2:

$$\left.\begin{aligned} E_2^{-m,k+m} &= H\big[E_1^{-m-1,k+m} \xrightarrow{d_1} E_1^{-m,k+m} \xrightarrow{d_1} E_1^{-m+1,k+m}\big] \\ &= E_\infty^{-m,k+m} = \operatorname{Gr}^W_{m+k} \mathbb{H}^k(X, \mathcal{K}^\bullet_{R\otimes\mathbb{Q}}). \end{aligned}\right\} \qquad \text{(III–10)}$$

iv) *The spectral sequence*

$${}_FE_1^{p,q} = \mathbb{H}^{p+q}(X, \operatorname{Gr}^p_F \mathcal{K}^\bullet_{\mathbb{C}}) \Longrightarrow \mathbb{H}^{p+q}(X, \mathcal{K}^\bullet_{\mathbb{C}})$$

degenerates at E_1; in particular the natural maps

$$\mathbb{H}^k(X, F^p\mathcal{K}^\bullet) \to \mathbb{H}^k(X, \mathcal{K}^\bullet_{\mathbb{C}})$$

are injective and

$$\operatorname{Gr}^p_F \mathbb{H}^k(X, \mathcal{K}^\bullet_{\mathbb{C}}) = \mathbb{H}^k(X, \operatorname{Gr}^p_F \mathcal{K}^\bullet_{\mathbb{C}});$$

v) *The spectral sequence for the filtered complex* $(\mathrm{Gr}^p_F(R\Gamma(X,\mathcal{K}^\bullet_{\mathbb{C}}),W)$ *degenerates at* E_2.

vi) *Referring to* (III–1) *we have an equality*

$$\chi_{\mathrm{Hdg}}(R\Gamma\mathcal{K}^\bullet) = \sum_k (-1)^k \left[\mathbb{H}^k(X,\mathcal{K}^\bullet)\right] \in K_0(\mathfrak{hs}_R). \quad \text{(III–11)}$$

Equivalently, with $a_X : X \to \mathrm{pt}$ *the constant map, like Prop. 2.33, we have*

$$\chi_{\mathrm{Hdg}}(R\Gamma\mathcal{K}^\bullet) = [(Ra_X)_*\mathcal{K}^\bullet] \in K_0(\mathfrak{hs}_R).$$

III) *Any morphism of mixed Hodge complexes of sheaves on* X *induces a homomorphism of mixed Hodge structures on the associated hypercohomology groups.*

Proof. I) is a direct consequence of the definitions. Given II) (i), III) follows. It suffices therefore to prove (i)–(vi).

Recall (Lemma 3.11) that on

$$E_1^{p,q}(K^\bullet_R \otimes \mathbb{Q}, W) = H^{p+q}(\mathrm{Gr}^W_{-p} K^\bullet_{R\otimes\mathbb{Q}}) = \mathbb{H}^{p+q}(X, \mathrm{Gr}^W_{-p} \mathcal{K}^\bullet_{R\otimes\mathbb{Q}})$$

the two direct filtrations and the inductive filtration induced by F coincide. By definition, F induces a Hodge structure of weight $-p+(p+q)=q$ on this term. The differential d_1 is compatible with F (see Lemma 3.11), and since d_1 is defined over $R\otimes\mathbb{Q}$ it commutes with complex conjugation and hence is compatible with $\bar{F}$. This implies that d_1 preserves the Hodge decomposition and hence is strictly compatible with the filtration F. This proves (ii).

We now consider the E_2-terms. By Theorem 3.12 (1) the three filtrations defined by F coincide and the resulting filtration is q-opposed to its complex conjugate, as before, so that we get a weight q-Hodge structure on $E_2^{p,q}(K^\bullet_R \otimes \mathbb{Q}, W)$. Now we prove the following Claim by induction on r.

Claim. For $r \geq 0$ the differentials of the spectral sequence $E_r(K^\bullet_R \otimes \mathbb{Q}, W)$ are strictly compatible with the inductive filtration $F = F_{\mathrm{ind}}$. They vanish for $r \geq 2$.

Indeed, for $r=0$ by formula (A–29) this means that the derivatives of the complex $\mathrm{Gr}^W_m K^\bullet_{\mathbb{C}} = R\Gamma(X, \mathrm{Gr}^W_m \mathcal{K}^\bullet_{\mathbb{C}})$ must be strictly compatible with the F-filtration, which holds by definition. For $r=1$ we just saw it. For $r \geq 2$ it suffices to show that $d_r = 0$. By induction, $E_r = E_2$ and by Theorem 3.12, the induction hypothesis implies that the three filtrations $F_{\mathrm{dir}}, F^*_{\mathrm{dir}}, F_{\mathrm{ind}}$ coincide on E_r and so (see Lemma 3.11) d_r preserves this filtration F. We just saw that the F-filtration on $E_2^{p,q} = E_r^{p,q}$ is q-opposed to the complex conjugate filtration and so d_r is a morphism of Hodge structures. But d_r maps $E_2^{p,q}$, a Hodge structure of weight q, to $E_r^{p+r,q-r+1}$ which has weight $q-r+1<q$ for $r>1$, and hence $d_r=0$.

Now we can complete the proof of the theorem. The Claim implies (iii). By Lemma 3.11 the F-filtration on $E_\infty^{p,q}$ induced by $H^{p+q}(K^\bullet_{\mathbb{C}})$ is the same as the one coming from the F-filtration on $E_2^{p,q} = E_\infty^{p,q}$ we just considered. So we have a Hodge structure of weight q on $\mathrm{Gr}^W_{-p} H^{p+q}(K_R^\bullet \otimes \mathbb{Q}) = \mathrm{Gr}^W_{-p} \mathbb{H}^{p+q}(\mathcal{K}^\bullet_{R\otimes\mathbb{Q}})$ and so, if we shift the W-filtration by $p+q$, we get a weight r-Hodge structure on the r-th graded parts graded $\mathrm{Gr}_r^{W[p+q]} = \mathrm{Gr}^W_{r-p-q}$. This proves (i). Theorem 3.12 (3) applies in our situation and this shows (iv) and (v).

We finally prove (vi). Start with the definition (III–1). By iii), taking in account the shift of the weight filtration when we pass to hypercohomology, we get

$$\begin{aligned}\chi_{\mathrm{Hdg}}(R\Gamma\mathcal{K}^\bullet) &= \sum_{k,m}(-1)^k\left[\mathbb{H}^k(X, \mathrm{Gr}^W_m \mathcal{K}^\bullet)\right]\\ &= \sum_{k,m}(-1)^k\left[\mathrm{Gr}^W_{k+m} \mathbb{H}^k(X, \mathcal{K}^\bullet)\right]\\ &= \sum_{k,m}(-1)^k\left[\mathrm{Gr}^W_m \mathbb{H}^k(X, \mathcal{K}^\bullet)\right] = \sum_k(-1)^k\mathbb{H}^k(X, \mathcal{K}^\bullet). \quad \square\end{aligned}$$

A morphism of mixed Hodge complexes of sheaves which is a bifiltered quasi-isomorphism of course induces an isomorphism of mixed Hodge structures on the hypercohomology groups. But this is even true for more general morphisms using Corollary 3.7:

Lemma-Definition 3.19. A morphism of mixed Hodge complexes of sheaves is called a **weak equivalence** if it is a quasi-isomorphism (but not necessarily a bi-filtered quasi-isomorphism). Weak equivalences induce isomorphisms of mixed Hodge structures on the hypercohomology of the complexes.

For later reference we record here how to produce the tensor product of R-mixed Hodge complexes. Let us first recall how one forms the tensor product of two bounded below (filtered) complexes $(K^\bullet, F)$ and $(L^\bullet, G)$. The tensor product complex $(K \otimes_R L)^\bullet$ is defined as follows

$$(K \otimes L)^n = \bigoplus_{i+j=n} K^i \otimes_R L^j, \quad d(x \otimes y) = dx \otimes y + (-1)^{\deg x} x \otimes dy$$

and the filtration $F \otimes G$ by

$$(F \otimes G)^m (K^\bullet \otimes L^\bullet) = \bigoplus_{i+j=m} F^i K^\bullet \otimes_R G^j L^\bullet.$$

This yields indeed a filtered complex denoted $(K^\bullet, F) \otimes (L^\bullet, G)$. Tensor products of bi-filtered complexes are defined similarly and denoted in the obvious way. We have the following result, whose easy proof we omit:

Lemma 3.20. 1) *Let there be given two R-mixed Hodge complexes of sheaves $\mathcal{K}^\bullet = (\mathcal{K}^\bullet_R, (\mathcal{K}^\bullet_{R\otimes\mathbb{Q}}, W), \alpha, (\mathcal{K}^\bullet_{\mathbb{C}}, W, F), \beta)$ and $\mathcal{L}^\bullet = (\mathcal{L}^\bullet_R, (\mathcal{L}^\bullet_{R\otimes\mathbb{Q}}, W), \alpha', (\mathcal{L}^\bullet_{\mathbb{C}}, W, F), \beta')$. The* ***tensor product*** *$\mathcal{K}^\bullet \otimes \mathcal{L}^\bullet$, given by*

$$((\mathcal{K}_R \otimes \mathcal{L}_R)^\bullet, (\mathcal{K}^\bullet_{R\otimes\mathbb{Q}}, W) \otimes (\mathcal{L}^\bullet_{R\otimes\mathbb{Q}}, W), \alpha \otimes \alpha', \\ (\mathcal{K}^\bullet_{\mathbb{C}}, W, F) \otimes (\mathcal{L}^\bullet_{\mathbb{C}}, W, F), \beta \otimes \beta')$$

is a mixed R-Hodge complex of sheaves on X.
2) *A similar assertion holds for mixed Hodge complexes $K^\bullet$ and $L^\bullet$. For the Hodge Grothendieck characteristics we have*

$$\chi_{\text{Hdg}}(K^\bullet \otimes L^\bullet) = \chi_{\text{Hdg}}(K^\bullet)\chi_{\text{Hdg}}(L^\bullet). \tag{III–12}$$

3) *The canonical morphism* (B–13)

$$R\Gamma(X, \mathcal{K}^\bullet) \otimes R\Gamma(X, \mathcal{L}^\bullet) \to R\Gamma(X, \mathcal{K}^\bullet \otimes \mathcal{L}^\bullet)$$

is a morphism of mixed Hodge complexes.

Example 3.21. Let X and Y be two topological spaces, and let $p : X \times Y \to X$, $q : X \times Y \to Y$ the two projections. If we start with a mixed R-Hodge complex of sheaves on X, say $\mathcal{K}^\bullet$ and a mixed Hodge complex of sheaves $\mathcal{L}^\bullet$ on Y, the tensor product of the two complexes $p^*\mathcal{K}^\bullet$ and $q^*\mathcal{L}^\bullet$ is the external product $\mathcal{K}^\bullet \boxtimes \mathcal{L}^\bullet$. In particular, the morphism (B–13) in this case becomes the morphism of mixed R-Hodge complexes

$$R\Gamma(X, \mathcal{K}^\bullet) \otimes R\Gamma(Y, \mathcal{L}^\bullet) \to R\Gamma(X \times Y, \mathcal{K}^\bullet \boxtimes \mathcal{L}^\bullet).$$

3.4 The Mixed Cone

We refer to Definition A.7 for the definition of the cone of a complex. We would like to construct the cone over a morphism of mixed Hodge complexes of sheaves as a mixed Hodge complexes of sheaves.

Theorem 3.22. *Let $\mathcal{K}^\bullet \xrightarrow{\phi} \mathcal{L}^\bullet$ be a morphism of mixed Hodge complexes of sheaves. We denote the weight and Hodge filtrations on $\mathcal{K}^\bullet$ by $W_\bullet(\mathcal{K}), F^\bullet(\mathcal{K}^\bullet)$ and similarly for $\mathcal{L}^\bullet$. The comparison morphisms for $\mathcal{K}^\bullet$ are the pseudo-morphisms α and β. The ones for $\mathcal{L}$ are given by α' and β'.*
1) *Let us put*

$$W_m\mathrm{Cone}^p(\phi_S) = W_{m-1}\mathcal{K}^{p+1}_S \oplus W_m\mathcal{L}^p_S, \quad S = R \otimes \mathbb{Q}, \quad \text{or } \mathbb{C}$$

and

$$F^r\mathrm{Cone}^p(\phi_{\mathbb{C}}) = F^r\mathcal{K}^{p+1}_{\mathbb{C}} \oplus F^r\mathcal{L}^p_{\mathbb{C}}.$$

Together with comparison morphisms given by

$$\begin{aligned}(\alpha, \alpha') &: \mathrm{Cone}^\bullet(\phi_R) \dashrightarrow \mathrm{Cone}^\bullet(\phi_{R\otimes\mathbb{Q}}, W)\\ (\beta, \beta') &: \mathrm{Cone}^\bullet(\phi_{R\otimes\mathbb{Q}}, W) \dashrightarrow \mathrm{Cone}^\bullet(\phi_{\mathbb{C}}, W, F)\end{aligned}$$

these data define the structure of a mixed Hodge complexes of sheaves on the cone $\mathrm{Cone}^\bullet(\phi)$. *We call this structure the **mixed cone**.*

2) *There is an exact sequence of mixed Hodge complexes of sheaves*

$$0 \to \mathcal{L}^\bullet \to \mathrm{Cone}^\bullet(\phi) \to \mathcal{K}^\bullet[1] \to 0$$

inducing a long exact sequence

$$\cdots \to \mathbb{H}^k(X, \mathcal{K}^\bullet) \to \mathbb{H}^k(X, \mathcal{L}^\bullet) \to \mathbb{H}^k(X, \mathrm{Cone}^\bullet(\phi)) \to \mathbb{H}^{k+1}(X, \mathcal{K}^\bullet) \to \cdots$$

of mixed Hodge structures with connecting homomorphism induced by ϕ. *In particular all maps are morphisms of mixed Hodge structures.*
3) *We have*

$$\chi_{\mathrm{Hdg}}(R\Gamma\,\mathrm{Cone}^\bullet(\phi)) = \chi_{\mathrm{Hdg}}(R\Gamma\mathcal{L}^\bullet) - \chi_{\mathrm{Hdg}}(R\Gamma\mathcal{K}^\bullet). \qquad \text{(III–13)}$$

Proof. 1) A morphism of pseudomorphisms consists of morphisms between the constituents of the chains which make up a pseudomorphisms and such that the obvious diagrams commute. This implies that each such diagram defines a morphism of cones or a quasi-isomorphism of cones or an inverse of such. In this way we get the pseudo-morphisms for the cones.

The map ϕ_R maps $W_m\mathcal{K}^\bullet_R$ to $W_m\mathcal{L}^\bullet_R$ and so on

$$\mathrm{Gr}^W_m \mathrm{Cone}^\bullet(\phi_R) = \mathrm{Gr}^W_{m-1} \mathcal{K}^\bullet_R[1] \oplus \mathrm{Gr}^W_m \mathcal{L}_R$$

the contribution of ϕ to the differential vanishes. So the preceding direct sum decomposition is a direct sum decomposition of complexes compatible with the F-filtration. Since both $\mathrm{Gr}^W_{m-1} \mathcal{K}^\bullet[1]$ and $\mathrm{Gr}^W_m \mathcal{L}^\bullet$ are Hodge complexes of sheaves of weight m, the direct sum $\mathrm{Gr}^W_m \mathrm{Cone}^\bullet(\phi)$ is. This completes the proof of 1).
2) This is a direct consequence of the definitions and the existence of an exact sequence for cones (formula (A–12)).
3) This follows from (III–6) and Lemma-Def. 2.35. □

Remark 3.23. If one would work with comparison morphisms in the derived category, as Deligne does, one gets diagrams which commute only *up to homotopy*. It follows that if one would use the same definition as above, the comparison maps for the cone would not commute with the derivative (remember that the derivative of the cone of a map involves the map itself). If one chooses an explicit homotopy it is still possible to define a representative for the mixed cone, but this really depends on the choice of the homotopy.

For this reason we have adapted Deligne's set-up. We work always with explicit representatives. In the geometric setting these representatives behave functorially which implies that in the geometric setting we automatically get morphisms of mixed complexes (of sheaves) in our sense which makes it possible to use the mixed cones in this situation.

Example 3.24. (RELATIVE COHOMOLOGY) Suppose that Y is a smooth subvariety of a compact Kähler manifold X with injection $i : Y \hookrightarrow X$. Recall

(Example 2.34) that for any smooth compact Kähler manifold Z we have introduced its Hodge-De Rham complex (equipped with the canonical De Rham-Godement marking)

$$\mathcal{H}dg^\bullet(Z) = \big(\underline{\mathbb{Z}}_Z, (\Omega_Z^\bullet, \sigma(\Omega_Z^\bullet)), \underline{\mathbb{Z}}_Z \hookrightarrow \Omega_Z^\bullet\big).$$

The relative cohomology $H^*(X,Y)$ can be viewed as the hypercohomology of

$$\mathcal{H}dg^\bullet(X,Y) := \mathrm{Cone}^\bullet\{\mathcal{H}dg^\bullet(X) \xrightarrow{i^*} i_*\mathcal{H}dg^\bullet(Y)\}[-1]$$

and hence carries a mixed Hodge structure making the long exact sequence for the pair (X,Y) an exact sequence of mixed Hodge structures. The Hodge Grothendieck characteristic of (X,Y) is the Hodge Grothendieck characteristic of the complex of global sections of $\mathcal{H}dg^\bullet(X,Y)$. Hence, using (III–13), (III–11) and Remark 2.35, we find

$$\chi_{\mathrm{Hdg}}(X,Y) = \chi_{\mathrm{Hdg}}(X) - \chi_{\mathrm{Hdg}}(Y).$$

Later, when we construct a mixed Hodge complex which computes the cohomology of possibly singular or non-compact algebraic varieties, the same construction can be applied when Y is any subvariety. See § 5.5

We next consider functoriality:

Lemma-Definition 3.25. Let $f : X \to Y$ be a continuous map and let $\mathcal{K}^\bullet$ be a mixed Hodge complex of sheaves on X. Defining

$$\begin{aligned}(Rf_*\mathcal{K}^\bullet)_R &= f_*\mathcal{C}_{\mathrm{Gdm}}(\mathcal{K}_R^\bullet) = Rf_*\mathcal{K}_R^\bullet\\ ((Rf_*\mathcal{K}^\bullet)_{\mathbb{Q}}, W) &= Rf_*(\mathcal{K}_{R\otimes\mathbb{Q}}^\bullet, W)\\ ((Rf_*\mathcal{K}^\bullet)_{\mathbb{C}}, W, F) &= Rf_*(\mathcal{K}_{\mathbb{C}}^\bullet, W, F)\end{aligned}$$

one obtains a mixed Hodge complex $Rf_*\mathcal{K}^\bullet$ of sheaves on Y (the comparison morphisms are the obvious ones), and there is an isomorphism of mixed Hodge structures

$$\mathbb{H}^p(X, \mathcal{K}^\bullet) \xrightarrow{\sim} \mathbb{H}^p(Y, Rf_*\mathcal{K}^\bullet).$$

The (easy) proofs of these assertions are left to the reader.

Example 3.26. Let $f : X \to Y$ be a holomorphic map between *compact* Kähler manifolds. Then there is a morphism of mixed Hodge complexes of sheaves on Y:

$$f^* : \mathcal{H}dg^\bullet(Y) \to Rf_*\mathcal{H}dg^\bullet(X)$$

which induces a long exact sequence of mixed Hodge structures

$$\to H^k(Y) \xrightarrow{f^*} H^k(X) \to \mathbb{H}^k(Y, \mathrm{Cone}^\bullet(f^*)) \to \dots .$$

3.5 Extensions of Mixed Hodge Structures

In this section we only consider integral mixed Hodge structures, but all of the results can easily be formulated and proven for R-Hodge structures. A mixed Hodge structure contains as part of the information the Hodge structures on the graded parts of the weight filtration. Two successive steps $W_{k-1} \subset W_k$ in the weight filtration define an extension of Gr_W^k by W_{k-1} and so the entire mixed Hodge structure can be considered as a successive extension of pure Hodge structures. So it is natural to study extensions in the category of mixed Hodge structures.

We have seen that the category of mixed Hodge structures is an abelian category (Corollary 3.9). We can therefore form the Yoneda Ext functor (see § A.2.6) $\mathrm{Ext}^n_{\mathrm{MHS}}(-,-)$. It is defined for $n \geq 1$ and for $n = 1$ will also be denoted $\mathrm{Ext}_{\mathrm{MHS}}$. As usual one puts $\mathrm{Ext}^0_{\mathrm{MHS}} = \mathrm{Hom}_{\mathrm{MHS}}$. By the general theory (see loc. cit.) the $\mathrm{Ext}^n_{\mathrm{MHS}}(A, B)$ are groups.

3.5.1 Mixed Hodge Extensions

The abelian group $\mathrm{Ext}_{\mathrm{MHS}}(A, B)$ is called the **group of mixed Hodge extensions** of A by B. Since congruences between 1-extensions are necessarily isomorphisms of mixed Hodge structures (by Cor. 3.7) the latter classifies isomorphism classes of extension between mixed Hodge structures.

Remark 3.27. A mixed Hodge structure on H is completely determined specifying a mixed Hodge structure on the free quotient $H/\operatorname{Tors}(H)$. In particular, first of all, the forgetful functor induces an isomorphism $E := \mathrm{Ext}_{\mathrm{MHS}}(\operatorname{Tors}(A), \operatorname{Tors}(B)) \simeq \mathrm{Ext}_{\mathrm{Abgrps}}(\operatorname{Tors}(A), \operatorname{Tors}(B))$. Secondly, since of course $E = \mathrm{Ext}_{\mathrm{Abgrps}}(A, B)$ there is a forgetful functor $\mathrm{Ext}_{\mathrm{MHS}}(A, B) \to E$ which can be shown to be a retraction for the natural exact sequence

$$0 \to E \to \mathrm{Ext}_{\mathrm{MHS}}(A, B) \to \mathrm{Ext}_{\mathrm{MHS}}(A/\operatorname{Tors}(A), B/\operatorname{Tors}(B)) \to 0.$$

So this sequence is split and there is no loss of information if we work with mixed Hodge structures on *torsion free* modules.

Definition 3.28. Let H be a mixed Hodge structure with $H_{\mathbb{Z}}$ torsion free. For $p \in \mathbb{Z}$ the **p-th Jacobian of** H is defined as

$$J^p(H) := H_{\mathbb{C}}/(F^p + H_{\mathbb{Z}}). \tag{III–14}$$

Since by (III–3) $F^0 H(p) = F^p H$ we have

$$J^p H \simeq J^0 \operatorname{Hom}(\mathbb{Z}, H(p)) = J^0 \operatorname{Hom}(\mathbb{Z}(-p), H). \tag{III–15}$$

Lemma 3.29. *If $W_{-1} H_{\mathbb{Q}} = H_{\mathbb{Q}}$ the group $J^p(H)$ is a Lie group.*

Proof. The condition implies that $F^0 H_{\mathbb{C}} \cap \overline{F^0 H_{\mathbb{C}}} = 0$ and hence $F^0 H_{\mathbb{C}}$ does not meet the image of $H_{\mathbb{Z}}$ in $H_{\mathbb{C}}$. In particularly, $H_{\mathbb{Z}}$ embeds discretely in $H_{\mathbb{C}}/F^p$. □

Example 3.30. Let H be a pure Hodge structure of weight $2m-1$. Then $J^m(H)$ is a compact complex torus. Indeed, we have a direct sum decomposition

$$H_{\mathbb{C}} := H_{\mathbb{Z}} \otimes \mathbb{C} = F^m \oplus \overline{F^m}$$

and any real element $x \in H_{\mathbb{Z}} \otimes \mathbb{R}$ belonging to one of these summands also belongs to the other one and so must be zero.

Theorem 3.31. *Let A and B be mixed Hodge structures with $A_{\mathbb{Z}}$ and $B_{\mathbb{Z}}$ torsion free.*

1) *There is a canonical isomorphism*

$$\mathrm{Ext}(A,B) \cong \mathrm{Hom}^W(A_{\mathbb{C}}, B_{\mathbb{C}}) / \mathrm{Hom}_F^W(A_{\mathbb{C}}, B_{\mathbb{C}}) + \mathrm{Hom}^W(A,B),$$

or, equivalently

$$\mathrm{Ext}(A,B) \cong W_0 \mathrm{Hom}(A,B)_{\mathbb{C}} / W_0 \cap F^0 \mathrm{Hom}(A,B)_{\mathbb{C}} + W_0 \mathrm{Hom}(A,B).$$

2) *Suppose that for some m we have $W_m B = B$ while $W_m A = 0$ (i.e. the weights of B are less than the weights of A, one says that A and B are* ***separated mixed Hodge structures****). There is a natural isomorphism of groups*

$$m : \mathrm{Ext}(A,B) \xrightarrow{\cong} J^0 \mathrm{Hom}(A,B)$$

given explicitly as follows. Let

$$E = [0 \to B \xrightarrow{\beta} H \xrightarrow{\alpha} A \to 0]$$

be an extension. Choose a ***retraction*** *$r : H \to B$, i.e. $r \circ \beta = \mathrm{id}_B$ and a section σ_F of $\alpha_{\mathbb{C}} : H_{\mathbb{C}} \to A_{\mathbb{C}}$ preserving the Hodge filtration. Then $m(E) \in J^0 \mathrm{Hom}(A,B)$ is represented by $r_{\mathbb{C}} \circ \sigma_F \in \mathrm{Hom}(A,B)_{\mathbb{C}}$.*

Proof. Since the separatedness implies that $W_0 \mathrm{Hom}(A,B) = \mathrm{Hom}(A,B)$, the second statement follows from the first, except for the explicit formula for m which we prove later.

Let

$$0 \to B \xrightarrow{\beta} H \xrightarrow{\alpha} A \to 0$$

be an extension of A by B. As $A_{\mathbb{Z}}$ is torsion free, the extension of the underlying $\mathbb{Z}$-modules splits. Let us choose a section

$$\sigma_{\mathbb{Z}} : A_{\mathbb{Z}} \to H_{\mathbb{Z}}$$

which preserves the weight filtration strictly. Any two such sections differ by an element of $\mathrm{Hom}_{\mathbb{Z}}(A_{\mathbb{Z}}, B_{\mathbb{Z}}) = \mathrm{Hom}(A,B)_{\mathbb{Z}}$ preserving weights, i.e by an element of $\mathrm{Hom}^W(A,B)$. The splitting defines an isomorphism of $\mathbb{Z}$-modules

$$f(\sigma_{\mathbb{Z}}) : B_{\mathbb{Z}} \oplus A_{\mathbb{Z}} \xrightarrow{\cong} H_{\mathbb{Z}}, \quad (b,a) \mapsto \beta(b) + \sigma(a)$$

such that $i_{\mathbb{Q}}$ sends the direct sum weight filtration to the weight filtration of $H_{\mathbb{Q}}$. Let us now choose a section

$$\sigma_{\mathbb{C}} : A_{\mathbb{C}} \to H_{\mathbb{C}}$$

which preserves the weight *and* the Hodge filtration strictly. This is possible, since for instance the exact sequence is compatible with the Deligne splitting and hence we can take any section that is compatible with it. Any two such sections differ by an element of $\mathrm{Hom}_F^W(A_{\mathbb{C}}, B_{\mathbb{C}})$. If we compare the corresponding isomorphism

$$f(\sigma_{\mathbb{C}}) : B_{\mathbb{C}} \oplus A_{\mathbb{C}} \xrightarrow{\cong} H_{\mathbb{C}}$$

with $f(\sigma_{\mathbb{Z}})$ we get the $\mathbb{C}$-linear automorphism $f(\sigma_{\mathbb{C}})^{-1} \circ (f(\sigma_{\mathbb{Z}}) \otimes 1)$ of $B_{\mathbb{C}} \oplus A_{\mathbb{C}}$, which in matrix form is given by

$$g(\phi) := \begin{pmatrix} 1_B & \phi \\ 0 & 1_A \end{pmatrix}.$$

The map $\phi : B \to A$ is just the difference of the two sections and hence preserves the weight filtration (both sections are strictly compatible with the weight filtrations and the weight filtration on $A_{\mathbb{Q}}$ is the one induced on A from the weight filtration on $H_{\mathbb{Q}}$). In general it *does not preserve the Hodge filtration.*

From the preceding arguments it follows that the class

$$[\phi] \in \mathrm{Hom}^W(A_{\mathbb{C}}, B_{\mathbb{C}}) / \mathrm{Hom}_F^W(A_{\mathbb{C}}, B_{\mathbb{C}}) + \mathrm{Hom}^W(A, B)$$

is a well defined invariant of the extension. If $[\phi] = 0$, the section $\sigma_{\mathbb{Z}}$ can be chosen in such a way that it preserves the weight and Hodge filtration and hence H is congruent to the direct sum mixed Hodge structure $A \oplus B$.

If a $\mathbb{C}$-linear map $\phi : A_{\mathbb{C}} \to B_{\mathbb{C}}$ preserves the weight filtration the automorphism g_ϕ of $B \oplus A$ preserves the weight filtration as well. The filtration on $B_{\mathbb{C}} \oplus A_{\mathbb{C}}$ defined by

$$F_\phi^\bullet := g_\phi(F^\bullet(B_{\mathbb{C}}) \oplus F^\bullet(A_{\mathbb{C}})) = F^\bullet(B_{\mathbb{C}}) + (1, \phi) F^\bullet(A_{\mathbb{C}}).$$

then induces a mixed Hodge structure, since on

$$\mathrm{Gr}_m^W g_\phi(B_{\mathbb{C}} \oplus A_{\mathbb{C}}) = g_\phi(\mathrm{Gr}_m^W B_{\mathbb{C}} \oplus \mathrm{Gr}_m^W A_{\mathbb{C}})$$

it induces the weight m Hodge structure which is the image under g_ϕ of the direct sum Hodge structure $\mathrm{Gr}_m^W B_{\mathbb{C}} \oplus \mathrm{Gr}_m^W A_{\mathbb{C}}$. So all classes $[\phi]$ occur.

The group structures are compatible: the Baer-sum of ϕ_i and ϕ_2 is the composition of pulling back along a diagonal and pushing out along a co-diagonal, i.e. if the two extensions are represented by the matrices $g(\phi_1)$ and $g(\phi_2)$, the Baer sum is represented by the matrix $g(\phi_1 + \phi_2)$.

Let us now come back to the explicit formula for the isomorphism m for a separated extension. We just note that the retraction r defines a section $\sigma_{\mathbb{Z}}$ by means of the formula

$$\sigma_{\mathbb{Z}}(a) = h - \beta \circ r(h), \quad h \in H \text{ with } \alpha(h) = a.$$

Since the extension is separated, both r and σ_F preserve automatically the weight filtration, so that by the preceding discussion, the extension is represented by the difference $\sigma_F - \sigma_{\mathbb{Z}}$. With $\sigma_F(a) - \sigma_{\mathbb{Z}}(a) = \beta(b)$ we find $b = r\circ\beta(b) = r\circ\sigma_F(a) - r(h) + r\circ\beta\circ r(h) = r\circ\sigma_F(a)$ which shows that indeed $r\circ\sigma_F$ represents the extension. □

Corollary 3.32. *If $V \to V'$ is a surjective morphism of mixed Hodge structures, then for any mixed Hodge structure H the induced map $\mathrm{Ext}_{\mathrm{MHS}}(H, V) \to \mathrm{Ext}_{\mathrm{MHS}}(H, V')$ is onto.*

Proof. This is basically true since Hom is right exact on free $\mathbb{Z}$-modules. Indeed, the induced map $\mathrm{Hom}^W(H, V)_{\mathbb{C}} \to \mathrm{Hom}^W(H, V'')_{\mathbb{C}}$ is surjective, inducing a surjective map between the quotients on both sides that give $\mathrm{Ext}(H, V)$, respectively $\mathrm{Ext}(H, V')$. □

Remark 3.33. If A and B are separated, the group $J^0 \mathrm{Hom}(A, B)_{\mathbb{C}}$ has the structure of a complex Lie group. Indeed, separateness is equivalent to saying that $\mathrm{Hom}(A, B)$ has only negative weights, i.e.

$$W_{-1} \mathrm{Hom}(A, B)_{\mathbb{Q}} = \mathrm{Hom}(A, B)_{\mathbb{Q}}$$

and the result follows upon applying Lemma 3.29 to the mixed Hodge structure $\mathrm{Hom}(A, B)$.

Examples 3.34. 1) For $m < n$ the group $\mathrm{Ext}_{\mathrm{MHS}}(\mathbb{Z}(m), \mathbb{Z}(n))$ is isomorphic to $\mathbb{C}/(2\pi\mathrm{i})^{n-m}\mathbb{Z}$, a twist of $\mathbb{C}^\times$,

2) If H is a Hodge structure of pure weight $2m - 1$ we have seen (III–15) that $J^m H \simeq J^0 \mathrm{Hom}_{\mathrm{MHS}}(\mathbb{Z}(-m), H)$ and hence

$$J^m H = \mathrm{Ext}_{\mathrm{MHS}}(\mathbb{Z}(-m), H),$$

a description which will turn out to be useful for an algebraic description of the Abel-Jacobi map in §. 7.1.2 A.

3) Let X be any smooth projective manifold. Take $A = \mathbb{Z}$ and $B = H^k(X, \mathbb{Z})(d)$ where d is chosen so that $k < 2d$ (for instance,if $k = 2m - 1$ is odd, one can take $d = m$). Then the weights are separated and by (III–15) we have

$$\begin{aligned}\mathrm{Ext}_{\mathrm{MHS}}(\mathbb{Z}, H^k(X, \mathbb{Z})(d)) &= J^0 \mathrm{Hom}(\mathbb{Z}, H^k(X, \mathbb{Z})(d)) \\ &\simeq J^d H^k((X, \mathbb{Z})) = H^k(X; \mathbb{C})/H^k(X) \oplus F^d H^k(X).\end{aligned}$$

3.5.2 Iterated Extensions and Absolute Hodge Cohomology

We return to $\mathrm{Ext}^n_{\mathrm{MHS}}$ for arbitrary n. As in the case of modules, given an exact sequence of mixed Hodge structures, we shall see below that there is a long exact sequence of Ext-groups and the above description of Ext^1 then implies that it is a right exact functor for $R = \mathbb{Z}$ and hence, as in the "classical" case, the higher Ext-groups vanish for $R = \mathbb{Z}$ or R a subfield of $\mathbb{C}$. As consequence of Lemma A.33 and Corollary 3.32 we have:

Proposition 3.35. *For any two mixed Hodge structures A and B, we have $\mathrm{Ext}^p_{\mathrm{MHS}}(A, B) = 0$ as soon as $p \geq 2$.*

In the remainder of this section we assume that our mixed (integral) Hodge structures are polarizable in the sense of Def. 3.1. We only consider morphisms which come from morphisms preserving *some* polarization. This leads to the abelian category consisting of mixed $\mathbb{Q}$-Hodge complexes with polarizable cohomology. Note that the polarizable $\mathbb{Q}$-Hodge complexes belong to this category, but have more structure. The usual construction of the cone provides the triangles which makes this category triangulated.

Given a bounded below complex $H^\bullet$ of mixed Hodge structures we can make it into a (normalized) mixed Hodge complex: the comparison morphisms are induced by the identity; the Hodge filtrations stay the same, but the weights have to be shifted by putting $(\phi W)_m H^k = W_{k+m} H^k$. In this process the boundary maps in the ϕW-gradeds becomes zero. So $G^\bullet = \mathrm{Gr}^{\phi W}_m H^\bullet$ is its own cohomology: $H^k(G^\bullet) = \mathrm{Gr}^W_{k+m} H^k$, which has a Hodge structure of weight $k + m$ by assumption. Moreover, the boundary maps being zero in $G^\bullet$, they are strictly compatible with the F-filtration. So the new complex is indeed a mixed Hodge complex. This process will be called **marking the complex of mixed Hodge structures** and we use ϕ to denote it. Beilinson [Beil86, 3.11], has shown:

Lemma 3.36. *Marking establishes an equivalence of triangulated categories*

$$\phi : D^{\mathrm{b}} \begin{pmatrix} \text{Graded polarizable} \\ \text{mixed Hodge structures} \end{pmatrix} \longrightarrow D^{\mathrm{b}} \begin{pmatrix} \text{normalized mixed Hodge} \\ \text{complexes with} \\ \text{polarizable cohomology.} \end{pmatrix}$$

To explain the notion of absolute Hodge cohomology, let us start with a Hodge structure H of weight $2k$. Hodge classes in H^k are precisely the morphisms of Hodge structure $\mathbb{Z}(-k) \to H^k$. If, instead, we have a complex $H^\bullet$ of mixed Hodge structures, we consider homomorphisms of *mixed* Hodge structures $\mathbb{Z} \to H^\bullet$. Any such morphism is determined uniquely by an element in $W^0 \cap F^0 H^\bullet$ in other words an element in the kernel of

$$\begin{aligned} H^\bullet \oplus W_0 H^\bullet_{\mathbb{Q}} \oplus W_0 \cap F^0 H^\bullet_{\mathbb{C}} &\xrightarrow{\delta} H^\bullet_{\mathbb{Q}} \oplus W_0 H^\bullet_{\mathbb{C}} \\ (x, y, z) &\mapsto (x - y, y - z). \end{aligned}$$

This is a surjective homomorphism and so, by the triangle of the cone (A–15) we can represent $\mathrm{Hom}_{\mathrm{MHS}}(\mathbb{Z}, H^\bullet)$ by $\mathrm{Cone}^\bullet(\delta)[-1]$. Hence $\mathrm{Ext}^k_{\mathrm{MHS}}(\mathbb{Z}, H^\bullet)$, the groups of the "$k$-th derived Hodge classes" in a bounded complex $H^\bullet$ of mixed Hodge structures can be calculated as the k-th cohomology groups of its normalized complex.

Beilinson generalizes this specific cone construction to normalized bounded mixed Hodge complexes in such a way that it is compatible with the marking. So start with a tent like (III–8). We first have to replace the weight filtration W by the backshifted weight filtration (see A.50) $\mathrm{Dec}\, W$ since under marking we are shifting the weight filtration. Then we have to generalize the above map δ coming from the identity to general comparison morphisms:

$$\begin{array}{ccc} K^\bullet \oplus (\mathrm{Dec}\, W)_0 K^\bullet_{\mathbb{Q}} \oplus ((\mathrm{Dec}\, W)_0 \cap F^0) K^\bullet_{C} & \xrightarrow{(\alpha,\beta)} & 'K^\bullet_{\mathbb{Q}} \oplus (\mathrm{Dec}\, W)_0 {}'K^\bullet_{\mathbb{C}} \\ (k, k_{\mathbb{Q}}, k_{\mathbb{C}}) & \mapsto & (\alpha_1 k - \alpha_2 k_{\mathbb{Q}}, \beta_1 k_{\mathbb{Q}} - \beta_2 k_{\mathbb{C}}). \end{array}$$

The **absolute Hodge cohomology** then is defined as

$$H^k_{\mathrm{Hodge}}(K^\bullet) := H^k\, \mathrm{Cone}^\bullet(\alpha, \beta)[-1].$$

One can verify that (in the *derived category*) $\mathrm{Cone}^\bullet(\alpha, \beta)$ indeed only depends on the data given by the polarizable mixed Hodge complex, i.e. the choices of $'K^\bullet_{\mathbb{Q}}$ and $'K^\bullet_{\mathbb{C}}$ are immaterial.

Example 3.37. Let $H^\bullet$ be a bounded complex of mixed Hodge structures with its corresponding normalized complex $\phi(H^\bullet)$. We saw that $\mathrm{Hom}_{\mathrm{MHS}}(\mathbb{Z}, H^\bullet) = \mathrm{Cone}^\bullet(\delta)[-1]$ and so

$$H^k_{\mathrm{Hodge}}(\phi(H^\bullet)) = H^k(\mathrm{Cone}^\bullet(\delta)[-1]) = \mathrm{Ext}^k_{\mathrm{MHS}}(\mathbb{Z}, H^\bullet) \qquad \text{(III–16)}$$

which shows the relation with the Ext-groups.

Pursuing the preceding a little further, we can consider the cohomology groups of any normalized mixed Hodge complex $K^\bullet$. These, by definition, admit mixed Hodge structures, so that the objects $\mathrm{Ext}^p_{\mathrm{MHS}}(\mathbb{Z}, H^q(K^\bullet))$ make sense. It is the $E_2^{p,q}$-term of the spectral sequence for the derived functor $R\,\mathrm{Hom}$ which reads

$$E_2^{p,q} = \mathrm{Ext}^p_{\mathrm{MHS}}(\mathbb{Z}, H^q(K^\bullet)) \Longrightarrow H^{p+q}_{\mathrm{Hodge}}(K^\bullet).$$

Since the higher Ext-groups vanish, we thus obtain a short exact sequence

$$0 \to \mathrm{Ext}^1_{\mathrm{MHS}}(\mathbb{Z}, H^{k-1}K^\bullet) \to H^k_{\mathrm{Hodge}} K^\bullet \to \mathrm{Hom}_{\mathrm{MHS}}(\mathbb{Z}, H^k K^\bullet) \to 0. \quad \text{(III–17)}$$

This sequence serves to relate absolute Hodge cohomology and Deligne cohomology. See §. 7.2. Suppose that $K^\bullet = \phi(H^\bullet)$, then $H^p(K^\bullet) = H^p(H^\bullet)$ and the sequence (III–17) becomes

$$0 \to \mathrm{Ext}^1_{\mathrm{MHS}}(\mathbb{Z}, H^{k-1}H^\bullet) \to \mathrm{Ext}^k_{\mathrm{MHS}}(\mathbb{Z}, H^\bullet) \to \mathrm{Hom}_{\mathrm{MHS}}(\mathbb{Z}, H^k H^\bullet) \to 0.$$

Note that while a higher Ext between mixed Hodge structures vanishes, this is no longer true for *complexes* of mixed Hodge structures.

Historical Remarks. The Deligne splitting is, as the name suggests, due to Deligne, but the published versions of it can be found in [C-K-S86] and [Mor]. The results in Sect. 3.2–3.4 are all due to Deligne ([Del71, Del73]). Extensions of (mixed) Hodge structures have been studied by Carlson [Car79, Car85b, Car87]. The last section explains Beilinson's results from [Beil86]. See also the article of Jannsen in [R-S-S].

Part II

Mixed Hodge structures on Cohomology Groups

4

Smooth Varieties

The main goal of this chapter is to show that there exists a functorial mixed Hodge structure on any of the cohomology groups of a smooth variety and which coincides with the classical Hodge structure if the variety is smooth and projective. To define this mixed Hodge structure, we first compactify the variety by a divisor whose singularities locally look like the crossing of coordinate hyperplanes. In § 4.1 we study the cohomology with respect to this compactification and we shall show in § 4.1–4.3 how to put weight and Hodge filtrations on the cohomology groups defining a mixed Hodge structure. The rational component of the Hodge De Rham complex which gives this Hodge structure can be given using so-called log structures which are treated in § 4.4 and which will be used in a decisive way in Chapter 11.

In § 4.5 we check that the mixed Hodge structure does not depend on our chosen compactification and that the construction is functorial. We also prove the theorem on the fixed part and show that for projective families over a smooth curve the Leray spectral sequence degenerates at E_2.

4.1 Main Result

Let U be a smooth complex algebraic variety. By [Naga] U is Zariski open in some compact algebraic variety X, which by [Hir64] one can assume to be smooth and for which $D = X - U$ locally looks like the crossing of coordinate hyperplanes. It is called a **normal crossing divisor**. If the irreducible components D_k of D are smooth, we say that D has **simple** or **strict normal crossings**.

Definition 4.1. We say that X is a **good compactification** of $U = X - D$ if X is smooth and D is a simple normal crossing divisor.

We return for the moment to the situation where $D \subset X$ is a hypersurface (possibly with singularities and reducible) inside a smooth n-dimensional complex manifold X and as above, we set

$$j : U = X - D \hookrightarrow X$$

A holomorphic differential form ω on U is said to have **logarithmic poles along** D if ω and $d\omega$ have at most a pole of order one along D. It follows that these holomorphic differential forms constitute a subcomplex $\Omega^\bullet_X(\log D) \subset j_*\Omega^\bullet_U$, the **logarithmic de Rham complex**

Suppose now that D has simple normal crossings, $p \in D$ and $V \subset X$ is an open neighbourhood with coordinates $(z_1, \dots, z_n)$ in which D has equation $z_1 \cdots z_k = 0$. On can show [Grif-Ha, p. 449]

$$\Omega^1_X(\log D)_p = \mathcal{O}_{X,p}\frac{dz_1}{z_1} \oplus \cdots \mathcal{O}_{X,p}\frac{dz_k}{z_k} \oplus \mathcal{O}_{X,p}dz_{k+1} \oplus \cdots \oplus \mathcal{O}_{X,p}dz_n,$$

$$\Omega^p_X(\log D)_p = \bigwedge^p \Omega^1_X(\log D)_p.$$

An essential ingredient in the proof of the following theorem is the residue map which is defined as follows. We set $D_k = \{z_k = 0\}$ and we let D' be the divisor on D_k traced out by D. Then writing $\omega = \eta \wedge (dz_k/z_k) + \eta'$ with η, η' not containing dz_k, the **residue map** can be defined as

$$\begin{aligned} \text{res} : \Omega^p_X(\log D) &\to \Omega^{p-1}_{D_k}(\log D') \\ \omega &\mapsto \eta\big|_{D_k}. \end{aligned}$$

As a special case we have the Poincaré residues $R_k : \Omega^1_X(\log D) \to \mathcal{O}_{D_k}$ which we shall use in § 11.1.1. As an aside, in § 4.2 we iterate this procedure to get residues for multiple intersections.

We can now formulate the main result of this chapter:

Theorem 4.2. *Let U be a complex algebraic manifold and let X be a good compactification, i.e. $D = X - U$ is a divisor with simple normal crossings. Then the following is true.*

1)
$$H^k(U; \mathbb{C}) = \mathbb{H}^k(X, \Omega^\bullet_X(\log D));$$

2) *The filtration W defined by*

$$W_m\Omega^p_X(\log D) = \begin{cases} 0 & \text{for } m < 0 \\ \Omega^p_X(\log D) & \text{for } m \geq p \\ \Omega^{p-m}_X \wedge \Omega^m_X(\log D) & \text{if } 0 \leq m \leq p. \end{cases}$$

induces in cohomology

$$W_m H^k(U; \mathbb{C}) = \operatorname{Im}\left(\mathbb{H}^k(X, W_{m-k}\Omega^\bullet_X(\log D)) \to H^k(U; \mathbb{C})\right),$$

a filtration which can be defined over the rationals. Together with the trivial filtration F on the complex $\Omega^\bullet_X(\log D)$ (see Example A.34.1) which in cohomology gives

$$F^p H^k(U, \mathbb{C}) = \operatorname{Im}\left(\mathbb{H}^k(X, F^p\Omega^\bullet_X(\log D)) \to H^k(U; \mathbb{C})\right)$$

these put a mixed Hodge structure on $H^k(U)$.

By Theorem 3.18, Part 2 of the above theorem would follow if we can put the structure of a mixed Hodge complex of sheaves on the complex $\Omega^\bullet_X(\log D)$; we postpone this to the next section (Proposition 4.11).

Part 1 is contained in the following Proposition, which we prove first.

Proposition 4.3. *The inclusion of complexes*

$$\Omega^\bullet_X(\log D) \to j_*\Omega^\bullet_U$$

is a quasi-isomorphism and induces a natural identification

$$H^k(U;\mathbb{C}) = \mathbb{H}^k(X, \Omega^\bullet_X(\log D)).$$

In other words, cohomology of U can be calculated using the log-complex.

Furthermore, the natural map $j_*\Omega^\bullet_U \to j_*s[\mathcal{C}^\bullet_{\mathrm{Gdm}}\Omega^\bullet_U] = Rj_*\Omega^\bullet_U$ *is a quasi-isomorphism inducing* $H^k(U;\mathbb{C}) = \mathbb{H}^k(X, Rj_*\underline{\mathbb{C}}_U)$.

Proof. The first assertion is a local calculation. We take for X is a polydisc Δ^n with coordinates $(z_1,\dots,z_n)$ and that $D = D_k$ is given as above by $z_1\cdots z_k = 0$. Then X and U are Stein manifolds (see Example B.17 1)) and hence $H^i(U,\Omega^j_U) = 0$ for all $i > 0$, $j \geq 0$. From Theorem B.18 it follows that the cohomology of U can be computed as the de Rham cohomology of the complex $\Omega^\bullet_U$:

$$H^q(U;\mathbb{C}) = H^q_{DR}(\Omega^\bullet_U) = H^q(\Gamma(U,\Omega^\bullet_U)).$$

It suffices therefore to show that $H^q(K^\bullet_{n,k}) \cong H^q(U;\mathbb{C})$, where

$$K^\bullet_{n,k} := \Gamma(\Delta^n, \Omega^\bullet_{\Delta^n}(\log D_k)).$$

In fact, if we put

$$R^1_{n,k} = \mathbb{C}\frac{dz_1}{z_1} \oplus \cdots \oplus \mathbb{C}\frac{dz_k}{z_k} \quad \text{(with the convention that } R^1_{n,0} = \mathbb{C}\text{)}$$

$$R^p_{n,k} = \bigwedge^p R^1_{n,k}$$

we shall prove by induction that the natural inclusions

$$\alpha_{n,k} : R^\bullet_{n,k} \to K^\bullet_{n,k}$$

are quasi-isomorphisms (here the differentials in the first complex are the zero maps). This completes the proof since on the one hand, the p-th cohomology of the first complex is exactly $R^p_{n,k}$, while on the other hand $R^p_{n,k}$ is the cohomology of U, which has the homotopy type of a product of k circles.

To show that $\alpha_{n,k}$ is a quasi-isomorphism consider the following diagram of complexes with exact rows

$$\begin{array}{ccccccccc}
0 \to & R^\bullet_{n,k-1} & \to & R^\bullet_{n,k} & \xrightarrow{\text{res}} & R^\bullet_{n-1,k-1}[-1] & \to 0 \\
 & \downarrow \alpha_{n,k-1} & & \downarrow \alpha_{n,k} & & \downarrow \alpha_{n-1,k-1} & \\
0 \to & K^\bullet_{n,k-1} & \to & K^\bullet_{n,k} & \xrightarrow{\text{res}} & K^\bullet_{n-1,k-1}[-1] & \to 0
\end{array}$$

If $\alpha_{n,k-1}$ and $\alpha_{n-1,k-1}$ are quasi-isomorphisms, then also $\alpha_{n,k}$ is a quasi-isomorphism by the five lemma. By the holomorphic Poincaré lemma, $\alpha_{n,0}$ is a quasi-isomorphism for all n, so by induction $\alpha_{n,k}$ is a quasi-isomorphism for all n, k.

To show the second assertion, consider the first spectral sequence for the derived functor for j_* (see Lemma-Definition A.46). It reads

$$E_1^{pq} = R^q j_* \Omega_U^p \Rightarrow R^{p+q} j_* \Omega_U^\bullet.$$

Since U is Stein, every point $x \in U$ has a basis of Stein open neighbourhoods V for which by Example B.17 1) $H^q(V, \Omega_U^p) = 0$ and hence

$$\left(R^q j_* \Omega_U^p\right)_p = \varinjlim_V H^q(V, \Omega_U^p) = 0.$$

It follows that this spectral sequence degenerates at E_1, i.e. $j_* \Omega_U^k \simeq R^k j_* \Omega_U^\bullet$ where the isomorphism is indeed induced by the natural homomorphism. By the holomorphic Poincaré-lemma $\Omega_U^\bullet$ is a resolution of $\underline{\mathbb{C}}_U$, and so the desired equality follows. □

Remark 4.4. We can introduce the subcomplex $\mathcal{E}_X^\bullet(\log D)$ of $j_* \mathcal{E}_U^p$ as in the holomorphic setting. Explicitly, $\mathcal{E}_X^p(\log D)$ is generated by local sections of the form $\frac{dz_{i_1}}{z_{i_1}} \wedge \cdots \wedge \frac{dz_{i_m}}{z_{i_m}} \wedge \beta$, where $m \leq p$ and β is a smooth $(p-m)$-form. The hypercohomology of this complex also computes the cohomology of U. This C^∞-complex is particularly useful when we consider its m-th graded piece with respect to the weight filtration. It is the complex $\mathrm{Gr}_m^W(\mathcal{E}_X^m(\log D) \wedge \mathcal{E}_X^\bullet$ shifted m places to the right. The complex of its global sections computes the hypercohomology of $\mathrm{Gr}_m^W(\Omega_X^\bullet(\log D))$. Explicitly

$$\mathbb{H}^k(\mathrm{Gr}_m^W(\Omega_X^\bullet(\log D))) = \frac{\{\alpha \in \Gamma(\mathcal{E}_X^m(\log D) \wedge \mathcal{E}_X^{k-m}) \mid d\alpha \in \Gamma(\mathcal{E}_X^{m-1}(\log D) \wedge \mathcal{E}_X^{k-m+2})\}}{d\Gamma(\mathcal{E}_X^m(\log D) \wedge \mathcal{E}_X^{k-m-1}) + \Gamma(\mathcal{E}_X^{m-1}(\log D) \wedge \mathcal{E}_X^{k-m+1})}.$$

4.2 Residue Maps

In this section we gather some facts on global residue maps which we shall use later. The set up is as in the previous section, so $D = D_1 \cup \cdots D_N$ is a simple normal crossing divisor inside a complex manifold X. We introduce

$$\begin{aligned} D_I &= D_{i_1} \cap D_{i_2} \cap \cdots \cap D_{i_m}, \quad I = \{i_1, \ldots, i_m\}; \\ D(I) &:= \sum_{j \notin I} D_I \cap D_j; \\ a_I &: D_I \hookrightarrow X \end{aligned}$$

and we set

$$\begin{aligned} D(0) &= X; \\ D(m) &= \coprod_{|I|=m} D_I, \quad m = 1, \ldots, N; \\ a_m &= \coprod_{|I|=m} a_I : D(m) \to X. \end{aligned}$$

Note that D_I is a submanifold of X of codimension $|I|$ and that $D(m)$ is the normalisation of the union of these submanifolds for $|I| = m$ fixed.

The goal is to define residues along D_I. So let $p \in D_I$. Then all m components D_i, $i \in I$ pass through p, but maybe more. We first need to choose local coordinates respecting in some sense the global enumeration of the components of D. Now choose coordinates $(U, z_1, \ldots, z_n)$ centred at p in such a way that $D_{i_j} = \{z_j = 0\}$ for $j = 1, \ldots, m$, and such that the remaining $k - m$ components of D are given by the equations $\{z_j = 0\}$, $j = m+1, \ldots, k$. Any local section ω of $\Omega_X^p(\log D)$ can then be written as

$$\omega = \frac{dz_1}{z_1} \wedge \cdots \wedge \frac{dz_m}{z_m} \wedge \eta + \eta'$$

where η has at most poles along components D_j, $j \notin I$, and η' is not divisible by the form $\frac{dz_1}{z_1} \wedge \cdots \wedge \frac{dz_m}{z_m}$. The restriction of η to D_I is independent of the chosen adapted local coordinates and so the map $\omega \mapsto \eta|_{D_I}$ generalizes and globalizes the previously locally defined residue maps. We note also that

$$d\omega = \frac{dz_1}{z_1} \wedge \cdots \wedge \frac{dz_m}{z_m} \wedge (-1)^m d\eta + d\eta'$$

which implies that the residue map is compatible with derivatives. Furthermore, if in the local description ω has weight $\leq m$, clearly $\eta|D_I$ is holomorphic. Let us collect this in a definition.

Definition 4.5. The **residue map**

$$\operatorname{res}_I : \Omega_X^\bullet(\log D) \to \Omega_{D_I}^\bullet(\log(D(I))[-m]$$

is locally defined by sending $\omega = [\frac{dz_1}{z_1} \wedge \cdots \wedge \frac{dz_m}{z_m} \wedge \eta + \eta'$ to $\eta|_{D(I)}$, where η' is not divisible by $\frac{dz_1}{z_1} \wedge \cdots \wedge \frac{dz_m}{z_m}$. The residue map restricts to

$$\operatorname{res}_I : W_m\Omega_X^\bullet(\log D) \to \Omega_{D_I}^\bullet[-m]. \qquad \text{(IV–1)}$$

Lemma 4.6. *The residue map* (IV–1) *is surjective and induces an isomorphism of complexes*

$$\operatorname{res}_m = \bigoplus_{|I|=m} \operatorname{res}_I : \operatorname{Gr}_m^W \Omega_X^\bullet(\log D) \xrightarrow{\cong} a_{m*}\Omega_{D(m)}^\bullet[-m]. \qquad \text{(IV–2)}$$

Proof. One constructs an inverse as follows. As before, fix an index set $I = \{i_1, \ldots, i_m\}$, $1 \leq i_1 < i_2 \cdots < i_m \leq N$. One defines

$$\rho_I : \Omega_X^p \to \mathrm{Gr}_m^W \Omega_X^{p+m}(\log D)$$
$$\rho_I(\beta) = \frac{dz_1}{z_1} \wedge \cdots \wedge \frac{dz_m}{z_m} \wedge \beta.$$

This map is well-defined, since, if $w_1, \ldots, w_n$ are other local coordinates with $D = \{w_1 = \cdots = w_k = 0\}$, the quotients z_i/w_i are holomorphic and also the forms $dz_i/z_i - dw_i/w_i$ are holomorphic, so that $\rho_I(\beta)$ in the w-coordinates differs from the expression in the z-coordinates by a form in $W_{m-1}\Omega_X^{p+m}(\log D)$ and so is zero in the quotient. Also, the elements of the form $\beta = z_{i_j}\beta'$, β' a local section of Ω_X^p, and $dz_{i_j} \wedge \beta''$, β'' a local section of Ω_X^{p-1}, map to zero so that the map ρ_I induces a map of complexes $\Omega_{D_I}^\bullet[-m] \to \mathrm{Gr}_m^W \Omega_X^\bullet(\log D)$ which can be assembled for $|I| = m$ to give a morphism of complexes

$$a_{m*}\Omega_{D(m)}^\bullet[-m] \to \mathrm{Gr}_m^W \Omega_X^\bullet(\log D).$$

This is clearly an inverse for the residue map. □

The E_1-term of the spectral sequence associated to the weight filtration on $H^k(U;\mathbb{C})$ by (A–29) is just

$$E_1^{-m,k+m} = \mathbb{H}^k(X, \mathrm{Gr}_m^W \Omega_X^\bullet(\log D))$$

which, by the above Lemma is isomorphic to $H^{k-m}(D(m);\mathbb{C})$. We want to describe d_1 by means of a suitable map between the cohomology groups on the $D(k)$. To this end we need to introduce various inclusion maps. With $I = (i_1, \ldots, i_m)$ and $J = (i_1, \ldots, \widehat{i_j}, \ldots, i_m)$ we set

$$\rho_j^I : D_I \hookrightarrow D_J$$
$$\rho_j^m = \bigoplus_{|I|=m} \rho_j^I : D(m) \hookrightarrow D(m-1).$$

$$\left.\begin{array}{l} \gamma_m = \bigoplus_{j=1}^m (-1)^{j-1} \left(\rho_j^m\right)_! : H^{k-m}(D(m))(-m) \\ \qquad\qquad \longrightarrow H^{k-m+2}(D(m-1))(-m+1) \end{array}\right\}. \qquad \text{(IV–3)}$$

The last equation employs the Gysin maps for ρ_j^m where we use the convention of (I–5) up to multiplication of both sides by the same power of $2\pi\mathrm{i}$. It is defined over the rationals. Over the complex numbers we could forget the Tate twists at the cost of neglecting a factor of $(2\pi\mathrm{i})$. However, a similar diagram holds in rational cohomology as we shall see later (Prop. 4.10). For that reason in the following proposition we don't leave out these twists.

Proposition 4.7. *For all $m \geq 1$ the following diagram is commutative*

$$\begin{array}{ccc} E_1^{-m,k+m}(W) & \xrightarrow{\mathrm{res}_m} & H^{k-m}(D(m),\mathbb{C})(-m) \\ \downarrow d_1 & & \downarrow -\gamma_m \\ E_1^{-m+1,k+m}(W) & \xrightarrow{\mathrm{res}_m} & H^{k-m+2}(D(m-1),\mathbb{C})(-m-1) \end{array}$$

Proof. We fix an index set $I = (i_1, \ldots, i_m)$ as above, a class $\eta \in H^{k-m}(D_I; \mathbb{C})$ which is the residue of a class $[\omega] \in \mathbb{H}^k(\mathrm{Gr}_m^W \Omega_X^\bullet(\log D))$. In fact, we then have

$$\mathrm{res}_I[\omega] = \eta.$$

As explained in Remark 4.4, we can take for ω a k-form on X which is C^∞ on U, which has logarithmic singularities along $\sum_{i\in I} D_i$, and such that $d\omega$ has weight $\leq m-1$. It is then fairly easy to see that $d_1[\omega] = [d\omega]$. Since $\gamma_m(\eta) = \sum_{j=1}^m (-1)^{j-1} \left(\rho_j^I\right)_! \eta$, it suffices to prove that

$$\mathrm{res}_J\, d\omega = -(-1)^{j-1} \left(\rho_j^I\right)_! \eta.$$

By (I–5) this amounts to showing that for all $\beta \in H^{2n-m+k}(D_J)$ one has

$$\frac{1}{2\pi \mathrm{i}} \int_{D_J} \mathrm{res}_J\, d\omega \wedge \beta = (-1)^k \int_{D_I} \eta \wedge (\rho_j^I)^*(\beta).$$

We compose

$$\mathrm{res}_J : \Omega_X^\bullet(\log D) \to \Omega_{D_J}^\bullet(\log(D(J))[-m+1]$$

with the residue map

$$\mathrm{res}_{i_k} : \Omega_{D_J}^\bullet(\log D(J)) \to \Omega_{D_I}^\bullet(\log D(I))[-1].$$

Since $dz_I = (-1)^{m-k} dz_J \wedge dz_{i_k}$, one gets

$$\mathrm{res}_{i_k} \circ \mathrm{res}_J = (-1)^{m-k}\, \mathrm{res}_I$$

so that

$$(-1)^{m-k} \int_{D_I} \mathrm{res}_I\, \omega \wedge (\rho_k^I)^* \beta = \int_{D_I} \mathrm{res}_{i_k} \left(\mathrm{res}_J\, \omega \wedge \beta\right).$$

We now use a result due to Leray ([Leray]):

Proposition 4.8 (Leray's residue formula). Let Y be a smooth hypersurface of a non-singular compact complex n-dimensional manifold X. Let $\theta \in \Gamma(\mathcal{E}_X^{2n-1}(\log Y))$ be such that $d\theta \in \Gamma(\mathcal{E}_X^{2n})$. Then

$$\int_Y \mathrm{res}(\theta) = \frac{1}{2\pi \mathrm{i}} \int_X d\theta.$$

Here $\mathrm{res}(\theta)$ is defined as in the case of holomorphic $(n-1)$-forms: if locally Y is defined by the equation $\{f = 0\}$, write $\theta = \dfrac{df}{f} \wedge \eta + \eta'$ where η' does not contain $\dfrac{df}{f}$ and where η is a $(2n-2)$-form which is locally C^∞; then $\mathrm{res}(\theta) = \eta | Y$.

We apply this to $X = D_J$, $Y = D_I$ and $\theta = \mathrm{res}_J\, \omega \wedge \beta$. We find

$$\int_{D_I} \mathrm{res}_{i_k} \left(\mathrm{res}_J\, \omega \wedge \beta\right) = \frac{1}{2\pi \mathrm{i}} \int_{D_J} d(\mathrm{res}_J\, \omega \wedge \beta) = (-1)^m \frac{1}{2\pi \mathrm{i}} \int_{D_J} \mathrm{res}_J\, d\omega \wedge \beta$$

which is exactly what we had to prove. □

4.3 Associated Mixed Hodge Complexes of Sheaves

In order to construct a mixed Hodge complex of sheaves computing the cohomology of $U = X - D$ we need to find a filtered complex over $\mathbb{Q}$ which over $\mathbb{C}$ is filtered quasi-isomorphic to $\Omega^\bullet_X(\log D)$ with the weight filtration. We are going to relate the weight filtration to the canonical filtration τ (see Example A.34) as an intermediate step to show that the weight filtration is defined over $\mathbb{Q}$. The same local computation sheds light on the integral structure as well. For that reason we treat this at the same time:

Lemma 4.9. 1) *The inclusion map*

$$(\Omega^\bullet_X(\log D), \tau) \to (\Omega^\bullet_X(\log D), W)$$

is a filtered quasi-isomorphism.
2) *There is a commutative diagram*

$$\begin{array}{ccc}
R^m j_* \underline{\mathbb{Z}}_U & \longrightarrow & R^m j_* \underline{\mathbb{C}}_U \\
\| \wr & & \| \wr \\
\mathrm{Gr}^\tau_m Rj_* \underline{\mathbb{Z}}_U & \longrightarrow & \mathrm{Gr}^W_m Rj_* \underline{\mathbb{C}}_U \\
\Big\downarrow r_m & & \| \wr \\
 & & \mathrm{Gr}^W_m(\Omega^\bullet(\log D)) \\
 & & \Big\downarrow r_m \\
a_{m*}\underline{\mathbb{Z}}_{D(m)}[-m](-m) & \xrightarrow{\tilde{\alpha}_m} & a_{m*}\underline{\mathbb{C}}_{D(m)}[-m]
\end{array}$$

where the map $\tilde{\alpha}_m$ is induced by the inclusion $\mathbb{Z}(-m) \hookrightarrow \mathbb{C}$ defining the Tate twist (III–3).
3) *The preceding commutative diagram defines a pseudo-morphism*

$$\alpha_m : a_{m*}\underline{\mathbb{Z}}_{D(m)}[-m](-m) \dashrightarrow \mathrm{Gr}^W_m \Omega^\bullet_X(\log D)$$

which gives the comparison morphism making the triple

$$\mathcal{K}_m := \left(a_{m*}\underline{\mathbb{Z}}_{D(m)}[-m](-m), (\mathrm{Gr}^W_m \Omega^\bullet_X(\log D), F), \alpha_m\right),$$

a weight-m Hodge complex of sheaves. Here F is the trivial filtration on the complex $\mathrm{Gr}^W_m(\Omega^\bullet_X \log D)$. The comparison isomorphism sends it to the complex $a_{m*}\Omega^\bullet_{D(m)}$ with its trivial filtration shifted by $-m$.
4) *Using the notation* (II–8) *we have*

$$\chi_{\mathrm{Hdg}}(R\Gamma(\mathcal{K}_m)) = (-1)^m \chi_{\mathrm{Hdg}} D(m) \cdot \mathbb{L}^m \in K_0(\mathfrak{hs}).$$

Proof. 1) and 2). It is easy to see that the inclusion is a filtered morphism. To see that it is a filtered quasi-isomorphism we consider the graded part

$$\left.\begin{aligned}\operatorname{Gr}^{\tau}_k[\Omega^\bullet_X(\log D)] &= H^k(\Omega^\bullet_X(\log D))\\ &\longrightarrow \operatorname{Gr}^W_k[\Omega^\bullet_X(\log D)] \xrightarrow{\overset{\text{qis}}{\sim}} (a_k)_*\underline{\mathbb{C}}_{D(k)}[-k].\end{aligned}\right\} \qquad \text{(IV–4)}$$

We have seen (Prop. 4.3) that replacing U by a small enough neighbourhood V of a point $Q \in D$ we have $\mathbb{H}^k(V, \Omega^\bullet_X(\log D)) = H^k(V - V \cap D; \mathbb{C})$. If $Q \in D(m)$ with $m < k$ both sides of (IV–4) are zero. If however $m \geq k$ we remark that since $(V - V \cap D)$ has the homotopy type of a product of m circles the left hand side has for stalk at Q a vector space of dimension $\binom{m}{k}$. On the other hand, the stalk at Q of $(a_k)_*\underline{\mathbb{C}}_{D(k)}$ consists of $\binom{m}{k}$ copies of $\mathbb{C}$ (corresponding to all possible I with $|I| = k$ such that $D_I \supset D(m) \ni Q$). So in all cases the remaining map in (IV–4) is a quasi-isomorphism.

Coming back to integral homology, $H_1(V)$ is freely generated by the m classes of loops γ_j around any of the m components of D and $(R^1 j_*\underline{\mathbb{C}}_V)_Q = \left(H^1(\Omega^\bullet_V(\log D \cap V))\right)_Q$ is freely generated by the classes of dz_k/z_k, $k = 1, \ldots, m$. Applying the residue map, we get the result for 1-cohomology from the residue formula $\int_{\gamma_j} dz_j/z_j = 2\pi \mathrm{i}$. Indeed, it says that the integral cohomology inside of the complex vector space $\left(R^1(a_{1*}\mathbb{C}_{D(1)})\right)_Q \cong H^1(a_1^{-1}V; \mathbb{C})$ is generated by the classes of $(1/2\pi\mathrm{i})(dz_j/z_j)$, $j = 1, \ldots, m$. For arbitrary rank k the result then also follows, since we have on the one hand $(R^k j_*\underline{\mathbb{C}}_V)_Q = \bigwedge^m (R^1 j_*\underline{\mathbb{C}}_V)_Q$ and on the other hand there is an isomorphism of sheaves of complex vector spaces

$$a_{k*}\underline{\mathbb{C}}_{D(k)} \cong \bigwedge^k a_{1*}\underline{\mathbb{C}}_{D(1)}.$$

3). The complex $\operatorname{Gr}^m_\tau Rj_*(\underline{\mathbb{Z}}_X))$ has only cohomology in degree m, which is

$$H^m(j_*\underline{\mathbb{Z}}_X) \xrightarrow[r]{\cong} a_{m*}\underline{\mathbb{Z}}_{D(m)}(-m) = R^m j_*\underline{\mathbb{Z}}_U.$$

Part 1 and 2 say that the residue map is a quasi-isomorphism of complexes which comes from a quasi-isomorphism on integral level, provided we take the correct identifications as stated. Since the hypercohomology of the complex $\operatorname{Gr}^W_m \Omega^\bullet_X(\log D)$ with the trivial filtration computes the cohomology of $D(m)$ with its induced Hodge structure, Example 2.34 states that we indeed get a Hodge complex. The twist by $-m$ forced upon us by the identifications, guarantees that we have a complex of weight m instead of weight 0.
4). The k-th cohomology group of the complex $R\Gamma\mathcal{K}_m$ is $H^{k-m}(D(m))(-m)$. Hence

$$\begin{aligned}\chi_{\text{Hdg}}(R\Gamma\mathcal{K}_m) &= \sum_k (-1)^k H^{k-m}(D(m))(-m)\\ &= \sum_k (-1)^{\ell+m} H^\ell(D(m))(-m)\\ &= (-1)^m \chi_{\text{Hdg}}(D(m)) \cdot \mathbb{L}^m. \qquad \square\end{aligned}$$

As a first consequence we find that Prop. 4.7 is valid over $\mathbb{Q}$ (since commutativity holds if it holds after tensoring with $\mathbb{C}$).

Proposition 4.10. *For all $m \geq 1$ the following diagram is commutative*

$$\begin{array}{ccc} {}_W E_1^{-m,k+m} & \xrightarrow{\mathrm{res}_m} & H^{k-m}(D(m);\mathbb{Q})(-m) \\ \big\downarrow d_1 & & \big\downarrow -\gamma_m \\ {}_W E_1^{-m+1,k+m} & \xrightarrow{\mathrm{res}_m} & H^{k-m+2}(D(m-1);\mathbb{Q})(-m+1). \end{array}$$

Here γ_m is the alternating sum of the Gysin homomorphisms (see (IV–3)*).*

As a second consequence of Lemma 4.9 we obtain the following description of the weight filtration:

$$\begin{aligned} W_m H^k(U;\mathbb{C}) &= \mathrm{Im}\left(\mathbb{H}^k(X, \tau_{m-k}\Omega_X^\bullet(\log D)) \to H^k(U;\mathbb{C})\right) \\ &= \mathrm{Im}\left(\mathbb{H}^k(X, \tau_{m-k} j_* \Omega_U^\bullet) \to H^k(U;\mathbb{C})\right). \end{aligned}$$

The last equality follows from the fact that a quasi-isomorphism between complexes is automatically a filtered quasi-isomorphism with respect to the canonical filtrations. Since the canonical filtration can be put on any complex, the weight filtration can be defined over $\mathbb{Q}$, replacing $Rj_*\underline{\mathbb{C}}_U$ by $\mathcal{C}^\bullet_{\mathrm{Gdm}}(\underline{\mathbb{Q}}_U)$ one then sets

$$W_m H^k(U;\mathbb{Q}) = \mathrm{Im}\left(\mathbb{H}^k(X, \tau_{m-k} j_* \mathcal{C}^\bullet_{\mathrm{Gdm}}(\underline{\mathbb{Q}}_U)) \to \mathbb{H}^k(X, j_*\mathcal{C}^\bullet_{\mathrm{Gdm}}(\underline{\mathbb{Q}}_U))\right).$$

We now have shown the main result of this chapter:

Proposition-Definition 4.11. *The following data form a mixed Hodge complex of sheaves on X, and is called the* ***Hodge-De Rham complex of*** (X, D), *denoted $\mathcal{H}dg^\bullet(X \log D)$.*

- *The complex $Rj_*\underline{\mathbb{Z}}_U = j_*\mathcal{C}^\bullet_{\mathrm{Gdm}}(\underline{\mathbb{Z}}_U)$ $(U = X - D)$;*
- *the complex $Rj_*\underline{\mathbb{Q}}_U$ with its canonical filtration and the obvious morphism*

$$\alpha : Rj_*\underline{\mathbb{Z}}_U \to Rj_*\underline{\mathbb{Q}}_U;$$

- *the complex $\Omega_X^\bullet(\log D)$ with the filtrations W, F and the filtered pseudo-morphism β defined by the following diagram*

$$\begin{array}{ccccccc} & & (Rj_*\Omega_U^\bullet, \tau) & & & (\Omega_X^\bullet(\log D), \tau)) & \\ & \nearrow & & \nwarrow & \swarrow & & \searrow \\ (Rj_*\underline{\mathbb{Q}}_U, \tau) \to (Rj_*\underline{\mathbb{C}}_U, \tau) & & & (j_*\Omega_U^\bullet, \tau) & & & (\Omega_X^\bullet(\log D), W) \end{array}$$

These complexes compute the cohomology of U and the filtrations (together with the trivial filtrations) induce on it the mixed Hodge structure announced in Theorem 4.2. For the Hodge-Grothendieck character of U and its Hodge-Euler polynomial we have

$$\chi_{\mathrm{Hdg}}(U) := \sum_{k\geq 0}(-1)^k \left[H^k(U)\right] = \sum_{m\geq 0}(-1)^m \chi_{\mathrm{Hdg}}(D(m))\cdot \mathbb{L}^m \quad \text{(IV–5)}$$

$$\begin{aligned} e_{\mathrm{Hdg}}(U) &= \sum_{k,p,q\geq 0}(-1)^k h^{p,q}[H^k(U)]u^p v^q \\ &= \sum_{m\geq 0}(-1)^m e_{\mathrm{Hdg}}(D(m)(uv)^m. \quad \text{(IV–6)} \end{aligned}$$

In particular, we have for all $p, q \geq 0$

$$\sum_{k\geq 0}(-1)^k h^{p,q}[H^k(U)] = (-1)^{p+q}\sum_{m\geq 0}(-1)^m h^{p-m,q-m}(D(m)).$$

For later reference, we state a result, the proof of which is left to the reader:

Lemma 4.12. *Let X and Y be smooth compact complex algebraic manifolds, $D \subset X$ a simple normal crossing divisor and $\pi : Y \to X$ a holomorphic map. We suppose that the inverse image E of D is either empty, all of Y or a simple normal crossing divisor on Y. In these three cases we put respectively*

$$\mathcal{H}dg^\bullet(Y \log E) = \begin{cases} \mathcal{H}dg^\bullet(Y) & \text{if } E = \varnothing \\ 0 & \text{if } E = Y \\ \mathcal{H}dg^\bullet(Y \log E) & \text{else.} \end{cases}$$

There is a canonical morphism

$$\pi^* : \mathcal{H}dg^\bullet(X \log D) \to R\pi_*\mathcal{H}dg^\bullet(Y \log E)$$

of mixed Hodge complexes of sheaves which induces a morphism of mixed Hodge structures

$$H^m(X - D) \to H^m(Y - E).$$

4.4 Logarithmic Structures

In this section we give an alternative description of the rational component of the Hodge-De Rham complex of (X, D), using the concept of logarithmic structure of Fontaine, Illusie and Kato [Ill94],[Kato88].

Definition 4.13. *Let $(X, \mathcal{O}_X)$ be an analytic space. A* ***pre-log structure*** *on X consists of a sheaf $\mathcal{M}$ of monoids on X together with a homomorphism $\alpha : \mathcal{M} \to \mathcal{O}_X$ of monoids. Here $\mathcal{O}_X$ is considered as a sheaf of monoids with the multiplication as its operation. The pair $(\mathcal{M}, \alpha)$ is called a* ***log structure*** *if $\alpha^{-1}(\mathcal{O}_X^*) \simeq \mathcal{O}_X^*$.*

Example 4.14. The **trivial log structure** on X is the pair $(\mathcal{O}_X^*, \mathcal{O}_X^* \hookrightarrow \mathcal{O}_X)$.

Let X be a complex manifold and $D \subset X$ a divisor with normal crossings on X. Define $\mathcal{M}_{X,D} = \mathcal{O}_X \cap j_*\mathcal{O}_U^*$ and let α be its inclusion in $\mathcal{O}_X$.

From now on we assume that D has simple normal crossings.

We let $\mathcal{M}_{X,D}^{\text{gp}}$ denote the sheaf of abelian groups associated to $\mathcal{M}_{X,D}$. It has the following universal property: there is a universal map $c : \mathcal{M}_{X,D} \to \mathcal{M}_{X,D}^{\text{gp}}$ and every homomorphism of monoid sheaves from $\mathcal{M}_{X,D}$ to a sheaf of groups on X factorizes uniquely over c. If $\mathcal{O}_X(*D)$ is the sheaf of germ of meromorphic functions on X with only poles along D (which is a sheaf of rings), then $\mathcal{M}_{X,D}^{\text{gp}}$ is the sheaf of its invertible elements. The choice of a local generator t for the ideal sheaf of D gives an isomorphism $\mathcal{O}_X(*D) \simeq \mathcal{O}_X[t^{-1}]$. Hence

$$\mathcal{M}_{X,D}^{\text{gp}}/\mathcal{O}_X^* \simeq a_*\underline{\mathbb{Z}}_{D(1)}.$$

Let $j : U = X - D \hookrightarrow X$. Consider the exponential map

$$\begin{aligned} \mathrm{e} : \mathcal{O}_X &\to \mathcal{M}_{X,D}^{\text{gp}} \\ f &\mapsto \exp(2\pi \mathrm{i} f). \end{aligned}$$

Its kernel is $\underline{\mathbb{Z}}_X = j_*\underline{\mathbb{Z}}_U$ and its cokernel is $a_*\underline{\mathbb{Z}}_{D(1)} \simeq R^1 j_*\underline{\mathbb{Z}}_U$. Hence, if we consider e as a complex of sheaves where $\mathcal{O}_X$ is placed in degree zero, it has the same cohomology as $\tau_{\leq 1} Rj_*\underline{\mathbb{Z}}_U$. The following construction provides us essentially with the exterior powers of $\mathrm{e} \otimes \mathbb{Q}$. It is a special case of the construction of Koszul complexes of a morphism by Illusie [Ill71, Sect. 4.3.1]. Define

$$K_p^q = \mathrm{Sym}_{\mathbb{Q}}^{p-q}(\mathcal{O}_X) \otimes \bigwedge_{\mathbb{Q}}^q (\mathcal{M}_{X,D}^{\text{gp}} \otimes_{\mathbb{Z}} \mathbb{Q}),$$

and $d : K_p^q \to K_p^{q+1}$ by

$$d(f_1 \cdots f_{p-q} \otimes y) = \sum_{i=1}^{p-q} f_1 \cdots f_{i-1} \cdot f_{i+1} \cdots f_{p-q} \otimes \mathrm{e}(f_i) \wedge y$$

for sections $f_1, \ldots, f_{p-q}$ of $\mathcal{O}_X$ and y of $\bigwedge_{\mathbb{Q}}^q (\mathcal{M}_{X,D}^{\text{gp}} \otimes_{\mathbb{Z}} \mathbb{Q})$. We get complexes

$$K_p^\bullet : 0 \to K_p^0 \xrightarrow{d} K_p^1 \to \cdots \to K_p^p \to 0,$$

and inclusions of complexes

$$K_p^\bullet \to K_{p+1}^\bullet, \; f_1 \cdots f_{p-q} \otimes y \mapsto 1 \cdot f_1 \cdots f_{p-q} \otimes y.$$

Theorem 4.15. *The map*

$$\phi_p : K_p^\bullet \to \Omega_X^\bullet(\log D)$$

given by

$$\phi_p(f_1 \cdots f_{p-q} \otimes y_1 \wedge \cdots \wedge y_q) = \frac{1}{(2\pi \mathrm{i})^q} \left(\prod_{i=1}^{p-q} f_i \right) \frac{dy_1}{y_1} \wedge \cdots \wedge \frac{dy_q}{y_q}$$

induces a quasi-isomorphism between $K_p^\bullet \otimes_{\mathbb{Q}} \mathbb{C}$ and $W_p \Omega_X^\bullet(\log D)$.

Proof. Due to [Ill71, Prop. 4.3.1.6] we have for $q \leq p$ that

$$H^q(K_p^\bullet) \simeq \mathrm{Sym}^{p-q}(\mathrm{Ker(e)}) \otimes \bigwedge^q(\mathrm{Coker(e)}) \simeq a_* \underline{\mathbb{Q}}_{D(q)}.$$

Moreover, $H^q(K_p^\bullet) = 0$ for $q > p$. The local representatives of these cohomology classes are mapped by ϕ_p to the local generators $\dfrac{1}{(2\pi \mathrm{i})^q} \dfrac{dz_{i_1}}{z_{i_1}} \wedge \cdots \wedge \dfrac{dz_{i_q}}{z_{i_q}}$ for the cohomology sheaves of $H^q(\Omega_X^\bullet(\log D))$. □

Corollary 4.16. *Let* $K_\infty^\bullet = \varinjlim_p K_p^\bullet$. *For* $m \in \mathbb{N}$ *define* $W_m K_\infty^\bullet$ *to be the image of* $K_m^\bullet$ *in* $K_\infty^\bullet$. *Then*

$$\varinjlim_n \phi_n : (K_\infty^\bullet, W) \otimes \mathbb{C} \to (\Omega_X^\bullet(\log D), W)$$

is a filtered quasi-isomorphism.

Corollary 4.17. *In the definition of the Hodge-De Rham complex of the pair* (X, D), *the rational component* $(Rj_* \underline{\mathbb{Q}}_U, \tau)$ *may be replaced by* $(K_\infty^\bullet, W)$. *This defines the same rational structure and weight filtration on the cohomology of* U.

Proof. We have a diagram

$$K_\infty^\bullet \xrightarrow[\sim]{\text{qis}} Rj_* j^* K_\infty^\bullet \xleftarrow[\sim]{\text{qis}} Rj_* \underline{\mathbb{Q}}_U.$$

Indeed, the second map is a quasi-isomorphism, because $K_{\infty|U}^\bullet$ is a resolution of $\underline{\mathbb{Q}}_U$; the first map is a quasi-isomorphism by the computations above. □

4.5 Independence of the Compactification and Further Complements

4.5.1 Invariance

Let us first look at what happens for a morphism $f : U \to V$ between two smooth varieties. It is possible to find smooth compactifications X of U and Y of V so that $D = X - U$ and $E = Y - V$ are divisors with simple normal crossings and such that f extends to a morphism $\bar{f} : X \to Y$. This can be done as follows. First you choose any compactifications of U and V with simple normal crossing divisors and then you take a suitable resolution of singularities of the closure of the graph of f. By Lemma 4.12 the morphism $\bar{f}$ induces a morphism of bi-filtered complexes $(\Omega_Y^\bullet(\log E), W, F) \to (Rf_* (\Omega_X^\bullet(\log D)), W, F)$ underlying a morphism $\mathcal{H}dg^\bullet(Y \log E) \to Rf_* \mathcal{H}dg^\bullet(X \log D)$. This in turn induces a morphism $f^* : H^k(V) \to H^k(U)$ preserving Hodge and weight filtration. This is therefore a morphism of mixed Hodge structures induced by the choice of the compactifications. Clearly, if f is biholomorphic, f^* is an isomorphism of mixed Hodge structures. Suppose now that X and Y are two compactifications of U and let Z be a resolution of the closure of the diagonal Δ of $U \times U$ inside $X \times Y$ such that Z is a good compactification of X as well. The two projections $Z \to X$ and $Z \to Y$ induce the identity on U. By the preceding remark, these then induce isomorphisms between the two mixed Hodge structure on $H^k(U)$ got by the compactification Z and the one got by the one by either X or Y. In particular, the mixed Hodge structure is independent of the compactification. In total, we have shown:

Proposition 4.18. *The mixed Hodge structure on $H^k(U)$, constructed in the previous section is independent of the choice of the compactification. Any morphism between smooth complex algebraic varieties $f : U \to V$ induces a morphism $f^* : H^k(V) \to H^k(U)$ of mixed Hodge structures. The latter comes from the morphism*

$$\mathcal{H}dg^\bullet(Y \log E) \to R\bar{f}_*\mathcal{H}dg^\bullet(X \log D)$$

induced by any extension $\bar{f} : X \to Y$ of f to good compactifications (Def. 4.1) (X, D) of U, respectively (Y, E) of V.

Example 4.19 (The mixed Hodge structure depends on the algebraic *structure).*

This example is due to Serre and is treated in detail in [Hart70]. We start out with an elliptic curve E and the $\mathbb{P}^1$-bundle X associated to the non-split rank two bundle V defined as an extension of the trivial line bundle by the trivial line bundle. The canonical trivial subbundle defines a section s of the $\mathbb{P}^1$-bundle, and we let U be its complement in X ; it is a $\mathbb{C}$-bundle over E. We claim that $U = \mathbb{C}^* \times \mathbb{C}^*$. Indeed, all sections of V meet s somewhere and so U does not contain compact submanifolds. On the other hand, pulling back $U \to E$ to the universal cover $\mathbb{C}$ of E trivializes this $\mathbb{C}$-bundle so that the total space becomes $\mathbb{C} \times \mathbb{C}$. The covering group $\mathbb{Z} \times \mathbb{Z}$ acts and since U has no compact submanifolds this action must be non-trivial on both factors and the quotient U is as claimed. But then also $\mathbb{P}^1 \times \mathbb{P}^1$ is a good compactification of U. We have $H^1(U) = \mathbb{Z} \oplus \mathbb{Z}$. It is easy to see that the restriction $H^1(X) \to H^1(U)$ is injective and hence an isomorphism. This shows that $H^1(U) \cong H^1(E) = W_1$ for this compactification, while for the second compactification $H^1(U)$ has pure weight 2: it is contained in $H^2(D)$, D the compactifying divisor (four copies of $\mathbb{P}^1$).

4.5.2 Restrictions for the Hodge Numbers

Proposition 4.20. *Let U be a smooth complex algebraic variety and let X be a good compactification of U. Then*

$$W_m H^k(U) = 0 \text{ for } m < k$$
$$W_k H^k(U) = \operatorname{Im}\left(H^k(X) \to H^k(U)\right).$$

The Hodge numbers $h^{p,q}$ of $H^k(U)$ can only be non-zero in the triangular region $p \leq k, q \leq k, p+q \geq k$.

Proof. The weight-m part is the image of $\mathbb{H}^k(X, \tau_{\leq m-k} Rj_* \underline{\mathbb{Q}}_U)$ inside the space $\mathbb{H}^k(X, Rj_* \underline{\mathbb{Q}}_U)$. Since $\tau_{\leq r} Rj_* \underline{\mathbb{Q}}_U = 0$ for $r < 0$ and $\tau_{\leq 0} Rj_* \underline{\mathbb{Q}}_U = \underline{\mathbb{Q}}_X$ placed in degree 0, the first two assertions follow. The last assertion follows from the fact that the spectral sequence for the weight filtration degenerates at the E_2-term and hence the rational Hodge structure $E_\infty^{-m,k+m} = \mathrm{Gr}_m^W H^k(U;\mathbb{Q})$ is a sub quotient of $H^{k-m}(D(m);\mathbb{Q})(-m)$. The Hodge numbers $h^{p,q}[H^{k-m}(D(m))]$ are zero if $p > k-m$ or $q > k-m$ so that, in view of the Tate shift, $h^{p,q}(\mathrm{Gr}_m^W H^k(U;\mathbb{Q})) = 0$ if $p > k$ or $q > k$. □

Remark 4.21. We shall see later (Corollary 6.30) that the Proposition remains true for any smooth compactification X of U.

Corollary 4.22. *Let Y be a smooth projective variety, V a smooth variety, $f : Y \to V$ a morphism, and $j : V \hookrightarrow X$ a smooth compactification of V. Then the subgroups $f^* H^k(V;\mathbb{Q})$ and $(j \circ f)^* H^k(X;\mathbb{Q})$ of $H^k(Y;\mathbb{Q})$ coincide.*

Proof. Because f^* and $(j \circ f)^*$ are both strictly compatible with the weight filtrations, it suffices to prove that the graded pieces have the same image. But $H^k(X;\mathbb{Q})$ is pure of weight k, so only the weight k-pieces matter. By the previous Proposition $\mathrm{Gr}_k^W H^k(X;\mathbb{Q}) \to \mathrm{Gr}_k^W H^k(V;\mathbb{Q})$ is onto and hence the image under restriction of these groups in $H^k(Y;\mathbb{Q})$ must be the same. □

4.5.3 Theorem of the Fixed Part and Applications

Recall (Theorem 1.38) that for smooth projective maps $f : V \to U$ the Leray spectral sequence

$$H^p(U, R^q f_* \mathbb{Q}) \Rightarrow H^{p+q}(V;\mathbb{Q})$$

degenerates at E_2. This implies in particular that the edge-homomorphism (A–30)

$$e_k(f) : H^k(V;\mathbb{Q}) \to H^0(U, R^k f_* \mathbb{Q})$$

is surjective.

Theorem 4.23 (Theorem of the fixed part). *Suppose that U is quasi-projective and that $f : V \to U$ is a smooth projective map. For any smooth compactification X of V, the natural restriction map composed with $e_k(f)$*

$$H^k(X;\mathbb{Q}) \to H^k(V;\mathbb{Q}) \xrightarrow{e_k(f)} H^0(U, R^k f_* \mathbb{Q})$$

is surjective. In other words, if we identify the right hand side with those k-classes on the fibre, invariant under monodromy, all of these are restrictions from classes on X.

Proof. Let $s \in U$ and let $Y = V_s$ be the fibre over s. Since $H^0(U, R^k f_* \mathbb{Q})$ can be identified with the invariant classes under the action of $\pi_1(U, s)$ on $H^k(Y;\mathbb{Q})$ it suffices to compare the images of $H^k(X;\mathbb{Q})$ and $H^k(U;\mathbb{Q})$ in $H^k(Y;\mathbb{Q})$. But by Corollary 4.22 these images are the same. On the other hand, the Invariant Cycle Theorem 1.40 tells us that the image of $H^k(U;\mathbb{Q})$ consists of the invariant classes. □

It follows that the Leray spectral sequence degenerates for a projective family over a one-dimensional base.

Theorem 4.24. *Let $f : X \to S$ a morphism between smooth projective varieties and let* $\dim(S) = 1$. *Then the first edge homomorphism* (A–30)

$$e_k = e_k(f) : H^k(X;\mathbb{Q}) \to E_2^{0,k}(f) = H^0(S; R^k f_* \mathbb{Q}).$$

is surjective, and the Leray spectral sequence for f degenerates at E_2.

Proof. The Leray spectral sequence for f degenerates precisely when e_k is surjective. We are going to show that this is always the case. Let $j : U \hookrightarrow S$ be the inclusion of the open set of regular values of f into S. Put $V = f^{-1}U$ and let $j^V : V \hookrightarrow X$ be the inclusion.

The edge homomorphisms $e_k = e_k(f)$, $e_k^V = e_k(f|_V)$ and the adjunction morphism

$$a_k : R^k f_* \underline{\mathbb{Q}}_X \to j_* j^* R^k f_* \underline{\mathbb{Q}}_X.$$

fit into the following commutative diagram

$$\begin{array}{ccc}
H^k(X,V;\mathbb{Q}) & & \\
\downarrow r & & \\
H^k(X;\mathbb{Q}) & \xrightarrow{e_k} & H^0(S; R^k f_* \underline{\mathbb{Q}}_X) \\
 & & \downarrow a_k \\
\downarrow j^V & & H^0(S, j_* j^* R^k f_* \underline{\mathbb{Q}}_X) \\
 & & \downarrow \cong \\
H^k(V;\mathbb{Q}) & \xrightarrow{e_k^V} & H^0(U, j^* R^k f_* \underline{\mathbb{Q}}_X).
\end{array}$$

The leftmost sequence is part of the long exact sequence for the pair (X, V).

The kernel of a_k is a sky-scraper sheaf supported on $S - U$. Its stalk at a point t can be described as follows. Choose small a enough disk Δ_t centred at t and let Δ_t^* be the punctured disk. Then at t the adjunction homomorphism can be identified with the restriction

$$(R^k j_*)_t : (R^k f_* \underline{\mathbb{Q}}_X)_t \simeq H^k(f^{-1}\Delta_t;\mathbb{Q}) \to H^k(f^{-1}\Delta_t^*;\mathbb{Q})^T = (j_* j^* R^k f_* \underline{\mathbb{Q}}_X)_t,$$

where the target is the subspace of invariants under the local monodromy T at t. Under these identifications, this is the same as the edge-homomorphism for the Leray-spectral sequence for $f|\Delta_t$. The map $(R^k j_*)_t$ comes from the restriction j_t^* figuring in long exact sequence for the pair $(f^{-1}\Delta_t, f^{-1}\Delta_t^*)$

$$\cdots \to H^k(f^{-1}\Delta_t, f^{-1}\Delta_t^*;\mathbb{Q}) \to H^k(f^{-1}\Delta_t;\mathbb{Q}) \xrightarrow{j_t^*} H^k(f^{-1}\Delta_t^*;\mathbb{Q}) \to \cdots.$$

It follows that

$$\mathrm{Ker}(a_k) = \bigoplus_t \mathrm{Ker}(a_k^{\sigma_t}) \simeq \mathrm{Im}\left[\bigoplus_t H^k(f^{-1}\Delta_t, f^{-1}\Delta_t^*;\mathbb{Q}) \to \bigoplus_t H^k(f^{-1}\Delta_t;\mathbb{Q})\right].$$

By excision $\bigoplus_t H^k(f^{-1}\Delta_t, f^{-1}\Delta_t^*;\mathbb{Q}) \simeq H^k(X,V;\mathbb{Q})$. Since the composition of restrictions $H^k(X,V;\mathbb{Q}) \to \bigoplus_t H^k(f^{-1}\Delta_t, f^{-1}\Delta_t^*;\mathbb{Q}) \to H^k(f^{-1}\Delta_t;\mathbb{Q})$ factors over the natural map $H^k(X,V) \xrightarrow{r} H^k(X)$ figuring in the long exact sequence of the pair (X,V), we deduce an exact sequence

$$H^k(X,U) \xrightarrow{e_k \circ r} H^0(S, R^k f_* \underline{\mathbb{Q}}_X) \xrightarrow{a_k} H^0(S, j_* j^* R^k f_* \underline{\mathbb{Q}}_X). \qquad \text{(IV–7)}$$

Let $\eta \in H^0(S; R^k f_* \underline{\mathbb{Q}}_X)$. By the Theorem of the Fixed Part (4.23) there is an element $\xi \in H^k(X;\mathbb{Q})$ with $e_k^V \circ j^V(\xi) = a_k(\eta)$. It follows that $e_k(\xi) - \eta \in \mathrm{Ker}(a_k)$ and hence, by the exact sequence (IV–7), is of the form $e_k(\theta)$, $\theta \in H^k(X;\mathbb{Q})$ so that $e_k(\xi - \theta) = \eta$. □

4.5.4 Application to Lefschetz Pencils

The Lefschetz hyperplane theorem (§ C.2.3) can be reformulated in terms of Hodge theory as follows.

Theorem 4.25. *Let X be an $(n+1)$-dimensional projective manifold and let $i : Y \hookrightarrow X$ be a smooth hyperplane section. Then*

1) *for $k < n$ the inclusion induces an isomorphism of weight k Hodge structures $i^* : H^k(X) \xrightarrow{\sim} H^k(Y)$;*
2) *for $k > n$ the Gysin maps induces an isomorphism of weight k Hodge structures $i_! : H^k(Y) \xrightarrow{\sim} H^{k+2}(X)(-1)$;*
3) *the rational middle cohomology splits as*

$$H^n(Y;\mathbb{Q}) = H^n_{\text{fixed}}(Y;\mathbb{Q}) \oplus H^n_{\text{var}}(Y;\mathbb{Q}), \quad H^n_{\text{fixed}}(Y;\mathbb{Q}) \underset{i^*}{\xrightarrow{\sim}} H^n(X;\mathbb{Q}); \qquad \text{(IV–8)}$$

this is a splitting preserving Hodge decompositions.

Consider now a Lefschetz pencil of hyperplanes $\{X_u\}_{u\in\mathbb{P}^1}$ of X with associated Lefschetz fibration

$$f : \tilde{X} = \mathrm{Bl}_B X \to \mathbb{P}^1,$$

where $B \subset X$ is the base locus of the pencil. Let $\Delta(f)$ be the critical locus, and let

$$j : U = \mathbb{P}^1 - \Delta(f) \hookrightarrow \mathbb{P}^1$$

be the inclusion. So $j^* R^n f_* \underline{\mathbb{Q}}_X$ is a locally constant sheaf. The splitting (IV–8) can be globalized over U. The subspaces $H^n_{\text{fixed}}(X_u; \mathbb{Q}) \subset H^n(X_u; \mathbb{Q})$ define a constant subsheaf $\mathbb{I}$ and the subspaces $H^n_{\text{var}}(Y; \mathbb{Q})$ define the subsheaf $\mathbb{V}$ of vanishing cohomology. By Cor. C.24 there is an orthogonal direct splitting

$$j^* R^n f_* \underline{\mathbb{Q}}_X = j^* R^n f_* \underline{\mathbb{Q}}_X = \mathbb{I} \oplus \mathbb{V}. \tag{IV–9}$$

In this situation, since the the local invariant cycle property holds (Cor. C.21), by Lemma C.13 the adjunction morphism $j^\sharp_n : R^n f_* \underline{\mathbb{Q}}_X \to j_* j^* R^n f_* \underline{\mathbb{Q}}_X$ is an isomorphism so that the above splitting can be used to study the Leray spectral sequence

$$E_2^{p,q}(f) = H^p(\mathbb{P}^1, R^q f_* \underline{\mathbb{Q}}_X) \Longrightarrow H^{p+q}(\tilde{X}; \mathbb{Q}).$$

Note that $R^n f_* \underline{\mathbb{Q}}_X$ definitely is not locally free. If for $k \neq n$ the adjunction maps

$$j^\sharp_k : R^k f_* \underline{\mathbb{Q}}_X \to j_* j^* R^k f_* \underline{\mathbb{Q}}_X \tag{IV–10}$$

are isomorphisms, the direct image sheaves $R^k f_* \underline{\mathbb{Q}}_X$ are locally constant, and conversely. By Cor. C.22 this is true for even n and "generically" true for odd n. We now have the following description of the terms in the Leray spectral sequence.

Theorem 4.26. *Let X be an $(n+1)$-dimensional projective manifold X and let $f : \tilde{X} \to \mathbb{P}^1$ a Lefschetz fibration and let Y be a smooth fibre. Suppose moreover that the adjunction morphism* (IV–10) *is an isomorphism for all $k = 0, \ldots, 2n$. Then the E_2-terms of the Leray spectral sequence for f have the following description.*

1) *For $m \neq n$ one has canonical isomorphisms of Hodge structures*

$$\begin{aligned} E_2^{0,m}(f) &\cong H^m(Y; \mathbb{Q}) \\ E_2^{2,m}(f) &\cong H^m(Y; \mathbb{Q})(-1) \\ E_2^{1,m}(f) &= 0. \end{aligned}$$

2) *For $m = n$, one has*

$$\begin{aligned} E_2^{0,n}(f) &\cong H^n_{\text{fixed}}(Y; \mathbb{Q}) \\ E_2^{2,n}(f) &\cong H^n_{\text{fixed}}(Y; \mathbb{Q})(-1) \\ E_2^{1,n}(f) &\cong H^1(\mathbb{P}^1, j_* \mathbb{V}). \end{aligned}$$

Proof. As noted before, the assertions for $m \neq n$ follow from the fact that our assumptions imply (Lemma C.13) that the sheaves $R^m f_* \underline{\mathbb{Q}}_X$ are locally constant, and hence constant.

The first assertion for $m = n$ follows from the fact that the n-th adjunction map is an isomorphism. The second assertion is dual to it as we now explain.

In view of the splitting (IV–9) it suffices to show that $H^2(\mathbb{P}^1, j_*\mathbb{V}) = 0$. By Theorem B.36 $H^0(\mathbb{P}^1, j_*\mathbb{V}^\vee) = H^0(U, \mathbb{V}^\vee)$ is dual to $H^2_c(U, \mathbb{V}) = H^2(\mathbb{P}^1, j_!\mathbb{V})$. It can be calculated by means of the exact sequence

$$0 \to j_!\mathbb{V} \to j_*\mathbb{V} \to \bigoplus_{t \in \Delta(f)} \mathbb{V}_t^T \to 0,$$

where Δ_t^* is a small punctured disk centred at t and $\mathbb{V}_t^T$ is the subspace of invariants in $\mathbb{V}_t$ under the local monodromy T. It follows that $H^2(\mathbb{P}^1, j_!\mathbb{V}) = H^2(\mathbb{P}^1, j_*\mathbb{V})$ is dual to $H^0(U, \mathbb{V}^\vee) = H^0(\mathbb{P}^1, j_*\mathbb{V})$ and this group indeed vanishes by the global invariant cycle theorem.

The last assertion follows directly from the definition of the sheaf of vanishing cohomology. □

By Theorem 4.24 the Leray spectral sequence degenerates, and so we have a decomposition $H^m(\tilde{X}; \mathbb{Q}) \simeq L^2 \oplus L^1/L^2 \oplus L^0/L^1$. Note that in general only L^2 is a subspace of $H^m(\tilde{X}; \mathbb{Q})$. In fact, since

$$L^2 = E_2^{2,m-2} = \operatorname{Im}[H^{m-2}(Y; \mathbb{Q})(-1) \xrightarrow{i_!} H^m(\tilde{X}; \mathbb{Q})],$$

it is a natural sub Hodge of $H^m(\tilde{X}; \mathbb{Q})$. As for L^1, we consider the quotient $L^0/L^1 = E_2^{0,m} = H^0(\mathbb{P}^1, R^m f_*\mathbb{Q})$. This is the subspace of $H^m(Y; \mathbb{Q})$ which is invariant under the global monodromy of the local system $\mathbb{V}$. By the Theorem of the Fixed Part 4.23 this is the image under restriction $H^m(X; \mathbb{Q}) \to H^m(Y : \mathbb{Q})$ and hence has a natural Hodge structure of weight m. Moreover, it follows that

$$L^1 = \operatorname{Ker}[H^m(\tilde{X}; \mathbb{Q})(-1) \xrightarrow{i^*} H^m(Y; \mathbb{Q})]$$

which describes L^1 as a sub Hodge structure as well. We can then put the quotient Hodge structures on the Leray quotients.

For $m = n + 1$ we dispose of the intersection pairing and we have natural identifications of the Leray quotients as a sub Hodge structures of $H^m(\tilde{X}; \mathbb{Q})$:

$$L^0/L^1 = (L^2)^\perp = (\operatorname{Ker} i^*)^\perp, \quad L^1/L^2 = \operatorname{Ker} i^* \cap (\operatorname{Im} i_!)^\perp.$$

Summarizing, we have:

Theorem 4.27. *Let X be an $(n+1)$-dimensional projective manifold X, let $f : \tilde{X} \to \mathbb{P}^1$, a Lefschetz fibration and let Y be any smooth fibre. Assume that the restriction $H^n(X) \to H^n(Y)$ is not an isomorphism (which, by Remark C.22 is the case if n is even and which is generically true for n odd). The Leray spectral sequence degenerates at E_2 and the Leray fibration on $H^m(\tilde{X}; \mathbb{Q})$ is a fibration of sub Hodge structures. For $m \neq n+1$ the resulting isomorphism of pure Hodge structures $H^m(\tilde{X}; \mathbb{Q}) \simeq L^2 \oplus L^0/L^1$ explicitly reads*

$$H^m(\tilde{X}; \mathbb{Q}) \simeq \begin{cases} H^{m-2}(Y; \mathbb{Q})(-1) \oplus H^m(Y; \mathbb{Q}), & \text{if } m \neq n, n+1, n+2, \\ H^{n-2}(Y; \mathbb{Q})(-1) \oplus H^n(Y; \mathbb{Q})_{\text{fixed}}, & \text{if } m = n, \\ H^n(Y; \mathbb{Q})_{\text{fixed}}(-1) \oplus H^{n+2}(Y; \mathbb{Q}), & \text{if } m = n+2. \end{cases}$$

For $m = n+1$ there is a decomposition $H^{n+1}(\tilde{X};\mathbb{Q}) = L^2 \oplus L^1/L^2 \oplus L^0/L^1$ into sub Hodge structures which reads

$$H^{n+1}(\tilde{X};\mathbb{Q}) = H^{n-1}(Y;\mathbb{Q})(-1) \oplus H^1(\mathbb{P}^1, j_*\mathbb{V}) \oplus H^{n+1}(Y;\mathbb{Q}).$$

Remark 4.28. It is amusing to compare these direct sum decompositions with the direct sum decompositions coming from the fact that $\tilde{X} = \mathrm{Bl}_B X$. Indeed, if $m \leq n$ the first decomposition coincides with the decomposition $H^{m-2}(B;\mathbb{Q})(-1) \oplus H^m(X;\mathbb{Q})$, while for $m > n+1$ one has to switch the two summands.

Historical Remarks. The construction of a mixed Hodge structure on a smooth variety given here is due to Deligne [Del71]. In this article most of the other results from this chapter can be found. For the results in § 4.5.3 we refer also to [Zuc76]. Degeneration of the Leray spectral sequence for Lefschetz pencils had been shown before, see [Katz73b, Th. 5.6.8].

5

Singular Varieties

In this chapter we shall put a functorial mixed Hodge structure on the cohomology groups of an arbitrary complex algebraic variety which in the smooth case coincides with the one defined in the previous chapter. The main idea is to express the cohomology of the variety in terms of cohomology groups of smooth compact varieties. To achieve this, we first take a variety X which is compact and contains our given variety U as a dense Zariski open subset. Then we define the notion of a simplicial resolution of the pair (X, D), where $D = X - U$ and deal with the mixed Hodge theory of simplicial varieties. These are introduced in § 5.1. Then, in § 5.1.3 and 5.2 we explain the construction of so-called cubical hyperresolutions of (X, D). These lead to simplicial resolutions with nice additional properties. Next, in § 5.3, we deal with the uniqueness and functoriality of the resulting mixed Hodge structure. Cup products and relative cohomology is discussed in § 5.4 and § 5.5 respectively.

It is crucial in this chapter that we allow varieties to be reducible.

5.1 Simplicial and Cubical Sets

5.1.1 Basic Definitions

The notion of a (co)-simplicial object starts from the standard p-simplex Δ_p, which is the convex hull in $\mathbb{R}^{p+1}$ of the $p+1$ standard unit-vectors

$$\Delta_p = \{(x_0, \dots, x_p) \mid x_i \geq 0, \sum_i x_i = 1\}.$$

Its boundary consists of the $(p-1)$-simplices $\Delta_p^q = \Delta_p \cap \{x_q = 0\}$, $q = 0, \dots, p$ inducing the embeddings $\delta^q : \Delta_{p-1} \to \Delta_p$, called the q-th **face maps**. Its vertices, the $p+1$ standard unit-vectors, can be identified with elements from the ordered set $\{0, \dots, p\}$ by the correspondence $i \iff e_i$. The standard p-simplex then corresponds to the ordinal $[p]$. In this way, the maps δ^q give examples of non-decreasing maps $[p-1] \to [p]$. Other examples of non-decreasing maps are coming from the **degeneration maps** $\sigma^q : \Delta_p \to \Delta_{p-1}$,

$q = 0, \ldots, p-1$ defined by $\sigma^q e_0 = e_0, \ldots, \sigma^q e_q = \sigma^q e_{q+1} = e_q, \sigma^q e_{q+2} = e_{q+1}, \ldots \sigma^q e_p = e_{p-1}$. This information can be captured in a diagram:

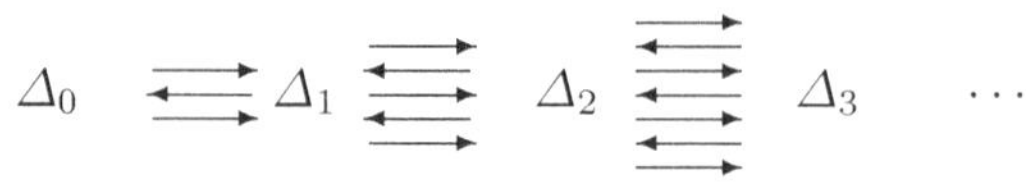

This is the first example of a co-simplicial set. A simplicial set $K_\bullet$ can be given by a diagram as above, but by reversing the arrows. A semi-simplicial set is given by a diagram with "face maps" only and for a cubical set we use cubes instead of simplices. We now give the formal definition.

Definition 5.1. 1) *The* ***simplicial category*** Δ *is the category with objects the ordered sets* $\{0, \ldots, n\}$, $n \in \mathbb{Z}_{\geq 0}$, *and with morphisms non-decreasing maps. If we only consider the strictly increasing maps we speak of the* ***semi-simplicial category*** $\triangle$. *The* ***cubical category*** *is the category* $\square$ *whose objects are the finite subsets of* $\mathbb{N}$ *and for which* $\mathrm{Hom}(I, J)$ *consists of a single element if* $I \subset J$ *and else is empty. We set*

$$[n] := \{0, \ldots, n\}.$$

The n*-truncated simplicial, semi-simplicial category, respectively cubical category is the full sub-category of the category* Δ_n, $\triangle_n$, *respectively* $\square$ *whose objects are the* $[k]$ *with* $k \in [n-1]$.

2) *A* ***simplicial, co-simplicial object*** *in a category* $\mathfrak{C}$ *is a contravariant functor* $K_\bullet : \Delta \to \mathfrak{C}$, *respectively a co-variant functor* $C^\bullet : \Delta \to \mathfrak{C}$. *A morphism between such objects is to be understood as a morphism of corresponding functors. Similarly we speak of a* ***semi-simplicial objects***, ***co-semi-simplicial objects***, ***cubical objects*** *and* ***co-cubical objects***. *We get an* n***-(co)simplicial object*** *by replacing* Δ *by* Δ_n *and similarly for* n***-(co)semi-simplicial object***. *Recalling that* δ_j *are the face maps, we set*

$$\begin{aligned} K_n &:= K_\bullet[n] \quad C^n = C^\bullet[n] \quad \textit{(the set of } n\textit{-simplices)} \\ d_j &= K(\delta^j), \quad d^j = C(\delta^j) \end{aligned}$$

Moreover, for a cubical object X *and* $I \subset \mathbb{N}$ *finite we write*

$$\begin{aligned} X_I &:= X(I) \\ d_{IJ} &:= X(I \hookrightarrow J) : X_J \to X_I, \quad I \subset J. \end{aligned}$$

So, a simplicial object $K_\bullet$ *in* $\mathfrak{C}$ *consists of objects* $K_n \in \mathfrak{C}$, $n = 0, \ldots,$ *and for each non-decreasing map* $\alpha : [n] \to [m]$, *there are morphisms* $d_\alpha : K_m \to K_n$. *For a co-simplicial object, just reverse the arrows:* $d^\alpha : C^n \to C^m$. *If moreover,* $\mathfrak{C}$ *is an additive category, we may put*

$$\delta_n := \sum_{j=0}^{n} (-1)^n d_j : K_n \to K_{n-1}, \quad \delta^n := \sum_{j=0}^{n} (-1)^n d^j : C^n \to C^{n+1}$$

thus defining a complex in $\mathfrak{C}$:

$$CK_\bullet := \{ \dots \xrightarrow{\delta_1} K_1 \xrightarrow{\delta_0} K_0 \}, \quad CC^\bullet := \{ C^0 \xrightarrow{\delta^0} C^1 \xrightarrow{\delta^1} \dots \}, \tag{V–1}$$

If S is any object in $\mathfrak{C}$ the **constant simplicial object** S is obtained by setting $S_n = S$ and taking the identity for the maps induced by face and degeneracy maps. An **augmentation** of a simplicial object to S is a morphism $K_\bullet \to S$ of simplicial objects. If $\mathfrak{C}$ is the category of topological spaces, we speak of a **simplicial space**, if $\mathfrak{C}$ is the category of complex algebraic varieties we speak of a **simplicial complex algebraic variety**. It should be clear what is meant by a **co-simplicial group, algebra, differential graded algebra** etc. For a simplicial abelian group $G_\bullet$, the complex $CG_\bullet$ is a chain complex, and for a co-simplicial abelian group $G^\bullet$ the complex $CG^\bullet$ is a co-chain complex

We define the **geometric realization** $|K_\bullet|$ of a simplicial space $K_\bullet$, using the convention that every non-decreasing map $f : [q] \to [p]$ has geometric realizations $|f| : \Delta_q \to \Delta_p$:

$$|K_\bullet| = \coprod_{p=0}^{\infty} \Delta_p \times K_p / R,$$

where the equivalence relation R is generated by identifying $(s, x) \in \Delta_q \times K_q$ and $(|f|(s), y) \in \Delta_p \times K_p$ if $x = K(f)y$ for all non-decreasing maps $f : [q] \to [p]$. The topology on $|K|$ is the quotient topology under R obtained from the direct product topology (note that the K_p are topological spaces by assumption). A semi-simplicial set has a geometric realization as well, using only strictly decreasing maps to describe the equivalence relation R. There is a natural augmentation

$$k : X_\bullet \to |X_\bullet| \tag{V–2}$$

defined by sending $x \in X_n$ to the equivalence class of (x, z_n), where z_n is the barycenter of Δ_n.

Examples 5.2 (of simplicial sets).

1) For any topological space X a **singular p-simplex** is a continuous map

$$\sigma : \Delta_p \to X.$$

These form the objects of the simplicial space $S_\bullet(X)$ of singular simplices in X. Any non-decreasing $f : [i] \to [j]$ seen as a map $f : \Delta_i \to \Delta_j$ induces a morphism $S(f) : S_j(X) \to S_i(X)$ by sending any j-simplex σ to the i-simplex $\sigma \circ f$.

2) Let A be a countable *ordered* set and let $\mathfrak{U} = \{U_\alpha\}_{\alpha \in A}$ be an open covering of a topological space X. For any subset I of A we let $U_I = \bigcap_{\alpha \in I} U_\alpha$. So, if $I \subset J$ there is an inclusion $d_{IJ} : U_J \hookrightarrow U_I$. If $j \in I$ we let $I(j) \subset I$ be the subset obtained by deleting the $j+1$-st element and we set $d_j = d_{I(j)I} : U_I \hookrightarrow U_{I(j)}$. The **nerve** of the covering $\mathfrak{U}$ is the semi-simplicial set $N(\mathfrak{U})_\bullet$ defined by

$$N(\mathfrak{U})_n := \coprod_{|I|=n+1} U_I, \quad d_j : N(\mathfrak{U})_n \to N(\mathfrak{U})_{n-1}, j = 0, \ldots, n.$$

The inclusions $U_I \hookrightarrow X$ define an augmentation $\epsilon(\mathfrak{U}) : N(\mathfrak{U})_\bullet \to X$.

3) Let $K_\bullet$ be a simplicial set. It induces a simplicial abelian group as follows. Its elements in degree q are the finite integral linear combination of q-simplices; morphisms are induced by those in $K_\bullet$. The associated chain complex is the **chain complex for** $K_\bullet$. Its homology groups $H_q(K_\bullet)$ are the **homology groups** of $K_\bullet$. Dualizing we define cohomology groups $H^q(K_\bullet)$. Replacing $\mathbb{Z}$ by any commutative ring R, one gets (co)simplicial R-modules leading to (co)chains and (co)homology groups with R-coefficients. For any topological space X the chain complex associated to the simplicial set $S_\bullet(X; R)$ is nothing but the singular chain complex (with values in R) whose homology and cohomology yields singular homology $H_\bullet(X; R)$, respectively singular cohomology $H^\bullet(X; R)$. See § B.1.1.

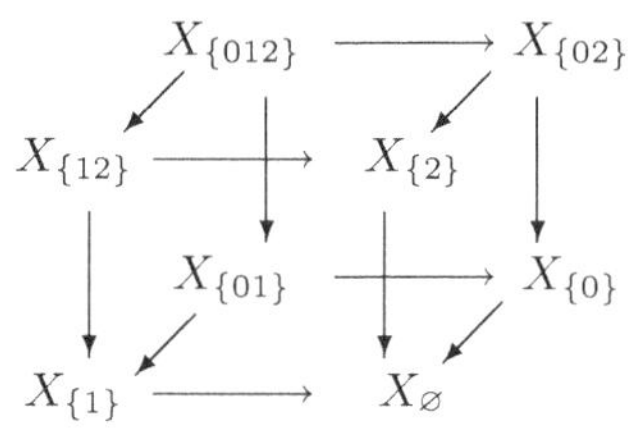

Fig. 5.1. A 3-Cubical variety

Examples 5.3 (of cubical varieties).

1) The nerve of a covering (Example 5.5.2) is in fact an A-cubical space.

2) Let Y be a variety with irreducible components $Y_0, \ldots, Y_n$. Put $Y_\varnothing = Y$ and $Y_I = \bigcap_{i \in I} Y_i$ for $I \subset [n]$ non-empty. The maps $d_{IJ} : Y_J \to Y_I$ are given by the inclusions. This defines an $(n+1)$-cubical variety.

3) Any $(k+1)$-cubical variety (X_I) can be considered as a morphism of k-cubical varieties $Y \to Z$ by putting $Z_I = X_I$ and $Y_I = X_{I \cup \{k\}}$ for $I \subset [k-1]$. In particular, a 1-cubical variety is the same as a morphism of varieties.

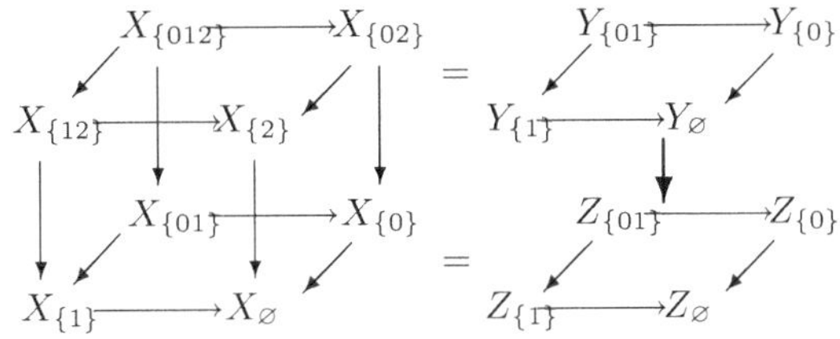

Fig. 5.2. A 3- cubical variety as a morphism between 2-cubical varieties

4) A generalization of the preceding: instead of the category of subsets of $[n-1]$ we may consider the category $\square_A$ of all finite subsets of a given set A and define A-cubical varieties as contravariant functors from $\square_A$ to varieties. Then for two finite sets A and B, the following notions are equivalent: $(A \sqcup B)$-cubical varieties, A-cubical objects in the category of B-cubical varieties and B-cubical objects in the category of A-cubical varieties. The reason is that $2^{A \sqcup B} = 2^A \times 2^B$.

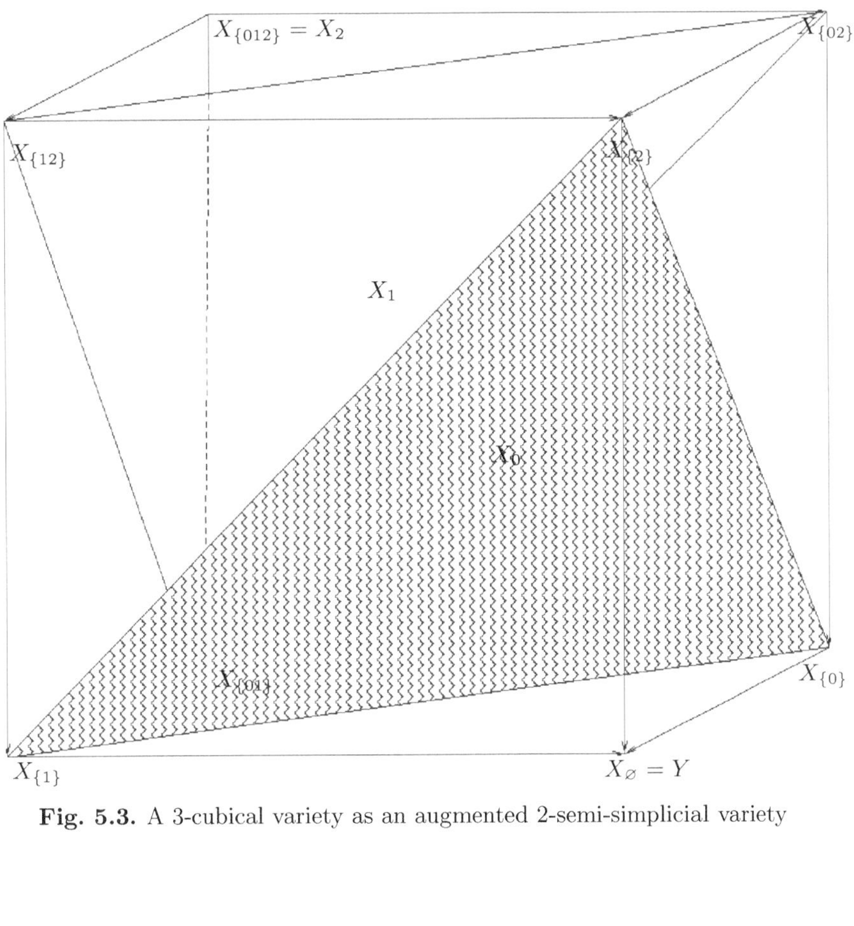

Fig. 5.3. A 3-cubical variety as an augmented 2-semi-simplicial variety

Remark 5.4. 1) Every semi-simplicial variety admits a unique augmentation to a point. A semi-simplicial variety augmented towards Y is just a semi-simplicial object in the category of Y-varieties.

2) Every $(n+1)$-cubical variety (X_I) gives rise to an augmented n-semi-simplicial variety $X_\bullet \to Y$ in the following way. We put

$$X_k = \coprod_{|I|=k+1} X_I, \quad k = 0, \ldots, n$$

and for each inclusion $\beta : [s] \to [r]$ and $I \subset [n]$ with $|I| = r+1$ writing $I = \{i_0, \ldots, i_r\}$, $i_0 < \ldots < i_r$, we let

$$X(\beta)|_{X_I} = d_{IJ}, \quad J = \beta(I) = \{i_{\beta(0)}, \ldots, i_{\beta(s)}\}.$$

For all $I \subset [n]$ we have a well-defined map $d_{\varnothing I} : X_I \to X_\varnothing = Y$. This is the desired augmentation. Note that this correspondence is functorial.

If $X = \{X_I\}$ is a cubical variety and $X_\bullet \to X_\varnothing$ its associated augmented semi-simplicial variety, the continuous map

$$|\epsilon| : |X_\bullet| \to X_\varnothing,$$

is called the **geometric realization of the cubical variety** X. For an A-cubical variety we can also directly describe it: the vertices of the cube $\square_A$, i.e. the *finite* subsets I of A are in one to one correspondence with the faces Δ_I of

$$\Delta_A = \{f : A \to [0,1] \mid \sum_{a \in A} f(a) = 1\};$$

they correspond to functions zero on $A - I$. Note that $\Delta_\varnothing$ is empty so that the augmenting variety does not play a role as indeed it should not. If $I \subset J$, there are inclusions $e_{IJ} : \Delta_I \to \Delta_J$ and together with the maps $d_{IJ} : X_J \to X_I$ they define the geometric realization as

$$[|X_\bullet| = \coprod_{I \subset A} \Delta_I \times X_I]/R,$$

where the equivalence relation R is generated by identifying $(f, d_{IJ}(x))$ and $(e_{IJ}(f), x)$.

5.1.2 Sheaves on Semi-simplicial Spaces and Their Cohomology

A **sheaf on a semi-simplicial space** is a semi-simplicial object in the category of pairs $(X, \mathcal{F})$ with X a topological space and $\mathcal{F}$ a sheaf on X, and whose morphisms are pairs $(f, f^\sharp) : (X, \mathcal{F}) \to (Y, \mathcal{G})$ with $f : X \to Y$ and $f^\sharp : \mathcal{G} \to f_* \mathcal{F}$ a sheaf homomorphism. More concretely, a sheaf $\mathcal{F}^\bullet$ on $X_\bullet$ consists of a family of sheaves $\mathcal{F}^k$ over X_k such that for increasing maps $\beta : [n] \to [m]$ we have sheaf morphisms $f^\sharp : \mathcal{F}^n \to X(\beta)^* \mathcal{F}^m$ satisfying $(f \circ g)^\sharp = f^\sharp \circ g^\sharp$. We can likewise consider **complexes of sheaves of abelian groups** $\mathcal{F}^{\bullet,\bullet}$ on $X_\bullet$ and **resolutions of a sheaf** on $X_\bullet$.

Examples 5.5. 1) The constant sheaf $\underline{G}_{X_\bullet}$, where G is an abelian group.

2) If $X_\bullet$ is a semi-simplicial complex analytic space the sheaves $\mathcal{O}_{X_n}$ define a sheaf $\mathcal{O}_{X_\bullet}$ on $X_\bullet$.

3) If $X_\bullet$ is a semi-simplicial complex manifold, for any $k \in N$, the sheaves $\Omega^k_{X_n}$ define $\Omega^k_{X_\bullet}$. They fit in the De Rham complex $\Omega^\bullet_{X_\bullet}$.

4) Let $\epsilon(\mathfrak{U}) : N(\mathfrak{U})_\bullet \to X$ be the augmented nerve of a covering $\mathfrak{U}$ of a topological space X as explained in Example 5.2.1). Let $\mathcal{F}$ be any sheaf on X. It defines a sheaf on the nerve: set $\mathcal{F}^I = \Gamma(U_I, \mathcal{F})$ and for $I \subset J$ let $\mathcal{F}^I \to \mathcal{F}^J$ be restriction maps induced by the inclusions d_{IJ}. This sheaf can be identified with the co-simplicial group $C^\bullet(\mathfrak{U}, \mathcal{F})$ where $C^n(\mathfrak{U}, \mathcal{F})$ are precisely the n-cochains with values in $\mathcal{F}$ and the associated complex $(C^\bullet(\mathfrak{U}, \mathcal{F}), \check{d})$ is the Čech-cochain complex with cohomology $H^q(\mathfrak{U}, \mathcal{F})$. Consider the double complex

$$C^{p,q}(\mathfrak{U}, \mathcal{F}) := C^q(\mathfrak{U}, \mathcal{C}^p_{\mathrm{Gdm}}(\mathcal{F})) \tag{V–3}$$

with the differentials in the p-direction coming from the Godement resolution and the differential in the q-direction the Čech-derivative. The associated simple complex neither computes $H^*(X, \mathcal{F})$ nor Čech-cohomology, but the two spectral sequences of the double complex (A–32) relate the two. In fact, the vertical rows are exact and the p-th row gives a resolution of $\Gamma(X, \mathcal{C}^p_{\mathrm{Gdm}}(\mathcal{F}))$ so that the first spectral sequence degenerates at E_2 and we have $'E_2^{p,0} = H^p(X, \mathcal{F})$:

$$'E_1^{p,q} = H^q(\mathfrak{U}, \mathcal{C}^p_{\mathrm{Gdm}}(\mathcal{F})) \implies H^{p+q}(s[C^{\bullet,\bullet}(\mathfrak{U}, \mathcal{F})] \simeq H^{p+q}(X, \mathcal{F}). \tag{V–4}$$

This spectral sequence is called the **Mayer-Vietoris spectral sequence**. On the other hand, for the second spectral sequence we have $''E_1^{p,q} = H^q(\mathfrak{U}, \mathcal{C}^p_{\mathrm{Gdm}}(\mathcal{F}))$ and if $\mathfrak{U}$ is acyclic with respect to the sheaves $\mathcal{C}^p_{\mathrm{Gdm}}(\mathcal{F})$, it degenerates at E_2 and gives $''E_2^{p,0} = H^p(\mathfrak{U}, \mathcal{F})$.

5) Let $\mathcal{F}^\bullet$ be a sheaf on $X_\bullet$. The Godement resolutions $\mathcal{C}^\bullet_{\mathrm{Gdm}}(\mathcal{F}^m)$ fit together to give a resolution $\mathcal{C}^\bullet_{\mathrm{Gdm}}(\mathcal{F}^\bullet)$ of the sheaf $\mathcal{F}^\bullet$.

Motivated by (V–4) we define the **cohomology** of a sheaf of abelian groups $\mathcal{F}^\bullet$ on a semi-simplicial space as follows. The abelian groups

$$F^{p,q} := \Gamma(X_q, \mathcal{C}^p_{\mathrm{Gdm}}(\mathcal{F}^q)) \tag{V–5}$$

form part of a double complex. As before, the differentials d' in the p-direction come from the Godement resolution, while now the differentials $d'' = \sum_{j=0}^n (-1)^n d^j$ in the q-direction are the differentials from the co-simplicial group $CF^{p,\bullet}$ which we introduced before (V–1). Define

$$H^k(X_\bullet, \mathcal{F}^\bullet) := H^k(sF^{\bullet,\bullet}). \tag{V–6}$$

In the special case where $X_\bullet$ is the nerve of an open covering $\mathfrak{U}$ of a topological space X and the sheaf is coming from a sheaf $\mathcal{F}$ on X, the double complex $F^{\bullet,\bullet}$ is the double complex (V–3). Hence (V–4) implies that

$$H^k(N(\mathfrak{U})_\bullet, \mathcal{F}) = H^k(X, \mathcal{F}). \tag{V–7}$$

Suppose that $\epsilon : X_\bullet \to Y$ is an augmentation and $\mathcal{F}^\bullet$ a sheaf on $X_\bullet$. The sheaves $\epsilon_* \mathcal{C}^q_{\mathrm{Gdm}}(\mathcal{F}^p)$ then form a double complex of sheaves on Y; its associated simple complex defines

$$R\epsilon_* \mathcal{F}^\bullet := s[\epsilon_* \mathcal{C}^\bullet_{\mathrm{Gdm}}(\mathcal{F}^\bullet)] \tag{V–8}$$

with k-th hypercohomology equal to $H^k(X_\bullet, \mathcal{F}^\bullet)$ as one readily verifies:

$$\mathbb{H}^k(Y, R\epsilon_* \mathcal{F}^\bullet) = H^k(X_\bullet, \mathcal{F}^\bullet). \tag{V–9}$$

There are natural adjunction maps of sheaves on Y, extending (B–24):

$$\epsilon^\sharp : \mathcal{G} \to R\epsilon_*(\epsilon^{-1}\mathcal{G})$$

Definition 5.6. [Del74, §5.3] An augmented semi-simplicial space $\epsilon : X_\bullet \to Y$ is said to be **of cohomological descent** if the natural map

$$\epsilon^\sharp : \underline{\mathbb{Z}}_Y \to R\epsilon_* \underline{\mathbb{Z}}_{X_\bullet}$$

is a quasi-isomorphism. In this case we have

$$\epsilon^* : H^q(Y) \xrightarrow{\sim} H^q(X_\bullet, \underline{\mathbb{Z}}_{X_\bullet}) \tag{V–10}$$

The last assertion is a consequence of (V–9).

The natural augmentation (V–2) is not always of cohomological descent (it is not always surjective), but we still have (see [Car85a, Th. 3.1]):

Proposition 5.7. *Let $k : X_\bullet \to |X_\bullet|$ be the natural augmentation. It induces isomorphisms*

$$H^k(|X_\bullet|, \underline{R}_{|X_\bullet|}) \xrightarrow{\sim} H^k(X_\bullet, \underline{R}_{X_\bullet}).$$

Proof (Sketch). Put $X = |X_\bullet|$ and $\mathcal{F} = \underline{R}_X$. For simplicity assume that $X_\bullet$ is n-semi-simplicial. Any subset I of $\{0, \ldots, n\}$ defines a face Δ_I with barycenter z_I. There is a natural map $p : X \to \Delta_n$ which is defined by sending $X_k \times \{z_I\}$, $|I| = k$ to the k-th vertex of Δ_n and extending affine linearly on $\{x\} \times \Delta_k$, $x \in X_k$ using the maps f_B. Using a metric on Δ_n where the edges have length 1, consider the open set U'_j of points having distance $1 - \frac{1}{2(n+1)}$ to the j-th vertex of Δ_n. Set $U_j = p^{-1}U'_j \subset X$. The covering $\mathfrak{U} = \{U_j \mid j = 0, \ldots, n\}$ has the property that U_I retracts to $p^{-1}z_I = X_{|I|}$. If $x \in X_k$, the class of (x, z_k) in X belongs to U_k and defines a map

$$j : X_\bullet \to N(\mathfrak{U})_\bullet$$

which is a homotopy equivalence of simplicial spaces and so defines an isomorphism $j^* : H^k(N(\mathfrak{U})_\bullet, \mathcal{F}) \xrightarrow{\sim} H^k(X_\bullet, j^*\mathcal{F})$ (here we use that we are working with constant sheaves). Moreover, by (V–7) the augmentation $\epsilon(\mathfrak{U})$ induces an isomorphism. Since $\epsilon(\mathfrak{U}) \circ j = k$ the result follows. □

5.1.3 Cohomological Descent and Resolutions

We now consider cubical and semi-simplicial *varieties*. An augmented semi-simplicial variety is of cohomological descent if this is the case for the underlying semi-simplicial topological space. For pairs we have the following notion.

Definition 5.8. [Del74, §5.3] Let X be a variety and D a closed subvariety of X, a **semi-simplicial resolution** of the pair (X, D) is a semi-simplicial variety $\epsilon : X_\bullet \to X$ augmented towards X such that all maps $X_k \to X$ are proper, X_k is smooth for all k, ϵ is of cohomological descent and the inverse image of D on each irreducible component X_k^i is either all of X_k^i, or empty, or a divisor with simple normal crossings on X_k^i.

Example 5.9. Suppose C is an algebraic curve with one nodal singularity P, with normalization $n : \tilde{C} \to C$ and $n^{-1}(P) = \{Q_0, Q_1\}$. Define $C_0 = \tilde{C}$, $C_1 = \{P\}$, $d_{0,1}(P) = Q_0$ and $d_{1,1}(P) = Q_1$. One obtains a 1–semi-simplicial space $C_\bullet$ with a natural augmentation to C given by n. This is a semi-simplicial resolution. Indeed, all maps occurring are finite so that $Rn_*\underline{\mathbb{Z}}_{C_\bullet}$ is the complex $[n_*\underline{\mathbb{Z}}_{\tilde{C}} \to \underline{\mathbb{Z}}_P]$ which resolves $\underline{\mathbb{Z}}_C$ since the sequence

$$0 \to \underline{\mathbb{Z}}_C \to n_*\underline{\mathbb{Z}}_{\tilde{C}} \to \underline{\mathbb{Z}}_P \to 0$$

is exact.

Definition 5.10. A cubical variety is said to be **of cohomological descent** (respectively a **cubical hyperresolution**) if its associated augmented semi-simplicial variety is of cohomological descent (respectively a semi-simplicial resolution).

Remark 5.11. For an n-cubical variety X let $\epsilon : X_\bullet \to X_\varnothing$ be its associated augmented semi-simplicial variety. We let $C^\bullet(X)$ denote the cone over the morphism $\underline{\mathbb{Z}}_{X_\varnothing} \to R\epsilon_*\underline{\mathbb{Z}}_{X_\bullet}$. Then $X_\bullet$ is of cohomological descent if and only if $C^\bullet(X)$ is acyclic. If $f : X \to Y$ is a morphism of n-cubical varieties, and Z the associated $(n+1)$-cubical variety, then Z is of cohomological descent if and only if $C^\bullet(Y) \to Rf_*C^\bullet(X)$ is a quasi-isomorphism.

Example 5.12. Using cubical varieties, we can rephrase and generalize Example 5.9 to any algebraic curve C. Let $\Sigma \subset C$ be the set of its singular points and $n : \tilde{C} \to C$ its normalization. Put $\tilde{\Sigma} = n^{-1}\Sigma$ and let $i : \Sigma \hookrightarrow C$, $\tilde{\imath} : \tilde{\Sigma} \hookrightarrow \tilde{C}$ the natural inclusions. The augmented semi-simplicial variety $C_\bullet \to C$ defined by the 2-cubical variety

$$\begin{array}{ccc} C_{01} = \tilde{\Sigma} & \xrightarrow{n|\tilde{\Sigma}} & \Sigma = C_1 \\ \downarrow{\scriptstyle \tilde{\imath}} & & \downarrow{\scriptstyle i} \\ C_0 = \tilde{C} & \xrightarrow{n} & C = C_\varnothing \end{array}$$

is of cohomological descent for the same reason as in Example 5.9.

There is another, more geometrical way to decide whether a cubical variety is of cohomological descent using the geometrical realization $|X_\bullet|$ of a semi-simplicial scheme $X_\bullet$ (see § 5.1.1). If $\epsilon : X_\bullet \to Y$ is an augmented semi-simplicial complex variety, there is an induced continuous map $|\epsilon| : |X_\bullet| \to Y$, the **geometric realization of the augmentation**. We have:

Proposition 5.13. *Let $\epsilon : X_\bullet \to Y$ be an augmented semi-simplicial complex variety. If $|\epsilon| : |X_\bullet| \to Y$ is proper and has contractible fibres, the augmented semi-simplicial complex variety is of cohomological descent. In particular*

$$\epsilon^* : H^q(Y) \xrightarrow{\sim} H^q(X_\bullet, \underline{\mathbb{Z}}_{X_\bullet}).$$

Proof. As to the first assertion, observe that it is local on Y in the complex topology and so we may take for Y an arbitrarily small neighbourhood of a fixed point. Put $X = |X_\bullet|$ and $e = |\epsilon|$. Prop. 5.7 together with formula (V–9) imply $\mathbb{H}^q(Y, R\epsilon_*\underline{\mathbb{Z}}_{X_\bullet}) \simeq H^q(X, Re_*\underline{\mathbb{Z}}_{X_\bullet}) \simeq H^q(X)$. On the other hand, $H^q(X) = H^q(Y)$ since the fibres of e are contractible. Hence ϵ induces an isomorphism $\mathbb{H}^q(Y, R\epsilon_*\underline{\mathbb{Z}}_{X_\bullet}) \xrightarrow{\sim} H^q(Y, \underline{\mathbb{Z}}_Y)$ and the complex $R\epsilon_*\underline{\mathbb{Z}}_{X_\bullet}$ is quasi-isomorphic to $\underline{\mathbb{Z}}_Y$. The formula for the cohomology in the global case then follows from (V–10). □

The nature of the fibres can often be decided by local topological considerations. In the preceding example (5.12) this is especially clear from Fig. 5.4 which describes $|\epsilon|$.

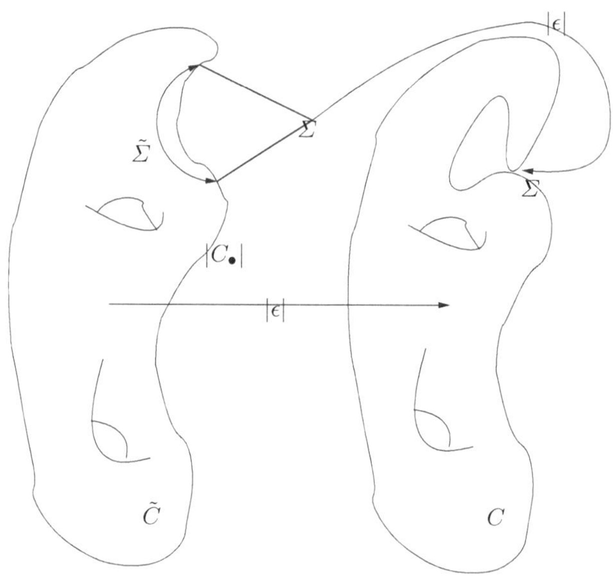

Fig. 5.4. The geometric realization of $\epsilon : C_\bullet \to C$

5.2 Construction of Cubical Hyperresolutions

In this section we will show that every pair (X, D) as above admits a cubical hyperresolution, hence a fortiori a semi-simplicial resolution. In fact we will show a stronger result, which takes the dimensions of the varieties appearing into account.

Definition 5.14. 1) A **proper modification** of a variety X is a proper morphism $f : \tilde{X} \to X$ such that there exists an open dense $U \subset X$ for which f induces an isomorphism $f^{-1}(U) \xrightarrow{\sim} U$. A **resolution** of X is a proper modification $f : \tilde{X} \to X$ for which $\tilde{X}$ is smooth.
2) The **discriminant** of a proper morphism $f : X \to S$ is the minimal closed subset $\Delta(f)$ of S such that f induces an isomorphism $X - f^{-1}(\Delta(f)) \to S - \Delta(f)$.

Remark 5.15. 1) Note that for a morphism $f : X \to S$ of irreducible varieties one has $\Delta(f) = S$ unless f is birational. Remember that we allow a variety to have several components. Suppose that all components of X have the same dimension, then the discriminant of f contains all components of S of dimension $< n$.
2) The notion of resolution that we use is a weaker one than Hironaka's. It is not assumed that the map f is a composition of blowing-ups with non-singular centres contained in the singular locus. A resolution in this weak sense may have a discriminant which contains regular points.

In 1996, Abramovich and de Jong [A-dJ] and, independently, Bogomolov and Pantev [B-P] have given rather short proofs of the following resolution theorem:

Theorem 5.16. *Let X be an (irreducible) algebraic variety and let D be a closed subset of X. Then there exists a resolution $f : \tilde{X} \to X$ which is a projective morphism and such that $f^{-1}(D)$ is a divisor with simple normal crossings on $\tilde{X}$.*

This weak resolution theorem suffices for our purposes.

Lemma-Definition 5.17. Let $f : \tilde{X} \to X$ be a proper modification with discriminant D. Define its **discriminant square** as the commutative diagram

$$\begin{array}{ccc} f^{-1}(D) & \xrightarrow{j} & \tilde{X} \\ \downarrow g & & \downarrow f \\ D & \xrightarrow{i} & X \end{array}$$

and let $Y_{\bullet}$ be the corresponding 2-cubical space. Then $Y_{\bullet}$ is of cohomological descent.

Proof. We can apply Prop. 5.13. Indeed, the fibre $|\epsilon|^{-1}(x)$, $x \in D$ is

$$\{x\} \cup f^{-1}(x) \cup f^{-1}(x) \times [0,1]/\sim$$

where $\{x\}$ gets identified with $f^{-1}(x) \times \{(0)\}$ and $y \in f^{-1}(x)$ with $(y,1) \in f^{-1}(x) \times \{(1)\}$. This yields the topological cone over the fibre $f^{-1}(x)$, which indeed is contractible onto its vertex. □

Remark 5.18. In the proof of the above lemma we only use that the restriction of the map f to the complement of $f^{-1}(D)$ is a homeomorphism. So we might weaken the notion of discriminant a little and still arrive at the same conclusion. This will be applied in the study of the discriminant hypersurface in the space of homogeneous polynomials of given degree in two variables, see Example 5.38.

Definition 5.19. The **discriminant** of a proper morphism $f : X \to S$ of cubical varieties is the smallest closed cubical subvariety D of S such that f induces isomorphisms $X_I - f^{-1}(D_I) \to S_I - D_I$ for all I.

This definition implies that we can form a discriminant square for a proper morphism between k-cubical varieties as we did for a morphism between ordinary varieties (Lemma-Definition 5.17). Such a square is a $(k+2)$–cubical variety and can be described as a morphism between discriminant squares for a morphism between $(k-1)$–cubical varieties. Since by Lemma-Def. 5.17 for $k = 1$, these are of cohomological descent, using Lemma 5.27 inductively proves:

Lemma 5.20. *The $(k+2)$–cubical variety defined by a discriminant square for a proper morphism between k–cubical varieties is of cohomological descent.*

Definition 5.21. Let $f : X_\bullet \to S_\bullet$ be a proper morphism of cubical varieties with discriminant $D_\bullet$ and let $T_\bullet$ be a closed cubical subspace of $S_\bullet$. Then we call f a **resolution** of $(S_\bullet, T_\bullet)$ if X_I is smooth, $f_I^{-1}(T_I)$ consists of certain components of X_I and divisors with simple normal crossings on some other components of X_I, and $\dim f_I^{-1}(D_I) < \dim S_I$ for all I.

Example 5.22. Let us consider 0-cubical varieties $f : X \to S$ with discriminant D and let $T = \varnothing$. Then the condition on the discriminant leads to a more restrictive notion than that of Definition 5.14. Indeed, if in the above definition X is supposed to be equidimensional, S must be equidimensional too, which is not necessarily the case in Def. 5.14 (but both definitions require X to be smooth and f to be bimeromorphic). Theorem 5.16 provides us with a resolution in the sense of Def. 5.21 (first take the disjoint union of the components and then apply the resolution component for component).

Definition 5.23. [G-N-P-P, I.2.6.1 on p. 10] Let $f : \tilde{X} \to X$ be a proper modification of an irreducible variety X and let $a : Y \to X$ be a dominant morphism. Then the **strict transform** of a under f is the diagram

$$\begin{array}{ccc} \tilde{Y} & \xrightarrow{b} & \tilde{X} \\ \downarrow{\scriptstyle \tilde{f}} & & \downarrow{\scriptstyle f} \\ Y & \xrightarrow{a} & X \end{array}$$

where one takes $U = X - \Delta(f)$, $\tilde{U} = f^{-1}(U)$, $V = a^{-1}(U)$, $\tilde{Y}$ is the closure of $\tilde{U} \times_U V$ in $\tilde{X} \times_X Y$, and $\tilde{f}$ and b are induced by the projections on the two factors.

Lemma-Definition 5.24. [G-N-P-P, I.2.6.2 on p. 10] Let X be an irreducible variety and for $r = 1, \ldots, n$ let $(f_r : X_r \to X)_r$ be a finite set of proper modifications of X. Then there is a minimal proper modification of X which dominates all f_r. It is denoted by $\sup_r(f_r : X_r \to X)$

Theorem 5.25. *Let S be an n-cubical variety and let T be a closed cubical subvariety. Then there exists a resolution $f : X \to S$ of (S, T).*

Proof. We follow the proof of [G-N-P-P, Thm I.2.6]. One defines an n-cubical set ΣS in the following way. For $I \subset [n-1]$ we let ΣS_I be the set of closed subspaces $S_{I,\alpha}$ of S_I for which there exists a $J \subset [n-1]$ containing I and an irreducible component of S_J such that $S_{I,\alpha}$ is the closure of its image under the morphism $S_J \to S_I$. Clearly for $I' \subset I$ we have a map $\Sigma S_I \to \Sigma S_{I'}$. Note that all $S_{I,\alpha}$ are irreducible. We need this cubical set in order to glue the resolutions constructed at each stage into a cubical variety. Indeed, ΣS_I contains the set of all components of S_I (if $J = I$) and the set of all subvarieties which are closures of images of components mapping to S_I under the maps $S_J \to S_I$ for $I \subset J$, $I \neq J$.

We build the cubical scheme X step by step, starting with $X_\varnothing$. For all $S_{\varnothing,\alpha} \in \Sigma S_\varnothing$ we choose a resolution $X_{\varnothing,\alpha} \to S_{\varnothing,\alpha}$ of $(S_{\varnothing,\alpha}, S_{\varnothing,\alpha} \cap T_\varnothing)$ and we let $f_\varnothing : X_\varnothing = \coprod X_{\varnothing,\alpha} \to S_\varnothing$. This resolves all components of $S_\varnothing$ (since $\Sigma S_\varnothing$ contains all components of $S_\varnothing$), and in addition contains smooth components lying over those proper subvarieties of $S_\varnothing$ which come from the cubical structure. This is the first step.

Now suppose that we have already defined X_J for all proper subsets J of I. For such J and $S_{I,\alpha} \in \Sigma S_I$ with image $S_{J,\beta} \in \Sigma S_J$ we have the dominant map $S_{I,\alpha} \to S_{J,\beta}$ and the proper modification $X_{J,\beta} \to S_{J,\beta}$. We let $W^J_{I,\alpha}$ be the strict transform of this pair and let $h_{I,\alpha} : W^J_{I,\alpha} \to S_{I,\alpha}$ be the natural map. Using the notation of Lemma-Def. 5.24, we then put $W_{I,\alpha} = \sup_{J \subset I}\left(W^J_{I,\alpha} \xrightarrow{h_{I,\alpha}} S_{I,\alpha}\right)$. Finally we let $X_{I,\alpha} \to W_{I,\alpha}$ be a resolution of $(W_{I,\alpha}, h^{-1}_{I,\alpha}(T_I))$ and put $X_I = \coprod_\alpha X_{I,\alpha}$. We have natural maps $f_I : X_I \to S_I$ and $X_I \to X_J$ for $J \subset I$. Constructed in this way, X_I is smooth, $f_I^{-1}(T_I)$ is of the desired form and $\dim f_I^{-1}(D_I) < \dim S_I$. Hence this procedure leads to a resolution of (S, T). □

Theorem 5.26. *For any variety X of dimension n and any Zariski closed subset T with dense complement there exists an $(n+1)$-cubical hyperresolution (X_I) of (X, T) such that* $\dim X_I \leq n - |I| + 1$.

Proof. We construct the hyperresolution step by step. Our induction hypothesis is, that after k steps we dispose of a $(k+1)$-cubical variety $X^{(k)}$ which is proper, of cohomological descent, with $X^{(k)}_{\varnothing} = X$ such that $X^{(k)}_I$ smooth for all non–empty $I \subset [k-1]$, $\dim X^{(k)}_I \leq n - |I| + 1$ for all $I \subset [k]$ and the inverse image of T in $X^{(k)}_I$ is a union of irreducible components of $X^{(k)}_I$ and a divisor with simple normal crossings.

The first step is to choose a resolution $\pi : \tilde{X} \to X$ of (X, T) with discriminant D, and to put

$$X^{(1)}_{\varnothing} = X,\ X^{(1)}_{\{0\}} = \tilde{X},\ X^{(1)}_{\{1\}} = D,\ X^{(1)}_{\{0,1\}} = \pi^{-1}D\ .$$

Suppose that we have performed k steps successfully. Consider $X^{(k)}$ as a morphism $f^{(k)} : Y^{(k)} \to Z^{(k)}$ of k-cubical varieties as in Example 5.3. Then Z_I is smooth for $I \neq \varnothing$. Let $T^{(k)}$ denote the inverse image of T in $Y^{(k)}$. We choose a resolution $\tilde{Y}^{(k)} \to Y^{(k)}$ of $(Y^{(k)}, T^{(k)})$ and consider the corresponding discriminant square

$$\left.\begin{array}{ccc} E^{(k)} & \to & \tilde{Y}^{(k)} \\ \downarrow & & \downarrow \\ D^{(k)} & \to & Y^{(k)} \end{array}\right\}, \qquad \text{(V–11)}$$

where $D^{(k)}$ is the discriminant of $\tilde{Y}^{(k)} \to Y^{(k)}$ and $E^{(k)}_I$ is the inverse image of $D^{(k)}_I$ in $\tilde{Y}^{(k)}_I$. Complete this square as follows

$$\begin{array}{ccccc} E^{(k)} & \to & \tilde{Y}^{(k)} & & \\ \downarrow & & \downarrow & \searrow & \\ D^{(k)} & \to & Y^{(k)} & \to & Z^{(k)} \end{array}$$

In this diagram the outer commutative square of k-cubical varieties can be considered as a $(k+2)$-cubical variety $X^{(k+1)}$. More precisely for all $I \subset [k-1]$ we let

$$Z^{(k)}_I = X^{(k+1)}_I,\ \tilde{Y}^{(k)}_I = X^{(k+1)}_{I \cup \{k\}},\ D^{(k)}_I = X^{(k+1)}_{I \cup \{k+1\}}\ E^{(k)}_I = X^{(k+1)}_{I \cup \{k, k+1\}}\ .$$

Note that $I \subset [k-1] \Rightarrow X^{(k+1)}_I = Z^{(k)}_I = X^{(k)}_I$ so in that case $\dim X^{(k+1)}_I \leq n - |I| + 1$. Moreover,

$$\dim \tilde{Y}^{(k)}_I = \dim Y^{(k)}_I = \dim X^{(k)}_{I \cup \{k\}} \leq n - |I|;$$

$$\dim D^{(k)}_I \leq \dim Y^{(k)}_I - 1 \leq n - |I| - 1$$

and

$$\dim E^{(k)}_I \leq \dim Y^{(k)}_I - 1 \leq n - |I| - 1\ .$$

So we conclude that $\dim X^{(k+1)}_I \leq n - |I| + 1$ for all $I \subset [k+1]$.

We finally have to check that $X^{(k+1)}$ is of cohomological descent. We need some technical preparations. First, for any cubical variety $X = \{X_I\}$ we set

$$C^\bullet(X) := \mathrm{Cone}^\bullet \left[\underline{\mathbb{Z}}_X \xrightarrow{\epsilon^\sharp} R\epsilon_* \underline{\mathbb{Z}}_{X_\bullet}\right],$$

where $\epsilon : X_\bullet \to X_\varnothing$ is the augmented semi-simplicial variety associated to X. Next we state a result of which the proof, a direct consequence of the definitions, is left to the reader.

Lemma 5.27. *Let X be a $(k+1)$-cubical variety and consider it as a morphism $f : Y \to Z$ of k-cubical varieties. Let $\mathcal{F}^\bullet$ be a sheaf complex on X restricting to sheaf complexes on Y and Z denoted by the same symbol. Then*

$$C^\bullet(X, \mathcal{F}^\bullet)[1] = \mathrm{Cone}^\bullet \left[C^\bullet(Z, \mathcal{F}^\bullet) \xrightarrow{C(f^\sharp)} Rf_* C^\bullet(Y, \mathcal{F}^\bullet)\right].$$

In particular, the left hand side is acyclic if and only if the map $C(f^\sharp)$ is a quasi-isomorphism.

Corollary 5.28. *Let X be a $(k+2)$-cubical variety and consider it as a commutative square of k-cubical varieties*

$$\begin{array}{ccc} Y & \xrightarrow{f} & Z \\ \downarrow{\scriptstyle a} & & \downarrow{\scriptstyle b} \\ T & \xrightarrow{g} & W. \end{array}$$

Then the cone over the natural map of complexes

$$C^\bullet(W)[1] \longrightarrow \mathrm{Cone}^\bullet \left[Rb_* C^\bullet(Z) \oplus Rg_* C^\bullet(T) \to R(g \circ a)_* C^\bullet(Y)\right]$$

is quasi-isomorphic to $C^\bullet(X)[2]$.

Proof. There is an induced square of complexes of sheaves on $X_\varnothing$

$$\begin{array}{ccc} R(g\circ a)_* C^\bullet(Y) & \xleftarrow{C(f^\sharp)} & Rg_* C^\bullet(Z) \\ \uparrow{\scriptstyle C(a^\sharp)} & & \uparrow{\scriptstyle C(b^\sharp)} \\ Rg_* C^\bullet(T) & \xleftarrow{C(g^\sharp)} & C^\bullet(W). \end{array}$$

The cone over the morphism of complexes $\mathrm{Cone}^\bullet C(g^\sharp) \to \mathrm{Cone}^\bullet(f^\sharp)$ given by $(C(b^\sharp), C(a^\sharp))$ is quasi-isomorphic to the complex $C^\bullet(X)[2]$. To conclude, we apply Lemma A.14. □

Continuation of the proof of Theorem 5.26. By Lemma 5.27, since $X^{(k)}$ is of cohomological descent, $C^\bullet(Z^{(k)}) \to Rf_*^{(k)} C^\bullet(Y^{(k)})$ is a quasi-isomorphism. By abuse of notation we omit the hyperdirect image functor and abbreviate this to

$$C^\bullet(Z^{(k)}) \xrightarrow[\sim]{\text{qis}} C^\bullet(Y^{(k)}). \tag{V–12}$$

Next, by Lemma 5.20, discriminant squares of resolutions of k-cubical varieties, if considered as $(k+2)$–cubical varieties, are of cohomological descent. Apply this to the discriminant square (V–11). Let W be the $(k+2)$–cubical variety it defines. Then $C^\bullet(W)$ is acyclic, and hence, by Cor. 5.28 we have

$$C^\bullet(Y^{(k)})[1] \xrightarrow[\sim]{\text{qis}} [C^\bullet(D^{(k)}) \oplus C^\bullet(\tilde{Y}^{(k)} Z)] \to C^\bullet(E^{(k)})$$

and hence, using (V–12), also

$$C^\bullet(Z^{(k)})[1] \xrightarrow[\sim]{\text{qis}} [C^\bullet(D^{(k)}) \oplus C^\bullet(\tilde{Y}^{(k)}) \to C^\bullet(E^{(k)})]$$

i.e., reasoning as before, $X^{(k+1)}$ is of cohomological descent. □

The proof can be adopted for certain cubical schemes ([G-N-P-P, proof of Thm. I.2.5]) and then yields:

Theorem 5.29. *Any cubical variety X admits a hyperresolution by a cubical cubical variety $Y = \{Y_{IJ}\}$ such that* $\dim Y_{IJ} \le \dim X - |I \times J| + 1$.

By example 5.3 4, we can view a cubical cubical variety Y as a cubical variety, say X' so that a resolution as in the above theorem gives rise to a morphism $f : X' \to X$ between cubical varieties. Since also a morphism between cubical varieties can be viewed as a cubical variety, this immediately gives

Corollary 5.30. *Let $X \to Y$ be a morphism of cubical varieties. Then there are cubical hyperresolutions $X' \to X$ and $Y' \to Y$ fitting in a commutative diagram of morphisms of cubical varieties*

$$\begin{array}{ccc} X' & \to & Y' \\ \downarrow & & \downarrow \\ X & \to & Y. \end{array}$$

5.3 Mixed Hodge Theory for Singular Varieties

5.3.1 The Basic Construction

Definition 5.31. 1) A **logarithmic pair** (X, D) is a smooth compact variety X together with a closed subvariety D such that for each irreducible component C of X, the intersection $C \cap D$ is either empty, or all of C, or a divisor with simple normal crossings on C.
2) A morphism of logarithmic pairs $f : (X, D) \to (X', D')$ is a morphism $f : X \to X'$ such that $f^{-1}(D') \subset D$.
3) A **semi-simplicial logarithmic pair** is defined to be a semi-simplicial object in the category of logarithmic pairs.

For each logarithmic pair (X, D) we can define the mixed Hodge complex of sheaves $\mathcal{H}dg^\bullet(X \log D)$ on X as the zero object on components of X which are contained in D and as usual (see Proposition 4.11) on the other components. For a morphism $f : (X, D) \to (X', D')$ we dispose of a morphism of mixed Hodge complexes of sheaves

$$f^* : \mathcal{H}dg^\bullet(X' \log D') \to Rf_*\mathcal{H}dg^\bullet(X \log D) .$$

Definition 5.32. The **Hodge-De Rham complex** for a semi-simplicial logarithmic pair $(X_\bullet, D_\bullet)$ with augmentation $\epsilon : X_\bullet \to Y$ is the mixed Hodge complex of sheaves

$$R\epsilon_*\mathcal{H}dg^\bullet(X_\bullet \log D_\bullet)$$

on Y obtained from the double complexes on X_r, $r \geq 1$ given by

$$\bigoplus_{p\geq 0}\bigoplus_{q\geq 0}(R\epsilon_q)_*\left[\mathcal{H}dg^p(X_r \log D_r)\right]_R, \quad R = \mathbb{Z}, \mathbb{Q}, \mathbb{C},$$

by taking at each level the associated single complex, and equipping this with filtrations W and F as follows:

$$W_m R\epsilon_*\left[\mathcal{H}dg^n(X_\bullet \log D_\bullet)\right]_R = \bigoplus_{q\geq 0}(R\epsilon_q)_* W_{m+q}\left[\mathcal{H}dg^{n-q}(X_q \log D_q)\right]_R, \; R = \mathbb{Q}, \mathbb{C}$$

and

$$F^p\Big[R\epsilon_*\left[\mathcal{H}dg^n(X_\bullet \log D_\bullet)\right]_{\mathbb{C}}\Big] = \bigoplus_{q\geq 0}(R\epsilon_q)_* F^p\left[\mathcal{H}dg^{n-q}(X_q \log D_q)\right]_{\mathbb{C}} .$$

The shift in the weight filtration is similar to the one in the construction of the mixed cone. Note that one has

$$\begin{aligned}\operatorname{Gr}^W_m R\epsilon_*\left[\mathcal{H}dg^\bullet(X_\bullet \log D_\bullet)\right]_R &= \bigoplus_{q\geq 0}(R\epsilon_q)_* \operatorname{Gr}^W_{m+q}\left[\mathcal{H}dg^\bullet(X_q \log D_q)\right]_R[-q]\\ &= \bigoplus_{q\geq 0}(R\epsilon_q \circ Ra_{m+q})_*\mathcal{H}dg^\bullet(D_q(m+q))[-m-2q](-m-q).\end{aligned}$$

In particular, we have

$$\left.\begin{aligned}&\chi_{\mathrm{Hdg}}(\operatorname{Gr}^W_m R\epsilon_*\mathcal{H}dg^\bullet(X_\bullet \log D_\bullet))\\ &\qquad = \textstyle\sum_{q\geq 0}(-1)^m \chi_{\mathrm{Hdg}}(D_q(m+q)) \cdot (-q-m)\end{aligned}\right\}. \quad \text{(V–13)}$$

If T is a closed subvariety of Y such that $\epsilon_q^{-1}(T) \subset D_q$ for all q, and $j : U = Y - T \to Y$ is the inclusion map, then one has a natural morphism of complexes

$$Rj_*\mathbb{Z}_U \to R\epsilon_*\left[\mathcal{H}dg^\bullet(X' \log D')\right]_{\mathbb{Z}}$$

which induces a homomorphism $H^k(U) \to \mathbb{H}^k(Y, [\mathcal{H}dg^\bullet(X' \log D')]_{\mathbb{Z}})$.

Suppose that one has a compact augmented semi-simplicial variety $\epsilon : X_\bullet \to Y$ which is of cohomological descent and a closed subspace T of Y such that $(X_\bullet, \epsilon_\bullet^{-1}(T))$ is a semi-simplicial logarithmic pair. Then, putting $D_\bullet = \epsilon_\bullet^{-1}(T)$, one has

$$\mathbb{H}^k(Y, R\epsilon_* \mathcal{H}dg^\bullet(X_\bullet \log D_\bullet)) \simeq H^k(|X_\bullet| - |D_\bullet|)$$

and this will put a mixed Hodge structure on the right hand side. If $(X_\bullet, D_\bullet)$ is a cubical hyperresolution of (Y,T) the right hand side is isomorphic to $H^k(Y - T)$ via the augmentation and so gets an induced mixed Hodge structure as well.

Given an algebraic variety U, we first choose an embedding $U \hookrightarrow Y$ of U as a dense Zariski open subset of a compact variety Y, and subsequently a cubical hyperresolution of the pair $(Y, Y - U)$. This shows the existence part in the next theorem.

Theorem 5.33. *Let U be a complex algebraic variety.*

i) *A mixed Hodge structure on the cohomology groups $H^k(U)$ is constructed as follows. Let Y be any compact variety containing U as a dense Zariski open subset with complement $T = Y - U$. Let $\epsilon : X_\bullet \to Y$ be a cubical hyperresolution of the pair (Y,T) and let $D_\bullet = \epsilon^{-1}T$. The mixed Hodge structure on the cohomology of U comes from considering it as the hypercohomology of the associated Hodge-De Rham complex on Y, i.e. the mixed Hodge complex of sheaves $R\epsilon_* \mathcal{H}dg^\bullet(X_\bullet \log D_\bullet)$. A different compactification together with an appropriate cubical hyperresolution gives the same mixed Hodge structure.*

ii) *Referring to III–1, for the Hodge character of U we have the following equality in $K_0(\mathfrak{hs})$*

$$\left.\begin{aligned} \chi_{\text{Hdg}}(U) :=& \textstyle\sum_{k \geq 0} (-1)^k \left[H^k(U)\right] \\ =& \textstyle\sum_{m,q \geq 0} (-1)^m \chi_{\text{Hdg}}(D_q(m+q)) \cdot (-m-q). \end{aligned}\right\} \qquad \text{(V–14)}$$

and hence for all $p, q \geq 0$ we have

$$\left.\begin{aligned} &\textstyle\sum_{k \geq 0} (-1)^k h^{p,q}[H^k(U)] \\ &\quad = (-1)^{p+q} \textstyle\sum_{r,m \geq 0} (-1)^m h^{p-m-r, q-m-r}(D_r(m-r)) \end{aligned}\right\} \qquad \text{(V–15)}$$

iii) *If $f : U \to V$ is a morphism, the induced homomorphism on cohomology is a morphism of mixed Hodge structures.*

iv) *For smooth varieties we obtain the same mixed Hodge structures as in Chap. 4.*

Proof. i) First note that, if one has a morphism $f : X_\bullet \to X'_\bullet$ of cubical varieties which are cubical hyperresolutions of $(Y, Y-U)$ and $(Y', Y'-U)$ respectively, then one has a morphism of mixed Hodge complexes of sheaves

$$f^* : R\epsilon'_* \mathcal{H}dg^\bullet(X'_\bullet \log D'_\bullet) \to Rf_* \circ R\epsilon_* \mathcal{H}dg^\bullet(X_\bullet \log D_\bullet)$$

and the induced map

$$f^* : \mathbb{H}^k(Y', R\epsilon'_* \mathcal{H}dg^\bullet(X'_\bullet \log D'_\bullet)) \to \mathbb{H}^k(Y, R\epsilon_* \mathcal{H}dg^\bullet(X_\bullet \log D_\bullet))$$

is a morphism of mixed Hodge structures and moreover a group isomorphism, hence it is an isomorphism of mixed Hodge structures. So two cubical hyperresolutions of pairs $(Y, Y-U)$ and $(Y', Y'-U)$ which are related by a morphism induce the same mixed Hodge structure on $H^k(U)$.

Now let $U \hookrightarrow Y'$ and $U \hookrightarrow Y''$ be two compactifications of U and let $\{X'_I\}$ and $\{X''_I\}$ be cubical hyperresolutions of $(Y', Y'-U)$ and $(Y'', Y''-U)$ respectively. First let Y be the closure of the diagonal of $U \times U$ inside $Y' \times Y''$. We have the diagram

$$\{X'_I\} \to Y' \leftarrow Y \to Y'' \leftarrow \{X''_I\}$$

and we will use theorem 5.29 to conclude that one has a diagram

$$\begin{array}{ccccccccc} \{X'_{IJ}\} & \to & \{Y'_I\} & \leftarrow & \{Y_I\} & \to & \{Y''_{IJ}\} & \leftarrow & \{X''_{IJ}\} \\ \downarrow & & \downarrow & & \downarrow & & \downarrow & & \downarrow \\ \{X'_I\} & \to & Y' & \leftarrow & Y & \to & Y'' & \leftarrow & \{X''_I\} \end{array}$$

where the top row is the cubical hyperresolution of the bottom row. Note that the cubical cubical varieties $\{X'_{IJ}\}$ can be considered as single cubical varieties as in Example 5.3. These are cubical hyperresolutions of $(Y', Y'-U)$ and $(Y'', Y''-U)$ respectively. Now we have related all the mixed Hodge structures by a chain of isomorphisms. This shows that the mixed Hodge structure on $H^k(U)$ does neither depend on the choice of compactification nor on the choice of a cubical hyperresolution.
ii) This follows from (III–11) and (V–13).
iii) Let $f : U \to V$ be a morphism. First we choose compactifications $U \subset Y$ and $V \subset Z$ such that f extends to $\overline{f} : Y \to Z$. Then we choose a cubical hyperresolution $\overline{f} : Y \to Z$ of the diagram $Y \to Z$ such that $\epsilon_Y : Y_\bullet \to Y$ and $\epsilon_Z : Z_\bullet \to Z$ are cubical hyperresolutions of (Y, U) and (Z, V) respectively. This results in a morphism of mixed Hodge complexes of sheaves

$$\overline{f}^* : (R\epsilon_Z)_* \mathcal{H}dg^\bullet(Z_\bullet \log E_\bullet) \to R\overline{f}_*(R\epsilon_Y)_* \mathcal{H}dg^\bullet(Y_\bullet \log D_\bullet)$$

where $D_\bullet = \epsilon_Y^{-1}(Y-U)$ and $E_\bullet = \epsilon_Z^{-1}(Z-V)$. This shows that the map $f^* : H^k(V) \to H^k(U)$ is a morphism of mixed Hodge structures.
iv) In the smooth case we can take a compactification Y of U such that $T = Y - U$ is a divisor with simple normal crossings on Y. Then Y, considered as a 0-cubical variety, is a cubical hyperresolution of (Y, T). The weight spectral sequence

$$E(R\Gamma(Y, R\epsilon_* \mathcal{H}dg^\bullet(X_\bullet \log D_\bullet)), W) \Rightarrow H(U)$$

has the form

$$\begin{aligned} E_1^{-m,k+m} &= \mathbb{H}^k(Y, \mathrm{Gr}_m^W R\epsilon_* \mathcal{H}dg^\bullet(X_\bullet \log D_\bullet)) \\ &= \textstyle\bigoplus_{q \geq 0} H^{k-m-2q}(D_q(m+q))(-m-q)\ .\square \end{aligned}$$

Example 5.34. Let D be a simple normal crossing divisor and let $D(m)$ be the disjoint union of m-fold intersection of components of D. A cubical resolution is given by $\{D_I\} \to D$, and hence

$$\chi_{\mathrm{Hdg}}(D) = \sum_{m \geq 1} (-1)^m \chi_{\mathrm{Hdg}}(D(m)). \tag{V–16}$$

5.3.2 Mixed Hodge Theory of Proper Modifications.

We first prove a fundamental Mayer-Vietoris type result for 2-cubical varieties.

Theorem 5.35. *Let*

$$\begin{array}{ccc} U & \to & Z \\ \downarrow & & \downarrow \\ Y & \to & X \end{array}$$

be a 2-cubical variety which is of cohomological descent. Then one has a long exact sequence of mixed Hodge structures

$$\ldots \to H^k(X) \to H^k(Y) \oplus H^k(Z) \to H^k(U) \to H^{k+1}(X) \to \ldots$$

For the Hodge characters one has

$$\chi_{\mathrm{Hdg}}(U) = \chi_{\mathrm{Hdg}}(Y) + \chi_{\mathrm{Hdg}}(Z) - \chi_{\mathrm{Hdg}}(X).$$

Proof. For simplicity we only treat the compact case. One may cover this 2-cubical variety by a similar diagram of cubical hyperresolutions

$$\begin{array}{ccc} U_\bullet & \to & Z_\bullet \\ \downarrow & & \downarrow \\ Y_\bullet & \to & X_\bullet \end{array}$$

We have the quasi-isomorphism

$$\mathcal{H}dg^\bullet(X_\bullet) \to \mathrm{Cone}^\bullet(\mathcal{H}dg^\bullet(Y_\bullet) \oplus \mathcal{H}dg^\bullet(Z_\bullet) \to \mathcal{H}dg^\bullet(U_\bullet))[-1]$$

The long exact sequence of the cone gives the desired sequence of mixed Hodge structures. □

Remark 5.36. If two subvarieties U_1 and U_2 of a given algebraic variety form an excisive couple (Def. B.4), then the inclusions define a 2-cubical variety $U_\bullet$ with $U_{12} = U_1 \cap U_2$ and with augmentation $U_\bullet \to U = U_1 \cup U_2$. The exact sequence is the Mayer-Vietoris sequence (Theorem B.6).

An example of such a situation is given by two closed subvarieties Y_1, Y_2 of a compact algebraic variety Y, since one can triangulate Y in such a way that Y_1 and Y_2 are subpolyhedra and these form excisive couples. We conclude

$$\chi_{\mathrm{Hdg}}(Y_1 \cup Y_2) = \chi_{\mathrm{Hdg}}(Y_1) + \chi_{\mathrm{Hdg}}(Y_2) - \chi_{\mathrm{Hdg}}(Y_1 \cap Y_2). \tag{V–17}$$

Corollary-Definition 5.37. *Let $f : \tilde{X} \to X$ be a proper modification with discriminant D. Put $E = f^{-1}(D)$. Let $g : f_{|E} : E \to D$ and let $i : D \to X$ and $\tilde{i} : E \to \tilde{X}$ denote the inclusions. Then one has a long exact sequence of mixed Hodge structures*

$$\ldots \to H^k(X) \xrightarrow{(f^*, i^*)} H^k(\tilde{X}) \oplus H^k(D) \xrightarrow{\tilde{i}^* - g^*} H^k(E) \to H^{k+1}(X) \to \ldots$$

It is called the ***Mayer-Vietoris sequence for the discriminant square*** *associated to f (see Definition-Lemma 5.17). One has*

$$\chi_{\mathrm{Hdg}}(\tilde{X}) = \chi_{\mathrm{Hdg}}(X) + \chi_{\mathrm{Hdg}}(E) - \chi_{\mathrm{Hdg}}(D).$$

Example 5.38. Let R_n be the vector space of homogeneous polynomials in two variables x, y of degree n with complex coefficients, and let P_n denote the associated projective space. For $i = 1, \ldots, [n/2]$ we have morphisms

$$d_{i,n} : P_i \times P_{n-2i} \to P_n$$

induced by the map $R_i \times R_{n-2i} \to R_n$ given by $(q, r) \mapsto q^2 r$. These morphisms are proper and generically injective. Let $S_{i,n} \subset P_n$ denote the image of $d_{i,n}$. The map $d_{m,2m} : P_m \to P_{2m}$ is an embedding, so $S_{m,2m} \simeq P_m$, whereas $d_{m,2m+1} : P_m \times P_1 \to P_{2m+1}$ is injective but not an immersion. However, still $d_{m,2m+1}$ induces an isomorphism on cohomology between $S_{m,2m+1}$ and $P_m \times P_1$. We are going to compute the mixed Hodge structures on $H^*(S_{i,n})$ for all $i \leq n/2$. The result is formulated as follows. Consider the cohomology ring $\mathbb{Q}[\lambda_i, \mu_{n-2i}] \simeq \mathbb{Q}[\lambda, \mu]/(\lambda^{i+1}, \mu^{n-2i+1})$ of $P_i \times P_{n-2i}$.
Claim: the map $d_{i,n}^* : H^*(S_{i,n}) \to H^*(P_i \times P_{n-2i})$ is injective and its image is the subalgebra generated by $2\lambda_i + \mu_{n-2i}$ and μ_{n-2i}^{n-2i}.

We prove this by descending induction on i and increasing induction on n. Clearly the claim holds for $n = 1, 2$ and for $i = [n/2]$. So let $n \geq 3$ and $i < [n/2]$. We have the Cartesian diagram

$$\begin{array}{ccc} P_i \times S_{1,n-2i} & \to & P_i \times P_{n-2i} \\ \downarrow{\scriptstyle d_{i,n}} & & \downarrow{\scriptstyle d_{i,n}} \\ S_{i+1,n} & \to & S_{i,n} \end{array}$$

in which the horizontal maps are inclusions, and $d_{i,n}$ induces a homeomorphism between $P_i \times P_{n-2i} - P_i \times S_{1,n-2i}$ and $S_{i,n} - S_{i+1,n}$. Hence the diagram represents a 2-cubical variety which is of cohomological descent (cf. Lemma 5.17) and by Theorem 5.35 we have the exact sequence

$$\ldots \to H^k(S_{i,n}) \to H^k(P_i \times P_{n-2i}) \oplus H^k(S_{i+1,n}) \xrightarrow{\gamma \times \beta} H^k(P_i \times S_{1,n-2i}) \to \ldots$$

We will show that the map $\gamma \times \beta$ is surjective. To see this, first observe that γ and β are $\mathbb{Q}$-algebra homomorphisms. The target is the degree k part of the subalgebra of $H^*(P_i \times P_1 \times P_{n-2i-2}) \simeq \mathbb{Q}[\lambda_i, \lambda_1, \mu_{n-2i-2}]$ generated by

$\lambda_i, 2\lambda_1 + \mu_{n-2i-2}$ and μ_{n-2i-2}^{n-2i-2}. The first two generators are in the image of γ whereas the last one is in the image of β. This shows that the long exact sequence above splits into short exact sequences and that $H^k(S_{i,n})$ is pure of weight k. Hence $H^*(S_{i,n})$ is a subalgebra of $H^*(P_i \times P_{n-2i})$, and its image equals $\gamma^{-1}(\operatorname{Im}\beta)$. Finally note that $\beta(2\lambda_{i+1} + \mu_{n-2i-2}) = 2\lambda_i + 2\lambda_1 + \mu_{n-2i-2} = \gamma(2\lambda_i + \mu_{n-2i})$ and $\gamma(\mu_{n-2i}^{n-2i}) = 0$. Hence $H^*(S_{i,n})$ contains $2\lambda_i + \mu_{n-2i}$ and μ_{n-2i}^{n-2i}. That these indeed generate $H^*(S_{i,n})$ now follows from a dimension count, which is left to the reader.

5.3.3 Restriction on the Hodge Numbers.

We prove some properties of the Hodge numbers of the mixed Hodge structure which we just defined.

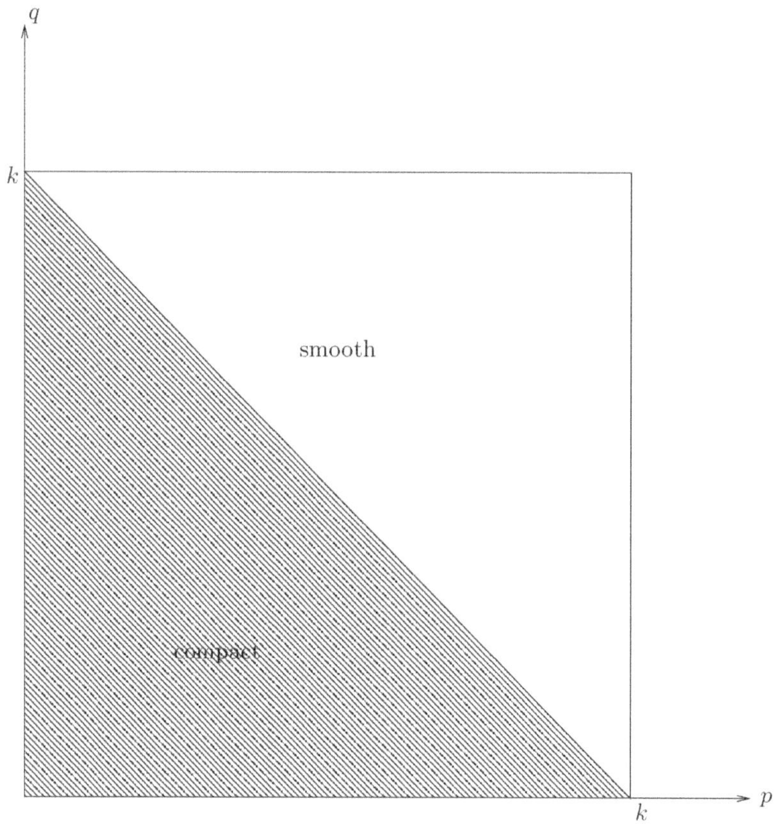

Fig. 5.5. The possible Hodge numbers when $k \leq n$

Theorem 5.39. *Let U be a complex algebraic variety of dimension n. Suppose that a Hodge number h^{pq} of $H^k(U)$ is non-zero. Then*

i) $0 \le p,q \le k$;
ii) *If* $k > n$ *then* $k - n \le p,q \le n$;
iii) *If* U *is smooth then* $p+q \ge k$;
iv) *If* U *is compact, then* $p+q \le k$.

Proof. Choose a compactification Y of U and a cubical hyperresolution $X_\bullet$ of $(Y, Y-U)$ such that $\dim X_r \le n-r$ for all r. This exists by Theorem 5.26.

Suppose $h^{pq}H^k(U)$ is non-zero. Then by the weight spectral sequence, $h^{pq}(H^{k-m-2r}(D_r(m+r))(-m-r))$ is non-zero for some m and some $r \ge 0$, i.e. $h^{p-m-r,q-m-r}(H^{k-m-2r}(D_r(m+r))$ is non-zero for some r,m. This implies $0 \le p,q \le k$. To prove the second statement, we use that $\dim D_r(m+r) \le n-m-2r$. The statement certainly holds for smooth compact varieties. Moreover $k > n$ implies that $k-m-2r > n-m-2r \ge \dim D_r(m+r)$, so if $h^{p-m-r,q-m-r}(H^{k-m-2r}(D_r(m+r))$ is non-zero and $k > n$ then $p-m-r$ is in the sub-interval $[k-m-2r-\dim D_r(m+r), \dim D_r(m+r)] \subset [k-n, n-m-2r]$ so

$$p \in [k-n+m+r, n-r] \subset [k-n, n] \ .$$

If U is smooth, we can use Proposition 4.20. If U is compact, $D_r = \varnothing$ for all r and $\mathrm{Gr}_m^W \mathcal{H}dg^\bullet(X_\bullet) = 0$ for $m > 0$. □

We want to reformulate this also in terms of weights. We first introduce the following concept:

Definition 5.40. Weight m **occurs** in a mixed Hodge structure (H, W, F) if $\mathrm{Gr}_m^W \ne 0$. The mixed Hodge structure is **pure of weight m** if m is the only weight which occurs.

Let X be an algebraic variety. We have the following table of weights which may occur.

Table 5.1. Table of weights on $H^k(X)$

	general	smooth	compact
$k \le n = \dim X$	$[0, 2k]$	$[k, 2k]$	$[0, k]$
$k \ge n$	$[2k-2n, 2n]$	$[k, 2n]$	$[2k-2n, k]$

Theorem 5.41. *Let* $f : Z \to U$ *be a surjective morphism of compact algebraic varieties. Then the induced map*

$$f^* : \mathrm{Gr}_k^W H^k(U) \to \mathrm{Gr}_k^W H^k(Z)$$

is injective for all $k \ge 0$.

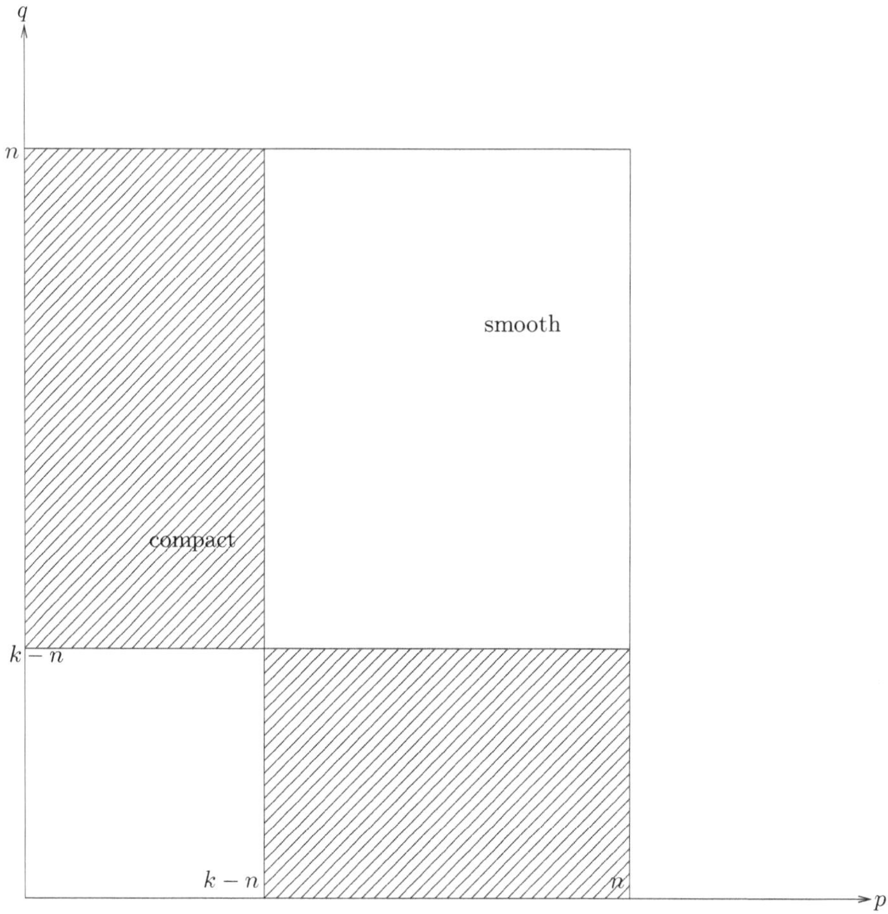

Fig. 5.6. The possible Hodge numbers when $k > n$

Proof. We proceed by induction on $\dim U$. The statement is certainly true for $\dim U = 0$. Also, it holds for U and Z smooth, by Theorem 2.29. Suppose the statement of the theorem holds for all U with $\dim U < k$. Let $f : Z \to U$ be as above with $\dim U = k$. We have a diagram

$$\begin{array}{ccccc} \tilde{Z} & \to & \tilde{U} & \leftarrow & D \\ \downarrow p & & \downarrow q & & \downarrow \\ Z & \to & U & \leftarrow & \Sigma \end{array}$$

where p and q are resolutions and the square on the right hand side is a discriminant square. As D is compact, $\mathrm{Gr}_k^W H^{k-1}(D) = 0$. Hence, by Theorem 5.35 we have the exact sequence

$$0 \to \mathrm{Gr}_k^W H^k(U) \to \mathrm{Gr}_k^W H^k(\Sigma) \oplus H^k(\tilde{U}) \to \mathrm{Gr}_k^W H^k(D) .$$

As $\dim \Sigma < k$ we get that $\mathrm{Gr}_k^W H^k(\Sigma) \to \mathrm{Gr}_k^W H^k(D)$ is injective. We conclude that $\mathrm{Gr}_k^W H^k(U) \to H^k(\tilde{U})$ is injective. As also $H^k(\tilde{U}) \to H^k(\tilde{Z})$ is injective, we get the injectivity of $\mathrm{Gr}_k^W H^k(U) \to H^k(\tilde{Z})$ which in turn implies the injectivity of $\mathrm{Gr}_k^W H^k(U) \to \mathrm{Gr}_k^W H^k(Z)$. □

Corollary 5.42. *Let $f : \tilde{X} \to X$ be a resolution of a compact algebraic variety X. Then for all $k \in \mathbb{N}$:*

$$W_{k-1}H^k(X) = \mathrm{Ker}[f^* : H^k(X) \to H^k(\tilde{X})] \ .$$

Proof. This follows from the injectivity of $\mathrm{Gr}_k^W H^k(X) \to H^k(\tilde{X})$. □

Corollary 5.43. *Let $u : Y \to X$ be a surjective morphism of compact algebraic varieties. Suppose that $H^k(Y)$ is pure of weight k. Then*

$$W_{k-1}H^k(X) = \mathrm{Ker}[u^* : H^k(X) \to H^k(Y)] \ .$$

Proof. There exists a diagram

$$\begin{array}{ccc} \tilde{Y} & \xrightarrow{\tilde{u}} & \tilde{X} \\ \downarrow{\scriptstyle g} & & \downarrow{\scriptstyle f} \\ Y & \xrightarrow{u} & X \end{array}$$

where g and f are resolutions of singularities. It induces a diagram

$$\begin{array}{ccc} H^k(\tilde{Y}) & \xleftarrow{\tilde{u}^*} & H^k(\tilde{X}) \\ \uparrow{\scriptstyle g^*} & & \uparrow{\scriptstyle f^*} \\ H^k(Y) & \xleftarrow{u^*} & H^k(X) \end{array}$$

in which the maps $\tilde{u}^*$ and g^* are injective. Hence $\mathrm{Ker}(u^*) = \mathrm{Ker}(f^*) = W_{k-1}H^k(X)$. □

5.4 Cup Product and the Künneth Formula.

We discuss mixed Hodge theoretic properties for the Künneth formula (Theorem B.7) for products of two complex algebraic varieties U and V.

Theorem 5.44 (Künneth respects mixed Hodge structures). *Let U and V be complex algebraic varieties. There is a natural isomorphism of mixed Hodge structures*

$$\bigoplus_{p+q=k} H^p(U;\mathbb{Q}) \otimes H^q(V;\mathbb{Q}) \to H^k(U \times V;\mathbb{Q}).$$

We have

$$\chi_{\mathrm{Hdg}}(U \times V) = \chi_{\mathrm{Hdg}}(U)\chi_{\mathrm{Hdg}}(V).$$

Before giving the proof, let us deduce that cup-products respect mixed Hodge structures. Indeed, taking $U = V$ and composing with the diagonal $\Delta : U \to U \times U$ we find:

Corollary 5.45. *Let U be a complex algebraic variety. Cup product*

$$H^i(U) \otimes H^j(U) \to H^{i+j}(U)$$

is a morphism of mixed Hodge structures.

Next we need to explain how the Künneth theorem is proved in topology. On ordered pairs of topological spaces the two functors

$$F : (X,Y) \mapsto S^\bullet(X \times Y)$$
$$G : (X,Y) \mapsto [S^\bullet(X) \otimes S^\bullet(Y)]$$

are related by the transpose of Alexander-Whitney homomorphism (B–7)

$$h = {}^tA : [S^\bullet(X) \otimes S^\bullet(Y)] \longrightarrow S^\bullet(X \times Y)$$

which is in fact a natural transformation from G to F. The Künneth formula essentially follows by showing that for all ordered pairs of topological spaces (X,Y) the transformation $h(X,Y)$ induces a homotopy equivalence

$$h(X,Y) : [S^\bullet(X) \otimes S^\bullet(Y)] \to S^\bullet(X \times Y).$$

What we have to bear in mind is that h is the realisation of the Künneth isomorphism on the level of singular chains, but we could have done the same on the level of Godement resolutions.
Proof of the theorem: We let X, Y be a compactification of U respectively V. We set $D := X - U$ and $E := Y - V$. We then construct a semi-simplicial resolution of $(X \times Y, D \times Y \cup X \times E)$ starting from given A-cubical hyperresolution $X(I)$ of (X,D) and a B-cubical hyperresolution $Y(J)$ of (Y,E). We want to construct a semi-simplicial variety on the first barycentric subdivision of $\square_A \times \square_B$. To do this, observe that its vertices correspond to pairs $\Delta_I \times \Delta_J$, $I \subset A$ and $J \subset B$ and any k simplex is determined by a unique flag $\Delta_{I_0} \times \Delta_{J_0} \subset \cdots \subset \Delta_{I_k} \times \Delta_{J_k}$ of length k. So to any such flag F corresponds a simplex Δ_F. Defining

$$(X \times Y)_F := X(i(F)) \times Y(j(F))$$

we get in a natural way a semi-simplicial variety with an augmentation

$$\epsilon_F : (X \times Y)_F \to X \times Y.$$

The maximal element in the flag F, $\Delta_{I_k} \times \Delta_{J_k}$ is denoted $\Delta_{i(F)} \times \Delta_{j(F)}$. As we observed, this corresponds to a vertex of the barycentric subdivision and the star of this vertex is the union of all simplices whose flags with maximal element corresponding exactly to this vertex. It follows that the geometric realization of $(X \times Y)_F$ is homeomorphic to the product $|X_\bullet| \times |Y_\bullet|$ of the geometric realizations of the semi-simplicial varieties associated to $X(I)$ and $Y(J)$. So we see that the above augmented semi-simplicial variety is of cohomological descent.

Let us now put

$$(X \times E \cup D \times Y)_F := X(i(F)) \times E(j(F)) \cup D(i(F)) \times Y(j(F)).$$

Making the identification

$$R(\epsilon(i(F) \times j(F)))_* \mathcal{H}dg^\bullet(X(i(F)) \times Y(j(F)) \log(X \times E \cup D \times F)_F) = \\ R(\epsilon_F)_* \mathcal{H}dg^\bullet((X \times Y)_F \log(X \times E \cup D \times F)_F),$$

we get a morphism of mixed Hodge complexes of sheaves

$$R\epsilon_* \mathcal{H}dg^\bullet(X_\bullet \log D_\bullet) \boxtimes R\epsilon_* \mathcal{H}dg^\bullet(Y_\bullet \log E_\bullet) \\ \longrightarrow R\epsilon_* \left(\mathcal{H}dg^\bullet(X \times Y)_F \log(X \times E \cup D \times Y)_F\right).$$

Using Proposition 3.21, on the level of mixed Hodge complexes this yields a morphism

$$R\Gamma\left(\epsilon_* \mathcal{H}dg^\bullet(X_\bullet \log D_\bullet)\right) \otimes R\Gamma\left(\epsilon_* \mathcal{H}dg^\bullet(Y_\bullet \log E_\bullet)\right) \\ \xrightarrow{h(X,D;Y,E)} R\Gamma\left(\mathcal{H}dg^\bullet(X \times Y)_F \log(X \times E \cup D \times Y)_F\right)$$

which is functorial in (X, D) and (Y, E) giving a natural transformation h between the functor

$$G(U, V) := R\Gamma\left(\epsilon_* \mathcal{H}dg^\bullet(X_\bullet \log D_\bullet)\right) \otimes R\Gamma\left(\epsilon_* \mathcal{H}dg^\bullet((Y_\bullet \log E_\bullet)\right)$$

which puts a mixed Hodge structure on the cohomology of the tensor product of the De Rham complexes for U and V, and the functor

$$F(U, V) := R\Gamma\left(\mathcal{H}dg^\bullet(X \times Y)_F \log(X \times E \cup D \times Y)_F\right)$$

which does the same for the cohomology of $U \times V$.

On the level of the (non-filtered) Godement resolutions of the constant sheaf $\mathbb{Q}$, and in the appropriate derived categories, these functors and the natural transformation h between them are the same as in the topological setting we discussed just before this proof. So h, by definition a morphism of mixed Hodge complexes, induces the Künneth isomorphism in cohomology.

This proof, combined with (III–12) also gives the formula for the Hodge characters. □

5.5 Relative Cohomology

5.5.1 Construction of the Mixed Hodge Structure

Let $f : X \to Y$ be a continuous map. The map $H^i(Y) \xrightarrow{f^*} H^i(X)$ fits in a long exact sequence

$$\cdots \to H^{i-1}(X) \to \tilde{H}^i(\mathrm{Cone}^\bullet(f)) \to H^i(Y) \xrightarrow{f^*} H^i(X) \to \cdots$$

where $\mathrm{Cone}^\bullet(f)$ is the mapping cone of f, cf. (B–37). Moreover, (Theorem B.22) if $f^\bullet : S^\bullet Y \to S^\bullet X$ is the map induced by f on the level of singular co-chains, then

$$\tilde{H}^i(\mathrm{Cone}^\bullet(f)) \simeq H^{i-1}(\mathrm{Cone}^\bullet(f^\bullet)) .$$

Now suppose that we have a morphism of varieties $f : U \to V$. We complete it to a diagram

$$\begin{array}{ccccc} U & \to & Y & \xleftarrow{\pi_Y} & Y_\bullet \\ \downarrow f & & \downarrow \overline{f} & & \downarrow \overline{f}_\bullet \\ V & \to & Z & \xleftarrow{\pi_Z} & Z_\bullet \end{array}$$

where Y and Z are compactifications of U and V and $\pi_Y : Y_\bullet \to Y$ and $\pi_Z : Z_\bullet \to Z$ are cubical hyperresolutions of $(Y, Y - U)$ and $(Z, Z - V)$ respectively. We let $D_\bullet = \pi_Y^{-1}(Y - U)$ and $E_\bullet = \pi_Z^{-1}(Z - V)$. Then we get a morphism of mixed Hodge complexes of sheaves

$$\overline{f}^* : (R\pi_Z)_* \mathcal{H}dg^\bullet(Z_\bullet \log E_\bullet) \to \overline{f}_*(R\pi_Y)_*\mathcal{H}dg^\bullet(Y_\bullet \log D_\bullet) .$$

We can use the mixed cone of $\overline{f}^*$ to put a mixed Hodge structure on the cohomology groups of $\mathrm{Cone}^\bullet(f)$; more precisely we have

$$\tilde{H}^k(\mathrm{Cone}^\bullet(f)) \simeq \mathbb{H}^{k-1}(Z, \mathrm{Cone}^\bullet(\overline{f}^*)) .$$

One obtains an exact sequence of mixed Hodge complexes of sheaves

$$0 \to R\overline{f}_*(R\pi_Y)_*\mathcal{H}dg^\bullet(Y_\bullet \log D_\bullet) \to \mathrm{Cone}^\bullet(\overline{f}^*) \to (R\pi_Z)_*\mathcal{H}dg^\bullet(Z_\bullet \log E_\bullet)[1] \to 0$$

which makes the long exact cohomology sequence into an exact sequence of mixed Hodge structures. The following is a special case:

Proposition 5.46. *Let V be a complex algebraic variety and $U \subset V$ a subvariety. The mixed Hodge structure on the cone of the inclusion $U \hookrightarrow V$ defines mixed Hodge structures on $H^k(V, U)$. One has*

$$\chi_{\mathrm{Hdg}}(V, U) := -\chi_{\mathrm{Hdg}}(\mathrm{Cone}^\bullet(\bar{f}^*)) = \chi_{\mathrm{Hdg}}(V) - \chi_{\mathrm{Hdg}}(U).$$

The long exact sequence in cohomology associated to the pair (V, U)

$$\cdots \to H^{k-1}(U) \to H^k(V, U) \to H^k(V) \to H^k(U) \to \ldots$$

is an exact sequence of mixed Hodge structures.

As to the weights, using Table 5.1, we conclude:

Corollary 5.47. *If U and V are smooth, $H^k(V,U)$ has at most weights in the range $[k-1,2k]$ and if U and V are compact, at most in the range $[0,k]$.*

Remark 5.48. We shall see below (Corollary 6.28) that if V is *compact and smooth* and $U \subset V$ is *open*, the group $H^k(V,U)$ has at most weights in the range $[k,2k]$.

For functoriality, the following is useful:

Observation 5.49. *For any commutative diagram of varieties (or 2-cubical variety)*

$$\begin{array}{ccc} U & \xrightarrow{f} & V \\ \downarrow{\scriptstyle g} & & \downarrow{\scriptstyle h} \\ U' & \xrightarrow{f'} & V' \end{array}$$

we obtain a morphism of mixed Hodge structures

$$(g,h)^* : H^*(\mathrm{Cone}^\bullet(f')) \to H^*(\mathrm{Cone}^\bullet(f)).$$

Corollary 5.50. *Consider a pair (X,T) where X is an algebraic variety and T a closed subvariety of X. Let $U = X - T \xrightarrow{j} X$ be the inclusion. There is a mixed Hodge structure on*

$$H^*_T(X) := H^*(X,U) = \tilde{H}^*(\mathrm{Cone}^\bullet(j))$$

such that the sequence

$$\ldots \to H^k_T(X) \to H^k(X) \to H^k(U) \to H^{k+1}_T(X) \to \ldots$$

becomes an exact sequence of mixed Hodge structures.

We can also look at triples (X,A,B) of a complex algebraic variety X with closed subvarieties $B \subset A$:

Corollary 5.51. *The inclusions $i : (A,B) \to (X,B)$ and $j : (X,B) \to (X,A)$ induce a long exact sequence of mixed Hodge structures*

$$\cdots \to H^k(X,A) \xrightarrow{j^*} H^k(X,B) \xrightarrow{i^*} H^k(A,B) \xrightarrow{\delta} H^{k+1}(X,A) \to \cdots .$$

5.5.2 Cohomology with Compact Support

Definition 5.52. Let U be an algebraic variety with compactification X. put $T = X - U$ and let $i : T \hookrightarrow X$ be the inclusion. The cohomology group $H^k_c(U)$ of U with compact supports is given a mixed Hodge structure through the isomorphism (see Cor. B.14)

$$H^k_c(U) \xrightarrow{\sim} H^k(X,T).$$

The **Hodge-Grothendieck character for compact support** is defined as

$$\chi^c_{\mathrm{Hdg}}(U) := \chi_{\mathrm{Hdg}}(X, T) = \chi_{\mathrm{Hdg}}(X) - \chi_{\mathrm{Hdg}}(T) \qquad \text{(V–18)}$$

with associated Hodge-Euler polynomial $P_{\mathrm{hn}} \circ \chi^c_{\mathrm{Hdg}}(U) = e_{\mathrm{Hdg}}(X) - e_{\mathrm{Hdg}}(T)$.

The invariance of the mixed Hodge structure follows from

Proposition 5.53. *Let $\pi : Y \to X$ be a proper modification with discriminant contained in T and let $E = \pi^{-1}(T)$ Then the natural map*

$$\pi^* : H^k(X, T) \to H^k(Y, E)$$

is an isomorphism of mixed Hodge structures.

Proof. We already know that π^* is a morphism of mixed Hodge structures. It remains to be shown that it is an isomorphism of groups. Let V_T be a closed tubular neighbourhood of T in X such that the inclusion of T into V_T is a homotopy equivalence. Let U_T be the interior of V_T. Put $V_E = \pi^{-1}(V_T)$ and $U_E = \pi^{-1}(U_T)$. Then

$$\begin{array}{c} H^k(X,T) \simeq H^k(X,V_T) \simeq H^k(X - U_T, V_T - U_T) \\ \qquad\qquad\qquad\qquad\qquad \downarrow \simeq \\ H^k(Y,E) \simeq H^k(Y,V_E) \simeq H^k(Y - U_E, V_E - U_E) \end{array}$$

because by the properness of π, the inclusion of E into V_E is also a homotopy equivalence.

Alternatively, observe that we are in the situation of a discriminant square, and that the associated 2-cubical variety is of cohomological descent according to Lemma-Definition 5.17. Consider this square as a morphism of pairs; then this induces an isomorphism on the cohomology of these pairs. □

As to weights, an immediate application of Corollary 5.47 gives:

Proposition 5.54. *The above mixed Hodge structure on $H^k_c(U)$ has at most weights in the interval $[0, k]$. Furthermore, the natural map $H^k_c(U) \to H^k(U)$ is a morphism of mixed Hodge structures. In fact, with X a compactification of U, it is the composition of the morphisms $H^k_c(U) \to H^k(X) \to H^k(U)$. In this set-up we have the exact sequence of mixed Hodge structures*

$$\ldots \to H^{k-1}(X) \to H^{k-1}(T) \to H^k_c(U) \to H^k(X) \to \ldots$$

where $T = X - U$.

The exact sequence for triples (Cor. 5.51) yield exact sequences of mixed Hodge structures for cohomology with compact support:

$$\cdots \to H^k_c(U - V) \to H^k_c(U) \to H^k_c(V) \to H^{k+1}_c(U - V) \to \cdots ,$$

where V is closed in U. Indeed, if $\bar{U}$ is a compactification of U, $\bar{V}$ the closure of V in $\bar{U}$, $S = \bar{U} - U$, $T = \bar{V} - V$ this is the exact sequence for the triple $(\bar{U}, \bar{V} \cup S, S)$. The additivity of the Hodge-Grothendieck characters (and hence for the Hodge-Euler polynomials) (V–18) in the setting of compact support follows immediately from this:

Proposition 5.55. *Let U be a complex algebraic variety which is the disjoint union of two locally closed subvarieties U_1 and U_2. Then*

$$\chi^c_{\mathrm{Hdg}}(U) = \chi^c_{\mathrm{Hdg}}(U_1) + \chi^c_{\mathrm{Hdg}}(U_2).$$

Remark 5.56. This has a motivic interpretation as follows. Let $K_0(\mathrm{Var})$ be the free abelian group on isomorphism classes of complex algebraic varieties modulo the so-called **scissor relations** where we identify the class $[X]$ of X with $[X - Y] + [Y]$ whenever $Y \subset X$ is a closed subvariety. The direct product being compatible with the scissor relation makes $K_0(\mathrm{Var})$ into a ring. Then there is a well defined ring-homomorphism

$$\chi^c_{\mathrm{Hdg}} : K_0(\mathrm{Var}) \to K_0(\mathfrak{hs})$$

extending the Hodge-Grothendieck characteristic.

We deduce the following theorem of Durfee [Du87] and Danilov and Khovanskii [D-K]:

Corollary 5.57. *Let X be an algebraic variety, which is the disjoint union of locally closed subvarieties $X_1, \ldots, X_m$. Then*

$$\chi^c_{\mathrm{Hdg}}(X) = \sum_{i=1}^{m} \chi^c_{\mathrm{Hdg}}(X_i)$$

and similarly for the Hodge-Euler polynomials.

Example 5.58. Let $T^n = (\mathbb{C}^*)^n$ be an n-dimensional algebraic torus. Then $e^c_{\mathrm{Hdg}}(T^1) = uv - 1$ so $e^c_{\mathrm{Hdg}}(T^n) = (uv-1)^n$. Consider an n-dimensional toric variety X. It is a disjoint union of T^n-orbits. Suppose that X has s_k orbits of dimension k. Then

$$e^c_{\mathrm{Hdg}}(X) = \sum_{k=0}^{n} s_k e^c_{\mathrm{Hdg}}(T^k) = \sum_{k=0}^{n} s_k (uv-1)^k.$$

If X has a pure Hodge structure (e.g. if X is compact and has only quotient singularities) then this formula determines the Hodge numbers of X.

Historical Remarks. In [Del74] Deligne defines a functorially mixed Hodge structure on the category of algebraic varieties. His treatment uses simplicial resolutions, while we base our treatment on the cubical version as introduced by Navarro Aznar and explained in [G-N-P-P].

6

Singular Varieties: Complementary Results

In § 6.1, following Arapura [Ara], we put a mixed Hodge structure on the Leray filtration. For technical reasons, this only works in the quasi-projective setting. For the most general statement we refer to Chapter 14, Corollary 14.14.

In § 6.2.3 and § 6.3 we return to the general setting and we study the behaviour of the mixed Hodge structure which we constructed in Chapter 5 under cup products and duality. As one may expect, this generalizes what we already have seen in the smooth compact situation (Chapt. 1 § 2.4). As an application, we prove the semi-purity of the link in § 6.2.3.

6.1 The Leray Filtration

We give a sketch of Arapura's proof [Ara] that the Leray filtration for a morphism $f : X \to Y$ between quasi-projective varieties is a spectral sequence of mixed Hodge structures. To start, note that the sheaves $R^q f_* \underline{\mathbb{Z}}_X$ are locally constant on the strata of a suitable finite analytic stratification of Y. We give these a name:

Definition 6.1. A sheaf $\mathcal{F}$ on a complex analytic space is **finitely constructible** if there is a finite stratification (see § C.1.1) by closed analytic subspaces such that $\mathcal{F}$ restricts to a locally constant sheaf on the open strata. We say that the filtration is *adapted* to $\mathcal{F}$.

For any finitely constructible $\mathcal{F}$ on X adapted to $X = X_m \supset \cdots \supset X_0$ we can construct a filtration adapted to the stratification, the so-called **skeletal filtration** as follows. With $k_\alpha : X - X_\alpha \hookrightarrow X$ the inclusion we set

$$\begin{aligned} \mathrm{Sk}^{\alpha+1}\mathcal{F} &:= (k_\alpha)_! k_\alpha^* \mathcal{F} \quad \alpha \geq 0 \\ \mathrm{Sk}^0 \mathcal{F} &:= \mathcal{F}. \end{aligned}$$

So $\mathrm{Sk}^{\alpha+1}\mathcal{F}$ is the same as $\mathcal{F}$ over the open set $X - X_\alpha$, but it is zero on the closed stratum X_α. The skeletal filtration is the decreasing filtration which, starting from $\mathcal{F}$ kills part of $\mathcal{F}$ on bigger and bigger closed substrata. Hence $\mathrm{Gr}^{\alpha}_{\mathrm{Sk}}\mathcal{F}$ is just $\mathcal{F}$ suitably restricted to the open stratum $X_\alpha - X_{\alpha-1}$. The spectral sequence defined by this filtration on the global sections is the **skeletal spectral sequence** $E_r^{\alpha,\beta}(X,\mathrm{Sk}\mathcal{F})$. This spectral sequence translates into the language of exact couples. Indeed, $H^q(X,\mathrm{Sk}^{\alpha+1}\mathcal{F}) = H^q(X,X_\alpha;\mathcal{F})$, and from the above interpretation of the gradeds $H^q(X,\mathrm{Gr}^{\alpha}_{\mathrm{Sk}}\mathcal{F}) = H^q(X_\alpha,X_{\alpha-1};\mathcal{F})$, so that Prop. A.41 implies

Proposition 6.2. *The skeletal spectral sequence is the spectral sequence for the bigraded exact couple* (D,E) *with* $D^{\alpha,\beta} = H^{\alpha+\beta}(X,X_{\alpha-1};\mathcal{F})$ *and* $E^{\alpha,\beta} = H^{\alpha+\beta}(X_\alpha,X_{\alpha-1};\mathcal{F})$.

If the Leray spectral sequence would be isomorphic to a skeletal spectral sequence for the sheaf $\underline{\mathbb{Q}}_X$ with respect to a filtration on X which is canonically related to f, we would have a geometric and functorial description of the Leray filtration. It turns out that this is not possible in general. To remedy this, we replace X by a quasi-projective variety X' which is the fibre product of X and a suitable affine variety Y' mapping to Y via the so-called **Jouanolou-trick**:

Theorem ([Jo, 1.5]). *Let X be a quasi-projective variety. There exists an* ***affinement for*** *X, i.e. an* affine *variety V and a morphism $h : V \to X$ whose fibres are isomorphic to the same complex affine space.*

Affinement behaves well with respect to fibre products: if $f : X \to Y$ is a morphism between quasi-projective varieties, $h : Y' \to Y$ an affinement of Y and $X' = X \times_Y Y'$ the fibre product, then the induced morphism $f' : X' \to Y'$ has the same fibres as f and is homotopy equivalent to $f : X \to Y$; if moreover V is an affinement of X' the canonical morphism $V \to Y'$ is homotopy equivalent to f.

The first assertion guarantees that the terms of the Leray spectral sequence for f and for f' are the same so that we may indeed replace f by its affinement f'. The second assertion can be paraphrased by saying that *any morphism between quasi-projective varieties is homotopic to a morphism between their affinements.* It is important for checking functoriality.

By the previous argument we may thus assume that Y is affine. Of course, constant sheaves are finitely constructible with respect to *any* stratification, and to capture the terms of the Leray-filtration, the main idea is to take a stratification of X which is the pull back of a stratification on Y which is cellular with respect to all of the direct images $R^q f_* \underline{\mathbb{Z}}_X$:

Definition 6.3. Let $\mathcal{F}$ be a finitely constructible sheaf adapted to a stratification $\{X_\alpha\}$. Put $X^o_\alpha = X_\alpha - X_{\alpha-1}$, $j_\alpha : X^o_\alpha \hookrightarrow X_\alpha$ and $\mathcal{F}_\alpha = (j_\alpha)_!(F|X^o_\alpha)$. We say that the stratification is **cellular** with respect to $\mathcal{F}$ if $H^q(X_\alpha,\mathcal{F}_\alpha) = 0$ unless $q = \alpha = \dim X_\alpha$.

This last property is crucial to show that the skeletal spectral sequence for the so constructed stratification is directly related to the spectral sequence for canonical filtration on the hyperdirect image $Rf_*\underline{\mathbb{Z}}_X$. Since this last spectral sequence is the Leray spectral sequence the result follows. The proof of this step is rather straightforward (see [Ara, Cor. 3.10 and Lemma 3.13]).

It thus remains to show that such cellular stratifications exist. This turns out to be a consequence of a central and non-trivial vanishing result due to Nori [Nori] who attributes it to Beilinson. See [Ara, Lemmas 3.4– 3.7].

Summarizing we then have:

Theorem 6.4. *Let $f : X \to Y$ be a morphism of quasi-projective varieties and let $h : Y' \to Y$ be an affinement. Let $f' : X' \to Y'$ be the fibre product, inducing $g : X' \to X$. Let $\mathcal{F}$ be a finitely constructible sheaf on X and let $\mathcal{F}' = f'^{-1}\mathcal{F}$. Then there exists a finite stratification of Y' by closed subvarieties Y'_k such that putting $X'_k = f^{-1}Y'_k$, the Leray spectral sequence for f and $\mathcal{F}$ is isomorphic to the spectral sequence for the bigraded exact couple (D, E) with $D = D_1 = H^{p+q}(X', X'_{p-1}; \mathcal{F}')$ and $E = E_1 = H^{p+q}(X'_p, X'_{p-1}; \mathcal{F}')$.*

This result explains the geometric nature of the Leray spectral sequence and implies the main result in Hodge theory we are after:

Theorem 6.5. *Let $f : X \to Y$ be a morphism between quasi-projective algebraic varieties. Then the Leray-spectral sequence $E_r(f) := E_r(f, \mathbb{Z}_X)$ for the constant sheaf $\mathbb{Z}_X$ is a spectral sequence of mixed Hodge structures and, in particular, the Leray filtration $L^\bullet[H^k(Y)]$ on $H^k(Y)$ is a functorial filtration of mixed Hodge structures. Functoriality means that given a morphism $h : Y' \to Y$, letting $f' : X' \to Y'$ be the fibre product inducing $g : X' \to X$, then, for all $r \geq 2$ the induced homomorphisms $h^* : E_r^{p,q}(f') \to E_r^{p,q}(f)$ are morphisms of mixed Hodge structures. In particular, the induced homomorphism $h^* : H^k(Y') \to H^k(Y)$ restricts to a morphism $L^s[H^k(Y')] \to L^s[H^k(Y')]$, $s \geq 0$ of mixed Hodge substructures.*

Proof (Sketch). As explained above, by Theorem 6.4 we may assume that Y is itself affine and admits a finite stratification by closed subvarieties Y_k such that the Leray spectral sequence is isomorphic to the spectral sequence for the bigraded exact couple

$$\left(H^{p+q}(X, X_{p-1}), H^{p+q}(X_p, X_{p+1})\right), \quad X_k = f^{-1}Y_k.$$

The maps in this couple come from the inclusions for triples $(X, A = X_p, B = X_{p-1})$. But the maps in the long exact sequence in cohomology for the pairs (X, A) and (A, B) are all morphisms of mixed Hodge structures. This proves the first assertion.

The functoriality of Arapura's construction is based on the functoriality of the skeletal filtration and the fact, pointed out just before the statement of the theorem, that a morphism between quasi-projective varieties is homotopic to a morphism between their affinements. Details are left to the reader. □

Remark. Functoriality holds in a more general form for commutative squares which are not necessarily fibre products. The following argument has been communicated to us by Alexei Gorinov.

Since functoriality holds for pullbacks, it remains to consider the case of two morphisms with the same base, i.e., suppose we have $f_i : X_i \to Y$, $i = 1, 2$ and $f : X_1 \to X_2$ such that $f_1 = f_2 \circ f$. By pulling everything back to an affinement of Y, we can assume that Y is affine. Now consider a stratification of Y that is cellular with respect to all direct images of the constant sheaf $\underline{\mathbb{Z}}_{X_i}$ under f_i, $i = 1, 2$. The morphism between the corresponding exact couples induces a morphism between the associated skeletal spectral sequences and hence respects the mixed Hodge structures by functoriality. Then apply Theorem 6.4.

For pairs we have:

Theorem 6.6 ([Ara, Thm. 4.1]). *Let $f : U \to V$ a morphism of quasi-projective varieties and $T \subset V$ a closed subvariety. Set $Z = f^{-1}(T)$ and let $j : U - Z \hookrightarrow U$, $j' : V - T \hookrightarrow V$ be the inclusions. The isomorphisms*

$$j'_! R^q f_* \mathbb{Z} \simeq R^q f(j_! \mathbb{Z})$$

induce a ***Leray spectral sequence for pairs*** *$E_r(f, T)$ converging to the relative cohomology of the pair (V, T). This is a spectral sequence of mixed Hodge structures in a functorial fashion.*

For cohomology with compact support, we take good compactifications X of U and Y of V respectively and a morphism $g : X \to Y$ extending f. Applying the theorem to the map of pairs $(X, X - U) \to (Y, Y - V)$ we deduce:

Corollary 6.7. *Let $f : U \to V$ a morphism of quasi-projective varieties. Then the Leray spectral sequence for cohomology with compact support $E_{r,c}(f)$ converging to cohomology of Y with compact support is a spectral sequence of mixed Hodge structures in a functorial fashion.*

We end by stating a non-trivial consequence of the proof of the preceding theorems

Theorem 6.8 ([Ara, Thm. 4.5]). *If $f : X \to Y$ is a* projective *morphism between quasi-projective varieties, then $H^i_c(Y, R^j f_* \mathbb{Z})$ has weights $\leq i + j$. In fact, the Hodge numbers $h^{p,q}[H^i_c(Y, R^j f_* \mathbb{Z})]$ are non-zero only for $p \geq i + j - \dim X$, $q \geq i + j - \dim X$ and $p + q \leq i + j$.*

6.2 Deleted Neighbourhoods of Algebraic Sets

6.2.1 Mixed Hodge Complexes

Let X be a complex algebraic variety and $Z \subset X$ a closed compact algebraic subset which contains the singular locus of X. An **algebraic neighbourhood** of Z in X is defined as $\alpha^{-1}([0, \delta])$ where $\delta > 0$ is sufficiently small and α is a **rug function**, i.e. a proper non-negative real algebraic function on a neighbourhood of Z in X with $\alpha^{-1}(0) = Z$. See [Du83]. If X is embedded in projective space and Z is smooth we can use the Fubini-Study metric to define α as the square of the distance function to Z; then an algebraic neighbourhood is obtained by intersecting a sufficiently small tubular neighbourhood of Z in projective space with X. A **deleted neighbourhood of Z in X** is defined as the complement of Z in an algebraic neighbourhood T of it in X. The boundary of an algebraic neighbourhood of Z in X is called its **link** in X. It is homotopy equivalent to a deleted neighbourhood.

Let $T \subset X$ be an algebraic neighbourhood of Z in X. We want to put a mixed Hodge structure on the cohomology of $T^* = T - Z$. We follow [Du83b]. Observe that deleted neighbourhoods behave well under proper modifications with discriminant contained in Z: for a discriminant square

$$\begin{array}{ccc} f^{-1}(D) & \xrightarrow{j} & \tilde{X} \\ \downarrow{\scriptstyle g} & & \downarrow{\scriptstyle f} \\ D & \xrightarrow{i} & X \end{array}$$

with $D \subset Z$ one has a one-to-one correspondence between deleted neighbourhoods of Z in X and of $f^{-1}(Z)$ in $\tilde{X}$. There always exists such a proper modification such that $\tilde{X}$ is smooth and $f^{-1}(Z) = E$ a divisor with simple normal crossings on $\tilde{X}$. Any two of these are dominated by a third one, so to obtain a well-defined mixed Hodge structure on $H^*(T^*)$ it suffices to deal with the case that $Z \subset X$ is a divisor with strict normal crossings and to show that there is a pull-back morphism in the case of one normal crossing situation dominating another. We will carry out the former, and leave the latter as an exercise to the reader.

So suppose that $D \subset X$ is a compact divisor with simple normal crossings, that T is an algebraic neighbourhood of D and $T^* = T - D$. Let $j : T^* \to T$ and $i : D \to T$ denote the inclusion maps. Then

$$H^*(T^*) \simeq \mathbb{H}^*(T, Rj_*\underline{\mathbb{Z}}_{T^*}) \simeq \mathbb{H}^*(D, i^*Rj_*\underline{\mathbb{Z}}_{T^*}).$$

The first isomorphism is a special case of (B–21), and the second one holds because D has a fundamental system of neighbourhoods all homotopy equivalent to T.

Let us further note that we have a resolution

$$0 \to \underline{\mathbb{Q}}_D \to (a_1)_*\underline{\mathbb{Q}}_{D(1)} \to (a_2)_*\underline{\mathbb{Q}}_{D(2)} \to \cdots$$

coming from the standard cubical hyperresolution $\epsilon : D_\bullet \to D$ of D where as before $D_\bullet$ is shorthand for the cubical variety $\{D_I\}$.

We use these remarks to construct a mixed Hodge complex of sheaves:

Theorem 6.9. *In the above setting, a mixed Hodge complex of sheaves $\mathcal{H}dg^\bullet(T^*)_{\mathbb{Z}}$ on T^* can be defined by setting*

– $\mathcal{H}dg^\bullet(T^*_{\mathbb{Z}}) := i^*Rj_*\underline{\mathbb{Z}}_{T^*}$;

- $\mathcal{H}dg^\bullet(T^*_{\mathbb{Q}}) := \big(i^* Rj_* \underline{\mathbb{Q}}_{T^*} \oplus \epsilon_* \underline{\mathbb{Q}}_{D_\bullet}\big)/\underline{\mathbb{Q}}_D$. *The first summand has only non negative weights and the second only non-positive ones; the embedding of* $\underline{\mathbb{Q}}_D$ *is given by* $(\alpha, -\beta)$ *with* α *the identification* $\tau_{\leq 0} i^* Rj_* \underline{\mathbb{Q}}_{T^*} = \underline{\mathbb{Q}}_D$ *and* β *the inclusion* $\underline{\mathbb{Q}}_D \to a_* \underline{\mathbb{Q}}_{D(1)}$.
- $\mathcal{H}dg^\bullet(T^*_{\mathbb{C}}) = \big(\Omega^\bullet_X(\log D) \oplus a_* \Omega^\bullet_{D_\bullet}\big) / \Omega^\bullet_X$ *on which the filtrations* W *and* F *are defined in the obvious way.*

The exact sequence

$$\mathcal{H}dg^\bullet(D) = W_0 \mathcal{H}dg^\bullet(^*) \to \mathcal{H}dg^\bullet(T^*) \to \mathcal{H}dg^\bullet(T^*)/W_0$$

gives rise to the long exact sequence of cohomology

$$\cdots \to H^k(D) \to H^k(T^*) \to H^{k+1}_D(T) \to H^{k+1}(D) \to \cdots \qquad \text{(VI–1)}$$

Proof. The proof is straightforward. We give some hints and remarks only. First of all, one might be tempted to take for the $\mathbb{Q}$-component the complex $i^* Rj_* \underline{\mathbb{Q}}_{T^*}$ together with its canonical filtration τ, but this does not work, as $\tau_{\leq 0} i^* Rj_* \underline{\mathbb{Q}}_{T^*} = \underline{\mathbb{Q}}_D$ does not give rise to a pure Hodge structure, unless D is itself smooth.

With the proposed modification, we obtain a mixed Hodge complex of sheaves essentially because

$$\mathrm{Gr}^m_W \mathcal{H}dg^\bullet(T^*_{\mathbb{Q}}) = \begin{cases} R^m j_* \underline{\mathbb{Q}}_{T^*}[-m] \simeq (a_m)_* \underline{\mathbb{Q}}_{D(m)}(-m)[-m] & \text{if } m > 0; \\ (a_{1-m})_* \underline{\mathbb{Q}}_{D(1-m)}[m] & \text{if } m \leq 0. \end{cases} \qquad \square$$

Remark 6.10. The mixed Hodge complex of sheaves $\mathcal{H}dg^\bullet(T^*)$ depends only on the first infinitesimal neighbourhood of D in X. In fact, the inclusion $D \subset X$ determines a logarithmic structure on D (called a **logarithmic embedding** of D) and all data of $\mathcal{H}dg^\bullet(T^*)$ can be constructed from this logarithmic structure. See [Ste95].

6.2.2 Products and Deleted Neighbourhoods

In this section we show first that the mixed Hodge structure on the deleted neighbourhood of an algebraic subvariety behaves well with respect to cup product: the mappings

$$H^k(T^*) \otimes H^\ell(T^*) \xrightarrow{\cup} H^{k+\ell}(T^*)$$

are morphisms of mixed Hodge structures. Here T is an algebraic neighbourhood of a divisor D with strict normal crossings in a smooth variety X, and D is compact. The mixed Hodge complex of sheaves $\mathcal{H}dg^\bullet(T^*)$ appears not to be suitable to define a cup product. We have to replace it by a sheaf complex which has also a multiplicative structure.

We first have the following

Lemma 6.11. *Let C be an irreducible component of $D(m)$ for some m. Then C is a smooth subvariety of X. Let $\mathcal{I}_C \subset \mathcal{O}_X$ denote its ideal sheaf. Then $\mathcal{I}_C \Omega_X^\bullet(\log D)$ is a subcomplex of $\Omega_X^\bullet(\log D)$.*

Proof. Let $P \in C$. Choose local holomorphic coordinates $(z_1, \ldots, z_n)$ on X centred at P such that $\mathcal{I}_{C,P} = (z_1, \ldots, z_k)\mathcal{O}_{X,P}$ and $\mathcal{I}_{D,P} = (z_1 \cdots z_l)\mathcal{O}_{X,P}$ for some $k \leq l \leq n$. For $\omega \in \mathcal{I}_C \Omega_X^p(\log D)_P$ write $\omega = \sum_{i=1}^m z_i \omega_i$ with $\omega_i \in \Omega_X^p(\log D)_P$ for $i = 1, \ldots, k$. Then

$$d\omega = \sum_{i=1}^m z_i(\frac{dz_i}{z_i} \wedge \omega_i + d\omega_i) \in \mathcal{I}_C \Omega_X^{p+1}(\log D)_P . \qquad \square$$

We denote the quotient complex $\Omega_X^\bullet(\log D)/\mathcal{I}_C \Omega_X^\bullet(\log D)$ by $\Omega_X^\bullet(\log D) \otimes \mathcal{O}_C$. We equip it with the filtrations W and F as a quotient of $\Omega_X^\bullet(\log D)$.

Theorem 6.12. *Let $i_C : C \to X$ and $j : X - D \to X$ be the inclusion maps. Then the complex $\Omega_X^\bullet(\log D) \otimes \mathcal{O}_C$ is quasi-isomorphic to $i_C^* Rj_* \underline{\mathbb{C}}_{X-D}$.*

Proof. We have an isomorphism $(\Omega_X^\bullet(\log D), W) \simeq (Rj_* \underline{\mathbb{C}}_{X-D}, \tau_{\leq})$ in the filtered derived category of bounded below complexes of sheaves of $\mathbb{C}$- vector spaces on X, which by restriction to C gives an isomorphism

$$(i_C^* \Omega_X^\bullet(\log D), W) \simeq (i_C^* Rj_* \underline{\mathbb{C}}_{X-D}, \tau_{\leq})$$

in the filtered derived category of bounded below complexes of sheaves of $\mathbb{C}$-vector spaces on C. It remains to be proven that the quotient map

$$(i_C^* \Omega_X^\bullet(\log D), W) \to (\Omega_X^\bullet(\log D) \otimes \mathcal{O}_C, W)$$

is a filtered quasi-isomorphism. To deal with this problem, note that for all $k \geq 0$ the Poincaré residue map

$$\mathrm{res}_k : \mathrm{Gr}_k^W \Omega_X^\bullet(\log D) \to (a_{k-1})_* \Omega_{D(k)}^\bullet[-k]$$

is an isomorphism of complexes (here $D(0) := X$). It has components

$$\mathrm{res}_I : \mathrm{Gr}_k^W \Omega_X^\bullet(\log D) \to \Omega_{D_I}^\bullet[-k]$$

where I is a subset of A of cardinality k and $D_I := \bigcap_{a \in I} D_a$. It follows that

$$i_C^* \mathrm{Gr}_k^W \Omega_X^\bullet(\log D) \simeq \bigoplus_{|I|=k} i_C^* \underline{\mathbb{C}}_{D_I}[-k] \simeq \bigoplus_{|I|=k} \underline{\mathbb{C}}_{D_I \cap C}[-k] .$$

Claim: the image of $\mathcal{I}_C \Omega_X^p(\log D) \cap W_k \Omega_X^p(\log D)$ under the map res_I coincides with $\mathcal{I} \Omega_{D_I}^{p-k} + d\mathcal{I}_C \wedge \Omega_{D_I}^{p-k-1}$.

Assuming the claim, we find that res_k induces an isomorphism

$$\mathrm{Gr}_k^W \Omega_X^\bullet(\log D) \otimes \mathcal{O}_C \simeq \bigoplus_{|I|=k} \Omega_{D_I \cap C}^\bullet[-k] .$$

Let us prove the claim. The map R_I presupposes an ordering of the set A of irreducible components of D. Write $I = \{i_1, \ldots, i_k\}$ with $i_1 < \ldots < i_k$ and choose local coordinates $(z_1, \ldots, z_n)$ on X centred at $P \in C$ such that D_{i_r} is defined near P by $z_r = 0$ for $r = 1, \ldots, k$ and $\mathcal{I}_{C,P}$ is generated by z_j for $j \in J$. Put $J_1 = J \cap \{1, \ldots, k\}$ and $J_2 = J - J_1$. Also suppose that D is defined near P by $z_1 \cdots z_l = 0$. Then $l \geq k$ and $J \subset \{1, \ldots, l\}$.

For $j \in J_2$ choose $\eta_i \in \Omega^{p-k-1}_{D_I,P}$ and $\zeta_i \in \Omega^{p-k}_{D_I,P}$ with lifts $\tilde{\eta}_i$ and $\tilde{\zeta}_i$ in $\Omega^{p-k-1}_{X,P}$ and $\Omega^{p-k}_{X,P}$ respectively. Let

$$\omega = \sum_{j \in J_2} \frac{dz_1}{z_1} \wedge \cdots \wedge \frac{dz_k}{z_k} \wedge (dz_j \wedge \tilde{\eta}_i + z_j \tilde{\zeta}_j) .$$

Then $\omega \in \mathcal{I}_C \Omega^p_X(\log D)_P \cap W_k \Omega^p_X(\log D)_P$ and

$$\mathrm{res}_I(\omega) = \sum_{j \in J_2} (dz_j \wedge \eta_j + z_j \zeta_j) .$$

Also remark that $\mathrm{res}_{I'}(\omega) = 0$ if $I \neq I' \subset A$ with $|I|' = k$. Hence we have an inclusion

$$\bigoplus_I (\mathcal{I}\Omega^{p-k}_{D_I,P} + d\mathcal{I}_C \wedge \Omega^{p-k-1}_{D_I,P}) \subset \mathrm{res}_k(\mathcal{I}_C \Omega^p_X(\log D)_P \cap W_k \Omega^p_X(\log D)_P) .$$

To prove the reverse inclusion, we let $\xi_i = \dfrac{dz_i}{z_i}$ if $1 \leq i \leq l$ and $\xi_i = dz_i$ if $i > l$. Also, for $B = \{b_1, \ldots, b_r\} \subset \{1, \ldots, n\}$ with $b_1 < \cdots < b_r$ we put $\xi_B = \xi_{b_1} \wedge \cdots \wedge \xi_{b_r}$. With this notation, $\Omega^p_X(\log D)_P$ is the free $\mathcal{O}_{X,P}$-module with basis the ξ_B with $|B| = p$. We have

$$\mathcal{I}_C \Omega^p_X(\log D)_P = \bigoplus_{|B|=p} \mathcal{I}_{C,P} \xi_B$$

and

$$W_k \Omega^p_X(\log D)_P = \bigoplus_{|B|=p} W_k \Omega^p_X(\log D)_P \cap \mathcal{O}_{X,P} \xi_B = \bigoplus_{|B|=p} J(B,k) \xi_B$$

where $J(B,k)$ is an ideal of $\mathcal{O}_{X,P}$ generated by square-free monomials.

For any B and any square-free monomial $z_E \in J(B,k)$ with $\mathrm{res}_I(z_E \xi_B) \neq 0$ one has $\{1, \ldots, k\} \subset B$ and $B \cap \{k+1, \ldots, l\} \subset E$. If moreover $z_E \in \mathcal{I}_{C,P}$ then $J_2 \cap E \neq \varnothing$. Choose $j \in J_2 \cap E$. If $j \in B$ then

$$z_E \xi_B = \pm z_E \frac{dz_j}{z_j} \wedge \xi_{B-\{j\}} = \pm dz_j \wedge z_{E-\{j\}} \xi_{B-\{j\}} \in d\mathcal{I}_C \wedge \Omega^{p-1}_X(\log D)_P$$

so $\mathrm{res}_I(z_E \xi_B) \in d\mathcal{I}_C \wedge \Omega^{p-k-1}_{D_I}$. On the other hand, if $j \notin B$ then

$$\mathrm{res}_I(z_E \xi_B) = z_j \mathrm{res}_I(z_{E-\{j\}} \xi_B) \in \mathcal{I}_C \Omega^{p-k}_{D_I} . \qquad \square$$

Corollary 6.13. $\mathbb{H}^k(C, \Omega_X^\bullet(\log D) \otimes \mathcal{O}_C) \simeq H^k(T_C - D; \mathbb{C})$ *where T_C is a tubular neighbourhood of C inside X.*

Corollary 6.14. *One has a cohomological mixed Hodge complex $\mathcal{H}dg^\bullet(C \log D)$ on C with*

$$(\mathcal{H}dg^\bullet(C \log D)_{\mathbb{Q}}, W) = (i_C^* Rj_* \underline{\mathbb{Q}}_{X-D}, \tau_{\leq})$$

and

$$(\mathcal{H}dg^\bullet(C \log D)_{\mathbb{C}}, W, F) = (\Omega_X^\bullet(\log D) \otimes \mathcal{O}_C, W, F) \ .$$

This defines a mixed Hodge structure on $H^k(U_C - D; \mathbb{C})$. Moreover, we have $W_0 \mathcal{H}dg^\bullet(\log D) \simeq \mathcal{H}dg^\bullet(C)$ so

$$W_k H^k(U_C - D) = \mathrm{Im}[H^k(C) \simeq H^k(U_C) \to H^k(U_C - D)] \ .$$

The data of all $\mathcal{H}dg^\bullet(D_I \log D)$ for $I \subset A$ give rise to a cohomological mixed Hodge complex on the cubical variety $D_\bullet$. We define

$$\mathcal{H}dg^\bullet(D \log D) = R\epsilon_* \mathcal{H}dg^\bullet(D_\bullet \log D) \ .$$

This is a cohomological mixed Hodge complex on D such that $\mathcal{H}dg^\bullet(D \log D)_{\mathbb{Q}} \simeq i_D^* Rj_* \underline{\mathbb{Q}}_{X-D}$. It gives a mixed Hodge structure on $H^k(T^*)$ where T is a tubular neighbourhood of D. The spectral sequence

$$E_1^{pq} = \mathbb{H}^q(D_p, \mathcal{H}dg^\bullet(D_p \log D)) \Rightarrow \mathbb{H}^{p+q}(D, \mathcal{H}dg^\bullet(D \log D))$$

can be considered as the Mayer-Vietoris spectral sequence corresponding to a covering of T^* by deleted neighbourhoods $T_{D_i} - D$.

Observe that we dispose of natural morphisms of cohomological mixed Hodge complexes $\mathcal{H}dg^\bullet(X \log D) \to \mathcal{H}dg^\bullet(D \log D)$ (which on cohomology induces the restriction mapping $H^k(X - D) \to H^k(T^*)$) and $\mathcal{H}dg^\bullet(D_\bullet) \to \mathcal{H}dg^\bullet(D \log D)$ (which on cohomology induces the restriction mapping $H^k(D) \simeq H^k(T) \to H^k(T^*)$). These morphisms induce a morphism

$$\mathcal{H}dg^\bullet(T^*) \to \mathcal{H}dg^\bullet(D \log D)$$

which is a weak equivalence.

There is a natural product

$$\Omega^\bullet(D \log D) \otimes \Omega^\bullet(D \log D) \to \Omega^\bullet(D \log D)$$

constructed as follows. Observe that

$$\Omega^\bullet(D \log D) \simeq \Omega_X^\bullet(\log D) \otimes_{\mathcal{O}_X} \epsilon_* \mathcal{O}_{D_\bullet}$$

where the first factor in the tensor product is a sheaf of differential graded algebras. Let us now also define the structure of a sheaf of differential graded algebras on $\epsilon_* \mathcal{O}_{D_\bullet}$. Recall that the cubical variety $D_\bullet$ depends on a chosen ordering $D_1, \ldots, D_N$ of the irreducible components of D which we fix from now on. The component D_I of the semi-simplicial set given by the ordered set $I = (i_0, \ldots, i_k)$, $i_k \in [1, \ldots, N]$ then comes with a sign $\epsilon(I)$, the sign of the permutation of I needed to put the elements of I in increasing order. The sheaf $\mathcal{O}_{D_I}$ as an $\mathcal{O}_D$-module is generated by the function e_I, the characteristic function of D_I multiplied by $\epsilon(I)$. To define a product structure its is sufficient to say what the product of e_I and e_J is. If I and J have one element in common, we may assume that this is the first element. We set

$$e_I * e_J = \begin{cases} 0 & \text{if } |I \cap J| \neq 1 \\ e_K,\ K = (i_0, I', J') & \text{if } I \cap J = i_0, I = (i_0, I'), J = (i_0, J'). \end{cases}$$

The multiplications on $\Omega^\bullet_X(\log D)$ and $\epsilon_* \mathcal{O}_{D_\bullet}$ then are both $\mathcal{O}_X$-linear and graded commutative, so they induce a graded-commutative multiplication on $\Omega^\bullet_D(\log D)$. It is left to the reader to verify that together with the differential this defines a sheaf of differential graded algebras. This multiplication is compatible with the multiplication on $i^* Rj_* \underline{\mathbb{Q}}_{T^*}$ which in turn induces the cup product on $H^*(T^*)$.

6.2.3 Semi-purity of the Link

The following is a weak version of the **semi-purity of the link**:

Theorem 6.15. *Let X be an algebraic variety of dimension n and let $Z \subset X$ be a compact subvariety of dimension s such that $X - Z$ is nonsingular. Let T^* be a deleted neighbourhood of Z in X. Then the mixed Hodge structure $H^q(T^*)$ has weights $\leq q+1$ if $q < n-s$ and weights $> q-1$ if $q \geq n+s$.*

Proof. Let $\pi : \tilde{X} \to X$ be a resolution of X such that $D = \pi^{-1}(Z)$ is a divisor with strict normal crossings on $\tilde{X}$. Without loss of generality we may assume that X is compact. Then $\tilde{X}$ is also compact. As the diagram

$$\begin{array}{ccc} D & \rightarrow & \tilde{X} \\ \downarrow & & \downarrow \pi \\ Z & \rightarrow & X \end{array}$$

is of cohomological descent, we have the long exact sequence

$$\cdots \to H^k(X) \to H^k(D) \oplus H^k(Z) \to H^k(\tilde{X}) \to H^{k+1}(X) \to \cdots .$$

Combine this with the long exact sequence (VI–1)

$$\cdots \to H^k(D) \to H^k(T^*) \to H^{k+1}_D(T) \to H^{k+1}(D) \to \cdots .$$

If $k < n-s$ then $H^{2n-k-1}(D)$ is pure of weight $2n-k-1$, so by duality $H^{k+1}_D(T)$ is pure of weight $k+1$. As $H^k(D)$ has weights $\leq k$ we find that $H^k(T^*)$ has weights $\leq k+1$. Hence its dual space

$$H^{2n-1-k}(T^*) \simeq \mathrm{Hom}(H^k(T^*); \mathbb{Q}(-n))$$

has weights $\geq 2n-k-1$. $\square$

Remark 6.16. In fact, we expect that stronger inequalities for the weights are valid:

$H^q(T^*)$ has a mixed Hodge structure with weights $\leq q$ if $q < n - s$ and with weights $> q$ if $q \geq n + s$.

This would follow if one were able to prove that

$$H^q_D(\tilde{X}) \hookrightarrow H^q(D) \text{ for } q \leq n - s$$

Remark 6.17. In the same notation as above, we also have the exact sequence of mixed Hodge structures

$$\cdots \to H^k(Z) \to H^k(T^*) \to H^{k+1}_Z(X) \to H^{k+1}(Z) \to \cdots . \qquad \text{(VI–2)}$$

This can be seen as follows: put

$$A := R\pi_*\mathcal{H}dg^\bullet(\tilde{X}) \qquad B := \mathcal{H}dg^\bullet(Z)$$
$$C := R\pi_*\mathcal{H}dg^\bullet(D_\bullet)) \quad E := \mathcal{H}dg^\bullet(\tilde{X}\log D)$$

We have a natural morphism

$$\mu : A \oplus B \to C \oplus E$$

such that $H^i(\mathrm{Cone}^\bullet(\mu)) \simeq H^{i+1}_Z(X)$. It restricts to

$$\lambda : A \oplus B \to C, \ \mu' : A \to C \oplus E$$

and $H^i(\mathrm{Cone}^\bullet(\lambda)) \simeq H^{i+1}(X)$, while $H^i(\mathrm{Cone}^\bullet(\mu')) \simeq H^i(T^*)$. The exact sequence (VI–1) results from the exact sequence

$$0 \to \mathrm{Cone}^\bullet(A \to D) \to \mathrm{Cone}^\bullet(A \oplus B \to C \oplus D) \to \mathrm{Cone}^\bullet(B \to C) \to 0$$

whereas the sequence (VI–2) results from

$$0 \to \mathrm{Cone}^\bullet(A \to C \oplus D) \to \mathrm{Cone}^\bullet(A \oplus B \to C \oplus D) \to B \to 0.$$

Remark 6.18. If $Z \subset X$ is a compact algebraic subset with neighbourhood T and $X - Z$ is smooth, then $M := \partial T$ is a compact oriented manifold, homotopy equivalent to a punctured neighbourhood U. Durfee and Hain [Du-H] have shown that the cup product $H^i(M) \otimes H^j(M) \to H^{i+j}(M)$ is a morphism of mixed Hodge structures. In the case of the link of an isolated singularity (X, x) of dimension n this implies that this cup product is the zero map if $i, j < n$ but $i + j \geq n$. Indeed, by strong semi-purity in that case the source of the cup product has weights at most $i + j$ whereas the target has weights at least $i + j + 1$.

This phenomenon has been used by McCrory [MC] and independently by Durfee, Steenbrink and Stevens, to give a description of the weight filtration of the link of a normal surface singularity in terms of Massey triple products.

Since the cup product $\alpha \cup \beta$ is zero for all $\alpha, \beta \in H^1(M;\mathbb{Q})$, the **Massey triple product** $\langle \alpha, \beta, \gamma \rangle$ can be defined for triples $\alpha, \beta, \gamma \in H^1(M;\mathbb{Q})$ as follows. Select 1-cochains f and g such that $df = \alpha \cup \beta$ and $dg = \beta \cup \gamma$. Then $\langle \alpha, \beta, \gamma \rangle$ is represented by $f \wedge \gamma + \alpha \wedge g$. See also the discussion in §9.4. The result is, that

$$W_0 H^1(M) = \{\alpha \in H^1(M) \mid \langle \alpha, \beta, \gamma \rangle = 0 \quad \text{for all } \beta, \gamma \in H^1(M)\}.$$

For links of isolated singularities in higher dimensions the weight filtration is not a topological invariant in general, as certain examples show (see [Ste-St]).

6.3 Cup and Cap Products, and Duality

6.3.1 Duality for Cohomology with Compact Supports

Next we show that for an algebraic variety U the cup product maps

$$H^i(U) \otimes H^j_c(U) \to H^{i+j}_c(U)$$

are morphisms of mixed Hodge structures. We deduce from this a duality result due to Fujiki [Fuj]. See Corollary 6.28.

Remark that, since $H^k_c(U) \to H^k(U)$ is a morphism of mixed Hodge structures, also the cup product on $H(U)$ automatically becomes a morphism of mixed Hodge structures. We knew this already (Corollary 5.45).

First assume that $U = X$, a compact algebraic variety and make use of the fact that cup product preserves mixed Hodge structures. Let us put a mixed Hodge structure on $H_j(X)$ using the transpose of the Kronecker homomorphism

$$H_j(X;\mathbb{Q}) \xrightarrow{\sim} \mathrm{Hom}_{\mathbb{Q}}(H^j(X;\mathbb{Q});\mathbb{Q}).$$

We then conclude:

Proposition 6.19. *For a* compact *variety* X

1) *the cap products*

$$H^i(X) \otimes H_j(X) \to H_{j-i}(X)$$

are morphisms of mixed Hodge structures;

2) *the Poincaré duality homomorphisms*

$$H^i(X)(n) \to H_{2n-i}(X), \quad n = \dim_{\mathbb{C}} X$$

are morphisms of mixed Hodge structures.

Now replace X by any complex algebraic variety U. There are cup products

$$H^i(U) \otimes H^j_c(U) \to H^{i+j}_c(U)$$

(see § B.1.2) and we want to show that these respect mixed Hodge structures.

We first suppose that U is *smooth* and choose a smooth compactification X of U such that $X - U = D$ is a divisor with simple normal crossings on X. We define

$$\mathcal{H}dg^\bullet(X, D) = \mathrm{Cone}^\bullet(\mathcal{H}dg^\bullet(X) \to \mathcal{H}dg^\bullet(D_\bullet)[-1]$$

where $D_\bullet$ is shorthand for the cubical variety $\{D_I\}$ associated to D. Its $\mathbb{C}$-component is denoted by

$$\Omega^\bullet_{X,D} = \mathrm{Cone}^\bullet(\Omega^\bullet_X \to a_* \Omega^\bullet_{D_\bullet}).$$

The above cup product now takes the shape

$$\mathbb{H}^i(X, \mathcal{H}dg^\bullet(X \log D)) \otimes \mathbb{H}^j(X, \mathcal{H}dg^\bullet((X, D)) \to \mathbb{H}^{i+j}(X, \mathcal{H}dg^\bullet(X, D))$$

which one would like to come from a morphism of sheaf complexes

$$\mathcal{H}dg^\bullet(X \log D) \otimes \mathcal{H}dg^\bullet(X, D) \to \mathcal{H}dg^\bullet(X, D) \ .$$

This appears not to be possible. However, we will construct a mixed Hodge complex of sheaves $\widetilde{\mathcal{H}dg}^\bullet(X, D)$ on X together with a quasi-isomorphism of mixed Hodge complexes of sheaves $\mathcal{H}dg^\bullet(X, D) \to \widetilde{\mathcal{H}dg}^\bullet(X, D)$ and a morphism

$$\mathcal{H}dg^\bullet(X \log D) \otimes \mathcal{H}dg^\bullet(X, D) \to \mathcal{H}dg^\bullet(X, D)$$

which realizes the cup product on cohomology. This will prove that the cup product under consideration is a morphism of mixed Hodge structures.

Set

$$\widetilde{\mathcal{H}dg}^\bullet(X, D) = \mathrm{Cone}^\bullet(\mathcal{H}dg^\bullet(X \log D) \to \mathcal{H}dg^\bullet(D \log D))[-1] \ .$$

Note that the inclusions $\mathcal{H}dg^\bullet(X) \to \mathcal{H}dg^\bullet(X \log D)$ and $\mathcal{H}dg^\bullet(D) \to \mathcal{H}dg^\bullet(D \log D)$ induce a morphism of cohomological mixed Hodge complexes

$$\beta : \mathcal{H}dg^\bullet(X, D) \to \widetilde{\mathcal{H}dg}^\bullet(X, D) \ .$$

Lemma 6.20. *The map induced by β on cohomology is an isomorphism.*

Proof. By excision, the map $H^k(X, D) \simeq H^k(X, U_D) \to H^k(X - D, U_D - D)$ is an isomorphism for all k. □

Remark 6.21. As this lemma is true also locally on X, we may even conclude that β is a quasi-isomorphism.

Corollary 6.22. *The cohomological mixed Hodge complexes* $\mathcal{H}dg^\bullet(X,D)$ *and* $\widetilde{\mathcal{H}dg}^\bullet(X,D)$ *determine the same mixed Hodge structure on* $H^k(X,D)$.

Proof. Indeed, β induces a morphism of mixed Hodge structures which is an isomorphism of vector spaces, hence an isomorphism of mixed Hodge structures. □

Now we proceed to the definition of the cup product on the level of complexes. Write

$$\tilde{\Omega}^\bullet_{X,D} = \widetilde{\mathcal{H}dg}^\bullet(X,D)_\mathbb{C} = \mathrm{Cone}^\bullet(\Omega^\bullet_X(\log D) \to \Omega^\bullet_D(\log D))[-1].$$

For each component C of $D_\bullet$ we have a natural cup product

$$\mu_C : \Omega^\bullet_X(\log D) \otimes_\mathbb{C} \Omega^\bullet_C \to \Omega^\bullet_X(\log D) \otimes \mathcal{O}_C\ .$$

These are the components of a cup product

$$\mu : \Omega^\bullet_X(\log D) \otimes_\mathbb{C} \Omega^\bullet_{X,D} \to \tilde{\Omega}^\bullet_{X,D}$$

which is compatible with the filtrations W and F. It now follows that the cup product maps

$$H^i(U) \otimes H^j_c(U) \to H^{i+j}_c(U)$$

are morphisms of mixed Hodge structures. This terminates the case where U is smooth.

Let us now extend this to arbitrary U. So we choose a compactification $U \subset X$, set $D = X - U$ and choose an A-cubical resolution $\{(X_I, D_I)\}$ of the pair (X,D). As in the final subsection of § 5.3, let $(X \times X)_F$ be the "flag-resolution" of $X \times X$. For simplicity, suppose that A is ordered and for $I = (i_0, \ldots, i_n) \subset A$ with $i_0 < \cdots < i_n$, we put $I_k = (i_0, \ldots, i_k$ and let $F(I)$ be the flag $(I_0 \times I_0, \cdots, I_n \times I_n)$ and embed X_I diagonally in $X_I \times X_I = X_{F(I)}$. This defines a morphism of semi-simplicial schemes over the diagonal embedding

$$\begin{array}{ccc} X_\bullet & \xrightarrow{\Delta_\bullet} & (X \times X)_\bullet \\ \downarrow & & \downarrow \\ X & \xrightarrow{\Delta} & X \times X. \end{array}$$

On $X_I \times X_I$ the complex $\mathcal{H}dg^\bullet(X_I \log D_I) \boxtimes \mathcal{H}dg^\bullet(X_I, D_I)$ pulls back under the diagonal embedding to $\mathcal{H}dg^\bullet(X_I \log D_I) \otimes \mathcal{H}dg^\bullet(X_I, D_I)$. The cup product defines the map

$$\mathcal{H}dg^\bullet(X_I \log D_I) \otimes \mathcal{H}dg^\bullet(X_I, D_I) \to \widetilde{\mathcal{H}dg}^\bullet(X_I, D_I).$$

As we have just seen, this is a pairing of mixed Hodge complexes of sheaves. This remains so after taking hyperdirect images under the augmentation. The hypercohomology of the resulting complexes yields the desired cup product pairing. We have shown:

Theorem 6.23. *Let U be a complex algebraic variety. The cup product pairings*

$$H^i(U) \otimes H^j_c(U) \to H^{i+j}_c(U)$$

are morphisms of mixed Hodge structures.

Remark 6.24. The restrictions of $\mathcal{H}dg^\bullet(X \log D)$, $\mathcal{H}dg^\bullet((X,D)$ and $\widetilde{\mathcal{H}dg}^\bullet(X,D)$ to $X-D$ are all equal to $\mathcal{H}dg^\bullet(X-D)$. Now consider the following situation: Y is a compact complex algebraic variety with closed subvarieties Z and W such that $Y-(Z\cup W)$ is smooth and $Z\cap W=\varnothing$. Then there is a cup product

$$H^i(Y-Z,W)\otimes H^j(Y-W,Z)\to H^{i+j}(Y,Z\cup W)$$

which is a morphism of mixed Hodge structures and induces a perfect duality if $i+j=2\dim(Y)$. The proof uses a proper modification $f: X\to Y$ such that X is smooth, f maps $X-f^{-1}(Z\cup W)$ isomorphically to $Y-(Z\cup W)$ and $D=f^{-1}(Z)$ and $E=f^{-1}(W)$ are divisors with simple normal crossings on X. By a gluing process one obtains cohomological mixed Hodge complexes $\mathcal{H}dg^\bullet(X\log D,E)$ etc. and a cup product map

$$\mathcal{H}dg^\bullet(X\log D,E)_{\mathbb{C}}\otimes\mathcal{H}dg^\bullet(X\log E,D)_{\mathbb{C}}\to\widetilde{\mathcal{H}dg}^\bullet(X,D\cup E)_{\mathbb{C}}$$

which is compatible with W and F. This answers a question raised by V. Srinivas.

We next want to reformulate our results in terms of Borel-Moore homology.

Lemma-Definition 6.25. *Borel-Moore homology gets a mixed Hodge structure through the isomorphisms*

$$H^{\mathrm{BM}}_k(U;\mathbb{Q})\xrightarrow{\sim}\mathrm{Hom}_{\mathbb{Q}}(H^k_c(U;\mathbb{Q});\mathbb{Q})$$

induced by the adjoint of the Kronecker pairing. It has at most weights in the interval $[-k,0]$.

The last assertion follows from Proposition 5.54.

Corollary 6.26. *The cap product pairings*

$$H^\ell(U)\times H^{\mathrm{BM}}_k(U)\to H^{\mathrm{BM}}_{k-\ell}(U)$$

are morphisms of mixed Hodge structure. In particular, with $[U]$ the fundamental class of $[U]$ in Borel-Moore homology (§ B.2.9), the Poincaré homomorphism

$$H^k(U)\xrightarrow{\cap[U]}H^{\mathrm{BM}}_{2n-k}(U)$$

is pure of type $(-n,-n)$ and an isomorphism if U is smooth.

6.3.2 The Extra-Ordinary Cup Product.

Let X be a smooth compact algebraic variety and let D be a divisor with simple normal crossings on X. The local cohomology groups $H^k_D(X) = H^k(X, X - D)$ are given a mixed Hodge structure using the mixed Hodge complex of sheaves

$$\mathcal{H}dg^\bullet(X) = \mathrm{Cone}^\bullet(\mathcal{H}dg^\bullet(X) \xrightarrow{u} \mathcal{H}dg^\bullet(X \log D))[-1]$$

but by excision we may as well take

$$\widetilde{\mathcal{H}dg}^\bullet(X) = \mathrm{Cone}^\bullet(\mathcal{H}dg^\bullet(D) \xrightarrow{v} \mathcal{H}dg^\bullet(D \log D))[-1] .$$

Observe that the morphisms $u_{\mathbb{C}}$ and $v_{\mathbb{C}}$ are injective, even after taking $\mathrm{Gr}_F \mathrm{Gr}^W$, so we have bi-filtered quasi-isomorphisms

$$(\mathcal{H}dg^\bullet(X)_{\mathbb{C}}, W, F) \to (\mathrm{Coker}(u_{\mathbb{C}}), W, F)[-1]$$

and

$$(\mathcal{H}dg^\bullet(X)_{\mathbb{C}}, W, F) \to (\mathrm{Coker}(v_{\mathbb{C}}), W, F)[-1] .$$

Moreover, the natural cup-product

$$\mathcal{H}dg^\bullet(X \log D)_{\mathbb{C}} \otimes_{\mathbb{C}} \mathcal{H}dg^\bullet(D)_{\mathbb{C}} \to \mathcal{H}dg^\bullet(D \log D)_{\mathbb{C}}$$

maps $\mathcal{H}dg^\bullet((X)_{\mathbb{C}} \otimes_{\mathbb{C}} \mathcal{H}dg^\bullet(D)_{\mathbb{C}}$ to $\mathcal{H}dg^\bullet(D)_{\mathbb{C}}$, so induces a cup product map

$$\mathrm{Coker}(u_{\mathbb{C}}) \otimes_{\mathbb{C}} \mathcal{H}dg^\bullet(D)_{\mathbb{C}} \to \mathrm{Coker}(v_{\mathbb{C}})$$

which is compatible with the filtrations W and F. Hence we conclude

Theorem 6.27. *The extra-ordinary cup product map*

$$H^i_D(X) \otimes H^j(D) \to H^{i+j}_D(X)$$

is a morphism of mixed Hodge structures.

Next let us consider the situation of a compact smooth variety Y and a closed subvariety T in Y. We want to extend the previous theorem to this situation. We will in fact reduce the general case to the normal crossing case as follows. First observe that $H^i_T(Y)$ has only weights $\geq i$. Indeed, we have the exact sequence

$$H^{i-1}(Y) \to H^{i-1}(Y - T) \to H^i_T(Y) \to H^i(Y)$$

and $H^i(Y)$ is pure of weight i; moreover, the cokernel of $H^{i-1}(Y) \to H^{i-1}(Y - T)$ has weights $\geq i$. We choose an embedded resolution of T in Y, i.e. a proper birational morphism $\pi : X \to Y$ such that $\pi^{-1}(T) = D$ is a divisor with simple normal crossings on X. We obtain a diagram

$$\begin{array}{ccccccc} H^{i-1}(Y-T) & \to & H^i_T(Y) & \to & H^i(Y) & \to & H^i(Y-T) \\ \downarrow\simeq & & \downarrow\pi^* & & \downarrow\pi^* & & \downarrow\simeq \\ H^{i-1}(X-D) & \to & H^i_D(X) & \to & H^i(X) & \to & H^i(X-D) \end{array}$$

and by weight considerations the composition

$$H^i(X) \to H^i(X-D) \simeq H^i(Y-T) \to H^{i+1}_T(Y)$$

in rational cohomology is the zero map. Hence we obtain short exact sequences of mixed Hodge structures

$$0 \to H^i_T(Y) \to H^i_D(X) \oplus H^i(Y) \to H^i(X) \to 0 \ .$$

The injectivity of $H^i(Y) \to H^i(X)$ implies that $H^i_T(Y) \to H^i_D(X)$ is injective for all i. In a similar way, but using the exact sequences of the pairs (Y,T) and (X,D) we find that $H^j(T) \to H^j(D)$ is injective. So the cup product maps fit in the commutative diagram with injective vertical maps

$$\begin{array}{ccc} H^i_T(Y) \otimes H^j(T) & \to & H^{i+j}_T(Y) \\ \downarrow & & \downarrow \\ H^i_D(X) \otimes H^j(D) & \to & H^{i+j}_D(X) \end{array}$$

where the bottom line is a morphism of mixed Hodge structures. Hence the top line is also a morphism of mixed Hodge structures. □

As a special case we have:

Corollary 6.28. *Let X be a compact smooth complex variety of pure dimension n and let T be a closed subvariety of X. Then for all k we have a non-singular pairing of mixed Hodge structures*

$$H^k_T(X) \otimes H^{2n-k}(T) \to H^{2n}_T(X) \cong \mathbb{Q}(-n) \ .$$

Consequently, $H^k_T(X)$ has at most weights in the range $[k, 2k]$.

The last clause follows since for a compact variety T the weights of $H^{2n-k}(T)$ are in the range $[0, 2n-k]$ and since we know already (Corollary 5.47) that $H^k_T(X)$ has weights in $[k-1, 2k]$.

Remark 6.29. The condition that X is compact can be weakened to "X smooth and T a compact subvariety of X". The duality shows that for X smooth the mixed Hodge structure on $H^i_T(X)$ depends only on T and $\dim X$, not on the analytic or algebraic structure of X.

Applying Corollary 5.47 to the pair (X, U), we deduce:

Proposition 6.30. *Let X be a smooth compactification of a smooth complex algebraic variety U. Then we have*

$$\begin{aligned} W_m H^k(U) &= 0 \quad \text{for } m < k \\ W_k H^k(U) &= \operatorname{Im}\left(H^k(X) \to H^k(U)\right) . \end{aligned}$$

We conclude this section with some results indicating how the dimension of the singular locus influences weights of the cohomology of the variety itself but also of the exceptional divisor of a "good " resolution of singularities.

Theorem 6.31. *Let X be an algebraic variety of dimension n. Let Z be an s-dimensional subvariety of X containing the singular locus Σ and let $\pi : \tilde{X} \to X$ be a resolution such that $\pi^{-1}(Z) = D$ is a divisor with normal crossings on $\tilde{X}$. Then*

- *for all $k \geq n+s$ one has $W_{k-1}H^k(D) = 0$;*
- *if moreover Z is compact, then $H^k(D)$ is pure of weight k for all $k \geq n+s$.*

Proof. We can find $(s+1)$ affine open subsets of X, say $U_0, \ldots, U_s$ whose union U covers Z. By Theorem C.14 each U_j has the homotopy type of a CW-complex of dimension $\leq n$ and so $H^k(U_j) = 0$ for $k > n$. Using the Mayer-Vietoris sequence one sees inductively that $H^k(U_0 \cup \cdots \cup U_t) = 0$ for $k > t+n$. Let $\tilde{U} = \pi^{-1}(U)$. Then the long exact sequence

$$\ldots \to H^k(U) \to H^k(\tilde{U}) \oplus H^k(Z) \to H^k(D) \to H^{k+1}(U) \to \ldots$$

shows that the map $H^k(\tilde{U}) \to H^k(D)$ is surjective for $k = n+s$ and an isomorphism for $k > n+s$ (as $H^k(Z) = 0$ in this range). Hence for $k \geq n+s$ the group $H^k(D)$ has weights $\geq k$ (as $\tilde{U}$ is smooth). So $W_{k-1}H^k(D) = 0$. If moreover Z is compact, then also D is compact and $H^k(D)$ has weights $\leq k$. Hence $H^k(D)$ is pure of weight k. □

Corollary 6.32. *In the situation of the previous theorem (with Z compact), $H^k_D(\tilde{X})$ is pure of weight k if $k < n-s$.*

Proof. This is just the statement dual to the second statement of the previous theorem. □

Theorem 6.33. *Let X be an algebraic variety with singular locus Σ and let $s = \dim(\Sigma)$. Then for $k > \dim(X) + s$ one has $W_{k-1}H^k(X) = 0$.*

Proof. Choose a resolution $\pi : (\tilde{X}, D) \to (X, \Sigma)$. When $k > \dim(X) + s$ we have the exact sequence

$$H^k(\tilde{X}) \to H^k(X) \to H^k(D)$$

because $H^k(\Sigma) = 0$. As $\tilde{X}$ is smooth, $W_{k-1}H^k(\tilde{X}) = 0$ and we have by Theorem 6.31 that $W_{k-1}H^k(D) = 0$. Hence $W_{k-1}H^k(X) = 0$. □

Historical Remarks.

Our treatment of the Leray spectral sequence is entirely based on [Ara], but we do not stress the motivic point of view.

The systematic study of cup and cap products as well as the various duality morphisms begun with Fujiki [Fuj]. The results in § 6.3 are slightly more general and the proof is simpler.

The Hodge structure on the link of a singularity is due to Durfee [Du83] and [Du-H]. Semi-purity of the link of isolated singularities was proven by Goresky and MacPherson [G-M82] using the decomposition theorem of Deligne, Beilinson, Bernstein and Gabber [B-B-D] which we discuss in a later chapter (theorem 14.42). The idea of a direct proof along the above lines is due to [G-N-P-P]. In [Nav], the decomposition theorem for the resolution map of isolated singularities has been proved.

7

Applications to Algebraic Cycles and to Singularities

Historically, one of the main motivations for the development of Hodge theory was the study of cycles. This certainly was one of the principal preoccupations of Sir William Hodge who stated his famous conjecture that algebraic cycles can be detected in cohomology by looking at the integral classes having pure Hodge type. In this chapter we shall explain this as well as Grothendieck's generalization. To state the latter requires certain subtle properties implied by the existence of functorial mixed Hodge structures on possibly singular and non-compact algebraic varieties derived in the previous chapters. This can be found in § 7.1. Intermediate Jacobians find their natural place in this section.

In § 7.2 we discuss the unified approach to cycle classes, due to Deligne and Beilinson. The Deligne-Beilinson groups are extensions of the integral (p, p) cohomology by the intermediate Jacobian and the cycle class map as well as the Abel-Jacobi map are subsumed in the Deligne cycle class map.

We treat Du Bois theory in § 7.3 and apply it to deduce vanishing theorems of Kodaira-Akizuki-Nakano type valid for singular spaces. Next, we prove the Grauert-Riemenschneider Vanishing Theorem for germs of isolated singularities to other cohomology groups. Finally some applications to Du Bois singularities are given.

7.1 The Hodge Conjectures

7.1.1 Versions for Smooth Projective Varieties

Let X be an n-dimensional smooth projective variety. We want to investigate rational Hodge substructures of the rational Hodge structure $H^m(X;\mathbb{Q})$ related to codimension c cycles Z on X for arbitrary m, not just for $m = 2c$. Recall (Prop. 5.46) that the exact sequence of relative cohomology for the pair $(X, U = X - Z)$

$$\cdots \to H^m_Z(X) \to H^m(X) \to H^m(U) \to \cdots \qquad \text{(VII–1)}$$

is a sequence of mixed Hodge structures. So the image of the first map is a Hodge substructure of $H^m(X)$, the group of classes supported on Z. Since torsion classes may be non-algebraic (see Remark 1.15), we pass to rational cohomology, and we say that, more generally, a rational Hodge substructure $H \subset H^m(X;\mathbb{Q})$ is **supported on** $Z \subset X$ if

$$H \subset \operatorname{Im}\{H^m_Z(X;\mathbb{Q}) \to H^m(X;\mathbb{Q})\}.$$

Taking all subvarieties of codimension c, the preceding observation implies that the **coniveau filtration**

$$N^c H^m(X;\mathbb{Q}) := \bigcup_{\operatorname{codim} Z \geq c} \operatorname{Im}\{(H^m_Z(X;\mathbb{Q}) \to H^m(X;\mathbb{Q}))\}$$

is a filtration by Hodge substructures. As an example of how this gives restrictions on the possible Hodge numbers, if Z is smooth of codimension c by Poincaré-Lefschetz duality $H^m_Z(X) \cong H^{m-2c}(Z)(c)$ so that its image H in $H^m(X;\mathbb{Q})$ has non-zero Hodge numbers only in the range $(c, m-c), \ldots, (m-c, c)$; this can be formalized as follows:

Definition 7.1. The **level** of a non-zero Hodge structure $V = \bigoplus V^{p,q}$ is the largest difference $|p-q|$ for which $V^{p,q} \neq 0$. On other words, a Hodge structure V of weight n has level at most $n-2p$ if and only if $F^pV = V$. For instance level 0 is only possible for even weight Hodge structures and means that it is pure of type (p,p), level 1 means that $H = H^{p-1,p} \oplus H^{p,p-1}$ etc.

The preceding discussion amounts to saying that for smooth Z of codimension c the image of $H^m_Z(X;\mathbb{Q})$ in $H^m(X;\mathbb{Q})$ has level at most $m-2c$. We want to explain how mixed Hodge theory can be used to show that this remains the case for singular Z. To do this we compare of the image in cohomology of $Z \hookrightarrow X$ with the image in cohomology under the composition $\tilde{Z} \to Z \hookrightarrow X$, where $\tilde{Z} \to Z$ is a resolution of singularities. The crucial observation here is the following result.

Lemma 7.2. *Assume that there are morphisms of complex algebraic varieties*

$$\tilde{Z} \xrightarrow{\sigma} Z \xrightarrow{i} X$$

with Z compact, $\tilde{Z}$ compact and smooth, X smooth and σ surjective. Then

$$\operatorname{Ker}\left(i^* : H^*(X) \to H^*(Z)\right) = \operatorname{Ker}\left((i \circ \sigma)^* : H^*(X) \to H^*(\tilde{Z})\right).$$

Proof. All these cohomology groups have mixed Hodge structures and it suffices to prove the equality on the graded pieces. Recall (Theorem 5.33) that the Hodge numbers $h^{p,q} = \dim H^m(T)^{p,q}$ vanish for $p+q > m$ if T is compact and for $p+q < m$ if T is smooth. Apply this first to the compact varieties Z and $\tilde{Z}$: the maps $\operatorname{Gr}^W_k(i^*)$ and $\operatorname{Gr}^W_k(\sigma^* \circ i^*)$ on m-cohomology are both zero for $k > m$ and so the two kernels live in weight $\leq m$, hence in weight m, since X is smooth. It suffices therefore to look at the m-th graded piece. Here the kernels are the same since the map

$$\operatorname{Gr}^W_m H^m(Z;\mathbb{Q}) \xrightarrow{\sigma^*} \operatorname{Gr}^W_m H^m(\tilde{Z};\mathbb{Q}) = H^m(\tilde{Z};\mathbb{Q})$$

induced by σ is injective (see Theorem 5.41). $\square$

We apply this to the situation of an embedding $i : Z \hookrightarrow X$ of a subvariety in a smooth projective variety and $\sigma : \tilde{Z} \to Z$ a resolution of singularities. By Proposition 1.19 the Gysin map associated to $f : \tilde{Z} \to X$

$$f_! : H^{k-2c}(\tilde{Z})(-c) \to H^m(X), \quad c = \operatorname{codim} Z$$

is a morphism of pure Hodge structures. In this situation we have:

Corollary 7.3. *In rational cohomology the subspace of $H^m(X)$ consisting of classes supported on Z coincides with the image of the Gysin map:*

$$\operatorname{Im}\left(H^m_Z(X;\mathbb{Q}) \to H^m(X;\mathbb{Q})\right) = \operatorname{Im}\left(H^{m-2c}(\tilde{Z};\mathbb{Q}) \xrightarrow{(\tilde{f})_!} H^m(X;\mathbb{Q})\right). \quad \text{(VII–2)}$$

Proof. The exact sequence (VII–1) together with lemma 7.2 yields a short exact sequence

$$H^m(X,Z;\mathbb{Q}) \to H^m(X;\mathbb{Q}) \xrightarrow{\tilde{f}^*} H^m(\tilde{Z};\mathbb{Q}).$$

Applying the Poincaré-duality isomorphism and renumbering the indices we get

$$H_m(U;\mathbb{Q}) \to H_m(X;\mathbb{Q}) \xrightarrow{\tilde{f}^!} H_{m-2c}(\tilde{Z};\mathbb{Q}).$$

Dualizing and applying the Kronecker-homomorphism (which is an isomorphism since we are working with rational coefficients) we find

$$\begin{aligned} \operatorname{Im}(f_!) &= \operatorname{Ker}\left(H^m(X;\mathbb{Q}) \to H^m(U;\mathbb{Q})\right) \\ &= \operatorname{Im}\left(H^m_Z(X;\mathbb{Q}) \to H^m(X;\mathbb{Q})\right) \end{aligned}$$

as claimed. □

As in the case of smooth Z, this implies that the Hodge structure on $\operatorname{Im}\{H^m_Z(X;\mathbb{Q}) \to H^m(X;\mathbb{Q})\}$ has the same Hodge numbers as one on the $(m-2c)$-cohomology group of a smooth compact variety and hence the level of the preceding Hodge structure is at most $m-2c$:

Corollary 7.4. *Let X be a smooth projective variety. Any rational Hodge substructure of weight m contained in $N^cH^m(X)$ has level at most $m-2c$.*

Now, one might conjecture that conversely $F^cH^m(X) \cap H^m(X;\mathbb{Q})$ is supported on a subvariety of codimension at least c, but this does not make sense, since $F^cH^m(X) \cap H^m(X;\mathbb{Q})$ itself is *not* a Hodge structure, as noted by Grothendieck in [Groth69]: it can have odd dimension (see [Lewis, 7.15] for details). Correcting this leads to the generalized Hodge conjecture, as improved by Grothendieck:

Conjecture 7.5 (GENERALIZED HODGE CONJECTURE $GHC(X,m,c)$). *Let X be a smooth projective variety. The largest rational Hodge substructure of $F^cH^m(X;\mathbb{C})\cap H^m(X;\mathbb{Q})$ is the union of all the rational Hodge substructures supported on codimension $\geq c$ subvarieties of X.*

Alternatively, for every $\mathbb{Q}$-Hodge substructure H' of $H^m(X;\mathbb{Q})$ of level at most $m-2c$, there exists a subvariety Z of X of codimension $\geq c$ such that the substructure H' is supported on Z.

Remark 7.6. The (classical) Hodge Conjecture 1.15 is the special case $m=2c$, with $H'=H^{2c}(X;\mathbb{Q})\cap\ H^{c,c}$.

7.1.2 The Hodge Conjecture and the Intermediate Jacobian

We have seen (Example 3.30) that to any Hodge structure H of odd weight $2m-1$ there is associated a complex torus

$$J(H)=H_{\mathbb{Z}}\backslash H_{\mathbb{C}}/F^m=H_{\mathbb{Z}}\backslash\overline{F^m},$$

the intermediate Jacobian of the associated Hodge structure. A special case arises for the cohomology groups of odd dimension of a smooth complex projective variety X of dimension n; we get **intermediate Jacobians**

$$J^m(X)=J(H^{2m-1}(X)),\ m=1,\ldots,n.$$

The first one, $J^1(X)$ is the usual Jacobian torus and $J^n(X)$ is the Albanese torus $\mathrm{Alb}(X)$. To define the Abel-Jacobi map, a dual description for the intermediate Jacobian is useful. Cup product pairing establishes a duality (modulo torsion) between $H^{2m-1}(X)$ and $H^{2n-2m+1}(X)(-n)$. Combined with Poincaré duality this yields the integration homomorphism

$$\alpha: H_{2n-2m+1}(X)\to H_{\mathrm{DR}}^{2n-2m+1}(X)^{\vee}$$

sending a cycle to integration over the cycle and so

$$J^m(X)\cong F^{n-m+1}H^{2n-2m+1}(X)^{\vee}/\alpha\big(H_{2n-2m+1}(X)\big).$$

Using this description we can define the Abel-Jacobi map on algebraic cycles of (complex) codimension m which are homologous to zero. If $Z=\partial c$ is such a cycle, integration over c defines an element of $H^{2n-2m+1}(X,\mathbb{C})^{\vee}$ and a different choice c' with $Z=\partial c'$ determines the same element in the intermediate Jacobian $J^m(X)$ since $c-c'$ gives a period integral. This gives the **Abel-Jacobi map**, a linear map

$$u_X^m: Z_{\mathrm{hom}}^m(X):=\left\{\begin{matrix}\text{codim } m\text{-cycles on } X\\ \text{homologous to zero}\end{matrix}\right\}\longrightarrow J^m(X). \qquad \text{(VII–3)}$$

The polarization on the Hodge structure of $H^{2m-1}(X)$ induces a non-degenerate bilinear form on the tangent space at 0 of the intermediate Jacobian which in general is indefinite, except when $m = 1$ (the ordinary Jacobian torus for divisors) or $m = \dim X - 1$ (the Albanese variety). In general, the subtorus associated to $N^1 H^{2m-1}(X)$ receives a definite polarization and hence is algebraic. We call it the **algebraic intermediate Jacobian**

$$J^m_{\text{alg}}(X) = J(N^1 H^{2m-1}(X)).$$

Likewise, we can consider the subtorus associated to the largest Hodge substructure of $F^{m-1}H^{2m-1}(X)$:

$$H^{2m-1}_{\text{Hdg}}(X) := \{\text{maximal rational Hodge substructure of level 1 contained in } H^{2m-1}(X),\}$$

the largest substructure on which the polarization restricts positively. So this torus, the **Hodge-theoretic intermediate Jacobian**

$$J^m_{\text{Hdg}}(X) := J\big(H^{2m-1}_{\text{Hdg}}(X)\big)$$

is also an abelian variety and contains $J^m_{\text{alg}}(X)$. The generalized Hodge conjecture amounts to saying that the two coincide.

We can now prove that the generalized Hodge conjecture is in fact equivalent to the classical Hodge conjecture. We start with the following simple

Observation 7.7. 1) *Conjecture $GHC(X, 2p, p-1)$ implies the classical Hodge conjecture for p-cycles on X.*
2) *If for every smooth curve C the Hodge classes in $H^1(C;\mathbb{Q})\otimes H^{2p-1}(X;\mathbb{Q}) \subset H^{2p}(C\times X;\mathbb{Q})$ come from algebraic cycles on $C\times X$, conjecture $GHC(X, 2p-1, p-1)$ holds.*

Proof. 1) Start with a rational Hodge class of type (p,p). The line V it spans is a rational Hodge structure of level 0 and hence of level ≤ 2. By $GHC(2p, p-1)$ there exists a codimension $(p-1)$-cycle $Z \subset X$ on which this Hodge structure, is supported. Let $\sigma : \tilde{Z} \to Z$ be a resolution of singularities. Then V is in the image of the Gysin-map $H^2(\tilde{Z};\mathbb{Q}) \to H^{2p}(X;\mathbb{Q})$. Because $H^2(\tilde{Z})$ carries a polarized Hodge structure, by semi-simplicity 2.12 the kernel of the Gysin-map is an orthogonal summand and its complement maps isomorphically onto the image of the Gysin map. In particular, V corresponds to a Hodge substructure $\tilde{V}$ of $H^2(\tilde{Z})$ of pure type $(1,1)$. Since the Hodge conjecture holds in this case, there is a divisor $\tilde{Y} \subset \tilde{Z}$ whose class spans $\tilde{V}$ and hence its image Y in X spans V.
2) By the Lefschetz hyperplane theorem for a smooth complete intersection curve $C \subset J^p_{\text{Hdg}}(X)$ the inclusion map induces a surjection on the level of H_1 and the Gysin map provides us with an injection $H^{2p-1}_{\text{Hdg}}(X) \hookrightarrow H^1(C)$ of pure Hodge structures, which, again by semi-simplicity gives a surjection $H^1(C) \twoheadrightarrow H^{2p-1}_{\text{Hdg}}(X)$ of Hodge structures and hence a Hodge class in the product $H^1(C)\otimes H^{2p-1}_{\text{Hdg}}(X)$. By assumption this class is supported on a cycle $\Gamma \subset C \times X$ and its image under projection $C \times X \to X$ supports all of $H^{2p-1}_{\text{Hdg}}(X)$ by construction. □

Remark 7.8. In fact, looking a bit more closely to the proof of 1), one can show that the converse of 2) holds.

Corollary 7.9. *If the classical Hodge conjecture is true for all smooth projective varieties, then the generalized Hodge conjecture is true.*

Proof. Note that $GHC(X, m, c)$ for all m and c follows if $GHC(X, m, c)$ is true in the two cases $m = 2p - 1, c = p$ and $m = 2p, c = p$. The last is the classical Hodge conjecture. By Observation 7.7, the first follows from the classical Hodge conjecture on the products $C \times X$, C a curve. □

7.1.3 A Version for Singular Varieties

The naive generalization of the Hodge conjecture to singular varieties turns out to be false. Indeed, there is a counterexample due to Bloch ([Jann, Appendix A]). There is a much more subtle generalization using 1-motives which is due to Barbieri-Viale [BaV02, BaV07]. Bloch's example as well as a later counterexample by Srinivas (cited in [BaV02]) still make sense in this formulation.

There is also a version for Borel-Moore homology which is much easier to formulate. To do this, we first reformulate the Hodge conjecture (for smooth projective varieties) for homology. If $Z \subset X$ is a subvariety of dimension d, we look at the image $\mathrm{Im}(H_m(Z;\mathbb{Q}) \to H_m(X;\mathbb{Q}))$, a rational Hodge substructure of weight $-m$. This Hodge structure has level $\leq m - 2d$, since the preceding image is the same as the image of $H_m(\tilde{Z};\mathbb{Q})$ in $H_m(X;\mathbb{Q})$ under the composition $\tilde{Z} \to Z \hookrightarrow X$. This leads to the **filtration by "niveau"**:

$$N_d H_m(X;\mathbb{Q}) := \bigcup_{\dim Z \leq d} \mathrm{Im}(H_m(Z;\mathbb{Q}) \to H_m(X;\mathbb{Q})),$$

where Z runs over all subvarieties of dimension $\leq d$. The subspace $N_d H_m(X;\mathbb{Q})$ is a rational Hodge substructure of $H_m(X;\mathbb{Q})$ of weight $-m$ and level $\leq m - 2d$ and hence contained in $F^{-d} H_m(X;\mathbb{C}) \cap H_m(X;\mathbb{Q})$.

Conjecture 7.10 (Generalized Hodge Conjecture (homological version)). *Let X be a smooth complex projective variety. The rational Hodge substructure $N_d H_m(X;\mathbb{Q})$ of $H_m(X;\mathbb{Q})$ coming from the cycles of dimension d is the largest rational Hodge substructure of $H_m(X;\mathbb{Q})$ contained in $F^{-d} H_m(X;\mathbb{C}) \cap H_m(X;\mathbb{Q})$.*

For arbitrary algebraic varieties (not necessarily smooth or compact) we can pass to Borel-Moore homology. Roughly speaking, we have to replace H_m^{BM} by its weight $-m$ part. According to Definition-Lemma 6.25 this is the lowest possible weight and so carries a pure Hodge structure of weight $-m$ so that modulo these changes the above homological version of the Hodge conjecture formally makes sense. What remains to be shown is that the part coming from the cycles has the correct level inside the lowest weight part. This is the content of the following Lemma.

Lemma 7.11. *Let U be a complex algebraic variety. The niveau d subspace $N_d H_m^{\mathrm{BM}}(U;\mathbb{Q}) \subset H_m^{\mathrm{BM}}(U;\mathbb{Q})$ intersects the lowest weight part of $H_m^{\mathrm{BM}}(U;\mathbb{Q})$ in a Hodge structure of level $\leq 2m-d$.*

Proof. We consider a compactification X of U and we set $D = X - U$. Fix a d-dimensional subvariety Z of U and let Y be its closure in X. First of all, strictness implies that

$$\begin{aligned}&\mathrm{Im}(W_{-m}H_m^{\mathrm{BM}}(Y;\mathbb{Q}) \to W_{-m}H_m^{\mathrm{BM}}(X;\mathbb{Q})) =\\ &\mathrm{Im}(H_m^{\mathrm{BM}}(Y;\mathbb{Q}) \to H_m^{\mathrm{BM}}(X;\mathbb{Q})) \cap W_{-m}H_m^{\mathrm{BM}}(X;\mathbb{Q}).\end{aligned}$$

Next, there is a commutative diagram of mixed Hodge structures

$$\begin{array}{ccc} H_{m-1}^{\mathrm{BM}}(D\cap Y) & \to & H_{m-1}^{\mathrm{BM}}(D) \\ \uparrow & & \uparrow \\ H_m^{\mathrm{BM}}(Z;\mathbb{Q}) & \to & H_m^{\mathrm{BM}}(U;\mathbb{Q}) \\ \uparrow & & \uparrow \\ H_m^{\mathrm{BM}}(Y;\mathbb{Q}) & \to & H_m^{\mathrm{BM}}(X;\mathbb{Q}), \end{array}$$

with all maps induced by suitable restrictions. Since in weight $(-m)$ the topmost line is identically zero, this shows that the niveau d subspace of $H_m^{\mathrm{BM}}(U;\mathbb{Q})$ in weight $(-m)$ is the restriction of $N_d W_{-m} H_m^{\mathrm{BM}}(X;\mathbb{Q})$.

Finally, we have to compare this with similar spaces in the homology of suitable resolutions. We choose a resolution of singularities $\sigma : \tilde{X} \to X$ such that the proper transform $\tilde{Y}$ of Y is smooth as well and we get a commutative diagram of mixed Hodge structures

$$\begin{array}{ccc} H_m^{\mathrm{BM}}(Z;\mathbb{Q}) & \to & H_m^{\mathrm{BM}}(X;\mathbb{Q}) \\ \uparrow & & \uparrow \\ H_m^{\mathrm{BM}}(\tilde{Z};\mathbb{Q}) & \to & H_m^{\mathrm{BM}}(\tilde{X};\mathbb{Q}). \end{array}$$

The image of the vertical arrows not only lands in the weight $(-m)$ part, but spans it. In fact this statement is dual to the assertion of Theorem 5.41. It follows that the niveau d subspace intersect the lowest weight subspace of $H_m^{\mathrm{BM}}(U;\mathbb{Q})$ exactly in the image of the niveau d subspace of $H_m^{\mathrm{BM}}(\tilde{X};\mathbb{Q})$.

The lower arrow in itself is a morphism of pure Hodge structures of weight $(-m)$ and $H_m^{\mathrm{BM}}(\tilde{Z};\mathbb{C}) = F^{-d}H_m^{\mathrm{BM}}(\tilde{Z};\mathbb{C})$, since Z has dimension d. So the niveau d subspace of $W_{-m}H_m^{\mathrm{BM}}(X;\mathbb{Q})$, which is the image inside $H_m^{\mathrm{BM}}(X;\mathbb{Q})$ of the niveau d subspace of $H_m^{\mathrm{BM}}(\tilde{X};\mathbb{Q})$, in fact belongs to $W_{-m} \cap F^{-d}H_m^{\mathrm{BM}}(X;\mathbb{C}) \cap H_m^{\mathrm{BM}}(X;\mathbb{Q})$. □

Motivated by this we abbreviate

$$\tilde{H}_m(U) := W_{-m}H_m^{\mathrm{BM}}(U;\mathbb{Q})$$

and call it the **pure part of the homology**. The generalized Hodge conjecture then reads:

Conjecture 7.12 (GENERALIZED HODGE CONJECTURE (homological version II)). *Let U be an algebraic variety. The rational Hodge substructure $N_d\tilde{H}_m(U)$ inside the pure part of the homology $\tilde{H}_m(U)$ coming from the cycles of dimension d is the largest rational Hodge substructure of $\tilde{H}_m(U)$ contained in $F^{-d}\tilde{H}_m(U;\mathbb{C}) \cap H_m^{\mathrm{BM}}(U;\mathbb{Q})$.*

As pointed out by Jannsen in [Jann], this conjecture follows as soon as the generalized Hodge conjecture for smooth projective varieties can be shown. The argument runs as follows.

The crucial ingredient is the semi-simplicity property for polarized pure Hodge structures (2.12). It implies that the kernel of the surjective morphism $\tilde{H}_m(\tilde{X}) \twoheadrightarrow \tilde{H}_m(U)$ is a direct factor and hence the largest rational Hodge substructure contained in F^{-d} of the source maps surjectively onto the largest rational Hodge substructure of F^{-d} of the target. So, if the Hodge conjecture holds for the source, a smooth projective variety, the former coincides with $N_d\tilde{H}_m(\tilde{X})$ and so the latter must coincide with $N_d\tilde{H}_m(U)$, completing the proof.

7.2 Deligne Cohomology

In this section X will be a smooth complex projective variety, although we occasionally allow smooth varieties which are not complete.

7.2.1 Basic Properties

In comparing the various fundamental classes it is natural to consider the fibre product of integral cohomology and $F^dH^{2d}(X;\mathbb{C})$ over complex cohomology. Deligne cohomology sets the stage for the comparison of fundamental classes on the level of complexes; a fibre product is just a kernel of a morphism and as explained in the Appendix (Example A.14) a kernel of a surjective morphism between complexes is quasi-isomorphic to the cone over this morphism (shifted to the right by 1). This leads to the following construction.

Lemma-Definition 7.13. 1) *Let R be any subring of $\mathbb{R}$ and consider $R(d) \subset \mathbb{C}$ as a complex in degree 0 mapped into the De Rham complex by the inclusion $\epsilon^d : R(d) \hookrightarrow \Omega_X^\bullet$. Denote the other natural inclusion by $\iota^d : F^d\Omega_X^\bullet \hookrightarrow \Omega_X^\bullet$. The shifted cone over the difference of the two is the* ***Deligne complex***

$$R(d)_{\mathrm{Del}} := \mathrm{Cone}^\bullet(\epsilon^d - \iota^d : R(d) \oplus F^d\Omega_X^\bullet \to \Omega_X^\bullet)[-1].$$

Equivalently, with Φ^d the composition of the inclusion of $R(d)$ in the De Rham complex followed by projection onto the quotient complex obtained by dividing out the subcomplex F^d, we have

$$R(d)_{\mathrm{Del}} := \mathrm{Cone}^\bullet\left(\Phi^d : R(d) \to \Omega_X^\bullet / F^d\Omega_X^\bullet\right)[-1].$$

This complex is quasi-isomorphic to the complex

$$0 \to R(d) \to \mathcal{O}_X \to \Omega^1_X \to \cdots \to \Omega^{d-1}_X \to 0. \qquad \text{(VII–4)}$$

2) *Its hypercohomology gives the* ***Deligne cohomology groups***

$$H^k_{\mathrm{Del}}(X, R(d)) := H^k(X, R(d)_{\mathrm{Del}}).$$

Similarly, if $Y \subset X$ *is closed, hypercohomology with support defines Deligne cohomology with support in* Y*, denoted*

$$H^k_Y(X, R(d)_{\mathrm{Del}}).$$

Examples 7.14. 1) For $d = 0$ we have the usual cohomology group $H^m(X; R)$.
2) For $d = 1$ and $A = \mathbb{Z}$, the 2-term Deligne complex is quasi-isomorphic to the sheaf $\mathcal{O}^*_X$ (placed in degree 1), i.e. we have

$$\mathbb{Z}(1)_{\mathrm{Del}} \xrightarrow[\sim]{\text{qis}} \mathcal{O}^*_X[-1]$$

and hence $H^m_{\mathrm{Del}}(X, \mathbb{Z}(1)) = H^{m-1}(X, \mathcal{O}^*_X)$. For $m = 2$ we get the Picard group and we see thus that in general the Deligne groups are only groups and not vector spaces.
3) We have

$$\mathbb{Z}(2)_{\mathrm{Del}} \xrightarrow[\sim]{\text{qis}} \{\mathcal{O}^*_X \xrightarrow{d\log} \Omega^1_X\}[-1].$$

And $H^2_{\mathrm{Del}}(X, R(2)) = H^1(\mathcal{O}^*_X \xrightarrow{d\log} \Omega^1_X)$, which for $R = \mathbb{Z}$ can be shown to be isomorphic to the group of isomorphism classes of line bundles equipped with a connection on X. See [Beil85].

The exact sequence of the cone (A–12) together with the second interpretation of the Deligne groups yields an exact sequence

$$H^{k-1}(X, \mathbb{Z}(d)) \to H^{k-1}(X; \mathbb{C})/F^d H^{k-1}(X; \mathbb{C}) \to H^k_{\mathrm{Del}}(X, \mathbb{Z}(d)) \to H^k(X, \mathbb{Z}(d)).$$

It can be rewritten as the short exact sequence

$$0 \to \frac{H^{k-1}(X; \mathbb{C})}{[H^{k-1}(X, \mathbb{Z}(d)) \oplus F^d H^{k-1}(X)]} \to H^k_{\mathrm{Del}}(X, \mathbb{Z}(d)) \to H^k(X, \mathbb{Z}(d)) \to 0.$$

We next use the interpretation of the Ext-groups for mixed Hodge structures from Example 3.34 (3) and obtain

$$\left.\begin{array}{r} 0 \to \mathrm{Ext}_{\mathrm{MHS}}(\mathbb{Z}, H^{k-1}(X, \mathbb{Z}(d)) \to H^k_{\mathrm{Del}}(X, \mathbb{Z}(d)) \to \\ \mathrm{Hom}_{\mathrm{MHS}}(\mathbb{Z}, H^k(X, \mathbb{Z}(d)) \to 0, \quad \forall k \le 2d. \end{array}\right\} \qquad \text{(VII–5)}$$

Comparing this with equation (III–17) in Chapter 3, we find the relation with a construction starting from the constant sheaf $\underline{\mathbb{Z}}_X(d)$:

Theorem 7.15. *Suppose that $k \le 2d$. Then we have*

$$H^k_{\mathrm{Del}}(X, \mathbb{Z}(d)) = H^k_{\mathrm{Hodge}}(\phi(\underline{\mathbb{Z}}_X(d))) = \mathrm{Ext}^k_{\mathrm{MHS}}(\mathbb{Z}, R\Gamma(\underline{\mathbb{Z}}_X(d))).$$

The Deligne complex behaves functorially with respect to algebraic morphisms and hence so do the Deligne groups.

There is a product structure on Deligne cohomology which comes from the multiplication on the level of complexes (VII–4)

$$\cup : \mathbb{Z}(p)_{\mathrm{Del}} \otimes \mathbb{Z}(q)_{\mathrm{Del}} \to \mathbb{Z}(p+q)_{\mathrm{Del}}$$

defined by

$$x \cup y = \begin{cases} x \cdot y & \text{if } \deg x = 0 \\ x \wedge dy & \text{if } \deg x > 0 \text{ and } \deg y = q \\ 0 & \text{otherwise.} \end{cases}$$

For $p = 0$ this gives back the usual cup-product in cohomology. For the proof of the following result we refer to [Es-V88, §1,§3],

Proposition 7.16. *The product on Deligne cohomology induces an anticommutative graded product structure on Deligne cohomology:*

$$\cup : H^k_{\mathrm{Del}}(X, R(p)) \otimes H^\ell_{\mathrm{Del}}(X, R(q)) \to H^{k+\ell}_{\mathrm{Del}}(X, R(p+q)).$$

The actions of the Deligne complex defined by the projection map combined with the usual products

$$\begin{aligned} R(p)_{\mathrm{Del}} \otimes R(q) &\longrightarrow R(p+q) \\ R(p)_{\mathrm{Del}} \otimes F^q(\Omega^\bullet) &\longrightarrow F^{p+q}(\Omega^\bullet) \\ R(p)_{\mathrm{Del}} \otimes \Omega^\bullet_X &\longrightarrow \Omega^\bullet_X \end{aligned}$$

induce a bi-graded action of the Deligne cohomology on the vector spaces $\bigoplus_q H^q_{\mathrm{DR}}(X)$, $\bigoplus_{p,q} H^q(X, R(p))$ and $\bigoplus_{p,q} F^p H^q(X)$ in such a way that the long exact sequence of the cone (see (A–12)*)*

$$\begin{aligned} \cdots \to H^{k-1}(X, R(d)) \oplus F^d H^{k-1}(X) \to H^{k-1}(X; \mathbb{C}) \to \\ H^k_{\mathrm{Del}}(X, R(d)) \to H^k(X, R(d)) \oplus F^d H^k(X) \to \cdots \end{aligned}$$

is compatible with these actions.

Example 7.17. Using Example 7.14 we have a pairing

$$\begin{array}{ccc} H^1_{\mathrm{Del}}(X, \mathbb{Z}(1)) \times H^1_{\mathrm{Del}}(X, \mathbb{Z}(1)) \to & & H^2_{\mathrm{Del}}(X, \mathbb{Z}(2)) \\ \| & & \| \\ H^1(\mathcal{O}^*_X) \times H^1(\mathcal{O}^*_X) & \to & \{\text{Line bundles with a connection}\} \end{array}$$

In [Beil85, §1-2], it is shown that this pairing $(f, g) \mapsto (L, \nabla)$ has the property that the curvature of ∇ is given by $d\log(f) \wedge d\log(g)$ and that the monodromy around a loop γ based $x \in X$ can be expressed as

$$\exp\left(\frac{1}{2\pi \mathrm{i}}\int_\gamma \log(f)d\log(g) - \log g(x)\int_\gamma d\log(f)\right).$$

When X is an open Riemann surface and γ a loop around a puncture y this number is the Tate symbol $(f,g)_y$ and hence in this case one has a natural mapping

$$K_2(X) \to H^2_{\mathrm{Del}}(X, \mathbb{Z}(2))$$

which maps the symbol $\{f,g\} := \sum_y (f,g)_y$ to the product $f \cup g$.

There is also a version of products in Deligne cohomology with supports. The product of two classes with support in two closed subvarieties Y_1 and Y_2 yields a class with support in the intersection $Y_1 \cap Y_2$, provided the two subvarieties meet properly. With the appropriate modifications, the preceding assertions then remain true.

Remark 7.18. Although the definition of the Deligne complex makes sense for non-compact complex algebraic manifolds, it is itself non-algebraic in nature, roughly because differential forms admitting arbitrary singularities at infinity are allowed. To remedy this, Beilinson has proposed to use forms with at most poles at infinity. So, as before, we let U be a smooth complex algebraic manifold and a smooth compactification $j : U \hookrightarrow X$ by means of a divisor D having simple normal crossings. Then we modify the definition of the Deligne complex by replacing $R(d)$ by the complex of sheaves $Rj_*R(d)$ on X, $F^d\Omega^\bullet_U$ by the complex $F^d\Omega^\bullet_X(\log D)$ and $\Omega^\bullet_X$ by $Rj_*\Omega^\bullet_X$. This yields the **Deligne-Beilinson complex**

$$R(d)_{\mathrm{DB}} := \mathrm{Cone}^\bullet\left(\epsilon^d - \iota^d : Rj_*R(d) \oplus F^d\Omega^\bullet_X(\log D) \to Rj_*\Omega^\bullet_X\right)[-1].$$

We shall only summarize its properties and refer to [Es-V88] for proofs. First of all, this complex restricted to U is quasi-isomorphic to the ordinary Deligne complex on U. Next, its hypercohomology groups

$$H^p_{\mathrm{DB}}(U, R(d)) := \mathbb{H}^p(X, R(d)_{\mathrm{DB}}),$$

the **Deligne-Beilinson groups** do not depend on the choice of the compactification. It follows that the inclusion $j : U \hookrightarrow X$ induces natural forgetful maps

$$H^p_{\mathrm{DB}}(U, R(d)) \to H^p_{\mathrm{Del}}(U, R(d)).$$

Secondly, as mentioned before, the complexes themselves behave well under morphisms $f : U \to V$ between smooth complex algebraic manifolds. In fact, if X, respectively Y is a smooth compactification of U, respectively V by means of simple normal crossing divisors such that f extends to a morphism $\bar{f} : X \to Y$, there is an induced morphism

$$\bar{f}^* : (R(d)_{\mathrm{DB}})_Y \to \bar{f}_*\left(R(d)_{\mathrm{DB}}\right)_X$$

which induces a homomorphism on the Deligne-Beilinson groups.

We also mention that a cup product on Deligne-Beilinson cohomology can be introduced (compatible with the forgetful maps above) for which the analogue of Proposition 7.16 is still true.

7.2.2 Cycle Classes for Deligne Cohomology

We recall § 2.4 that for any irreducible subvariety Y of codimension d in a compact algebraic manifold X, we have defined an integral fundamental class $\mathrm{cl}(Y) \in H^{2d}(X, \mathbb{Z}(d))$ which under the inclusion

$$\epsilon_d : \mathbb{Z}(d) \hookrightarrow \mathbb{C}$$

maps to the image of the Hodge class $\mathrm{cl}_{\mathrm{Hdg}}(Y) \in F^d H^{2d}(X;\mathbb{C})$ in $H^{2d}(X;\mathbb{C})$. This image has pure type (p,p).

In local cohomology on the one hand we have the Thom class $\tau(Y) \in H_Y^{2d}(X, \mathbb{Z}(d))$ which maps to $\mathrm{cl}(Y)$ when we forget the support. On the other hand we have the Thom-Hodge class $\tau_{\mathrm{Hdg}}(Y) \in \mathbb{H}_Y^{2d}(X, F^d\Omega_X^\bullet)$ mapping to $\mathrm{cl}_{\mathrm{Hdg}}(Y)$ upon forgetting the support.

Looking at place $k = 2d$ in the preceding long exact sequence for a cone, written for cohomology with supports in a closed subvariety $Y \subset X$ leads to the definition of the cycle class in Deligne cohomology as we shall now explain.

Proposition 7.19. *Let $Y \subset X$ be a codimension d subvariety. There is a unique* ***Thom-Deligne class***

$$\tau_{\mathrm{Del}}(Y) \in H_Y^{2d}(X, \mathbb{Z}_{\mathrm{Del}}(d))$$

which maps to the Thom class $\tau(Y) \in H_Y^{2d}(X, \mathbb{Z}(d))$ and the Thom-Hodge class $\tau_{\mathrm{Hdg}}(Y) \in H_Y^{2d}(X, F^d\Omega_X^\bullet)$. Forgetting supports, we have the ***Deligne class****, the fundamental class $\mathrm{cl}_{\mathrm{Del}}(Y) \in H_{\mathrm{Del}}^{2d}(X, \mathbb{Z}(d))$ in Deligne cohomology. Under the maps induced by the two projections $\mathbb{Z}(d)_{\mathrm{Del}} \to \mathbb{Z}(d)$, respectively $\mathbb{Z}(d)_{\mathrm{Del}} \to F^d\Omega_X^\bullet$ it maps to the usual fundamental class, respectively the Hodge fundamental class.*

Proof. Consider the exact sequence

$$\begin{aligned} \to H_Y^{2d-1}(X;\mathbb{C}) &\to H_Y^{2d}(X, \mathbb{Z}_{\mathrm{Del}}(d)) \\ &\to H_Y^{d}(X, \mathbb{Z}(d)) \oplus H_Y^{2d}(X, F^d\Omega_X^\bullet) \to H_Y^{2d}(X;\mathbb{C}). \end{aligned}$$

The group $H_Y^{2d-1}(X;\mathbb{C})$ vanishes, being dual to $H_{2\dim Y+1}(Y;\mathbb{C})$. So Deligne cohomology can be expressed as a fibre product

$$H_Y^{2d}(X, \mathbb{Z}_{\mathrm{Del}}(d)) = H_Y^{2d}(X, \mathbb{Z}(d)) \times_{H_Y^{2d}(X;\mathbb{C})} H_Y^{2d}(X, F^d\Omega_X^\bullet).$$

Since the two Thom classes have the same image in $H_Y^{2d}(X;\mathbb{C})$ there is a unique Deligne-Thom class with the stated properties. □

We want to relate the Deligne-Thom class to the Abel-Jacobi map. As a first step, we prove:

Lemma 7.20. *The Deligne cohomology group $H_{\mathrm{Del}}^{2d}(X, \mathbb{Z}(d))$ fits in an exact sequence of abelian groups*

$$0 \to J^d(X) \to H_{\mathrm{Del}}^{2d}(X, \mathbb{Z}(d)) \to H^{d,d}(X, \mathbb{Z}(d)) \to 0$$

relating the intermediate Jacobian and the Hodge groups.

Proof. Recall the definition of the Deligne cohomology as the twisted cone over the difference of the two inclusions $\mathbb{Z}(d) \hookrightarrow \mathbb{C}$ and $F^p\Omega^\bullet_X \hookrightarrow \Omega^\bullet$. The coboundary map in the exact sequence of the cone is just the map induced by (minus) this difference (Examples A.14) and so in particular we find that the kernel of

$$H^{2d}(X,\mathbb{Z}(d)) \oplus F^d H^{2d}(X;\mathbb{C}) \to H^{2d}(X;\mathbb{C})$$

is exactly the Hodge group

$$H^{d,d}(X,\mathbb{Z}(d)) := H^{2d}(X,\mathbb{Z}(d)) \cap (\epsilon^d)^{-1} H^{d,d}(X).$$

The exact sequence of the cone then gives a short exact sequence

$$\begin{aligned} 0 \to H^{2d-1}(X;\mathbb{C})/\{F^d H^{2d-1}(X) + H^{2d-1}(X,\mathbb{Z}(d))\} \to H^{2d}_{\mathrm{Del}}(X,\mathbb{Z}(d)) \to \\ \to H^{d,d}(X,\mathbb{Z}(d)) \to 0 \end{aligned}$$

and since ϵ^d is multiplication by $(2\pi \mathrm{i})^d$, the first term is the d-th intermediate Jacobian. □

As the next step, we explain how to describe the Abel-Jacobi map in an algebraic fashion. Let Z be an algebraic cycle homologous to zero with support $|Z|$. Consider the exact cohomology sequence for cohomology with support in $|Z|$. It is an exact sequence of mixed Hodge structures as we have seen before (§ 5.5). A portion of its reads

$$\cdots \to H^{2m-1}_{|Z|}(X) \to H^{2m-1}(X) \to H^{2m-1}(X - |Z|) \to H^{2m}_{|Z|}(X) \to \dots.$$

The first group in this sequence is zero, while the last group is pure of type (m,m) (it is the free $\mathbb{Z}$-module generated by the components of Z) and contains the Thom-class $\tau(Z)$ of Z. Since it maps to zero under the last map of the preceding sequence, we thus find an extension of $H^{2m-1}(X)$ by $\mathbb{Z}\tau(Z) = \mathbb{Z}(-m)$. By Example 3.34 1, this thus defines an element in the intermediate Jacobian $J^m(X)$. A calculation shows that this is the Abel-Jacobi image of Z. For more details the reader may consult [Es-V88, § 7] where the following more general result is proven.

Theorem 7.21. *The Abel-Jacobi map* (VII–3) $u_X : \mathcal{Z}^d_{\mathrm{hom}}(X) \to J^d(X)$ *fits in the following commutative diagram*

$$\begin{array}{ccccccccc} 0 \to & \mathcal{Z}^d_{\mathrm{hom}}(X) & \to & \mathcal{Z}^d(X) & \to & \mathcal{Z}^d(X)/\mathcal{Z}^d_{\mathrm{hom}} & \to 0 \\ & \downarrow{\scriptstyle u_X} & & \downarrow{\scriptstyle \mathrm{cl}_{\mathrm{Del}}} & & \downarrow{\scriptstyle \mathrm{cl}} & \\ 0 \to & J^d(X) & \to & H^{2d}_{\mathrm{Del}}(X,\mathbb{Z}(d)) & \to & H^{d,d}(X,\mathbb{Z}(d)) & \to 0. \end{array}$$

7.3 The Filtered De Rham Complex And Applications

7.3.1 The Filtered De Rham Complex

Let X be a complex algebraic variety and $X_\bullet$ a cubical hyper-resolution. Let $\epsilon : X_\bullet \to X$ be the associated augmented simplicial variety. The isomorphism class of the mixed Hodge complex $\epsilon_* \mathcal{K}^\bullet_{\mathrm{DR}}(X_\bullet)$ in the derived bi-filtered category of complexes of $\mathcal{O}_X$-modules heavily depends on the choice of the hyperresolution. For instance, the graded objects with respect to the weight filtration give the cohomology of all of the smooth constituents of the hyperresolution. If X is compact and if we forget the weight filtration we obtain the De Rham complex $R\epsilon_*\Omega^\bullet_{X^\bullet}$. By definition its hypercohomology computes the cohomology of X together with its Hodge filtration. This is no longer true when X ceases to be compact, although we still have $\mathbb{H}^q(X, R\epsilon_*\Omega^\bullet_{X^\bullet}) = H^q(X;\mathbb{C})$. Surprisingly, on the level of sheaves of differential forms a certain uniqueness result still holds as shown by Du Bois and which we quote without proof from [DuB]:

Theorem 7.22. *Let X be a complex algebraic variety and let $\epsilon : X_\bullet \to X$, $\epsilon' : X'_\bullet \to X$ be two cubical hyperresolutions related by a morphism of cubical varieties $f : X'_\bullet \to X_\bullet$ in the sense that*

$$\begin{array}{ccc} X'_\bullet & \xrightarrow{\ f\ } & X_\bullet \\ & {\scriptstyle \epsilon'}\searrow \quad \swarrow{\scriptstyle \epsilon} & \\ & X & \end{array}$$

is commutative. Then the canonical map

$$R\epsilon_*\Omega^p_{X_\bullet} \to R\epsilon'_*\Omega^p_{X'_\bullet}$$

which obtained by applying $R\epsilon_*$ to $\Omega^p_{X_\bullet} \xrightarrow{f^*} f_*\Omega^p_{X'_\bullet} \to Rf_*\Omega^p_{X'_\bullet}$ is a quasi-isomorphism.

It follows that the complex $R\epsilon_*\Omega^\bullet_{X_\bullet}$ equipped with the trivial filtration considered in the derived filtered category of complexes of coherent $\mathcal{O}_X$-modules is uniquely determined by X. This object is called the **filtered De Rham complex** and denoted $(\tilde{\Omega}^\bullet_X, F)$. By abuse of notation, we shall write

$$\tilde{\Omega}^p_X := \mathrm{Gr}^p_F\, \tilde{\Omega}^\bullet_X[p].$$

Explicitly, fixing some cubical hyperresolution $X^\bullet$, the de Rham complex is given by

$$\tilde{\Omega}^k_X := \bigoplus_{p+q=k} (\epsilon_q)_*\Omega^p_{X_q}$$

with differential the sum $d + d''$ of the ordinary differentiation d coming from the individual complexes $\Omega^\bullet_{X_q}$ and d'' coming from the Čech differential

$$\sum_{i=0}^{q+1}(-1)^i d^*_{i,q+1} : (\epsilon_q)_*\Omega^p_{X_q} \to (\epsilon_{q+1})_*\Omega^p_{X_{q+1}}$$

We shall have occasion to compare this filtered De Rham complex with the De Rham complex constructed from the sheaf Ω^1_X of **Kähler differentials** on X. These sheaves are defined locally as follows. Suppose that $i : U \hookrightarrow V$ is a local chart exhibiting U as the locus of zeroes of finitely many functions f_j, $j = 1, \ldots, k$ defined on a smooth manifold V. Let $\mathcal{I}_V$ be the ideal they define and $d\mathcal{I}_V$ the $\mathcal{O}_U$-module generated by their differentials. Then on U we define the sheaf of Kähler differentials as as

$$\Omega^1_U := \Omega^1_V / (\mathcal{I}_V \Omega^1_V + d\mathcal{I}_V).$$

Clearly, this is a coherent $\mathcal{O}_U$-module. That this is independent of the chosen embedding needs some verification. Indeed, we have

$$\Omega^1_U = \mathcal{I}_\Delta / \mathcal{I}^2_\Delta, \quad \Delta \subset U \times U \text{ the diagonal.}$$

See [Hart77] for the algebraic situation and [Gr-R77] for analytic varieties. By definition Ω^p_U is the p-exterior wedge (over $\mathcal{O}_U$) of Ω^1_U, and by construction the resulting sheaves Ω^p_X are $\mathcal{O}_X$-coherent. The usual d-operator induces natural derivations $d : \Omega^p_X \to \Omega^{p+1}_X$ yielding the **Kähler De Rham complex**

$$\mathcal{O}_X \xrightarrow{d} \Omega^1_X \xrightarrow{d} \cdots \xrightarrow{d} \Omega^n_X. \tag{VII–6}$$

The augmentation $\epsilon : X^\bullet \to X$ induces a homomorphisms of complexes

$$\varphi^\bullet : \Omega^\bullet \to \tilde{\Omega}^\bullet.$$

If we give both the trivial filtration, this morphism is unique in the filtered derived category of complexes of coherent $\mathcal{O}_X$-sheaves.

Examples 7.23. 1) Let Y be a smooth variety and E a divisor of Y with simple normal crossings. We let $\mathcal{I}_E$ be the ideal sheaf of E inside Y. The sheaf Ω^1_E of Kähler differentials now fits in an exact sequence

$$0 \to \mathcal{I}_E/\mathcal{I}^2_E \xrightarrow{d} \Omega^1_Y \otimes \mathcal{O}_E \to \Omega^1_E \to 0,$$

so it is a locally free $\mathcal{O}_Y$-module. The components of E are smooth and as in example 5.3 1), the unions E_q of the q-fold intersections define a simplicial resolution $a_\bullet : E_\bullet \to E$ of E. There is a natural map

$$\Omega^q_E \to (a_0)_* \Omega^q_{E_1}$$

whose kernel consists of those germs of q-forms on E that are supported on the singular locus E_{sing} of E (since a_0 is an isomorphism away from the singular locus and the right hand side is torsion free). These germs constitute the **torsion q-forms** Tors^q_E on E. Using the identification

$$\Omega^q_E = \Omega^q_Y / (\mathcal{I}_E \Omega^q_Y + d\mathcal{I}_E \wedge \Omega^{q-1}_Y)$$

one easily verifies that the torsion equals

$$\mathrm{Tors}_E^q = \mathcal{I}_E \Omega_Y^q(\log E)/(\mathcal{I}_E \Omega_Y^q + d\mathcal{I}_E \wedge \Omega_Y^{q-1}). \tag{VII–7}$$

On the level of $a_* \Omega^\bullet$ we have the Čech differentials d_i''. We thus get a sequence

$$0 \to \mathrm{Tors}^q \to \Omega_E^q \to (a_0)_* \Omega_{E_1}^q \xrightarrow{d_0''} (a_1)_* \Omega_{E_2}^q \xrightarrow{d_1''} \cdots$$

This sequence is exact as we show by a local computation. Indeed, if $Y = \mathbb{C}^n$ and E is given by $z_1 \cdots z_k = 0$, the preceding sequence splits into a direct sum of $\binom{n}{p}$ sequences, one for each monomial form dz_I, $I \subset \{1, \ldots, n\}$. Each of these is then a Koszul-type exact sequence of the form

$$0 \to \prod_{j \in J} z_j A \to A \to \bigoplus_{j \in J} A/(z_j) \to \bigoplus_{j,k \in J} A/(z_j, z_k) \cdots$$

where $J = \{1, \ldots, k\} - I$ and $A = \mathcal{O}_{\mathbb{C}^n}$. This shows that up to quasi-isomorphism we have

$$\tilde{\Omega}_E^\bullet = \Omega_E^\bullet / \mathrm{Tors} = \Omega_Y^\bullet / \mathcal{I}_E \Omega_Y^\bullet(\log E),$$
(all equipped with the trivial filtration).

2) Let $f : Y \to X$ be a morphism of complex algebraic varieties which is a homeomorphism. Since the property of being of cohomological descent is a topological one, any cubical hyperresolution for Y induces one for X by composing with f. Hence $\tilde{\Omega}_X^\bullet = f_* \tilde{\Omega}_Y^\bullet$. In the category of birational morphisms with target X which are homeomorphisms, there is a maximal object, the **weak normalization** $n' : X^{\mathrm{wn}} \to X$ of X. To compute the filtered De Rham complex, it is therefore sufficient to compute it for its weak normalization. For instance, we always have

$$H^0(\tilde{\Omega}_X^\bullet) \xrightarrow{\sim} n'_* \mathcal{O}_{X^{\mathrm{wn}}}.$$

Let us next look at a curve X. Uni-branch singularities get resolved by the weak normalization procedure and the full normalization $n : \tilde{X} \to X$ pulls apart the different branches. If $\Sigma = X_{\mathrm{sing}}$ the obvious exact sequence

$$0 \to n'_* \mathcal{O}_{X^{\mathrm{wn}}} \to n_* \mathcal{O}_{\tilde{X}} \to \underline{\mathbb{C}}_\Sigma \to 0$$

together with the defining sequence

$$0 \to n_* \mathcal{O}_{\tilde{X}} \xrightarrow{d} \underline{\mathbb{C}}_\Sigma \oplus n_* \Omega_{\tilde{X}}^1 \to 0$$

combine to show that the filtered De Rham complex of the curve X is given by

$$0 \to n'_* \mathcal{O}_{X^{\mathrm{wn}}} \xrightarrow{d} n_* \Omega_{\tilde{X}}^1 \to 0$$

together with its trivial filtration.

Next, we state some important properties of the De Rham complex.

Proposition 7.24. *Let X be a complex algebraic variety. Then*

1) *$\tilde{\Omega}^\bullet_X$ is a resolution of the constant sheaf $\underline{\mathbb{C}}_X$;*
2) *the differentials of the graded complex $\mathrm{Gr}_F\, \tilde{\Omega}^\bullet_X$ are $\mathcal{O}_X$-linear and the cohomology sheaves of these complexes are $\mathcal{O}_X$-coherent;*
3) *if X is compact, the spectral sequence of hypercohomology*

$$E_1^{p,q} = \mathbb{H}^{p+q}(X, \mathrm{Gr}_F^p\, \tilde{\Omega}^\bullet_{\mathbb{C}}) \longrightarrow H^{p+q}(X; \mathbb{C})$$

degenerates at E_1 and $E_1^{pq} = \mathrm{Gr}_F^p\, H^{p+q}(X; \mathbb{C})$.

These properties follow more or less immediately from the fact that the filtered De Rham complex is obtained by forgetting the weight filtration in the full complex $\mathcal{K}^\bullet_X$. The degeneration of the spectral sequence is a consequence of Theorem 3.18.

We next study the functorial properties of the De Rham complex. Let $f : Y \to X$ be a morphism of complex algebraic varieties. We have seen (Theorem 5.29) that we can find a cubical hyperresolution $f_\bullet : Y_\bullet \to X_\bullet$ of the diagram $f : Y \to X$. Pulling back differential forms, one obtains a morphism of filtered complexes

$$f^* : \tilde{\Omega}^\bullet_X \to Rf_* \tilde{\Omega}^\bullet_Y.$$

Its cone is the De Rham complex for f:

$$\tilde{\Omega}^\bullet_f := (\mathrm{Cone}^\bullet f^*, \text{trivial filtration})$$

and its hypercohomology computes the cohomology of the mapping cone of f. If X and Y are compact, the spectral sequence for the Hodge filtration degenerates at E_1 and $\mathrm{Gr}_F^p\, H^m(\mathrm{Cone}^\bullet(f)) = H^m(\mathrm{Gr}_F^p\, \tilde{\Omega}^\bullet_f)$. If $f : Y \hookrightarrow X$ is a closed embedding, we write this also as $\tilde{\Omega}^\bullet_{(X,Y)}$. Its hypercohomology computes now the relative cohomology $H^*(X, Y)$ with Hodge gradeds given by $H^q(\tilde{\Omega}^p_{(X,Y)})$.

Example 7.25. Suppose that X is a complex algebraic variety and $j : \Sigma \hookrightarrow X$ a closed subvariety such that $U = X - \Sigma$ is smooth. Then the De Rham complex of the pair (X, Σ) can be calculated as follows. Choose a resolution $\pi : Y \to X$ of X which is a biholomorphism away from $E := \pi^{-1}\Sigma$ and such that E is a divisor with simple normal crossings. The natural map

$$\pi^* : \tilde{\Omega}^\bullet_{(X,\Sigma)} \to R\pi_* \tilde{\Omega}^\bullet_{(Y,E)}$$

is a quasi isomorphism. This can be seen as follows. Let $i : E \hookrightarrow Y$ be the inclusion, $\pi' = \pi|E$ and $k = \pi \circ i = j \circ \pi'$. The 2-cubical variety

$$\begin{array}{ccc} E & \xrightarrow{\ i\ } & Y \\ \downarrow{\scriptstyle \pi'} & \searrow^{k} & \downarrow{\scriptstyle \pi} \\ \Sigma & \xrightarrow{\ j\ } & X \end{array}$$

is of cohomological descent since by assumption Σ contains the discriminant locus of π. It follows that the morphism

$$\tilde{\Omega}^\bullet_X \xrightarrow{(i^*,-\pi^*)} \operatorname{Cone}^\bullet\left(Rj_*\tilde{\Omega}^\bullet_\Sigma \oplus R\pi_*\tilde{\Omega}^\bullet_Y \xrightarrow{\pi'^*+j^*} Rk_*\tilde{\Omega}^\bullet_E\right)[-1]$$

is a quasi-isomorphism. Using Example A.14 3) then translates this into the desired quasi-isomorphism.

On the other hand, the De Rham complex of the embedding $i : E \hookrightarrow Y$ can be computed as follows. We have seen that $\tilde{\Omega}^\bullet_E = \Omega^\bullet_E/\operatorname{Tors}$. As the map

$$i^*\Omega^\bullet_Y \to \Omega^\bullet_E/\operatorname{Tors}$$

is surjective, its cone is filtered quasi isomorphic to the kernel of i^* (see Example A.14 2). Hence, by equation (VII–7) above, we have

$$\tilde{\Omega}^\bullet_{(Y,E)} = \mathcal{I}_E\Omega^\bullet_Y(\log E).$$

In passing, we note that this complex only computes the relative cohomology of (Y, E) if Y is compact. If this is not the case, one has to compactify (Y, E), say into (Y', E') with Y' smooth, E' a divisor such that $E' \cap Y = E$ and such that $D := Y - Y'$ together with E' forms a divisor having simple normal crossings. The same argument as before then shows that the cohomology of (Y, E) can be computed as the hypercohomology of the complex

$$\mathcal{I}_{E'}\Omega^\bullet_{Y'}(\log(E' + D)).$$

The spectral sequence for the trivial filtration degenerates at E_1 and the gradeds of the Hodge filtration on cohomology are given by the cohomology of the gradeds of the trivial filtration on this complex.

7.3.2 Application to Vanishing Theorems

The idea that topological vanishing theorems together with Hodge theory give analytic vanishing results is due to Kollár and Esnault-Viehweg (see for instance [Es-V92] for references and further explanations)

To give an idea of this method let us discuss how it can be used to show the Kodaira Vanishing Theorem [Kod53] which states that for any ample line bundle L on a *smooth* projective manifold one has

$$H^p(X, L^{-1}) = 0, \quad \text{for } p < n = \dim X.$$

Suppose first that L has a section vanishing simply along a divisor $H \subset X$ with simple normal crossings. This implies that the Hodge filtration on $H^p(X - H; \mathbb{C})$ is obtained from De Rham complex $\Omega^\bullet_X(\log H)$ with its trivial filtration. Since $X - H$ is affine, by Theorem C.14, we have $H^k(X - H; \mathbb{C}) = 0$ for $k > n$. The Hodge gradeds of these cohomology groups are $H^{k-p}(\Omega^p_X(\log H))$ and so these vanish for $k > n$ as well. In particular $H^{k-n}(K_X \otimes L) = 0$ for $k > n$ which, by Serre duality, is equivalent to Kodaira Vanishing. Next, for some $N \geq 1$, L^N will have a section vanishing simply along a smooth divisor H and one considers the N-cyclic covering

$$f : Y \to X$$

ramified exactly along H. The line bundle f^*H then has a section vanishing simply along the divisor $H' = f^{-1}H$ and $Y - H'$ is affine so that now $H^k(Y, K_Y \otimes \mathcal{O}_Y(H')) = 0$. Since $K_Y \otimes \mathcal{O}_Y(H') = f^*(K_X \otimes L)$, the result follows from the Leray spectral sequence for f.

This idea can be used without much difficulty to prove the Akizuki-Nakano vanishing theorem [A-N]:

Theorem 7.26 (Akizuki-Nakano). *Let X be a smooth complex projective variety and L an ample line bundle on X. Then $H^p(X, \Omega_X^q \otimes L) = 0$ for $p + q > n$.*

Instead of giving the proof, we shall see that it follows also from our main result Theorem 7.29.

We next want to show how purely local results can be derived from global vanishing theorems.

Theorem 7.27 (Global-to-local principle). *Suppose that $f : X \to Y$ is a morphism between projective varieties, q a natural number and $\mathcal{F}$ a coherent sheaf on X with the property that $H^q(X, \mathcal{F} \otimes f^*L) = 0$ for all ample line bundles L on Y. Then $R^q f_* \mathcal{F} = 0$.*

Proof. Let L be sufficiently ample so that $R^q f_* \mathcal{F} \otimes L$ is generated by sections and $H^i(Y, R^j f_* \mathcal{F} \otimes L) = 0$ for $i > 0$ and all $j \geq 0$. The Leray spectral sequence

$$E_2^{i,j} = H^i(Y, R^j f_* \mathcal{F} \otimes L) \Rightarrow H^{i+j}(X, \mathcal{F} \otimes f^*L)$$

then degenerates at E_2 so that $H^0(Y, R^q f_* \mathcal{F} \otimes L) = H^q(X, \mathcal{F} \otimes f^*L)$. But the latter space vanishes by assumption so that $R^q f_* \mathcal{F} = 0$. □

This principle can be used to derive statement b) from a) in the following result due to Grauert and Riemenschneider [Gr-Rie]:

Theorem 7.28 (Grauert-Riemenschneider). *Let X be a compact n-dimensional complex algebraic variety, $\pi : Y \to X$ a proper modification with Y smooth and L an ample line bundle on X. Then*

a) $H^q(Y, K_Y \otimes \pi^*L) = 0$ *for* $q > 0$;
b) $R^q \pi_* \Omega_Y^n = 0$ *for* $q > 0$.

This idea will be used in the proof theorem of the main theorem below which generalizes the Grauert-Riemenschneider result as we shall see.

We now go to singular varieties and use the filtered De Rham complex instead of the logarithmic De Rham complex. The main result generalizing all of the above is as follows:

Theorem 7.29 ([G-N-P-P, Ste85a]). *Let X be a compact complex algebraic variety of dimension n and let L be an ample line bundle on X. Let $(\tilde{\Omega}^\bullet_X, F)$ be the filtered De Rham complex of X (so F is the trivial filtration). Then*

a) *the hypercohomology groups $\mathbb{H}^m(X, \mathrm{Gr}^p_F \tilde{\Omega}^\bullet_X \otimes L)$ vanish for $m > n$;*
b) *the cohomology sheaf $H^m(\mathrm{Gr}^p_F \tilde{\Omega}^\bullet_X)$ vanishes for $m < p$ or $m > n$.*

Guillen-Navarro Aznar-Puerta-Steenbrink vanishing theorem

Remark. For X smooth, the complex $\mathrm{Gr}^p_F \tilde{\Omega}^\bullet_X$ is just the single sheaf Ω^p_X placed in degree p so that its m-th hypercohomology group is $H^{m-p}(X, \Omega^p_X)$ and a) in the preceding Theorem is just the Akizuki-Nakano vanishing theorem. On the other end of the extreme, to calculate $F^n \tilde{\Omega}^\bullet_X$ it suffices to take a smooth proper modification $\pi : Y \to X$ and take $\pi_* \Omega^n_Y$ viewed a complex concentrated in degree n. We thus obtain the vanishing result of Grauert and Riemenschneider.

Proof (of Theorem 7.29). We reduce the theorem to the following relative version.

Proposition 7.30. *Let X be a compact complex algebraic variety of dimension n and let L be an ample line bundle on X. Let $i : \Sigma \hookrightarrow X$ be a closed subvariety such that the complement $X - \Sigma$ is smooth. Then*

a) *the hypercohomology groups $\mathbb{H}^m(X, \mathrm{Gr}^p_F \tilde{\Omega}^\bullet_{X,\Sigma} \otimes L)$ vanish for $m > n$;*
b) *the cohomology sheaf $H^m(\mathrm{Gr}^p_F \tilde{\Omega}^\bullet_{X,\Sigma})$ vanishes for $m < p$ or $m > n$.*

To see that this Proposition implies Theorem 7.29, we use the p-graded piece of the F-filtration of the exact sequence of the cone (Appendix A, formula A–12)

$$0 \to i_* \tilde{\Omega}^\bullet_\Sigma \to \tilde{\Omega}^\bullet_{X,\Sigma} \to \tilde{\Omega}^\bullet_X[1] \to 0,$$

once in hypercohomology after tensoring with $\mathcal{O}_X(L)$ and once in cohomology. The result of Theorem 7.29 then follows from Prop. 7.30 by induction on the dimension of X. □

Proof (of Prop. 7.30). To start, observe that we have already computed the filtered De Rham complex $\tilde{\Omega}^\bullet_{X,\Sigma}$ in Example 7.25. We use the same notation employed there. So $\pi : Y \to X$ is a proper smooth modification and $E = \pi^{-1}\Sigma$ is a divisor with simple normal crossings and $\tilde{\Omega}^\bullet_{X,\Sigma} = \mathcal{I}_E \Omega^\bullet_Y(\log E)$ so that

$$\mathbb{H}^m(X, \mathrm{Gr}^p_F \tilde{\Omega}^\bullet_{X,\Sigma} \otimes L) = H^{m-p}(Y, \mathcal{I}_E \Omega^p_Y(\log E) \otimes \pi^* L)$$

and

$$H^m(\mathrm{Gr}^p \tilde{\Omega}^\bullet_{X,\Sigma}) = R^{m-p}\pi_* \mathcal{I}_E \Omega^p(\log E).$$

It follows that it is sufficient to show the following two assertions:

a′) $H^q(Y, \mathcal{I}_E \Omega^p_Y(\log E) \otimes \pi^* L) = 0$ for $p + q > n$, b') $R^q\pi_*\mathcal{I}_E\Omega^p_Y(\log E) = 0$ for $p+q>n$.

By the global to local principle, it suffices to prove a'. Instead, we prove the dual statement. To formulate it, introduce the rank n-bundle

$$\mathcal{V} := T_X(-\log E),$$

by definition the dual of $\Omega^1(\log E)$. Since $\Omega^n_Y(\log E) = \Omega^n_Y \otimes \mathcal{O}(E)$, the isomorphism

$$\Lambda^p\mathcal{V} \otimes \Lambda^n\mathcal{V}^\vee \xrightarrow{\sim} \Lambda^{n-p}\mathcal{V}^\vee$$

then yields

$$\Omega^p_Y(\log E)^\vee \otimes K_Y \simeq \Omega^{n-p}_Y(\log E) \otimes \mathcal{O}_Y(-E).$$

It follows that Serre duality takes the form

$$H^p(Y, \Omega^q_Y(\log E) \otimes \pi^* L^{-1})^\vee = H^{n-p}(Y, \Omega^{n-q}_Y(\log E) \otimes \mathcal{O}_Y(-E) \otimes \pi^* L).$$

Hence, since $\mathcal{I}_E = \mathcal{O}_Y(-E)$, to prove a') we are reduced to proving

$$H^q(Y, \Omega^p_Y(\log E) \otimes \pi^* L^{-1}) = 0, \qquad p+q<n.$$

To show this, we reduce first to the case when L is very ample. This goes as follows. Let L^N be very ample. Choose a section of L^N vanishing along a smooth hyperplane H which is transverse to all mappings $E_{i_1}\cap\cdots\cap E_{i_p} \to X$, where the E_j are the components of E. Let $f : X' \to X$, respectively $g : Y' \to Y$ be the N-cyclic covering branched along H, respectively $H' = \pi^{-1}H$. Then Y' is smooth and π induces a proper modification $\pi' : Y' \to X'$ fitting in a commutative diagram

$$\begin{array}{ccc} Y' & \xrightarrow{g} & Y \\ \downarrow{\scriptstyle \pi'} & & \downarrow{\scriptstyle \pi} \\ X' & \xrightarrow{f} & X. \end{array}$$

Moreover if $\Sigma' = f^{-1}\Sigma$, the complement $X' - \Sigma'$ is smooth and now $Y' - g^{-1}H'$ is affine. Put $E' = g^{-1}E = (\pi')^{-1}\Sigma'$ and $L' = f^*L$ and assume that the groups $H^q(Y', \Omega^p_{Y'}(\log E') \otimes \pi'^* L'^{-N})$ vanish for $p+q<n$. Since $E+H'$ is a divisor with simple normal crossings, a calculation (see for instance Lemma 3.16 in [Es-V92]) shows that

$$g_*\Omega^p_{Y'}(\log E') = \bigoplus_{i=0}^{N} \Omega^p_Y(\log E) \otimes \pi^* L^{-i}.$$

Since $(\pi')^*L' = g^*\pi^*L$, the Leray spectral sequence for g then gives

$$0 = H^q(Y, g_*\Omega^p_{Y'}(\log E') \otimes \pi^*(L^{-N})) = \bigoplus_{i=0}^{N} H^q(Y, \Omega^p_Y(\log E) \otimes \pi^*(L^{-N-i}))$$

and the desired vanishing then follows. For the remainder of the proof we thus may indeed replace Y', X', π', L' by Y, X, π, L thereby assuming that

L is very ample with a section vanishing along a hypersurface H with the required transversality.

We complete the proof by induction on the dimension. Indeed $H' = \pi^{-1}H$ is smooth and $H' \cup E$ is a divisor with simple normal crossings as is $D = E \cap H'$, and then the induction hypothesis implies

$$H^q(H', \Omega^p_{H'}(\log D) \otimes (\pi|H)^* L^{-1}) = 0, \quad \text{for } p + q < n - 1. \qquad \text{(VII–8)}$$

Next, since $\pi^* L^{-1}$ is the ideal sheaf of H', the groups $H^q(Y, \Omega^p_Y(\log E) \otimes \pi^* L^{-1})$ occur in the long exact sequence in cohomology of the tautological sequence

$$0 \to \mathcal{I}_{H'} \to \mathcal{O}_Y \to \mathcal{O}_{H'} \to 0$$

tensored by $\Omega^p_Y(\log E)$. So it suffices to show that the maps

$$H^q(Y, \Omega^p_Y(\log E)) \xrightarrow{c_{pq}} H^q(H', \Omega^p_Y(\log E) \otimes \mathcal{O}_{H'}) \qquad \text{(VII–9)}$$

are isomorphisms for $p + q < n - 1$ and an injection for $p + q = n - 1$.

Next, we observe that the map π maps D onto $S = \Sigma \cap H$ while the smooth complement $H' - D$ maps isomorphically onto $H - S$. So the pair $(Y - E, H' - D)$ maps isomorphically onto $(X - \Sigma, H - S)$. By the version of the Lefschetz theorem stated in Remark C.16, the restriction maps

$$H^k(X - \Sigma) \to H^k(H - S)$$

for the latter are isomorphisms in the range $k < n - 1$ and injective if $k = n - 1$. These maps are morphisms of mixed Hodge structures and so, taking the graded parts of the Hodge filtration, the maps

$$a_{pq} : H^q(Y, \Omega^p_Y(\log E)) \to H^q(H', \Omega^q_{H'}(\log D))$$

are likewise isomorphisms for $p + q < n - 1$ and injective for $p + q = n - 1$. These maps factor as follows

$$H^q(Y, \Omega^p_Y(\log E)) \xrightarrow{c_{pq}} H^q(H', \Omega^p_Y(\log E) \otimes \mathcal{O}_{H'}) \xrightarrow{b_{pq}} H^q(H', \Omega^p_{H'}(\log D)).$$

The first map c_{pq} is the map (VII–9). The second map is induced by the long exact sequence

$$0 \to \Omega^{p-1}_{H'}(\log D) \otimes \mathcal{O}_{H'}(-H') \to \Omega^p_Y(\log E) \otimes \mathcal{O}_{H'} \to \Omega^p_{H'}(\log D) \to 0.$$

To show that this indeed is an exact sequence of locally free $\mathcal{O}_{H'}$-sheaves, by standard linear algebra considerations it suffices to do this for $p = 1$ where it follows from a local calculation which we omit.

Since $\mathcal{O}_{H'}(-H')$ is the restriction of $\pi^* L^{-1}$ to H', the long exact sequence in cohomology together with the induction hypothesis (VII–8) then show that b_{pq} is an isomorphism for $p + q < n - 1$ and injective for $p + q = n - 1$. Since the same is true for $a_{pq} = b_{pq} \circ c_{pq}$, we have this result for c_{pq} as well, which is what we wanted to prove. □

7.3.3 Applications to Du Bois Singularities

Recall the notion of a Cohen-Macaulay singularity from § 2.5. We use the criterion (see [S-T, Thm 1.14]) :

$$(X,x) \text{ Cohen-Macaulay } \iff H^k_{\{x\}}\mathcal{O}_X = 0, \quad k = 0,\dots,\dim X - 1.$$

The dualizing complex $\omega_X^\bullet$ for general complex spaces X (see § 2.5) is an ingredient in Grothendieck's local duality theorem ([Groth67, Thm. 6.3]):

Theorem 7.31 (Local Grothendieck duality theorem). *Let X be a complex space of pure dimension n and let $\mathcal{F}$ be a coherent sheaf which is locally free on the smooth locus of X. For every $x \in X$ there is a canonical isomorphism*

$$\operatorname{Ext}^q_{\mathcal{O}_{X,x}}(\mathcal{F}_x, \omega^\bullet_{X,x}) \xrightarrow{\sim} H^{n-q}_{\{x\}}(X,\mathcal{F})^\vee.$$

As a consequence, (X,x) is Cohen-Macaulay if and only if the dualizing complex only has cohomology in degree 0. This sheaf is the **dualizing sheaf** ω_X. We say that the singularity is **Gorenstein** if $\omega_{X,x} = \mathcal{O}_{X,x}$. Local complete intersections are always Gorenstein.

For isolated singularities (X,x), the local cohomology groups can be calculated by means of a **good resolution** $\pi : Y \to X$. By definition this means that X is a contractible Stein space, Y is smooth and the exceptional divisor $E = \pi^{-1}x$ has simple normal crossings. In this setting there is another useful duality theorem (see [Kar, Theorem 3.2]):

Theorem 7.32 (Local version of Serre's duality theorem). *Let $\pi : (Y,E) \to (X,x)$ be a good resolution. Let $\mathcal{F}$ be a locally free sheaf on Y. Then for $q < n$ the local cohomology groups $H^q_E(Y,\mathcal{F})$ are finite dimensional and there is a canonical duality isomorphism*

$$H^q_E(Y,\mathcal{F}) \xrightarrow{\sim} H^{n-q}(E, (\mathcal{F}^\vee \otimes \Omega^n_Y)|E)^\vee.$$

Corollary 7.33. *Let $\pi : (Y,E) \to (X,x)$ be a good resolution of an isolated n-dimensional singularity x. Then*

$$H^q(Y,\mathcal{O}_Y) \cong H^{q+1}_{\{x\}}(X,\mathcal{O}_X), \quad q = 0,\dots,n-2.$$

Proof. By Grauert-Riemenschneider vanishing (Theorem 7.28), the space $H^{n-q}(E,\Omega^n_Y|E)$ vanishes for $q < n$ since it is the stalk of $R^{n-q}\pi_*\Omega^n_Y$ at x. By local Serre duality, the latter space is dual to $H^q_E(Y,\mathcal{O}_Y)$. So, by the exact sequence in local cohomology, setting $U = Y - E = X - \{x\}$, the restriction maps $H^q(Y,\mathcal{O}_Y) \to H^q(U,\mathcal{O}_U)$ are isomorphisms for $q < n-1$. Then the exact sequence in local cohomology for (X,x) together with the fact that X is Stein gives the isomorphisms $H^q(\mathcal{O}_U) \xrightarrow{\sim} H^{q+1}_{\{x\}}(X,\mathcal{O}_X)$. □

It follows that the numbers $\dim H^p(Y, \mathcal{O}_Y)$ are invariants of the singularity for $p = 1, \ldots, n-2$. As to $p = n-1$, denoting by L^2 the "square integrable sections", one has ([Kar, Prop. 4.2]):

$$\dim H^{n-1}(Y, \mathcal{O}_Y) = \dim \left(H^0(U, \Omega_U^n) / L^2(U, \Omega_U^n) \right). \qquad \text{(VII–10)}$$

There is one further invariant, the δ-invariant which measures how far the singularity is from being normal, where we say that a singularity is **normal**, respectively **weakly normal** if its is isomorphic to its normalization, its weak normalization, respectively. It is defined as follows:

$$\delta(X, x) := \dim_{\mathbb{C}}(\mathcal{O}_{\widetilde{X,x}} / \mathcal{O}_{X,x})$$

where the tilde denotes the normalization. Traditionally, the **geometric genus** is related to these invariants by means of the defining formula

$$(-1)^n p_g(X, x) := -\delta(X, x) + \sum_{q=1}^{n-1} (-1)^{q+1} \dim_{\mathbb{C}} H^q(Y, \mathcal{O}_Y).$$

For a normal Cohen-Macaulay singularity only $H^{n-1}(Y, \mathcal{O}_Y)$ can be non-zero so that

$$p_g(X, x) = \dim H^{n-1}(Y, \mathcal{O}_Y). \qquad \text{(VII–11)}$$

We say that a normal singularity is **rational** if it is Cohen-Macaulay and if p_g vanishes. Equivalently, for a good resolution $f : Y \to X$ we have $H^q(\mathcal{O}_Y) = 0$ for all $q > 0$. Another way of saying this is that the higher direct images $R^i f_* \mathcal{O}_Y$, $i > 0$ all vanish. In fact this inspires the definition for possibly non-isolated rational singularities: a variety X has **rational singularities** if for every resolution of singularities $f : Y \to X$ one has $R^i f_* \mathcal{O}_Y = 0$, $i > 0$. One can show ([K-K-M-S, pp. 50–51]) that it suffices to verify this assertion for one resolution, for instance, for a good resolution in case we have an isolated singularity. This shows that the definitions are indeed compatible.

By (VII–10) and (VII–11) above, a Cohen-Macaulay singularity is rational if all n-forms on U extend to square integrable ones. Examples of such singularities include quotient singularities and toroidal singularities. Note that a quotient singularity need not be Gorenstein; this is only the case for quotients by a subgroup of $\mathrm{SL}(n)$.

A class of singularities that has become important in Mori's approach to the classification of higher dimensional varieties (see for example [C-K-M]) is the class of **canonical singularities**. To define these, recall that on a normal complex algebraic variety the canonical divisor K_X is the divisor of a meromorphic n-form on X, where $n = \dim X$. Equivalently, with $i : X_{\mathrm{reg}} \hookrightarrow X$ the inclusion of the smooth locus into X, we have

$$\omega_X := \mathcal{O}_X(K_X) = i_* \Omega^n_{X_{\mathrm{reg}}}.$$

Note that for a Cohen-Macaulay singularity the sheaf ω_X is indeed the dualizing sheaf so that this notation is consistent.

Now X has canonical singularities if first of all a multiple, say rK_X of K_X is a Cartier divisor and secondly, if for every resolution of singularities $f: Y \to X$ we have $f^*(rK_X) = rK_Y + \sum r_i E_i$ where $r_i \geq 0$ and E_i runs over the exceptional divisors. More generally, if the numbers $a_i = r_i/r$ always satisfy $a_i \geq -1$, respectively $a_i > -1$, we say that X has **log terminal** respectively **log canonical singularities**. The smallest positive integer r such that rK_X is Cartier is called the **index** of the singularity. By [Reid, 3.6] one can always locally achieve index 1 by a finite surjective morphism $X' \to X$. Gorenstein singularities are examples of index 1 singularities.

Coming back to the filtered De Rham complex, one may introduce the **Du Bois invariants**

$$b^{p,q}(X,x) := \dim H^q(\tilde{\Omega}^p_{X,x}).$$

For isolated singularities one can take for X a contractible Stein space, so that $H^q(\mathcal{O}_{X,x}) = 0$ whenever $q > 0$. The natural morphism

$$\varphi^0 : \mathcal{O}_X \to \tilde{\Omega}^0_X$$

then induces isomorphisms in cohomology if

i) $\mathcal{O}_X \xrightarrow{\sim} H^0(\tilde{\Omega}^\bullet_X))$, i.e. X is weakly normal, and
ii) $b^{0,q}(X,x) = 0$ for $q > 0$.

This motivates the following definition.

Definition 7.34. A variety X has **Du Bois singularities** if $\varphi^0 : \mathcal{O}_X \xrightarrow[\sim]{\text{qis}} \tilde{\Omega}^0_X$. A point $x \in X$ is a Du Bois singularity if φ^0_x is an isomorphism. So X is Du Bois if all its points are du Bois.

Examples 7.35. 1) As to curve singularities, we calculated the De Rham complex explicitly and so it follows that a curve is Du Bois if and only if it is weakly normal, that is all branches are smooth and the tangent directions are all distinct.

2) Rational singularities are Du Bois. We give here a sketch of the proof from [Kov]. Suppose that a complex space X has at most rational singularities. So, if $f: Y \to X$ is a resolution of singularities, the induced map $\mathcal{O}_X \to Rf_*\mathcal{O}_Y$ (of complexes of $\mathcal{O}_X$-modules) is a quasi-isomorphism. In particular, this map has a left inverse in the appropriate derived category. Since the preceding map factors over the natural map $h : \mathcal{O}_X \to \tilde{\Omega}^0_X$, also this latter map has a left inverse. Now we have the following

Splitting principle: *If the natural map $h : \mathcal{O}_X \to \tilde{\Omega}^0_X$ has a left inverse, X has only Du Bois singularities.*

To show this, we first observe that the assertion is local and so we may assume that X is algebraic. Next, the assumption on h is stable under taking hyperplanes. On the other hand, a cubical resolution for X induces one for its generic hyperplane H. This implies that the filtered De Rham complex $\tilde{\Omega}^0_H$ is quasi-isomorphic to $\mathcal{O}_H \otimes \tilde{\Omega}^0_X$. So, if X has points that are not Du Bois, a generic hyperplane section will have the same property. In proving the assertion, we may therefore assume that X is Du Bois except maybe at one isolated point x. So h is a quasi-isomorphism except maybe at x. The assumption on h implies in particular that the induced map on the level of the local (hyper)cohomology groups

$$h_x : H^k_x(X, \mathcal{O}_X) \to \mathbb{H}^k_x(X, \tilde{\Omega}^0_X)$$

is injective for all k. We shall show that these maps are surjective as well. Here we use the assumption that X is algebraic. We choose a compactification $\bar{X}$ with complement a divisor D and we use (see Prop. 7.24(iii)) that the spectral sequence of hypercohomology associated to the filtered De Rham complex on $\bar{X}$ degenerates at E_1. In particular the map $H^k(\bar{X}, \mathbb{C}) \to \mathbb{H}^k(\bar{X}, \tilde{\Omega}^0_{\bar{X}})$ is surjective. Since it factors over $H^k(\bar{X}, \mathbb{C}) \to H^k(\bar{X}, \mathcal{O}_{\bar{X}})$, the map $h : H^k(\bar{X}, \mathcal{O}_{\bar{X}}) \to \mathbb{H}^k(\bar{X}, \tilde{\Omega}^0_{\bar{X}})$ is surjective as well. Now put $Z = D \cup \{x\}$ and consider the commutative diagram

$$\begin{array}{ccccccc}
H^{k-1}(X - \{x\}, \mathcal{O}_X) & \to & H^k_Z(\bar{X}, \mathcal{O}_{\bar{X}}) & \to & H^k(\bar{X}, \mathcal{O}_{\bar{X}}) & \to & H^k(X - \{x\}, \mathcal{O}_X) \\
\downarrow & & \downarrow{\scriptstyle h_Z} & & \downarrow{\scriptstyle h} & & \downarrow \\
\mathbb{H}^{k-1}(X - \{x\}, \tilde{\Omega}^0_X) & \to & \mathbb{H}^k_Z(\bar{X}, \tilde{\Omega}^0_{\bar{X}}) & \to & \mathbb{H}^k(\bar{X}, \tilde{\Omega}^0_{\bar{X}}) & \to & \mathbb{H}^k(X - \{x\}, \tilde{\Omega}^0_X).
\end{array}$$

In this diagram the leftmost and the rightmost vertical maps are isomorphisms for all k. We have seen that the map h is onto. It follows that h_Z must be onto. In particular $H^k_x(\bar{X}, \mathcal{O}_{\bar{X}}) \to \mathbb{H}^k_x(\bar{X}, \tilde{\Omega}^0_{\bar{X}})$ must be onto, and by excision this proves that indeed $H^k_x(X, \mathcal{O}_X) \to \mathbb{H}^k_x(X, \tilde{\Omega}^0_X)$ is onto.
To complete the proof, we invoke a general

Localization principle: *Suppose that $h : \mathcal{F}^\bullet \to \mathcal{G}^\bullet$ is a morphism of complexes of sheaves on X that is a quasi-isomorphism everywhere except maybe at $x \in X$. Suppose moreover that h induces for all k isomorphisms $\mathbb{H}^k_x(\mathcal{F}^\bullet) \to \mathbb{H}^k_x(\mathcal{G}^\bullet)$. Then h is also a quasi-isomorphism at x.*

To see this, note that replacing $\mathcal{G}^\bullet$ by a quasi-isomorphic complex, we may assume (see § A.1) that h is injective with quotient the mapping cone $\mathcal{Q}^\bullet = \mathrm{Cone}^\bullet(h)$. By assumption, $\mathcal{Q}^\bullet$ is everywhere acyclic except maybe at x. The assumption on $\mathbb{H}^k_x$ implies that $\mathbb{H}^k_x(\mathcal{Q}^\bullet) = 0$ and then the exact sequence for local cohomology yields $\mathbb{H}^k(X, \mathcal{Q}^\bullet) \xrightarrow{\sim} \mathbb{H}^k(X - \{x\}, \mathcal{Q}^\bullet) = 0$. The spectral sequence $H^i(X, H^j(\mathcal{Q}^\bullet)) \Longrightarrow \mathbb{H}^{i+j}(X, \mathcal{Q}^\bullet)$ degenerates since $H^i(\mathcal{Q}^\bullet)$ is supported on x and so for all k we have $0 = \mathbb{H}^k(X, \mathcal{Q}^\bullet) = H^0(X, H^k(\mathcal{Q}^\bullet)) = H^k(\mathcal{Q}^\bullet)_x$, as desired.

3) Log terminal singularities are Du Bois. To show this, following again [Kov], we first observe that, since the statement is local, we may assume that our singular variety X admits a finite cover $f : X' \to X$ with canonical singularities of index one [Reid, 3.6]. Suppose that we can show that X' is Du Bois. Then X is Du Bois as well: we claim that $\mathcal{O}_X \to Rf_*\mathcal{O}_{X'}$ has a left inverse so that we can apply the above splitting principle. Indeed, $R^if_*\mathcal{O}_{X'} = 0$ for $i > 0$ since f is finite and the (normalized) trace map splits $\mathcal{O}_X \to f_*\mathcal{O}_{X'}$. So it remains to show that canonical index 1 singularities

U are Du Bois. Let $f : V \to U$ be a resolution of singularities. Then, by definition, $f^*\omega_U$ is a subsheaf of ω_V and so $f_*\omega_V = \omega_U$. By the Grauert-Riemenschneider vanishing result (Theorem 7.28(ii)), $R^i f_*\omega_V = 0$ for $i > 0$. It follows that $Rf_*\omega_V$ is quasi-isomorphic to ω_U. So the inclusion $f^*\omega_U \to \omega_V$ yields

$$Rf_*f^*\omega_U \to Rf_*\omega_V \xrightarrow[\sim]{\text{qis}} \omega_U$$

which, away from the singularities, and hence everywhere, is a left inverse for $\omega_U \to Rf_*f^*\omega_U$. Hence, upon tensoring with ω_U^{-1} this gives a left inverse for $\mathcal{O}_U \to Rf_*\mathcal{O}_V$. The conclusion again follows on applying the splitting principle.

As to log canonical singularities, we have a partial result due to Kovács ([Kov, Theorem K]) which is too technical to reproduce here.

4) Gorenstein Du Bois surface singularities have been classified. In addition to the rational singularities we have the following list:
- the simple elliptic and the cusp singularities ([Ste83]);
- ordinary double curve singularities (locally of the form $xy = 0$), the pinch point ($xy^2 = z^2$ and three "degenerate cusps" with equations $xyz = 0$ (the ordinary triple point), $z^2+x^2y^2 = 0$ and $z^2+y^3+x^2y^2 = 0$. See [Str].

5) Non-Gorenstein Du Bois singularities abound. Take a smooth curve with a suitably ample line bundle on it and blow down the zero section. The resulting singularity is such an example. See [DuB, Prop. 4.13].

6) See [Is85], [Is86], [Is87] for more on Du Bois singularities.

Historical Remarks. The statement of the (original) Hodge conjecture can be found in [Ho50]. Grothendieck's generalization is stated in [Groth69], while the version for singular varieties can be found in [Jann]. For a nice overview of what is known for the Hodge conjecture see [Lewis].

Griffiths introduced the intermediate Jacobian in [Grif68], the relation with Deligne cohomology is due to Deligne. See [Es-V88] and [ElZ-Z]. For a rather complete overview of Deligne-Beilinson cohomology see [Es-V88].

The vanishing results in § 7.3.2 are largely due to Guillén, Navarro Aznar, Pascual-Gainza and Puerta [G-N-P-P]. We present Steenbrink's simplified proof from [Ste85a].

The filtered De Rham complex has been introduced by Du Bois [DuB] and motivates the terminology "Du Bois singularity". § 7.3.3 contains further historical remarks on singularities.

Part III

Mixed Hodge Structures on Homotopy Groups

8

Hodge Theory and Iterated Integrals

In this chapter we give Hain's construction of a mixed Hodge structure on homotopy groups. His results are explained in § 8.2 after a first section in which we collect some basic material from homotopy theory that we need later on. A central result in this section is the Borel-Serre theorem which (under suitable assumptions) relates the homotopy groups of a topological space to the homology of its loop space. Loop spaces are no longer finite dimensional manifolds and so we can not hope to put directly a mixed Hodge structure on their cohomologygroups. However, dually, the cohomology of a loop space can be calculated by means of an integration procedure which associates to an ordered set of forms on a differential manifold a single form on its loop space, which is called an "iterated integral". It is explained in § 8.3.

In § 8.4 and § 8.5 we explain how to deal with the fundamental group of a smooth complex projective variety. The mixed Hodge structure given there depends on the choice of base points.

The construction for the higher homotopy groups for a simply connected smooth complex projective variety is carried out in § 8.6 and § 8.7. It proceeds along the following lines. The starting point is a theorem due to Chen which states that the cohomology of the loop space of X can be calculated by means of iterated integrals and this can be done through an algebraic construction on the de Rham complex on X itself, the "bar construction". Starting with weight and Hodge filtrations on the De Rham complex of X we thus naturally get similar filtrations on its bar constructions.

What complicates the story is that the (duals of the) homotopy groups are not given by the full cohomology of the loop space, but rather by its "indecomposables". These are defined by means of the multiplicative structure of the cohomology ring. Therefore the usual cup product on the De Rham complex on X is needed to capture the indecomposables in the cohomology of the loop space of X. This leads to multiplicative mixed Hodge complexes which incorporates these ingredients on the level of complexes. However, we also need a good product on the level of rational complexes. The classical cup-product on the level of cochains is not the correct ingredient, since one needs a (graded) commutative product. This is explained in § 8.6, where we construct the De Rham complex and its bar construction over the *rationals* making use of a certain rationally defined complex introduced by Sullivan and which does have a graded commutative product. We then have all the ingredients to put the structure of a mixed multiplicative Hodge complex on the De Rham complex leading to the desired mixed Hodge structure on the cohomology of the loop space and on its indecomposables.

We will not give full proofs for all of Hain's results. We discuss fundamental groups of smooth projective varieties and homotopy groups of smooth projective varieties and only outline how to adapt the latter to the case of arbitrary complex algebraic varieties. Also, the complementary results stated in § 8.2 will not be proved. However, all the ingredients necessary to understand Hain's proofs of these results can be found in this chapter.

8.1 Some Basic Results from Homotopy Theory

We start by recalling the definition and the basic properties of the homotopy groups. For any two pairs of topological spaces (X, A) and (Y, B) we use the notation $[(X, A), (Y, B)]$ for the set of homotopy classes of maps $X \to Y$ sending A to B (any homotopy is supposed to send A to B as well). Let $I = [0, 1]$ be the unit interval. Fixing a point s on the k-sphere S^k, we have

$$\pi_k(X, x) = [(I^k, \partial I^k), (X, x)] = [(S^k, s), (X, x)];$$

There is a natural product structure on these sets (divide I^k in two and use the first map on one half and the second map on the other half). This makes $\pi_k(X, x)$ into a group, which turns out to be abelian for $k \geq 2$.

We next relate the $(k+1)$-st homotopy group of any pathwise connected topological space X to the k-th cohomology group of the loop space. We denote the path space of X, equipped with the compact-open topology by PX, and the loop space of loops based at x, with the induced topology by $P_x X$.

One has natural isomorphisms

$$\pi_{k+1}(X, x) \xrightarrow{\sim} \pi_k(P_x X, e_x) \quad k \geq 0,$$

where e_x is the constant loop at x. The isomorphism is obtained by viewing a map $[I^{k+1}, \partial I^{k+1}] \to (X, x)$ as a parametrized map $f : [I^k, \partial I^k] \times [I, \partial I] \to (X, x)$, yielding the map $I^k \to P_x X$ sending $u \in I^k$ to the loop $t \mapsto f(u, t)$.

Homotopy and homology are related through the **Hurewicz homomorphism**

$$h_k : \pi_k(Y, y) \to H_k(Y),$$

defined by associating to the class of a map $f : S^k \to Y$ the image under f_* of a generator of $H_k(S^k)$. The following important result tells us when the Hurewicz homomorphism actually is an isomorphism:

Theorem 8.1 (Hurewicz theorem). *Suppose that (X, x) is $(k-1)$-connected, i.e. $\pi_s(X, x) = 1$, $s = 0, \ldots, k-1$. Then h_k is an isomorphism.*

The map dual to h_k is a homomorphism $H^k(Y;\mathbb{Q}) \to \mathrm{Hom}(\pi_k(Y),\mathbb{Q})$. The left hand side can be huge. But in fact the homomorphism factors over a quotient of $H^k(Y;\mathbb{Q})$, the indecomposables of degree k, obtained by dividing out the subspace generated by products of two or more factors of positive degree. This follows from the fact that the cohomology ring of a sphere has no non-trivial cup products. For later use, we need a name for the quotient just considered.

Definition 8.2. – Let R be a ring and A an R-algebra with unit 1. An **augmentation** is an R-algebra homomorphism $\epsilon : A \to R$ sending 1 to 1 and its kernel is called the **augmentation ideal** IA.
– If $R = k$ is a field, $\epsilon : A \to k$ an augmentation, the k-vector space

$$QA = (IA)/(IA)^2 = IA \otimes_A k$$

is called the space of the **indecomposables** of A. It can be identified with the k-vector space with basis a minimal set of generators of A as a k-algebra.

Example 8.3. i) Let G be a group, and let $R[G]$ be its group ring with R-coefficients. *The* augmentation $\epsilon : R[G] \to R$ is the R-homomorphism defined by $\epsilon(r) = r$, $\epsilon(g) = 1$. In this case the augmentation ideal J is generated by the elements $g - 1$, $g \in G$. If $G = \pi_1(X, x)$, the fundamental group of a topological space, its group ring can be identified with $H_0(P_xX)$: indeed, the path components of the loop space P_xX are the homotopy classes of loops based at x. This shows that this ring can be very big indeed. If however the fundamental group is finitely presented, the quotients $\mathbb{Z}\pi_1(X,x)/J^k$ by the successive powers of the augmentation ideal J are finitely generated. So it makes sense to ask whether in this case these quotients have a mixed Hodge structure.
ii) Let Y be a connected topological space and $A = H^*(Y;\mathbb{Q})$ its cohomology-algebra. The inclusion of a point $y \hookrightarrow Y$ induces an augmentation $H^*(Y;\mathbb{Q}) \to \mathbb{Q}$ whose kernel is exactly the reduced cohomology having only strictly positive degrees. So pointed connected spaces give augmented cohomology algebras. An example is provided by the loop space $Y = P_xX$ of a simply connected pair (X, x).

So, our discussion provides us with a morphism

$$Qh^k : QH^k(Y;\mathbb{Q}) \to \mathrm{Hom}(\pi_k(Y,y),\mathbb{Q}). \qquad \text{(VIII–1)}$$

This map is rarely an isomorphism, but if Y is a loop space, we can apply a result due to Borel and Serre. Its formulation requires yet another notion from homotopy theory which we briefly discuss. Details can be found in [Wh].

Example 8.4. An H**-space** is a pointed space (Y, y) equipped with a continuous map $h : Y \times Y \to Y$ such that the two inclusions $z \mapsto (z, y)$, $z \mapsto (y, z)$

composed with h are homotopic to the identity. Such a map is called a **multiplication**. If this multiplication is associative up to homotopy and if moreover there exist a map $i : (Y, y) \to (Y, y)$ which serves as an inverse up to homotopy, we say that (Y, y) is a **group like H-space**.

The standard example of a group like H-space is the loop space P_xX of the loops at x (take the constant loop at x as the base point in P_xX and composition of loops for the multiplication map). Another example is a topological group with identity as base point and group multiplication as the multiplication map.

Now we can formulate the result we are after (see [Mil-Mo]):

Theorem 8.5 (THEOREM OF BOREL AND SERRE). *Let Y a connected H-space such that $\pi_k(Y, y)$ is finitely generated. Then the map Qh^k (VIII–1) is an isomorphism.*

Corollary 8.6. *Let X be a simply connected topological space such that $\pi_{s+1}(X)$ is finitely generated. Then we have an isomorphism*

$$Qh^s : QH^s(P_xX; \mathbb{Q}) \xrightarrow{\sim} \operatorname{Hom}(\pi_s(P_xX), \mathbb{Q}) \xrightarrow{\sim} \operatorname{Hom}(\pi_{s+1}(X, x), \mathbb{Q}).$$

We can indeed apply this to algebraic varieties:

Lemma 8.7. *The fundamental group of a complex algebraic variety is finitely presented. For simply connected complex algebraic varieties the higher homotopy groups are finitely generated.*

Proof. Any compact algebraic variety admits a triangulation and hence its fundamental group is finitely presented. A non compact algebraic variety can be compactified by a divisor whose components meet transversally. We can then triangulate the compactification in such a way that the compactifying divisor becomes a subcomplex. The complement is thus a finite union of (open) simplices and hence the fundamental group is again finitely presented.

The higher homotopy groups of a finite cell complex in general are *not* finitely generated, but for simply connected spaces with finitely generated homology (such as complex algebraic varieties) this still holds. This is a non-trivial result in homotopy theory. See e.g. [Span, p. 509]. □

We next explain the extra structure which is present on the cohomology of H-spaces such as P_xX, that of a Hopf algebra, whose definition we now give:

Definition 8.8. Let k be a field and A a k-algebra containing k. Suppose that $\epsilon : A \to k$ is an augmentation. A **co-multiplication** is an algebra homomorphism $\Delta : A \to A \otimes_k A$ and it makes A into a **Hopf algebra** if it is associative in the sense that $(\Delta \otimes 1) \circ \Delta = (1 \otimes \Delta) \circ \Delta$, and if the augmentation satisfies the rule $(\epsilon \otimes \mathrm{id}_A) \circ \Delta = \mathrm{id}_A = (\mathrm{id}_A \otimes \epsilon) \circ \Delta$. If in addition A is graded and the multiplication and the co-multiplication preserve the grading we say that A is **a graded Hopf algebra**. If in addition the multiplication $A^p \otimes A^q \to A^{p+q}$ satisfies $xy = (-1)^{pq} yx$, we say that the the Hopf algebra is **graded commutative**.

Example 8.9. Let Y be any connected H-space with multiplication h. Suppose that the cohomology $H^*(Y)$ of Y is finitely generated. Then one has a Künneth decomposition $H^*(Y \times Y) \cong H^*(Y) \otimes H^*(Y)$ and hence a co-product

$$\mu : H^*(Y) \xrightarrow{h^*} H^*(Y \times Y) \simeq H^*(Y) \otimes H^*(Y).$$

To verify that $H^*(Y)$ with the usual cup product multiplication m and co-product μ is a Hopf algebra, one uses that the multiplication m is related to the diagonal embedding $\delta : X \hookrightarrow X \times X$ as follows

$$m : H^*(Y) \otimes H^*(Y) \cong H^*(Y \times Y) \xrightarrow{\delta^*} H^*(Y).$$

Returning to a general graded commutative Hopf algebra A, the co-product Δ induces a new co-product $\Delta^{(\mathrm{J})}$ by composing it with the operator

$$\begin{aligned} \mathrm{J} : A \otimes A &\to A \otimes A \\ (a \otimes b) &\mapsto (-1)^{pq} b \otimes a, \quad a \in A^p,\, b \in A^q. \end{aligned}$$

The **co-bracket** is the map

$$\Delta - \Delta^{\mathrm{J}} : A \to A \otimes A. \qquad \text{(VIII–2)}$$

This co-bracket descends to the indecomposables and makes QA into what is called a **Lie co-algebra**.

In particular, by Example 8.9, for any group-like H-space Y the indecomposables $QH^*(Y)$ form a co-algebra equipped with a co-bracket. For such Y there is a dual procedure leading to a bracket

$$\pi_s(Y, y) \times \pi_t(Y, y) \xrightarrow{[\,,]} \pi_{s+t}(Y, y)$$

as follows. For simplicity of notion, we shall use $u \cdot v$ for the product $h(u, v)$, $u, v \in Y$. Let there be given maps $f : (I^s, \partial I^s) \to (Y, y)$ and $g : (I^t, \partial I^t) \to (Y, y)$, then the product map $f \cdot g : (I^s, \partial I^s) \times (I^t, \partial I^t) = (I^{s+t}, \partial I^{s+t} \to (Y, y)$ is given by $(f \cdot g)(u, v) = f(u) \cdot g(v)$. Since Y is group-like, for every $f : (I^s, \partial I^s) \to (Y, y)$ there is an inverse map $f^{-1} : (I^s, \partial I^s) \to (Y, y)$ defined by $f^{-1}(u) = f(u)^{-1}$. The bracket is just the commutator:

$$[f, g] := f \cdot g \cdot f^{-1} \cdot g^{-1}.$$

That this bracket is graded commutative and that the Jacobi identity holds is not trivial. See [Wh]. In the special case where $Y = P_x X$, (X, x) simply connected, the bracket becomes

$$\pi_{s+1}(X, x) \times \pi_{t+1}(X, x) \to \pi_{s+t+1}(X, x),$$

the **Whitehead product**. Under the duality given by Cor. 8.6 the co-product on $QH^*(P_x X; \mathbb{Q})$ is indeed dual to the Whitehead product:

Proposition 8.10. *Let (X, x) be a connected simply connected space all of whose homotopy groups are finitely generated. Then the Hurewicz isomorphisms induce an isomorphism of graded Lie co-algebras*

$$QH^*(P_xX;\mathbb{Q})[-1] \xrightarrow{\sim} \bigoplus_{s\geq 1} \mathrm{Hom}(\pi_s(X,x);\mathbb{Q}),$$

where the co-bracket is dual to the Whitehead product.

We close this section by discussing the concept of a **(Hurewicz) fibration**. This is a continuous surjective map between topological spaces $p : E \to B$ which has the homotopy lifting property: given a map $g : X \to E$, every homotopy of $p \circ g$ can be lifted to a homotopy of g. For such a fibration any two fibres are homotopy equivalent [Span, p.101] and with $e \in E$ the base point and F the typical fibre, one has the homotopy exact sequence ([Span, p. 377])

$$\cdots \pi_n(F, e) \to \pi_n(E, e) \to \pi_n(B, p(e)) \to \pi_{n-1}(F, e) \cdots \qquad \text{(VIII–3)}$$

Although fibrations look rather special, up to homotopy all maps are fibrations. To explain this, we first recall that $f : (X, x) \to (Y, y)$ is a **homotopy equivalence** if f has an inverse g up to homotopy, i.e. $f \circ g$ and $g \circ f$ are homotopic to the identity. Clearly homotopy equivalences induce isomorphisms on homotopy groups. The converse holds whenever X and Y are CW complexes. This result is due to Whitehead [Span, p. 405]).

It is now quite easy to see how one can functorially replace any continuous map $f : X \to Y$ by a Hurewicz fibration. Using the path space PY of continuous paths in Y, the total space of the fibration is

$$E_f = \{(x, \gamma) \in X \times PY \mid \gamma(0) = f(x)\}$$

and the map $\pi_f : E_f \to Y$ given by sending a pair (x, γ) to the endpoint $\gamma(1)$ of γ gives it the structure of a fibration. The map $s : X \to E_f$ which sends x to the pair (x, constant path at $f(x)$) is a homotopy equivalence so that indeed $f : X \to Y$ may be replaced by the fibration $\pi_f : E_f \to Y$. The **homotopy fibre** $E_f(y)$ of f above y by definition is the fibre of π_f above y. Its homotopy type depends only on the path component to which y belongs. So, if we start with a Hurewicz fibration over a path connected space, any fibre is homotopy equivalent to the homotopy fibre.

8.2 Formulation of the Main Results

Our ultimate aim is to put a mixed Hodge structure on the homotopy groups of any complex algebraic variety. As explained below, there is a crucial difference between the fundamental group and the higher homotopy groups due to the fact that the fundamental group in general is not abelian in contrast to the higher homotopy groups. There are also technical difficulties in the case of higher homotopy if one does not assume that the variety is simply connected. The main result is now

Theorem 8.11. *Let X be a complex algebraic variety with base point $x \in X$. Then the following holds.*

1) *For each $s \geq 0$, the finitely generated $\mathbb{Z}$-module $\mathbb{Z}\pi_1(X,x)/J^{s+1}$ has a natural mixed Hodge structure, functorial with respect to base point preserving morphisms of varieties. Moreover, the product in this ring is a morphism of mixed Hodge structures;*
2) *In case X is simply connected, the higher homotopy groups $\pi_k(X,x)$ carry natural mixed Hodge structures functorial with respect to morphisms of simply connected varieties and independent of base points. The Whitehead products are morphisms of mixed Hodge structure.*
3) *In case X is simply connected, the Hurewicz maps $H_k(X) \to \pi_k(X)$ are morphisms of mixed Hodge structure.*

Remark 8.12. The hypothesis that (X,x) be simply connected is too restrictive. In order to have a mixed Hodge structure on the higher homotopy groups of a complex variety (X,x), it is sufficient that (X,x) be **nilpotent** in the sense that $\pi_1(X,x)$ is nilpotent and, moreover, the natural action of $\pi_1(X,x)$ on $\pi_k(X,x) \otimes \mathbb{Q}$ is unipotent for each $k \geq 2$. Examples include path connected H-spaces, since by [Span, Chap. 7, Thm. 10], such a space has an abelian fundamental group with trivial action on higher homotopy groups.

There are some further results which complete the assertions of the preceding theorem. For this we need to pass to rational coefficients. First of all, the filtration of the group-ring $\mathbb{Q}\pi_1(X,x)$ by powers of its augmentation ideal J defines an inverse system

$$\{\mathbb{Q}\pi_1(X,x)/J^{s+1} \mid s = 0,1,2,\dots\}$$

whose limit is the ***J*-adic completion**

$$\mathbb{Q}\widehat{\pi_1(X,x)} := \varprojlim \mathbb{Q}\pi_1(X,x)/J^{s+1}.$$

This $\mathbb{Q}$ vector space is in general not finite dimensional and we need to modify the definition of a mixed Hodge structure as follows.

Definition 8.13. Let $V = \varprojlim V_k$ where each V_k is a finite dimensional $\mathbb{Q}$-vector space. A **pro-mixed Hodge structure** on V consists of mixed Hodge structures on each V_k such that the linear maps $V_\ell \to V_k$, $\ell > k$ are morphisms of mixed Hodge structures.

Similarly one defines the notion of an **ind-mixed Hodge structure** on $V = \varinjlim V_k$.

With this in mind, we have

Complement (I). *The mixed Hodge structures of Theorem 8.11 1) induces a pro-mixed Hodge structure on* $\widehat{\mathbb{Q}\pi_1(X,x)}$. *The product map for this algebra is a morphism of mixed Hodge structures. With the standard pro-mixed Hodge structure on the tensor product* $\widehat{\mathbb{Q}\pi_1(X,x)} \otimes_{\mathbb{Q}} \widehat{\mathbb{Q}\pi_1(X,x)}$, *also the diagonal map sending* $g \in \widehat{\mathbb{Q}\pi_1(X,x)}$ *to* $g \otimes g$ *is a morphism of mixed Hodge structures.*

It is also natural to consider the maps in the long exact sequence of a Hurewicz fibration:

Complement (II). *The homotopy exact sequence* (VIII–3) *for a Hurewicz fibration* $p : E \to B$ *of complex algebraic varieties is a sequence of mixed Hodge structures if the fibre* F *and the base* B *are simply connected (this implies that the total space* E *is simply connected), or more generally, if* E, B *are simply connected and* F *is nilpotent (see Remark 8.12).*

Example 8.14. As an application, we look at the Hopf fibration $\mathbb{C}^{n+1} - \{0\} \to \mathbb{P}^n$ with fibre $\mathbb{C}^*$. Since complex projective space is simply connected, when we apply the homotopy exact sequence to the Hopf fibration, it shows that $\pi_2(\mathbb{P}^n)$ is cyclic with one generator, while for $m \geq 3$ we have $\pi_m(\mathbb{P}^n) \cong \pi_m(S^{2n+1})$. By [Span, Chap. 9, § 7] these groups are torsion, except the infinite cyclic group $\pi_{2n+1}(S^{2n+1})$. So for Hodge theory, only this group and $\pi_2(\mathbb{P}^n)$ are interesting and they must have a pure Hodge structure. The coboundary $\pi_2(\mathbb{P}^n) \to \pi_1(\mathbb{C}^*)$ is an isomorphism of mixed Hodge structures and so the natural mixed Hodge structure on the latter is pure of type $(-1,-1)$ as well. As for the Hodge structure on $\pi_{2n+1}(\mathbb{P}^n) \cong \pi_{2n+1}(\mathbb{C}^{n+1} - 0)$, let us look the Hurewicz homomorphism

$$\pi_{2n+1}(\mathbb{C}^{n+1} - 0) \to H_{2n+1}(\mathbb{C}^{n+1} - 0) \cong \mathbb{Z}.$$

Since by loc. cit. the ℓ-th homotopy group of the k-sphere is trivial for $\ell < k$, by Theorem 8.1 this is an isomorphism. The group on the right admits a mixed Hodge structure which can be calculated by identifying $\mathbb{C}^{n+1} - 0$ as the complement of the exceptional divisor together with the hyperplane at infinity inside the blow up of $\mathbb{P}^{n+1}$ in a point outside the hyperplane at infinity. The Gysin sequence (Prop 1.19) shows that $H_{2n+1}(\mathbb{C}^{n+1} - 0) = \mathbb{Z}(n+1)$. So this gives the Hodge structure on $\pi_{2n+1}(\mathbb{C}^{n+1} - 0)$.

For the relative situation there is the useful

Complement (III). *Let* $f : (X,x) \to (Y,y)$ *be a morphism of complex algebraic varieties. Suppose that* Y *is (path) connected and that the monodromy action of* $\pi_1(Y,y)$ *on the each cohomology group of the homotopy fibre* $E_f(y)$ *of* f *is unipotent. Then these cohomology groups* $H^k(E_f(y))$ *carry natural mixed Hodge structures; if moreover the homotopy fibre is simply connected or, more generally nilpotent, the higher homotopy groups* $\pi_k(E_f(Y))$ *of the homotopy fibre also carry natural mixed Hodge structures. The monodromy representation of the group algebra* $\mathbb{Q}\pi_1(Y,y)$ *on the cohomology* $H^*(E_f(y))$ *of the homotopy fibre is a representation of mixed Hodge structures.*

Remark. The monodromy hypothesis is rarely satisfied when we have a family of smooth algebraic varieties since the monodromy acts semi-simply (see 10.13). But the preceding result can also be used if Y is simply connected, even when f is not a fibre bundle. For Y a punctured disk Hain has shown that there is a "limit" mixed Hodge structure on the cohomology (respectively homotopy) of the homotopy fibre. See [Hain86].

8.3 Loop Space Cohomology and the Homotopy De Rham Theorem

8.3.1 Iterated Integrals

Motivated by the isomorphism from the Borel-Serre theorem (Corollary 8.6) it is natural to ask for a De Rham theorem for the cohomology of path spaces. The basic idea is that, although loop spaces in general are infinite dimensional, one can still define differential forms on them, do integration and derive an analogue of De Rham's theorem using so called plots instead of coordinate charts.

Definition 8.15. – A **plot** into a set Y is a map of an open convex subset of $\mathbb{R}^n$ into Y (n is arbitrary);
– A **differentiable space** is a set Y together with a collection of plots into Y such that the following properties hold:
i) Compatibility: for any plot $p : U \to Y$ and any smooth map $f : U' \to U$, $U' \subset \mathbb{R}^m$, convex, also $p \circ f$ is a plot;
ii) Constant maps from open convex subsets in $\mathbb{R}^n$ are plots;
iii) Glueing: a map from an open convex set $U \subset \mathbb{R}^n$ to Y is a plot as soon as U can be covered by plots.

Standard examples of such differentiable spaces are finite dimensional differentiable manifolds, as well as their path spaces. In the first case the plots are the usual differentiable maps of open convex subspaces in $\mathbb{R}^n$ to Y. In the last case, we observe first that any map $f : U \to PY$ induces a **suspension** $Sf : I \times U \to Y$, upon setting $Sf(t,u) = f(u)(t)$ (one views $f(u)$ as an actual path in Y). Then for $U \subset \mathbb{R}^n$, we declare $f : U \to PY$ to be a plot if its suspension is.

Definition 8.16. A differential form ω on a differentiable space Y is given by a differential form ω_p on U for each plot $p : U \to Y$. These should satisfy the compatibility rule that for any smooth map $f : U' \to U$, $U' \subset \mathbb{R}^m$ convex, the differential form $\omega_{p\circ f}$ on the plot $p \circ f$ coincides with $f^*\omega_p$.

One can then add, multiply and differentiate differential forms plot-wise, defining a De Rham complex on Y denoted $E_{\mathrm{DR}}(Y)$. If Y is a smooth manifold this contains and is quasi-isomorphic to the usual De Rham complex, and hence computes cohomology of Y. This is no longer true however in general. For path spaces PX of simply connected differentiable manifolds X with finite dimensional De Rham cohomology Chen has shown that the cohomology of $Y = PX$ is the cohomology of a certain subcomplex of the De Rham complex $E_{\text{DR}}(Y)$, the complex of iterated integrals.

To define an iterated integral, we first look at the behaviour of an r-form ω on X when pulled back under the suspension $S(p) : I \times U \to X$ of a plot $p : U \to PX$. One has a canonical decomposition

$$S(p)^*\omega = \beta(t,u) + dt \wedge \gamma(t,u), \quad (t,u) \in I \times U,$$

where β and γ contain no dt.

Definition 8.17. Let there be given an *ordered* collection of differential forms ω_i of degree $k_i \geq 1$, $i = 1, \dots, s$ on X. The associated **elementary iterated integral**

$$\int \omega_1\omega_2\cdots\omega_s \in E^{\ell}_{\text{DR}}(PX), \quad \ell = \sum_{i=1}^{s}(k_i - 1)$$

is the ℓ-form which on the plot $p : U \to PX$ is given by

$$\int\cdots\int_{0\leq t_1\leq\cdots\leq t_s\leq 1} \gamma_1(t_1,u)\wedge\gamma_2(t_2,u)\cdots\wedge\gamma_s(t_s,u)dt_1dt_2\dots dt_s,$$

where $S(p)^*\omega_i = \beta_i(t,u) + dt\wedge\gamma_i(t,u)$ is the canonical decomposition considered previously. An **iterated integral** on X is a differential form on its path space $P(X)$ which is a (finite) linear combination of a constant and elementary iterated integrals. Notation:

$$\int E_{\text{DR}}(X) \subset E_{\text{DR}}(PX).$$

Since an iterated integral is a differential form on path space, it has an derivative. To calculate it we introduce the involution

$$\mathrm{J} : E_{\text{DR}}(X) \to E_{\text{DR}}(X), \quad \mathrm{J}\omega = (-1)^p\omega, \quad \omega \text{ a } p \text{ form on } X$$

and then (see [Chen76, §2.1])

$$\begin{aligned} d\textstyle\int \omega_1\cdots\omega_s &= d'\textstyle\int\omega_1\cdots\omega_s + d''\textstyle\int\omega_1\cdots\omega_s \\ d'\textstyle\int \omega_1\cdots\omega_s &= \textstyle\sum_{i=1}^{s}\int(-1)^{i+1}\mathrm{J}\omega_1\cdots\mathrm{J}\omega_{i-1}(\mathrm{J}\omega_i\wedge\omega_{i+1})\omega_{i+2}\cdots\omega_s \\ d''\textstyle\int \omega_1\cdots\omega_s &= \textstyle\sum_{i=1}^{s}\int(-1)^{i}\mathrm{J}\omega_1\cdots\mathrm{J}\omega_{i-1}d\omega_i\omega_{i+1}\cdots\omega_s. \end{aligned}$$

So the iterated integrals form a *subcomplex* of the De Rham complex of PX.

The following properties are easily verified using the definitions.

Lemma 8.18. a) *(Functoriality) If $f : X \to Y$ is a smooth map between differentiable manifolds, one defines the pull back of an iterated integral*

$$\int f^*(\omega_1\omega_2\cdots\omega_s) := \int f^*\omega_1 f^*(\omega_2)\cdots f^*(\omega_s) \in E_{\mathrm{DR}}(PX)$$

and for 1-forms ω_i, $i = 1, \dots, s$ we have

$$\int_{f\circ\gamma} \omega_1\omega_2\cdots\omega_s = \int_\gamma f^*(\omega_1\omega_2\cdots\omega_s), \quad \gamma \in PX.$$

b) *The following relations hold when we restrict iterated integrals to P_xX:*

$$\begin{aligned}
\textstyle\int_\gamma dh\omega_1\omega_2\cdots\omega_s &= \textstyle\int_\gamma (h\omega_1)\omega_2\cdots\omega_s - h(x)\int_\gamma \omega_1\cdots\omega_s;\\
\textstyle\int_\gamma \omega_1\cdots\omega_{i-1}dh\omega_i\cdots\omega_s &= \textstyle\int_\gamma \omega_1\cdots\omega_{i-1}(h\omega_i)\omega_{i+1}\cdots\omega_s\\
&\quad \textstyle- \int_\gamma \omega_1\cdots(h\omega_{i-1})\omega_i\cdots\omega_s;\\
\textstyle\int_\gamma \omega_1\cdots\omega_s dh &= \textstyle h(x)\int_\gamma \omega_1\cdots\omega_s - \int_\gamma \omega_1\cdots\omega_{s-1}(h\omega_s).
\end{aligned}$$

8.3.2 Chen's Version of the De Rham Theorem

Chen's homotopy De Rham theorem states that the cohomology of the subcomplex given by iterated integrals already computes $H^*_{\mathrm{DR}}(PX)$. We need a slightly more general version using the parlance of differential graded algebras. We recall the definition:

Definition 8.19. Let k be a field. A graded k- algebra $A = \bigoplus_{p\geq 0} A^p$ is a **differential graded algebra** if there is a k-derivation $d : A \to A$, $dA^p \subset A^{p+1}$, i.e. d is k-linear and satisfies the Leibniz-rule $d(xy) = (dx)y + (-1)^p x dy$, $x \in A^p, y \in A^q$. Moreover, one should have $d\circ d = 0$. If in addition A is a Hopf algebra (see Def. 8.8) and the co-multiplication is a morphism of differential graded algebras we say that A is a **differential graded Hopf algebra**. If A is a differential graded algebra with $A^0 = k$, we say that A is **connected**.

The standard example of a commutative differential graded algebra with connected cohomology is the De Rham-algebra of k-valued differential forms on a path connected differentiable manifold X, where k is a subfield of $\mathbb{C}$:

$E_{\mathrm{DR}}(X;k)$: the differential graded algebra of k-valued differential forms on X.

If A is a differential graded subalgebra of $E_{\mathrm{DR}}(X;k)$, we set

$\int A(X,x)$: the subalgebra of $E_{\mathrm{DR}}(P_xX;k)$ formed by the iterated path integrals of forms on X belonging to $A \subset E_{\mathrm{DR}}(X;k)$.

With these notations, we can now formulate Chen's De Rham theorem in the simply connected case [Chen76, §2.3]:

Theorem 8.20 (Chen's Homotopy De Rham Theorem). *Let X be a connected and simply connected manifold, all of whose rational cohomology groups are finite dimensional. Let A be a differential graded subalgebra of the algebra of differential forms with the property that the inclusion in the full De Rham algebra is a quasi-isomorphism. Then the cohomology of the complex $\int A(X,x)$ computes the cohomology of the loop space P_xX with coefficients k.*

8.3.3 The Bar Construction

The formulas for the derivative of an iterated integral and the formulas from Lemma 8.18 motivate the bar and the reduced bar construction on the De Rham algebra. For simplicity of notation, we take $k = \mathbb{R}$ so that we only treat the real De Rham algebra. First note that the choice of a base point $x \in X$ defines an augmentation $E_{\rm DR}(X) \to \mathbb{R}$.

The bar construction makes in fact sense for any augmented differential graded algebra (A, ϵ) with augmentation ideal $IA = \operatorname{Ker} \epsilon$ as we explain now. We first introduce

$$B^{-s,t}A = \text{degree } t \text{ elements inside } \bigotimes^{s} IA.$$

Traditionally one denotes the element $\omega_1 \otimes \cdots \otimes \omega_s$ of $B^{-s,t}A$ by $[\omega_1|\ldots|\omega_s]$. We make $B^{-\bullet,\bullet}A$ into a double complex by setting

$$\begin{aligned}
d' &: B^{-s,t}A \longrightarrow B^{-s+1,t}A,\\
&[\omega_1|\cdots|\omega_s] \mapsto \sum_{i=1}^{s}(-1)^{i+1}[\mathrm{J}\omega_1|\cdots|\mathrm{J}\omega_{i-1}|\mathrm{J}\omega_i \wedge \omega_{i+1}|\omega_{i+2}|\cdots|\omega_s]\\
d'' &: B^{-s,t}A \longrightarrow B^{-s,t+1}A,\\
&[\omega_1|\cdots|\omega_s] \mapsto \sum_{i=1}^{s}(-1)^{i}[\mathrm{J}\omega_1|\cdots|\mathrm{J}\omega_{i-1}|d\omega_i|\omega_{i+1}|\cdots|\omega_s].
\end{aligned}$$

The **bar construction** BA of A is the associated single complex. It has extra structure. There always is the co-product

$$\begin{aligned}
\Delta : BA &\to BA \otimes BA\\
[\omega_1|\cdots|\omega_r] &\mapsto \textstyle\sum_{i=0}^{r}[\omega_1|\cdots|\omega_i] \otimes [\omega_{i+1}|\cdots|\omega_r].
\end{aligned}$$

and the product

$$\begin{aligned}
BA \otimes BA &\xrightarrow{\wedge} BA\\
[\omega_1|\cdots|\omega_r] \otimes [\omega_{i+1}|\cdots|\omega_{r+s}] &\mapsto \textstyle\sum_{\sigma} \operatorname{sign}(\sigma)[\omega_{\sigma(1)}|\cdots|\omega_{\sigma(r+s)}],
\end{aligned}$$

where σ runs over all shuffles of type (r,s) and the sign of such a shuffle is determined by giving ω_i the weight $-1 + \deg(\omega_i)$. Let us recall that a

shuffle σ of type (r,s) is a permutation of $\{1,\ldots,n=r+s\}$ such that $\sigma^{-1}(1)<\cdots<\sigma^{-1}(r)$ and $\sigma^{-1}(r+1)<\cdots<\sigma^{-1}(r+s)$. In general these two structures don't give a differential graded Hopf algebra, except when the algebra $A^\bullet$ is graded commutative. We leave this verification to the reader.

Returning to the De Rham complex, the structure of its bar complex depends on the chosen base point $x\in X$ and we therefore use the notation $BE_{\mathrm{DR}}(X,x)$ in this case. We can now explain the degree convention. The map $[\omega_1|\ldots|\omega_s]\mapsto\int\omega_1\cdots\omega_s$ maps an element in the bar construction of the De Rham complex to a differential form of degree $\sum_{i=1}^s \deg\omega_i - s$ and so the total degree on the bar complex coincides with this degree. In fact, the definitions are such that the following statement becomes a tautology:

Lemma 8.21. *Iterated integration induces a map of differential graded algebras*

$$BE_{\mathrm{DR}}(X,x)\to E_{\mathrm{DR}}(P_xX).$$

The three relations from Lemma 8.18 motivate the construction of a subcomplex of the bar construction which can be defined for any augmented differential graded algebra. Indeed, we look at the subspace of BA generated by elements of the form

$$\left\{\begin{array}{l}[dh|\omega_1|\cdots|\omega_s]-[h\omega_1|\omega_2|\cdots|\omega_s]+\epsilon(h)[\omega_1|\cdots|\omega_s]\\ [\omega_1|\cdots|\omega_s|dh]-\epsilon(h)[\omega_1|\cdots|\omega_s]+[\omega_1|\cdots|h\omega_s],\\ [\omega_1|\cdots|\omega_{i-1}dh|\omega_i|\cdots|\omega_s]-[\omega_1|\cdots|\omega_{i-1}|h\omega_i|\cdots|\omega_s]\\ \qquad\qquad\qquad\qquad +[\omega_1|\cdots|h\omega_{i-1}|\omega_i|\cdots|\omega_s]\end{array}\right. \tag{VIII–4}$$

where $h\in A^0$. The quotient by this subcomplex is called the **reduced bar construction** $\bar{B}A$. The equivalence class of $[\omega_1|\cdots|\omega_r]$ in the reduced bar construction is traditionally denoted $(\omega_1|\cdots|\omega_r)$.

Remark 8.22. For later use we need a slight generalization of this construction for a (not necessarily commutative) differential graded algebra A which is zero in degrees <0. It involves also a right A-module M and a left A-module N. Instead of IA (which does not make sense since there is no augmentation), we use the positive degree part A^+ and we introduce

$$T^{-s,t}(M,A,N):=\text{degree } t \text{ elements of } [M\otimes\otimes^s A^+\otimes N].$$

The previous $B^{-s,t}$ is the special case where $M=N=k$, considered A-modules through the augmentation $\epsilon: A\to k$. This motivates the notation $m[a_1|\cdots|a_s]n$ for the element $m\otimes a_1\otimes\cdots\otimes a_s\otimes n$ in $T^{-s,t}(M,A,N)$.

Note that the formulas (VIII–4) make sense in this setting, viewing $h\in A^0$ as acting from the right on $m\in M$ and from the left on $n\in N$:

$$\left\{\begin{array}{l}m[dh|a_1|\cdots|a_s]n-m[ha_1|a_2|\cdots|a_s]n+mh[a_1|\cdots|a_s]n\\ m[a_1|\cdots|a_s|dh]n-m[a_1|\cdots|a_s]hn+m[a_1|\cdots|ha_s]n,\\ m[a_1|\cdots|a_{i-1}dh|a_i|\cdots|a_s]n-m[a_1|\cdots|a_{i-1}|ha_i|\cdots|a_s]n\\ \qquad\qquad\qquad\qquad +m[a_1|\cdots|ha_{i-1}|a_i|\cdots|a_s]n\end{array}\right.$$

These relations generate a subcomplex and the quotient is the **reduced bar construction** $\bar{B}(M, A, N)$. The equivalence class of $m[a_1|\cdots|a_s]n$ is commonly denoted by $m(a_1|\cdots|a_s)n$. It is a differential graded algebra with product denoted $\wedge$ coming from the shuffle product.

As an *example* consider the path space $P_{x,y}X$ of paths from x to y. There are two augmentations $\epsilon_x, \epsilon_y : E_{\mathrm{DR}} \to \mathbb{R}$ and we can have two structures on $\mathbb{R}$ as a $E_{\mathrm{DR}}(X)$ module, denoted $\mathbb{R}_x$, respectively $\mathbb{R}_y$. The corresponding reduced bar is $\bar{B}(\mathbb{R}_x, E_{\mathrm{DR}}, \mathbb{R}_y)$. Chen's theorem actually is more elaborate and also states that the cohomology of this complex computes $H^*(P_{x,y}; \mathbb{R})$. We shall see later (§ 8.6) that there is a rational differential graded algebra $\mathsf{A}_{\mathbb{Q}}$ with augmentations $\mathsf{A}_{\mathbb{Q}} \to \mathbb{Q}$ depending on the base point such that the reduced bar complex computes $H^*(P_{x,y}; \mathbb{Q})$. So for $k = \mathbb{R}$ or $k = \mathbb{Q}$ we thus have the complexes $\bar{B}(k_x, \mathsf{A}_{\mathbb{Q}}, k_y)$ computing $H^*(P_{x,y}X; k)$.

Coming back to the reduced bar construction for the De Rham complex, note that it is inspired by the relations for the iterated integrals. So by construction, the integration map factors over $\bar{B}E_{DR}(X, x)$. In fact (see [Chen77]) we have:

Theorem 8.23. *Let X be a simply connected manifold all whose homotopy groups are finitely generated. If A is a subcomplex of the De Rham complex such that its inclusion into the full De Rham complex is a quasi-isomorphism, then the differential graded algebra morphisms*

$$BA \to \bar{B}A \to \int A(X, x) \hookrightarrow \int E_{\mathrm{DR}}(X, x)$$

are all quasi-isomorphisms. Moreover, the resulting isomorphism between the cohomology algebras

$$H^*BA \xrightarrow{\sim} H^*(P_xX; \mathbb{R})$$

is a Hopf algebra isomorphism.

8.3.4 Iterated Integrals of 1-Forms

To define the associated iterated integral of 1-forms, we need no plots, since the corresponding differential form on PX now has degree 0. So, given an ordered set of s k-valued one-forms $\omega_1, \omega_2, \ldots, \omega_s$, the associated iterated integral is the function

$$\int \omega_1\omega_2\cdots\omega_s : PX \to k$$

given by the formula

$$\int_\gamma \omega_1\omega_2\cdots\omega_s = \underset{0\le t_1\le\cdots\le t_s}{\int\cdots\int} f_1(t_1)f_2(t_2)\cdots f_s(t_s)dt_1\cdots dt_s,$$

where γ is any path and

$$\gamma^*\omega_i = f_i(t)dt.$$

An **integrated integral of length** $\leq s$ is a k-linear combination of a constant and iterated integrals associated to at most s one-forms.

Continuing with the case of 1-forms, let us remark that a function $f : PX \to k$ has differential zero if and only if the composite $f \circ p : U \to k$ is constant for any plot $p : U \to PX$, i.e. f is constant on path components of PX. Loops based at x in the same homotopy class form a single path component and so $\int \omega_1\omega_2 \cdots \omega_s : P_xX \to k$ belongs to $H^0(E_{\mathrm{DR}}(PX))$ if and only if the value of this iterated integral is constant on loops that are homotopic. If $s = 1$ this is equivalent to $d\omega_1 = 0$ which motivates the following definition.

Definition 8.24. An iterated integral of 1-forms on *loop space* P_xX is **closed** or a **homotopy functional** if its value does not change under homotopies preserving the base point x.

8.4 The Homotopy De Rham Theorem for the Fundamental Group

The fundamental group can also be studied using iterated integrals of one-forms. First we need a relation with a ordinary integrals. To formulate it, let us observe that *closed* $\mathbb{R}$-valued iterated integrals on P_xX can also be extended to functions on the group ring

$$\mathbb{R}\pi_1(X, x) \to \mathbb{R}$$

(and similarly if we replace $\mathbb{R}$ by $\mathbb{C}$). The following relation plays a central role in the sequel.

Lemma 8.25. *Let* $a = (\alpha_1 - 1)(\alpha_2 - 1)\cdots(\alpha_r - 1)$. *Then*

$$\int_a \omega_1 \cdots \omega_s = \begin{cases} 0 & \text{if } r > s \\ \int_{\alpha_r} \omega_r \cdots \int_{\alpha_1} \omega_1 & \text{if } r = s. \end{cases}$$

Proof. Define the simplex $\Delta(I^r) = \{(t_1, \ldots, t_r) \in I^r \mid 0 \leq t_1 \leq \cdots \leq t_r \leq 1\}$ and divide the unit interval into r equal subintervals $I_k = [\frac{k-1}{r}, \frac{k}{r}]$. Let α_i, $i = 1, \ldots, r$ be closed paths based at x and write

$$\begin{aligned} \gamma &= \alpha_1 * \cdots * \alpha_r \\ a_k &= (\alpha_1 - 1)\cdots(\alpha_k - 1)\alpha_{k+1}\cdots\alpha_r,\ k = 1\ldots, r \\ \gamma^*\omega_k &= f_k(t)dt,\ k = 1\ldots, s. \end{aligned}$$

By induction one shows

$$\int_{a_k} \omega_1 \cdots \omega_s = \sum \int_{I_{i_1} \times \cdots I_{i_s} \cap \Delta(I^s)} f_1(t_1)\cdots f_s(t_s)dt_1 \cdots dt_s,$$

where the sum is taken over all choices of non-decreasing s-tuples of integers $(i_1, \dots, i_s)$ containing the set $\{1, \dots, k\}$. For $k > s$ the sum is zero and for $k = s = r$ there is only one summand. Fubini completes the proof. □

We can now formulate the sought for De Rham theorem for the fundamental group.

Theorem 8.26. *Let A be a differential graded subalgebra of the De Rham algebra $E_{\mathrm{DR}}(X)$ of a manifold X. Consider closed iterated integrals which are homotopy functionals and the tautological integration homomorphism*

$$\begin{array}{ccc} H^0(\int A(X,x)) & \longrightarrow & H^0(P_x X; \mathbb{R}) \\ \| & & \| \\ \{\text{closed iterated integrals}\} & \longrightarrow & \mathrm{Hom}_{\mathbb{Z}}(\mathbb{Z}\pi_1(X,x), \mathbb{R}) \\ \int \omega_1 \cdots \omega_s & \longmapsto & \{[\gamma] \mapsto \int_\gamma \omega_1 \cdots \omega_s\}. \end{array}$$

Suppose that the inclusion of A into the De Rham complex induces an isomorphism on H^1 and an injection on H^2. Suppose also that $\pi_1(X,x)$ is finitely generated. Recalling that J is the augmentation ideal, any closed iterated integral of length s vanishes on elements in J^r when $r > s$. Integration induces therefore a homomorphism

$$\left\{ \begin{matrix} \text{closed iterated integrals} \\ \text{of length} \le s. \end{matrix} \right\} \longrightarrow \mathrm{Hom}_{\mathbb{Z}}(\mathbb{Z}\pi_1(X,x)/J^{s+1}, \mathbb{R}).$$

This is an isomorphism.

Proof (sketch). The first assertion is Lemma 8.25.

For the second assertion we first make some observations. The crucial remark is the fact, that the transport matrix of a flat connection does not change under homotopies preserving the endpoints. See Lemma B.42. Let us briefly recall the construction. Suppose that we have a piece-wise smooth path $\gamma : I \to X$ and a fixed $(n \times n)$-matrix ω of smooth 1-forms. We consider this matrix as a connection matrix of a connection ∇ on the trivial bundle $\mathbb{R}^n \times X$. Recall (Lemma-Def. B.41) that parallel transport along the path γ is given by the transport map

$$\mathbf{w} = \mathbf{v}\, T(\gamma),$$

where $T(\gamma)$ is the following (convergent) series of iterated integrals evaluated on γ

$$T(\gamma) = 1 + \int_\gamma \omega + \int_\gamma \omega\omega + \int_\gamma \omega\omega\omega + \cdots$$

The iterated integrals appearing in this series in fact are matrix-valued functionals instead of just real or complex valued. For flat connections the transport map does not depend on homotopies preserving endpoints

(Lemma B.42). So in this case the entries of the transport matrix are homotopy functionals.

We can now prove the second assertion. We try to find iterated integrals as coefficients in an upper triangular connection matrix of a flat connection on a trivial bundle $V \times X$. To get the right number of such, we carry this out for the finite dimensional real vector space $V = \mathbb{R}\pi_1(X, x)/J^{s+1}$. A representation of the fundamental group in a vector space V defines a vector bundle with a flat connection on it as explained in Example B.40. Here right multiplication defines an action of the fundamental group $G = \pi_1(X, x)$ on V. The resulting representation ρ has the property that $(\rho([\alpha]) - I)^{s+1} = 0$. This is because $GJ^i \subset J^{i+1}$ and so preserves the filtration

$$V = V^0 = J^0/J^{s+1} \supset V^1 = J^1/J^{s+1} \supset \cdots V^s = J^s/J^{s+1} \supset 0,$$

the induced action on the graded quotients being trivial.

Now, as in § B.3, let us form the associated flat bundle

$$E = (V \times \text{universal cover of } X)/G, \quad (v, x) \cdot g = (vg, g^{-1}x).$$

The subspaces V^i define flat subbundles E^i and we can easily find a smooth trivialization of E which trivializes at the same time all the subbundles E^i as bundles together with the flat connection. This implies that the connection form of the flat connection on E takes its values in the the Lie-algebra $\mathfrak{g}$ of endomorphisms of V that preserve the flag $V = V^0 \supset V^1 \cdots \supset V^s \supset 0$. Its connection form ω is a $\mathfrak{g}$-valued 1-form. So for any smooth path γ in X, setting $\gamma^*\omega = A(t)dt$, we have the desired nilpotency $A(t)^{s+1} = 0$ and hence the entries of A consist indeed of iterated integrals of length $\leq s$. Also, since the connection is flat, the value $T(\gamma)$ depends only on the homotopy class of γ. If we let W be the space of iterated integrals which are homotopy functionals, the transport matrices therefore give an element

$$T \in W \otimes \mathrm{End}(V).$$

The monodromy of the local system E is a homomorphism

$$\mathbb{R}\pi_1(X, x) \to \mathrm{End}(V),$$

which by construction factors over $V = \mathbb{R}\pi_1(X, x)/J^{s+1}$ and induces the canonical injection

$$\mathbb{R}\pi_1(X, x)/J^{s+1} = V \hookrightarrow \mathrm{End}(V), \quad g \mapsto A_g = (v \mapsto v \cdot g).$$

So T in fact takes values in $W \otimes V$. Now, quite formally, integration induces a map $f : W \to V^\vee$, while T viewed as an element in $W \otimes V$ defines $g : V^\vee \to W$ by contraction. The map $f \circ g$ is dual to the map $V \to V$ induced by integration $\gamma \mapsto \int_\gamma T$, and it is the identity by construction. From $f \circ g = \mathrm{id}$ we derive that the integration map f must be surjective. Since it is obviously injective on iterated integrals that are homotopy functionals, this completes the proof. □

Remark 8.27. The theorem implies that the space of closed iterated integrals is dual to the J-adic completion $\widehat{\mathbb{R}\pi(X,x)}$ of the the group ring of the fundamental group (over $\mathbb{R}$) with respect to the augmentation ideal J.

8.5 Mixed Hodge Structure on the Fundamental Group

The main result is:

Theorem 8.28. *Let X be a smooth complex projective algebraic variety and $x \in X$. Then, for each $s \geq 0$, the finitely generated $\mathbb{Z}$-module $\mathbb{Z}\pi_1(X,x)/J^{s+1}$ has a natural mixed Hodge structure.*

Remark. The general case of complex algebraic varieties can be treated using the machinery of cohomological descent for simplicial varieties. For the latter we refer to [Hain87].

Proof (of theorem 8.28 – sketch only). The point of departure is the de Rham Theorem 8.26. For brevity, introduce the notation $B_s(X,x)$ for the the real vector space of iterated integrals over 1-forms and of length $\leq s$ and $H^0B_s(X,x)$ for the closed iterated integrals of length $\leq s$.[1]

We put a *weight filtration* on the spaces of iterated integrals by setting

$$W_kB_s(X,x) = \begin{cases} B_k(X,x) & \text{if } k \leq s \\ B_s(X,x) & \text{if } k \geq s. \end{cases}$$

The difficulty is to show that it is defined over the rationals. This we do later.

A *Hodge filtration* is defined starting with the usual Hodge filtration coming from the type decomposition of complex one forms, where a form is in F^1 if it is of pure type $(1,0)$ and in F^0 otherwise. Explicitly, F^p is spanned as a complex vector space by the complex-valued iterated integrals over forms $\omega_1 \cdots \omega_s$, $\omega_j \in F^{p_j}$ with $\sum p_j \geq p$.

We need to show that the weight and Hodge filtration define a mixed Hodge structure. To start with, we make the following

Observation 8.29. *There is a natural isomorphism*

$$\begin{aligned} \mathbb{R} \oplus H^1_{\mathrm{DR}}(V) &\to H^0(B_1(X),x) \\ (c,[\omega]) &\mapsto c + \textstyle\int \omega. \end{aligned}$$

To see this, note that the integral of a one form ω depends on homotopy classes of the paths if and only if ω is closed and it evaluates to zero on closed paths whenever ω is exact.

We now compare closed iterated integrals of length s with elements inside $\otimes^s H^1_{\mathrm{DR}}(X)$:

---

[1] $B_s(X,x)$ is the image of $B^{-s,s}E_{\mathrm{DR}}(X,x)$ under the integration map and $H^0B_s(X,x)$ is the image of the d-closed forms in $B^{-s,s}E_{\mathrm{DR}}(X,x)$.

Proposition 8.30. 1) *The leading term of an iterated integral*

$$\sum_{|J|=s} a_J \int \omega_{j_1} \cdots \omega_{j_s} + \text{iterated integrals of smaller length} \qquad \text{(VIII–5)}$$

defines an element in $\bigotimes^s E^1(X)/dE^0(X)$, *and hence a homomorphism*

$$p : B_s(X,x) \to \bigotimes^s E^1(X)/dE^0(X).$$

2) *The homomorphism* p *is zero on iterated integrals of lower lengths and when restricted to homotopy functionals it induces a homomorphism*

$$\bar{p} : H^0(B_s(X,x)) \to \bigotimes^s H^1_{\mathrm{DR}}(X)$$

with kernel $H^0 B_{s-1}(X,x)$. *Hence we have an exact sequence*

$$0 \to H^0 B_{s-1}(X,x) \to H^0 B_s(X,x) \xrightarrow{\bar{p}} \bigotimes^s H^1_{\mathrm{DR}}(X). \qquad \text{(VIII–6)}$$

Proof. 1) By Lemma 8.25, the value of the leading term on $(\alpha_1 - 1)\cdots(\alpha_s - 1)$ is equal to $\int_{\alpha_s} \omega_s \cdots \int_{\alpha_1} \omega_1$. This evaluates to zero for all choices of closed paths $\alpha_1, \ldots, \alpha_s$ if and only if at least one of the ω_i is exact as we shall now verify by induction. For $s = 1$ the function $f(y) := \sum a_{j_1} \int_x^y \omega_{j_1}$ is well defined and $df = \sum a_{j_1}\omega_{j_1}$.

Now assume that $s > 1$ and choose a basis $[\theta_i]$ for the finite dimensional subspace of $E^1(X)/dE^0(X)$ generated by the ω_{j_k} that appear in the leading term of (VIII–5) and write

$$\begin{aligned} \textstyle\sum_{|J|=s} a_J[\omega_{j_1}] \otimes \cdots \otimes [\omega_{j_r}] &= \textstyle\sum_{|J|=s} A_J[\theta_{j_1}] \otimes \cdots \otimes [\theta_{j_s}] \\ &= \textstyle\sum_i [\theta_i] \otimes \sum A_J[\theta_{j_2}] \otimes \cdots \otimes [\theta_{j_s}]. \end{aligned}$$

Consider now for each $(s-1)$-tuple of closed paths the 1-form modulo differentials

$$[\theta] = \sum_i [\theta_i] \otimes \sum A_J \int_{\alpha_2} \theta_{j_2} \cdots \int_{\alpha_s} \theta_{j_s}.$$

By assumption $[\theta]$ evaluates to zero over any closed path and hence must be zero. But the $[\theta_i]$ form a basis for the space we are working in and hence all coefficients must be zero. By the induction hypothesis the corresponding multiform must be zero and hence also the multiform we started with must vanish.

This completes the proof of 1); in particular we dispose of a well defined homomorphism

$$p : B_s(X,x) \to \bigotimes^s E^1(X)/dE^0(X).$$

2) Since any iterated integral of length $\leq s-1$ evaluates to zero on $a = (\alpha_1 - 1)\cdots(\alpha_s - 1)$ (apply Lemma 8.18 again), it follows that p is zero on the subspace $B_{s-1}(X,x)$.

Let us restrict p to closed integrals. Observe that for a homotopy functional $\int \omega_{j_1}\cdots\omega_{j_s}$ the value over $a = (\alpha_s - 1)\cdots(\alpha_1 - 1)$ does not change when we deform the loops α_j, keeping endpoints fixed. Hence the product $\int_{\alpha_s}\omega_s \cdots \int_{\alpha_1}\omega_1$ does not change under such deformations. By Stokes' theorem, if the ω_{j_k} are not exact, there are loops α_k, $k \neq i$ such that $\int_{\alpha_k}\omega_{j_k} \neq 0$. So $\int_{\alpha_i}\omega_{j_i} = \int_a \omega_{j_1}\cdots\omega_{j_s}/\prod_{k\neq i}\int_{\alpha_k}\omega_{j_k}$ then does not depend on the homotopy class of the loop α_i and so ω_{j_i} is closed. So p maps closed integrals to closed forms. Let us determine those integrals that map to an exact form under p, that is

$$p(I) = d\left(\sum b_{k_1,\dots,k_{s-1}}\eta_{k_1}\otimes\cdots\otimes\eta_{k_{i-1}}\otimes f_{k_i}\otimes\eta_{k_{i+1}}\otimes\cdots\otimes\eta_{k_{s-1}}\right)$$

where the η_j are 1-forms and f_j is a function. If we now evaluate I, since the η_j are closed, we only have iterated integrals of the form

$$\int\left(\sum b_{k_1,\dots,k_{s-1}}\eta_{k_1}\otimes\cdots\otimes\eta_{k_{i-1}}\otimes df_{k_i}\otimes\eta_{k_{i+1}}\otimes\cdots\otimes\eta_{k_{s-1}}\right).$$

These can be reduced to iterated integrals of smaller length using the previously established formulas from Lemma 8.18.

Since, as we have seen, p maps iterated integrals of length $\leq s-1$ to zero this proves (VIII–6) . □

Corollary 8.31. *The weight filtration is defined over* $\mathbb{Q}$.

Proof. This is obviously true for $s = 1$ and for the induction we use that the image of $\bar{p}$ is defined over $\mathbb{Q}$, since it is the kernel of the map

$$\begin{aligned}\textstyle\sum_{i=1}^{s-1} c_i: \quad & \otimes^s H^1(X) \to \textstyle\sum_{i+j=s-2}\otimes^i H^1(X)\otimes H^2(X)\otimes\otimes^j H^1(X)\\ & c_i(z_1\otimes\cdots\otimes z_s) = z_1\otimes\cdots\otimes z_i\otimes(z_i\wedge z_{i+1})\otimes z_{i+2}\otimes\cdots\otimes z_s.\end{aligned}$$ □

Completion of the proof of Theorem 8.28. The Hodge filtration induces the correct Hodge filtration on the subspace inside $\bigotimes^s H^1(X;\mathbb{C})$. To complete the proof we recall (Criterion 3.10) that two filtrations on a rational vector space induce the structure of a rational mixed Hodge structure, if it is the middle term of an exact sequence of rational bi-filtered vector spaces whose morphisms strictly preserve the filtrations, and which moreover induce the structure of a rational mixed Hodge structure on the two extremes of the exact sequence.

This finishes the sketch of the proof that $\mathbb{Z}\pi_1(X,x)/J^{s+1}$ carries a mixed Hodge structure. From the preceding construction it should be clear that it is functorial. □

The space $H^0(\int A(X,x))$ of closed iterated integrals is a direct limit of the spaces $H^0(B_s(X,x))$ of length $\leq s$ closed iterated integrals. Moreover, these spaces define the weight spaces on $H^0(\int A(X,x))$. By Remark 8.27 this real space is the $\mathbb{R}$-dual of $\varprojlim \mathbb{Z}\pi_1(X,x)/J^{s+1}$ and hence recalling Definition 8.13 we have:

Corollary 8.32. *There is a pro-mixed Hodge structure on* $\widehat{\mathbb{Q}\pi_1(X,x)}$ *compatible with the mixed* $\mathbb{Q}$*-Hodge structure on each of the* $\mathbb{Q}\pi_1(X,x)/J^{s+1}$ *from Theorem 8.28.*

Remark 8.33. The mixed Hodge structure depends on base points and in certain cases this can be understood geometrically through a Torelli type theorem as shown by Hain [Hain87b] and Pulte [Pul]:
Let (X,x) *and* (Y,y) *be two pointed compact Riemann surfaces. Suppose that there is a ring isomorphism*

$$\varphi : \mathbb{Z}\pi_1(X,x)/J(X,x)^3 \to \mathbb{Z}\pi_1(Y,y)/J(Y,y)^3$$

inducing an isomorphism of mixed Hodge structures. Then there is an isomorphism $f : X \to Y$ *with* $f(x) = y$ *with the possible exception of two points* x *on* X

8.6 The Sullivan Construction

As explained in the introduction to this chapter, we need a graded commutative product structure on a rationally defined algebra computing the cohomology. Such an algebra has been found by Sullivan in [Sull] using the complex of rational polynomial forms. Let us briefly explain how this works. Let us first look at a simplicial complex K (§ B.1). Consider a collection of p-forms whose coefficients with respect to barycentric coordinates are polynomials with rational coefficients, one on each simplex of K with the obvious demand on compatibility that the form associated to a simplex in the boundary of a given simplex is the restriction of the form on the entire simplex. Such collections form a $\mathbb{Q}$-vector space $A^p(K)$ and differentiation and wedge product on each simplex defines the structure of a **rational** differential graded algebra $\mathsf{A}(K) = \bigoplus_p A^p(K)$. The integration map is a map of cochain complexes $\mathsf{A}(K) \to C^\bullet(K;\mathbb{Q})$, where the right hand side is the complex of $\mathbb{Q}$-valued simplicial cochains. It induces an algebra isomorphism $H^*\mathsf{A}(K) \xrightarrow{\sim} H^*(K;\mathbb{Q})$. One shows that this is compatible with subdivisions and so, if X is triangulable, this construction solves our problem.

One can generalize this to arbitrary *topological spaces* X by using the simplicial set $S_\bullet(X)$ whose p-simplices are the singular p-simplices $\sigma : \Delta_p \to X$ introduced in Example 5.2.1. Each non decreasing map $f : [q] \to [p]$ has as its geometric realisation an affine map $|f| : \Delta_q \to \Delta_p$ and hence there are induced maps $|f|^* : \mathsf{A}(\Delta_p) \to \mathsf{A}(\Delta_q)$ on forms with $\mathbb{Q}$-polynomial coefficients. To construct $\mathsf{A}(S_\bullet X)$ we take a sequence of $\mathbb{Q}$-polynomial forms ω_σ indexed by singular simplices σ, where ω_σ is a form on the associated simplex Δ_p. These sequences (ω_σ) should satisfy the compatibility relations $|f|^*\omega_\sigma = \omega_{f(\sigma)}$ for all non-decreasing maps f, as explained before. For X a *differentiable manifold* one can instead use the simplicial set $S_\bullet^\infty$ of smooth singular simplices and rational polynomial forms on them:

Definition 8.34. Let X be a differentiable manifold. The **Sullivan algebra** or **Sullivan De Rham-complex** $\mathsf{A}(X)_\mathbb{Q}$ is the differential graded algebra

$$\mathsf{A}(X)_\mathbb{Q} := \mathsf{A}(S_\bullet^\infty X).$$

The basic result now is (see [Halp, 15.19]):

Theorem 8.35. *Let X be differentiable manifold and let $\mathsf{A}(X)_\mathbb{Q}$ be the Sullivan De Rham complex of rational polynomial forms on the simplicial set of smooth singular simplices and let $\mathsf{A}_\infty(X)$ be the complex of smooth forms on the same simplicial set. Then*

1) *The inclusion $\mathsf{A}(X)_\mathbb{Q} \otimes \mathbb{R} \hookrightarrow \mathsf{A}_\infty(X)$ is a quasi-isomorphism;*
2) *the natural map $E_{\mathrm{DR}} \to \mathsf{A}_\infty(X)$, which results from pulling back a global form via a smooth singular simplex, is a quasi-isomorphism;*
3) *the integration map (sending the De Rham complex, Sullivan's complex and the complex $\mathsf{A}_\infty(X)$ to singular cohomology) induces the usual De Rham isomorphism*
$$H^*_{\mathrm{DR}}(X) \xrightarrow{\sim} H^*(\mathsf{A}_\infty(X)).$$
4) *For any $\mathbb{Q}$-algebra k, using k-*valued forms *this generalizes to*
$$H^*_{\mathrm{DR}}(X;k) \xrightarrow{\sim} H^*(\mathsf{A}_\infty(X);k) \xleftarrow{\sim} H^*(\mathsf{A}(X)_\mathbb{Q} \otimes_\mathbb{Q} k) \xrightarrow{\sim} H^*(X;k).$$

The bar construction is algebraic in nature. Moreover, as can easily be seen [Hain87, 1.1.1], a quasi-isomorphism $A^\bullet \to A'^\bullet$ between two augmented differential graded algebras $A^\bullet, A'^\bullet$ over a field k with $H^0(A^\bullet) \cong H^0(A'^\bullet) \cong k$ a induces a quasi-isomorphism $BA^\bullet \to BA'^\bullet$ between the two bar constructions. We deduce:

Corollary 8.36. *Let X be a simply connected smooth manifold. Let k be a $\mathbb{Q}$-algebra and let $A^\bullet$ be an augmented commutative k-differential graded algebra together with a quasi-isomorphism*

$$f : A^\bullet \to \mathsf{A}(X)_\mathbb{Q} \otimes_\mathbb{Q} k,$$

from $A^\bullet$ to the Sullivan-algebra of X. Then f induces an isomorphism of k-Hopf algebras
$$H^*(BA^\bullet) \to H^*(P_xX;k).$$

Remark 8.37. There is a logarithmic version of this construction for the cohomology of a smooth but not necessarily compact algebraic variety U. Recall (§ 4.1) that the weight filtration comes from a weight filtration on a subcomplex of the complexified De Rham complex, the complex of forms with logarithmic poles. To have the weight filtration over $\mathbb{Q}$ we have shown (4.3) that this filtered complex is quasi-isomorphic to the Godement resolution of the constant sheaf with its canonical filtration. We would like to replace the complex of logarithmic forms with its weight filtration by a quasi-isomorphic filtered differential graded algebra which is already defined over the rationals. Unfortunately, the canonical filtration τ on the Sullivan complex $\mathsf{A}_{\mathbb{Q}}(U)$ leads to sheaves that are not fine. To be precise, write $U = X - D$, with X smooth and D a strict normal crossing divisor. The complex of sheaves $(j_*\mathsf{A}_{\mathbb{Q}}, \tau)$ associated to the filtered presheaf $V \mapsto (\mathsf{A}_{\mathbb{Q}}(U \cap V), \tau)$ are no longer fine. But there is a Čech-ist approach developed by Hain [Hain87, 5.6] which solves this problem and extends the above approach to the non-compact smooth setting.

8.7 Mixed Hodge Structures on the Higher Homotopy Groups

In order to apply the bar construction to mixed Hodge theory, one needs to show that the bar construction of a differential graded algebra which is a mixed Hodge complex is again a mixed Hodge complex.

Definition 8.38. Let k be a subfield of $\mathbb{R}$. A mixed (commutative) k-Hodge complex

$$\mathbf{A} := \left((A_k, W), (\mathsf{A}_{\mathbb{C}}, W, F), \beta_A : (A, W) \otimes \mathbb{C} \xrightarrow[\sim]{\text{qis}} (A_{\mathbb{C}}, W)\right)$$

built on A_k is a **multiplicative mixed Hodge complex** if A_k and $A_{\mathbb{C}}$ are (commutative) differential graded algebras and the comparison isomorphism β_A is a morphism of filtered differential graded algebras.

If we regard $(k, \mathbb{C})$ as a k-mixed Hodge complex by giving k pure Hodge type $(0,0)$, an augmentation $\epsilon : \mathbf{A} \to k$ is required to be a morphism of k-mixed Hodge complexes. We say then that $(\mathbf{A}, \epsilon)$ is an augmented multiplicative k-mixed Hodge complex.

So, if $(\mathbf{A}, \epsilon)$ is an augmented multiplicative k-mixed Hodge complex there is a splitting of mixed Hodge complexes $\mathbf{A} = \operatorname{Ker} \epsilon \oplus k$.

Recall the bar construction of a differential graded algebra A:

$$B^n(A) = \bigoplus_{s \geq 1} \big[\underbrace{IA \otimes \cdots \otimes IA}_{s}\big]^{n+s},$$

where IA is the augmentation ideal. We see that $B^n\mathbf{A}$ inherits weight and Hodge filtrations from $\mathbf{A}$. However, the weight filtration does *not* give the correct weight filtration on $B\mathbf{A}$. Indeed, in general there appears a shift when we pass from the weight filtration on a mixed Hodge complex to that of its cohomology (Theorem 3.18). Here, making this shift on the level of complexes, i.e. on $B\mathbf{A}$, will produce the correct weight filtration in cohomology (which computes homotopy) if one uses the **bar-weight filtration** given by the differential graded subalgebras

$$(BW)_k\mathbf{A} = \bigoplus_{s\geq 0} W_{k-s}\big[\underbrace{I\mathbf{A}\otimes\cdots\otimes I\mathbf{A}}_{s}\big]^{\bullet+s}.$$

The basic technical result [Hain87, 3.2.1], then is

Proposition 8.39. *If* $\mathbf{A}$ *is a connected augmented multiplicative mixed* k*-Hodge complex, the bar construction* $B\mathbf{A}$ *with induced Hodge filtration and bar-weight filtration is a mixed Hodge complex. If* $\mathbf{A}$ *is commutative, the shuffle-product preserves both filtrations and* $B\mathbf{A}$ *is a multiplicative mixed Hodge complex. Moreover, the co-product is a morphism of multiplicative mixed Hodge complexes.*

Proof. Let us first write down the E_0- and E_1-terms for the bar-weight filtration

$$\begin{aligned}
{}_{BW}E_0^{p,q} &= \mathrm{Gr}^{BW}_{-p}(B^{p+q}\mathbf{A}) = \bigoplus_{s\geq 0}\mathrm{Gr}^W_{-p-s}\big[\underbrace{I\mathbf{A}\otimes\cdots\otimes I\mathbf{A}}_{s}\big]^{p+q+s},\\
&= \bigoplus_{s\geq 0} {}_W E_0^{p+s,q}\big[\underbrace{I\mathbf{A}\otimes\cdots\otimes I\mathbf{A}}_{s}\big]^{\bullet+s}\\
{}_{BW}E_1^{p,q} &= H^{p+q}(\mathrm{Gr}^{BW}_{-p}B\mathbf{A}) = \bigoplus_{s\geq 0} H^{p+q}(\mathrm{Gr}^W_{-p-s}\big[\underbrace{I\mathbf{A}\otimes\cdots\otimes I\mathbf{A}}_{s}\big]^{\bullet+s})\\
&= \bigoplus_{s\geq 0} {}_W E_1^{p+s,q}\big[\underbrace{I\mathbf{A}\otimes\cdots\otimes I\mathbf{A}}_{s}\big]^{\bullet+s}.
\end{aligned}$$

Unravelling the definitions, what one needs to prove is that the differential d_0 on the E_0-term for the bar-weight filtration is strict with respect to the F-filtration and that the E_1-terms carry pure Hodge structures.

Now $I\mathbf{A}$ underlies a mixed Hodge complex and hence, by Lemma 3.20, so does its s-th tensor product. Since d_0 is strictly compatible with the F-filtration on the ${}_WE_0$ terms by definition of a mixed Hodge complex, the same hold then for d_0 on the E_0-terms of the bar-weight filtration.

Next, since the s-th tensor product of $I\mathbf{A}$ underlies a mixed Hodge complex, by definition each of the summands of the E_1-terms carries a pure Hodge structure of weight q and so does ${}_{BW}E_1^{p,q}$ completing the proof. □

Corollary 8.40. *The space of the indecomposables* $QH^*(B\mathbf{A})$ *in the cohomology of the bar construction of a connected augmented multiplicative mixed Hodge complex* $\mathbf{A}$ *carries a* k*-mixed Hodge structure. Moreover, the co-bracket* (VIII–2) *is a morphism of mixed Hodge structures.*

Proof. From the general results on mixed Hodge complexes (Theorem 3.18), it follows that the cohomology of the bar construction $B\mathbf{A}$ carries a mixed k-Hodge structure. It also follows that the product in the algebra $B\mathbf{A}$ induces a morphism of mixed Hodge structures on $H^*(B\mathbf{A})$ as well as on the augmentation ideal $J = \mathrm{Ker}(H^*(B\mathbf{A}) \to k)$ so that J^2 is a sub mixed Hodge structure and $J/J^2 = QH^*(B\mathbf{A})$ receives the structure of a k-mixed Hodge structure. Since the co-product on $H^*(B\mathbf{A})$ is a morphism of mixed Hodge structures (Prop. 8.39), the same holds for the co-bracket on its decomposables (make use of the description of the co-bracket (VIII–2)). □

Corollary 8.41. *The homotopy groups of a simply connected smooth complex projective variety carry a functorial mixed Hodge structure. The Whitehead products are morphisms of mixed Hodge structure.*

Proof. We have seen (Example 2.34) that there is a pure $\mathbb{Q}$-Hodge complex of weight 0 which computes the integral cohomology of X. It consists of the two complexes

$$(R\Gamma(X,\mathbb{Z}) \otimes \mathbb{Q}, (R\Gamma(X,\Omega^\bullet), \text{trivial filtration}))$$

together with the comparison isomorphisms induced by the injection $\mathbb{C} \hookrightarrow \Omega^\bullet$. To get a multiplicative complex we must choose a concrete realization of the complex $R\Gamma(X,\mathbb{Z})\otimes\mathbb{Q}$ which is a commutative differential graded algebra. We take the Sullivan algebra $\mathsf{A}_\mathbb{Q}$ of $\mathbb{Q}$-polynomial differential forms together with the trivial filtration. The latter respects multiplication of differential forms and hence this gives a $\mathbb{Q}$-augmented multiplicative mixed Hodge complex built on $\mathsf{A}_\mathbb{Q}$. Choosing a base point $x \in X$ gives $B\mathsf{A}$ the structure of an augmented commutative differential graded Hopf algebra and Corollary 8.40 then states that the indecomposables $QH^*(B\mathsf{A})$ admit a mixed Hodge structure. The theorem of Borel-Serre (8.6) tells us that the degree s-piece of this space is dual to the homotopy group $\pi_{s+1}(X,x)$ which therefore also gets a mixed Hodge structure.

By Theorem 8.10 the dual of the Whitehead product is the co-bracket on the decomposables in the cohomology of the loop space. So, again by Corollary 8.40, it is a morphism of mixed Hodge structures. □

Remark. As in Remark 8.37, one can extend the preceding constructions to smooth not necessarily compact complex projective varieties.

For possibly singular complex algebraic varieties, we have put a mixed Hodge structure on the cohomology starting from the Hodge- De Rham complexes of sheaves on the smooth varieties of a suitable cubical hyperresolution. Taking global sections of the Godement resolution then gives the De Rham complex whose cohomology gets the induced mixed Hodge structure. Replacing the De Rham complex by a suitable logarithmic version of the Sullivan De Rham complex we can even extend the preceding constructions to this situation as well. The details can be found in [Hain87].

Theorem 8.42. *The mixed Hodge structure on the homotopy group $\pi_s(X,y)$, $s \geq 2$ is independent of the chosen base point $y \in X$.*

Proof. Note that for a simply connected and connected space X there is a canonical isomorphism

$$\phi = \phi_\gamma : \pi_s(X,y) \to \pi_s(X,x)$$

by choosing any path γ from x to y. More precisely, the map

$$\begin{aligned} \Phi : P_{x,y}X \times P_y X &\to P_x X \\ (\gamma,\alpha) &\mapsto \gamma * \alpha * \gamma^{-1} \end{aligned}$$

induces a graded homomorphism $QH^*(P_xX) \to QH^*(P_yX)$ which corresponds to the dual of ϕ under the Borel-Serre isomorphism (Prop. 8.10).

We need a suitable multiplicative mixed Hodge complex that computes $H^*(P_{x,y}X)$ such that Φ induces a morphism of mixed Hodge structures. Now recall that we have discussed a generalization of the bar-construction in Remark 8.22. We use the notation employed there and we use in particular the notation of the example. It is fairly obvious that the differential graded algebra $\bar{B}(k_x, \mathsf{A}_\mathbb{Q} \otimes k, k_y)$ $k = \mathbb{Q}, \mathbb{R}$ are the ingredients needed. Our differential graded algebras $\bar{B}(k_x, \mathsf{A}_\mathbb{Q} \otimes k, k_x)$ and $\bar{B}(k_y, \mathsf{A}_\mathbb{Q} \otimes k, k_y)$ compute the cohomology of P_xX respectively P_yY and the product

$$\begin{aligned} \bar{B}(k_x, \mathsf{A}_\mathbb{Q} \otimes k, k_x) &\to \bar{B}(k_x, \mathsf{A}_\mathbb{Q} \otimes k, k_y) \otimes \bar{B}(k_y, \mathsf{A}_\mathbb{Q} \otimes k, k_y) \\ (\omega_1|\cdots|\omega_s) &\mapsto \textstyle\sum_{0\leq i<j<s} \pm(\omega_1|\cdots|\omega_i) \wedge (\omega_s|\cdots|\omega_{j+1}) \otimes (\omega_{i+1}|\cdots|\omega_j) \end{aligned}$$

relates these. Now, the cohomology of the left hand side computes $H^*(P_xX;k)$, while the right hand computes the cohomology of $P_{x,y}X \times P_yX$. Since $H^0(P_{x,y}X;k) = k$, projection on the appropriate summand of the Künneth decomposition then gives

$$H^k(P_xX;k) \to H^0(P_{x,y}X;k) \otimes H^k(P_yX;k) \cong H^k(P_yX;k).$$

On the level of indecomposables this is exactly the dual of the map ϕ. Since the map Φ^* clearly preserves Hodge and weight filtrations, the above map ϕ^* and its dual ϕ must be maps of mixed Hodge structures. □

Finally we look at the Hurewicz homomorphism:

Theorem 8.43. *For a simply connected algebraic variety the Hurewicz homorphism*

$$h_k : \pi_k(X,x) \to H_k(X)$$

is a morphism of mixed Hodge structures.

Proof. Consider the suspension map

$$\begin{aligned} s : (I, \partial I) \times P_xX &\to (X,x) \\ (t,\gamma) &\longrightarrow \gamma(t). \end{aligned}$$

It induces the map $s_* : H_s(P_xX) \cong H_s(P_xX) \otimes H_1(I, \partial I) \to H_{s+1}(X)$ fitting in the commutative diagram

$$\begin{array}{ccc} \pi_{s+1}(X,x) & \xrightarrow{h_{s+1}} & H_{s+1}(X) \\ \| & & \uparrow s_* \\ \pi_s(P_xX, e_x) & \xrightarrow{h_s} & H_s(P_x). \end{array}$$

Now dually this gives

$$\begin{array}{ccc} \mathrm{Hom}(\pi_{s+1}(X,x),\mathbb{Q}) & \xleftarrow{h^{s+1}} & H^{s+1}(X;\mathbb{Q}) \\ \| & & \downarrow s^* \\ \mathrm{Hom}(\pi_s(P_xX, e_x),\mathbb{Q}) & \xleftarrow[Qh^s]{\sim} & QH^s(P_x), \end{array}$$

where s^* comes from the integration map $\mathsf{A}^{s+1}(X) \to \mathsf{A}^s(P_xX)$ (an $(s+1)$-form α is mapped to its iterated integral $\int \alpha$, viewed as an s-form on P_xX). It follows that on the level of differential graded algebras the Hurewicz map h^{s+1} is induced by the map

$$\begin{aligned} \mathsf{A}^{s+1}_{\mathbb{Q}}(X) &\to \bar{B}\mathsf{A}^{s+1}_{\mathbb{Q}}(X) \\ \alpha &\mapsto (\alpha). \end{aligned}$$

This map therefore obviously preserves weight and Hodge filtrations. □

Historical Remarks. Most of the results in this chapter are due to Hain ([Hain87]). It uses the approach to homotopy De Rham theorems through iterated integrals as initiated by K.T. Chen [Chen76] who built on topological results of Adams [Adams] (in the simply connected case) as well as Stallings [Stal] (for the fundamental group).

9

Hodge Theory and Minimal Models

This chapter is devoted to Sullivan's theory of the minimal model and Morgan's construction of a mixed Hodge structure on the homotopy groups using minimal models. A priori this mixed Hodge structure might differ from Hain's. But in fact, for the *higher* homotopy groups the two are equal (as communicated to us by Hain). However, since the base point is absent in Morgan's construction for the *fundamental group*, it cannot be the same as Hain's. On the other hand, Morgan's construction is more powerful since it gives a mixed Hodge structure on the cohomology level of the constituents of the (rational) Postnikov tower (which contains all information from rational homotopy).

We briefly sketch Morgan's construction in case of a simply connected compact Kähler manifold, omitting his construction for the mixed Hodge structure of the fundamental group. See also Remark 9.25.

A rough outline of Morgan's construction goes as follows. Recall (Theorem 8.35) that for any polyhedron X, Sullivan has shown that the differential graded algebra $\mathsf{A}(X)$ of polynomial forms with rational coefficients computes $H^*(X;\mathbb{Q})$. The idea is that this differential graded algebra contains sufficient information to reconstruct the *rational* homotopy type of X. For simply connected spaces this just means that we can compute $\pi_k \otimes \mathbb{Q}$, $k > 1$, but it is more complicated to define what is meant by $\pi_1 \otimes \mathbb{Q}$. In the simply connected case, treated in § 9.2 and § 9.3, the main result is that for $k \geq 2$ the rational homotopy group $\pi_k \otimes \mathbb{Q}$ is canonically dual to the indecomposables in degree k of a certain differential graded algebra $\mathsf{MA}(X))$ canonically associated to $\mathsf{A}(X)$, the minimal model. In our case, the fact that the cohomology of X is finite dimensional implies that the model $\mathsf{MA}(X))$ is also finite dimensional. There is a mixed Hodge structure on this algebra compatible with the differential and the wedge product. This is the content of Morgan's main result, stated in § 9.3, but whose proof we only sketch. It follows that the indecomposables carry a mixed Hodge structure.

In § 9.4.1 Sullivan's construction for the fundamental group is explained with an application to Kähler manifolds in § 9.4.3: the real De Rham fundamental group is determined by the cup-product form on H^1. This turns out to impose severe restrictions on the possible fundamental groups of Kähler manifolds.

9.1 Minimal Models of Differential Graded Algebras

We start with the basic definitions. Let us work over a fixed field k. We shall work with **connected** differential graded k-algebras A (i.e. $A^0 = k$) so that the augmentation ideal equals

$$I(A) = A^+ = \bigoplus_{p>0} A^p$$

and the indecomposables $Q(A) = I(A)/I(A)^2$ can be viewed as the k-space generated by a minimal set of generators. An algebra A is said to be **1-connected** if it is connected and if $H^1(A) = 0$.

We shall build an extension of A by adjoining the elements in degree n as follows. Let V be a k-vector space and let $\Lambda_n V$ be the free graded commutative algebra with unit generated by V (so that $\Lambda_n V$ is the polynomial algebra on V if n is even and the exterior algebra if n is odd).

Definition 9.1. Let A be a differential graded k-algebra and let V be a finite dimensional k-vector space. A linear map $\varphi : V \to A^{(n+1)}$ with $d \circ \varphi = 0$ determines a **Hirsch extension in degree** n. This is the algebra $A \otimes \Lambda_n V$ made into a differential graded algebra by placing V in degree n and by extending the differential upon setting $dx = \varphi(x)$ when $x \in V$. This differential graded algebra is denoted

$$A \otimes_\varphi \Lambda_n V.$$

The Hirsch-extension is **decomposable** if the image of φ is decomposable, i.e. $\varphi(V) \subset A^+ \wedge A^+ = I(A)^2$.

Remark 9.2. Clearly, the indecomposables of a decomposable Hirsch extension are just the direct sum of V with the indecomposables of A.

A minimal differential graded algebra is built from k by successive decomposable Hirsch extensions. For a given differential graded algebra A a **minimal model** MA is a minimal differential graded algebra which is quasi-isomorphic to A and which maps to A. Formally:

Definition 9.3. 1) A differential graded algebra M is called **minimal** if

- $\mathsf{M}^0 = k$ (i.e. it is connected);
- $d\mathsf{M} \subset \mathsf{M}^+ \wedge \mathsf{M}^+ = I\mathsf{M}^2$, i.e. d is decomposable. Equivalently, d induces the zero map on the indecomposables $Q\mathsf{M}$;
- There is a **series** for M, i.e. an increasing union of differential graded sub-algebras
$$k = \mathsf{M}_0 \subset \mathsf{M}_1 \subset \mathsf{M}_2 \subset \cdots \mathsf{M}$$
such that $\mathsf{M}_n \subset \mathsf{M}_{n+1}$ is a Hirsch extension.

2) A **minimal model** $\mathsf{M}A$ for a given differential graded algebra A is a minimal differential graded algebra $\mathsf{M}A$ together with a quasi-isomorphism of differential graded algebras $f : \mathsf{M}A \to A$.

If M is minimal and 1-connected, there is a **canonical series** for M by letting M_n be the subalgebra generated by elements in degrees $\leq n$. In this case $\mathsf{M}_n \subset \mathsf{M}_{n+1}$ is indeed a Hirsch extension of degree n as we shall explain. We have a representation $\mathsf{M}_{n+1} = \mathsf{M}_n \otimes \Lambda_{n+1} V$ as vector spaces with V the vector space of indecomposables of M in degree $n+1$. Since d is decomposable, and since there are no degree 1 elements, for any $v \in V$ the derivative dv is a linear combination of elements that are products of indecomposables in degrees at most n, i.e. $dv \in \mathsf{M}_n$. Conversely we have:

Theorem 9.4. *Any differential graded algebra A which is 1-connected has a minimal model $\mathsf{M}A$ with $\mathsf{M}^1(A) = 0$. If $H^*(A)$ is a finite dimensional k-vector space, then the minimal model $\mathsf{M}A$ is a finitely generated k-algebra.*

Sketch of the proof. We put $\mathsf{M}_0 = k$, $f_0 : k \to A$ the canonical map and we assume inductively that we have constructed successive Hirsch extensions $\mathsf{M}_0 \subset \cdots \subset \mathsf{M}_m$ and maps of differential graded algebras $f_j : \mathsf{M}_j \to A$, $j = 0, \ldots, m$ which are isomorphisms in cohomology of degree $\leq j$ and injections in degree $m+1$. We want to construct the next step as a Hirsch extension of M_m. Consider $\mathrm{Cone}^\bullet(f_m)$. The exact sequence of the cone (A–12) together with the inductive assumptions show that $H^i(\mathrm{Cone}^\bullet(f_m)) = 0$ for $i \leq m+1$. We put $V = H^{m+2}(\mathrm{Cone}^\bullet(f_m))$ and we define $f_{m+1} : V \to A^{m+1}$ by choosing a section of

$$\{(m+2)\text{-cocycles in } \mathrm{Cone}^\bullet(f_m)\} \longrightarrow H^{m+2}(\mathrm{Cone}^\bullet(f_m))$$

and then projecting onto the M-summand. Explicitly, choose, linearly in $w \in V$, cocycle representatives $(m_w, a_w) \in \mathsf{M}_m^{m+2} \oplus A^{m+1}$. Hence $dm_w = 0$, $f_m(m_w) = da_w$ and the class of m_w is w. Then we define $f_{m+1}(w) = a_w$ and $\mathsf{M}_{m+1} = \mathsf{M}_m \otimes_{f_{m+1}} \Lambda_{m+1} V$. The map $f_{m+1} : \mathsf{M}_{m+1} \to A$ equals f_m on M_m and f_{m+1} on V and is extended multilinearly. We define $dw = m_w$ in M_{m+1} so that $d(dw) = dm_w = 0$ and $f_m(dw) = f_m(m_w) = da_w = df_{m+1}w$ so that f_{m+1} is a map of differential graded algebras. The verification that f_{m+1} induces an isomorphism in cohomology up to degree $\leq m+1$ and an injection in degree $m+2$ is left to the reader. For details, we refer to [Grif-Mo, Ch. IX]. □

Example 9.5. Let A be the De Rham algebra of $\mathbb{P}^n$. Clearly $\mathsf{M}_2(A) = P(u)$, the polynomial algebra on a generator of degree 2. Since $u^{n+1} = dv_{2n+1}$, we find

$$\mathsf{M}A = P(u) \otimes_{\mathrm{id}} \Lambda_{2n+1}[v_{2n+1}].$$

Next, we need to discuss unicity of the minimal model. One uses the concept of **homotopic** differential graded algebras. To explain this, let $k[t, dt]$ be the tensor product $k[t] \otimes \Lambda(dt)$ where we place t in degree 0 and dt in degree 1. The differential is the obvious one, sending $p(t)$ to $p'(t)dt$ and dt to 0. For $k = \mathbb{R}$ one can view this as the algebra of differential forms on the real line with polynomial coefficients. A **homotopy** between two differential graded algebra homomorphisms $f, g : A \to B$ is a map of differential graded algebras $H : A \to B \otimes k[t, dt]$ such that $f = e(0) \circ H$, $g = e(1) \circ H$. Here $e(x) : B \otimes k[t, dt] \to B$ is the evaluation map $t \mapsto x \in k$, $dt \mapsto 0$.

In loc. cit. one finds the unicity statement for minimal models:

Theorem 9.6. *Let A be a 1-connected differential graded algebra and let $f : \mathsf{M}A \to A$, $f' : \mathsf{M}'A \to A$ two minimal models for A with $\mathsf{M}A^1 = \mathsf{M}'A^1 = 0$. There exist an isomorphism $\iota : \mathsf{M}A \xrightarrow{\sim} \mathsf{M}'A$ such that f and $f' \circ \iota$ are homotopic. The isomorphism ι is determined up to a homotopy by this condition.*

Remark 9.7. The minimal model $\mathsf{M}A$ has no generators in degree 1. This implies that the map ι induces a *unique* isomorphism between the indecomposables. To explain this, recall that the indecomposables in degree n in a minimal model is the vector space, say V_n, used to build the n-th step of the filtration as a Hirsch-extension. The map ι being unique on the successive quotients of the canonical filtration induces a unique map on the quotients of the induced filtration on decomposables. But this filtration is split by degree and the degree n piece is exactly V_n. So ι is indeed unique on decomposables. We say that the space of indecomposables is a **homotopy invariant**.

In the non-simply connected case this notion has to be adapted. Without giving the details here, let us only mention that the appropriate notion is that of **lomotopy** as introduced by Halperin [Halp, 11.19]. We refer to loc. cit. for the unicity statements in the non simply-connected case.

9.2 Postnikov Towers and Minimal Models; the Simply Connected Case

Recall [Span, p. 426], that for any $n \geq 1$ and any group π (abelian if $n > 1$), there exists a CW complex $K(\pi, n)$ which is unique up to homotopy and has exactly one non-vanishing homotopy group, $\pi_n = \pi$. Recall also that the path space $PK(\pi, n+1)$ fibres over $K(\pi, n+1)$ with fibre $K(\pi, n)$. Given a path connected base space B with base point b and a fibration $E \to B$ over it, we say that the fibration is **principal** if the action of $\pi_1(B, b)$ on the fibre is trivial up to homotopy. This is in particular the case if the base is simply connected. So $PK(\pi, n+1) \to K(\pi, n+1)$ is a principal fibration if $n \geq 1$, the **universal principal fibration with fibre** $K(\pi, n)$. Indeed, any principal fibration over B with fibre $K(\pi, n)$ is obtained by pulling back this universal fibration by means of a map $f : B \to K(\pi, n+1)$, unique up to homotopy. By [Span, p. 447] this map is classified by its **obstruction class** $e(f) \in H^{n+1}(B; \pi)$, i.e. there is a bijection

$$e : [B, K(\pi, n+1)] \xleftrightarrow{1:1} H^{n+1}(B; \pi).$$

In the remainder of this chapter we shall restrict ourselves to a simply connected topological space X. The diagram

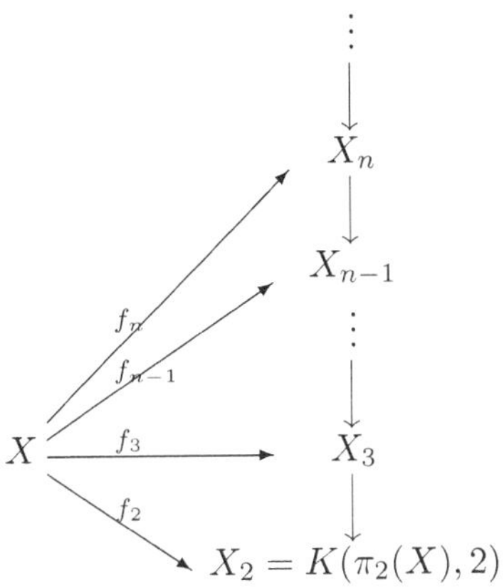

is called a **Postnikov tower** of X if the following conditions are satisfied

1. X_n has zero homotopy groups in degrees $> n$;
2. $X_n \to X_{n-1}$ is a principal fibration with fibre $K(\pi_n(X), n)$;
3. f_n induces a isomorphisms $\pi_k(X) \xrightarrow{\sim} \pi_k(X_n)$ for $k \leq n$.

A Postnikov tower is inductively built as a tower of principal fibrations starting from $K(\pi_2, 2)$ and stage X_n is built from X_{n-1} by specifying a characteristic element $e_{n+1} \in H^{n+1}(X_{n-1}; \pi_n(X))$. Such towers exist (loc. cit, p. 444). Moreover, taking the limit of the inverse system we get a space $X' = \lim_n X_n$ with the same homotopy type as X. So we can recover X up to homotopy from its Postnikov tower.

Next we need the concept of a **rational Postnikov tower**. This tower encodes the information in the $\mathbb{Q}$-vector spaces $\pi_k(X) \otimes \mathbb{Q}$, $k \geq 2$. To do this, consider the CW complex $K(\mathbb{Q}, n)$. So, if $\pi_2(X) \otimes \mathbb{Q} \cong \mathbb{Q}^s$ we can start with $(X_2)_{\mathbb{Q}} = K(\pi_2(X) \otimes \mathbb{Q}, 2) = \prod^s K(\mathbb{Q}, 2)$. Then one inductively replaces each fibration $X_n \to X_{n-1}$ in the construction of the Postnikov tower by the corresponding fibration $(X_n)_{\mathbb{Q}} \to (X_{n-1})_{\mathbb{Q}}$ using the characteristic element

$$\begin{aligned}(e_{n+1})_{\mathbb{Q}} \in H^{n+1}((X_{n-1})_{\mathbb{Q}}; \pi_n(X) \otimes \mathbb{Q}) = \\ \mathrm{Hom}_{\mathbb{Q}}\left((\pi_n X \otimes \mathbb{Q})^{\vee}, H^{n+1}((X_{n-1})_{\mathbb{Q}}; \mathbb{Q})\right).\end{aligned}$$

For details see [Grif-Mo, Chapter VII].

The basic idea of Sullivan's theory is that from this rational Postnikov tower, one can inductively build a minimal model $\mathsf{MA}(X)$ for the Sullivan algebra $\mathsf{A}(X)$ by making a Hirsch extension with V the dual of $\pi_n(X) \otimes \mathbb{Q}$ and φ any $\mathbb{Q}$-linear map

$$\varphi : (\pi_n(X) \otimes \mathbb{Q})^{\vee} \to \{\text{closed forms in } A^{n+1}((X_{n-1})_{\mathbb{Q}})\}$$

which induces the characteristic element

$$(e_{n+1})_{\mathbb{Q}} \in \mathrm{Hom}_{\mathbb{Q}}\left((\pi_n X \otimes \mathbb{Q})^{\vee}, H^{n+1}((X_{n-1})_{\mathbb{Q}}; \mathbb{Q})\right).$$

The statement of the following theorem follows quite directly from this.

Theorem 9.8 ([Sull]). *Let X be a simply connected polyhedron with finite-dimensional rational cohomology $H^*(X;\mathbb{Q})$. Then the canonical series of the minimal model $\mathsf{MA}(X)$ of the Sullivan algebra $\mathsf{A}(X)$ has the following properties.*

1) $\mathsf{M}_j\mathsf{A}(X)$ *is the minimal algebra of the Sullivan algebra of the j-stage of the rational Postnikov tower;*
2) *The space of indecomposables of $\mathsf{MA}(X)$ in degree n is canonically dual to $\pi_n(X)\otimes\mathbb{Q}$ and the characteristic element $(e_{n+1})_{\mathbb{Q}}$ gets identified with the map*

$$Q\mathsf{MA}(X)^n = (\pi_n(X)\otimes\mathbb{Q})^\vee \to H^{n+1}(\mathsf{M}_{n-1}\mathsf{A}(X))$$

induced by the n-th step extension of the minimal model;

3) *The Whitehead product*

$$\sum_{i+j=k} (\pi_i X\otimes\mathbb{Q})\otimes(\pi_j X\otimes\mathbb{Q}) \to \pi_{k-1}X\otimes\mathbb{Q}$$

is dual to the map induced by

$$d: Q^{k-1}\mathsf{MA}(X) \to (Q\mathsf{MA}(X)\wedge Q\mathsf{MA}(X))^k .$$

Example 9.9. We computed the minimal model of the De Rham algebra of $\mathbb{P}^n$ in Example 9.5. It follows that

$$\pi_k(\mathbb{P}^n)\otimes\mathbb{Q} = \begin{cases} 0 & k\neq 2, 2n+1 \\ \mathbb{Q} & k = 2, 2n+1. \end{cases}$$

9.3 Mixed Hodge Structures on the Minimal Model

The basic result is the following theorem together with its ensuing corollary.

Theorem 9.10 ([Mor]). *Let X be a simply connected smooth complex variety. The minimal model $\mathsf{MA}(X)$ of the Sullivan algebra admits a mixed Hodge structure with the following properties*

1) *The defining morphism $f:\mathsf{MA}(X)\to\mathsf{A}(X)$ induces a morphism of mixed Hodge structures in cohomology;*
2) *The differential and product structure of $\mathsf{MA}(X)$ are morphisms of mixed Hodge structures;*
3) *The mixed Hodge structure is well-defined and functorial only up to homotopy.*

Corollary 9.11. *Let X be a simply connected smooth complex variety.*

1) *The rational homotopy groups carry a functorial mixed Hodge structure and the Whitehead products are morphisms of mixed Hodge structure;*

2) *The rational cohomology rings of the stages X_n in the rational Postnikov tower carry a mixed Hodge structure and the maps in this tower $X \to X_n$ and $X_{n+1} \to X_n$ induce morphisms of mixed Hodge structures in rational cohomology;*
3) *The rational invariants*

$$(e_{n+1})_{\mathbb{Q}} : [\pi_n(X) \otimes \mathbb{Q}]^{\vee} \to H^{n+1}(X; \mathbb{Q})$$

are morphisms of mixed Hodge structures.

Proof (of the corollary). Use the fact (see Remark 9.7) that in the simply connected case the indecomposables form a homotopy invariant of the minimal algebra, and then use Theorems 9.8 and 9.10. □

Remark 9.12. 1) It can be shown (D. Hain, letter to the authors) that the mixed Hodge structure on the homotopy groups as constructed by Hain (see the previous Chapter) is the same as the one found by Morgan.
2) Morgan's constructions can be modified so as to apply to cubical schemes and thus there are results similar to the previous two theorems valid for arbitrary complex algebraic varieties. See [Nav] for details.

Before giving a sketch of the proof of Theorem 9.10, we need to introduce one of the basic ingredients in Morgan's proof:

Definition 9.13. Let k be a field contained in $\mathbb{R}$. A **k-mixed Hodge diagram** consists of a (biregularly) filtered differential k-algebra (A, W) and a bi-filtered differential $\mathbb{C}$-algebra (E, W, F) together with a comparison morphism, which is a quasi-isomorphism of filtered differential algebras

$$(A, W) \otimes_k \mathbb{C} \xrightarrow[\beta]{\overset{\text{qis}}{\sim}} (E, W)$$

such that

1) if we use β to put a real structure on the terms of the W-spectral sequence, the inductive F-filtration (§ 3.2) on the terms ${}_W E_1^{p,q}(E)$ is q-opposed to its complex conjugate;
2) the differentials of the W-spectral sequence are strictly compatible with the inductive F-filtration.

To compare two such diagrams we introduce two more concepts:

1) An **elementary equivalence** between mixed Hodge diagrams $((A, W), \beta, (E, W, F))$ and $((A', W), \beta', (E', W, F))$ consists of a commutative diagram

$$\begin{array}{ccc} A_{\mathbb{C}} & \xrightarrow{\beta} & E \\ \big\downarrow{\scriptstyle f_{\mathbb{C}}} & & \big\downarrow{\scriptstyle g} \\ A'_{\mathbb{C}} & \xrightarrow{\beta'} & E' \end{array}$$

of quasi-isomorphisms of differential algebras together with a homotopy H between $g \circ \beta$ and $\beta' \circ f_{\mathbb{C}}$ such that $f_{\mathbb{C}}$ and H are compatible with W and g is compatible with F and W.

2) An **equivalence** is a finite string of elementary equivalences, possibly with arrows in both directions.

Remark. 1. Let $(A,W),\beta,(E,W,F)$ be a mixed Hodge diagram. Putting the conjugate filtration $\bar{F}$ on $\bar{E}$ defines the conjugate diagram $((A,W),\bar{\beta},(\bar{E},W,\bar{F}))$.

2. The condition 2) for a mixed Hodge diagram implies that the inductive F-filtration on the terms of the W-spectral sequence coincides with the first and second indirect filtrations (Theorem 3.12). There is therefore no danger of confusion when we speak in the sequel of *the* F-filtration on these terms. Condition 1) implies that $H^{p+q}(\mathrm{Gr}^W_{-p} E)$ has a pure Hodge structure of weight q and hence

$$(A,(A,W),\mathrm{id},(E,W,F),\beta)$$

is a mixed k-Hodge complex; because β is compatible with the multiplication, it is even a *multiplicative* mixed Hodge complex in the sense of Def. 8.38. Moreover, equivalent diagrams give quasi-isomorphic multiplicative mixed k-Hodge complexes. So the above concept can be considered as a refinement. In particular, the filtrations $\mathrm{Dec}\,W$ and F induce the structure of a mixed Hodge structure on the cohomology $H^*(A)$.

Examples 9.14. 1) Let X be a smooth projective variety. The De Rham algebra $E_{\mathrm{DR}}(X)$ and the usual Hodge filtration on its complexification defines a real mixed Hodge diagram by classical Hodge theory (Chap. 2). The comparison morphism is the identity in this case.

2) Let U be a smooth algebraic variety, X a good compactification of U with inclusion $j: U \hookrightarrow X$. Let D_j, $j \in J$ be the components of $D = X - U$. We define $E_{\mathrm{DR}}(X,D)$ be the differential graded algebra generated by the $E_{\mathrm{DR}}(X)$ and symbols θ_j of degree 1 with $d\theta_j = \omega_j$, a closed smooth 2-form on X with support in a tubular neighbourhood U_j of D_j such that its class in $H^2_{D_j}(U_j) = H^2(U_j, U_j - D_j)$ is the Thom class. The W-filtration counts the number of θ_j. For E we take the smooth De Rham complex $E(X \log D)$ with logarithmic forms along D with the usual weight and Hodge filtration (see Remark 4.4). The comparison morphism

$$\beta : E_{\mathrm{DR}}(X,D) \otimes \mathbb{C} \to E(X \log D) \otimes \mathbb{C}$$

can be defined as soon as a 1-form β_j with logarithmic singularities along D has been constructed with $d\beta_j = \omega_j$. See [Mor, Lemma 3.2]. Indeed, we then put $\beta(\theta_j) = \beta_j$. The fact that we do obtain a real mixed Hodge diagram is a restatement of the main results in Chap. 4. See [Mor, §3]. We call $(E_{\mathrm{DR}}(X,D),W),\beta,(E(X \log D),W,F))$ a **Hodge-De Rham diagram** for (X,D). Any other compactification of U and other choices for the forms θ_j or β_j lead to equivalent mixed Hodge diagrams. So Hodge-De Rham diagrams are defined up to equivalences.

Proof (Sketch of the proof of Theorem 9.10).
Step 1. The complex bigraded minimal model. We consider more generally 1-connected real differential graded algebras A with finite dimensional cohomology. We assume that these fit into a mixed Hodge diagram $((A,W),\beta,(E,W,F))$ so that the cohomology $H^*(E)\cong H^*(A)\otimes\mathbb{C}$ admits a real mixed Hodge structure. Since the real minimal model $f:\mathsf{M}A\to A$ is a quasi-isomorphism, also $H^*(\mathsf{M}A)$ receives a real mixed Hodge structure. On the complex cohomology we therefore have a canonical bigrading, the Deligne splitting (Lemma-Def. 3.4). This bigrading can be lifted to a bigraded minimal differential graded algebra which has certain extra properties which make it unique up to homotopies, the *bigraded minimal model* of $A\otimes\mathbb{C}$.

Definition 9.15. 1) A differential graded algebra M has a **compatible bigrading** if

$$M=\bigoplus_{0\le r,s} M^{r,s},\qquad M^{0,0}=A^0=k,$$

such that the wedge product and the d-operator are of type $(0,0)$.
2) A **morphism** from a differential graded algebra M with compatible bigrading to a mixed Hodge diagram $\mathsf{D}=(A,W),\beta,(E,W,F)$ consists of a diagram

$$\begin{array}{ccccc} & & M & & \\ & \swarrow^{\psi} & \downarrow^{\rho} & \searrow^{\psi'} & \\ E & \xleftarrow{\beta} & A\otimes\mathbb{C} & \xrightarrow{\bar\beta} & \bar E \end{array}$$

and homotopies $H:M\to E\otimes k[t,dt]$ and $H':M\to\bar E\otimes k[t,dt]$ from $\beta\circ\rho$ to $\bar\beta\circ\rho$ such that (as usual, $\mathrm{Dec}\,W$ denotes the filtration W backshifted as in Remark A.50):

$$\begin{aligned}&\rho(M^{r,s})\subset(\mathrm{Dec}\,W)_{r+s}A,\\ &\psi(M^{r,s})\subset R^{r,s}E,\ \psi'(M^{r,s})\subset L^{r,s}\bar E\\ &H^*(M^{r,s})\subset(\mathrm{Dec}\,W)_{r+s}E\otimes k[t,dt],\ H'(M^{r,s})\subset(\mathrm{Dec}\,W)_{r+s}\bar E\otimes k[t,dt].\end{aligned}$$

Here

$$\begin{cases} R^{r,s}E=(\mathrm{Dec}\,W)_{r+s}\cap F^rE\\ L^{r,s}\bar E=(\mathrm{Dec}\,W)_{r+s}\bar E\cap\bar F^q\bar E+\sum_{i\ge 2}(\mathrm{Dec}\,W)_{r+s-i}\bar E\cap\bar F^{r-i+1}\bar E.\end{cases}$$

If in addition $\rho:M\to A\otimes\mathbb{C}$ is a minimal model for $A\otimes\mathbb{C}$, we say that $\mathsf{M}(\mathsf{D}):=(M,\psi,\rho,\psi',H,H')$ is a **bigraded minimal model** for D.

The main result concerning existence and uniqueness of bigraded models is as follows:

Theorem 9.16 ([Mor, §6]). *Any mixed Hodge diagram $\mathsf{D}=((A,W),\beta,(E,W,F))$ has a bigraded minimal model $\mathsf{M}(\mathsf{D})$. The bigrading induced on the cohomology $H^*(\mathsf{D})$ by its (own) mixed Hodge structure agrees with the* *bigrading induced on the cohomology of the minimal model* $\mathsf{M}\mathsf{D}$. *Equivalent mixed Hodge diagrams give rise to isomorphic bigraded minimal models. The isomorphism is unique up to a homotopy compatible with the bigradings.*

The bigraded minimal model for a such a mixed Hodge diagram D for simplicity will be written $\mathsf{M}(A \otimes \mathbb{C})$ and the morphism

$$\rho : \mathsf{M}(A \otimes \mathbb{C}) \to A \otimes \mathbb{C}$$

will be called the **complex minimal model**. The bigrading on $\mathsf{M}(A \otimes \mathbb{C})$ define weight and Hodge filtrations:

$$W_m(\mathsf{M}(A \otimes \mathbb{C})) := \bigoplus_{r+s \le m} \mathsf{M}(A \otimes \mathbb{C})^{r,s}$$
$$F^k(\mathsf{M}(A \otimes \mathbb{C}) := \bigoplus_{r \ge k} \mathsf{M}(A \otimes \mathbb{C})^{r,s}.$$

A restatement of Theorem 9.16 in terms of these is:

Corollary 9.17. *The bigraded morphism*

$$\mathsf{M}(A \otimes \mathbb{C}) \xrightarrow{\beta \circ \rho} E$$

sends the filtrations induced by (W, F) *to the filtrations* $(\operatorname{Dec} W, F)$ *on* E.

We finish this first step by applying the above to the geometric situation: let U be a smooth complex algebraic variety with compatible compactification (X, D). We conclude from the previous corollary that the minimal model

$$\mathsf{M}(X; \mathbb{C}) := \mathsf{M}E_{\mathrm{DR}}(X; \mathbb{C})$$

of the complex De Rham forms gets a bigrading through the choice of a Hodge-De Rham diagram for (X, D). Moreover, different choices differ by automorphisms homotopic to the identity.

Step 2. Weight filtrations and mixed Hodge structures on the real model. Still in the general situation of a Hodge diagram as in step 1, we want to find a weight filtration W on the *real* minimal model $\mathsf{M}E$ such that

$$\rho : (\mathsf{M}E, W) \to (E, \operatorname{Dec} W)$$

is a filtered algebra morphism. This is an example of a more general concept:

Definition 9.18. 1) A filtration W on a minimal algebra M is called **minimal** if both d and the product are strictly compatible with W;
2) the filtration on a filtered algebra (A, W) **passes to the minimal model** if there exists a minimal model $\rho : \mathsf{M} \to A$ for A together with a minimal filtration on M such that ρ is compatible with the filtrations. We call $(\mathsf{M}, W) \xrightarrow{\rho} (A, W)$ a **filtered minimal model**.

By [Mor, Cor. (7.6)], filtered minimal models are unique up to isomorphisms, themselves unique up to homotopies preserving the filtrations.

As an example, the weight filtration on $A \otimes \mathbb{C}$ passes to $\mathsf{M}(A \otimes \mathbb{C})$ and we need to see that it passes to $\mathsf{M}A$. This is possible thanks to:

Principle (of deforming the field of definition [Mor, Thm. (7.7)]). *If (A, W) is a filtered k-algebra ($k \subset \mathbb{C}$ a subfield) and $(A, W) \otimes \mathbb{C}$ passes to the minimal model, then also (A, W) does. Moreover, if two minimal filtrations on the same minimal model have the property that over $\mathbb{C}$ the identity is homotopic to a filtered isomorphism, then the same is true over k.*

It follows that the weight filtration on the real minimal model passes to the minimal model.

Now we have almost all the ingredients needed to put a real mixed Hodge structure on the real minimal model of a mixed Hodge diagram $((A, W), \beta, (E, W, F))$. In fact, because the comparison morphism β is a quasi-isomorphism, there is an isomorphism

$$\mathsf{M}(\beta) : \mathsf{M}(A \otimes \mathbb{C}) \to \mathsf{M}E$$

which is well defined up to homotopy and which may be assumed to be an isomorphism of filtered minimal models (since β induces a quasi-isomorphism $(A, \operatorname{Dec} W) \otimes \mathbb{C} \to (E, \operatorname{Dec} W)$). Moreover such a filtered isomorphism $\mathsf{M}(\beta)$ is well-defined up to homotopy compatible with the filtrations. One then can complete this step to arrive at the following result:

Theorem 9.19 ([Mor, Theorem (8.6)]). *Suppose that A is 1-connected and that $H^*(A)$ is finite dimensional. If A fits into a Hodge diagram, any filtered isomorphism $\mathsf{M}(\beta)$ as above defines a real mixed Hodge structure on $\mathsf{M}A$ such that d and the product are morphisms of mixed Hodge structures. The map induced by $\rho : \mathsf{M}A \to A$ in cohomology is a map of mixed Hodge structures, where we put a mixed Hodge structure on $H^*(A)$ by viewing a mixed Hodge diagram as a mixed Hodge complex.*

As an aside, the proof consists of an easy induction argument, exploiting the construction of the minimal model.

Step 3. Rational structures. Here we go back to the construction of the $\mathbb{Q}$-multiplicative mixed Hodge complexes which give the rational cohomology of a smooth algebraic manifold U (§ 8.6) and refine them to a Hodge diagram as follows. Let (X, D) be a good compactification for U and choose a C^1-triangulation for X such that D becomes a subcomplex. We let $\mathsf{A}(X)$ be the Sullivan complex of $\mathbb{Q}$-polynomial forms with respect to this triangulation and likewise, we let $\mathsf{A}_\infty(X)$ be the version with piece-wise smooth forms. We next define an algebra $\mathsf{A}(X, D)$ built from $\mathsf{A}(X)$ in a similar way as $E_{\mathrm{DR}}(X, D)$ is built from $E_{\mathrm{DR}}(X)$ (Example 9.14 2)). We then define $\mathsf{A}_\infty(X, D)$ starting from $\mathsf{A}_\infty(X)$. We refer to [Mor, §2] for the definition of the arrows in the following diagram and for the proof that is commutative up to homotopy

$$\begin{array}{ccccc}
\mathsf{A}(X, D) \otimes \mathbb{R} & \hookrightarrow & \mathsf{A}_\infty(X, D) & \hookleftarrow & E_{\mathrm{DR}}(X) \\
\downarrow{\scriptstyle\psi} & & \downarrow{\scriptstyle\psi_\infty} & & \downarrow{\scriptstyle\psi_{\mathrm{DR}}} \\
\mathsf{A}(U) \otimes \mathbb{R} & \hookrightarrow & \mathsf{A}_\infty(U) & \hookleftarrow & E_{\mathrm{DR}}(U)
\end{array}$$

The principle of deforming the field of definition shows that the W-filtration on the rational algebra $\mathsf{A}(U)$ passes to the minimal model $\mathsf{MA}(U)$ and the inclusions on the bottom line of the previous diagram, after tensoring with $\mathbb{C}$, become compatible with the weight and Hodge filtrations. Since these define the Hodge-De Rham diagram up to quasi-isomorphisms, the mixed Hodge structure on the complex minimal model comes from a rational weight filtration which induces the rational weight filtration on $H^*(\mathsf{MA}(U)) \cong H^*(U;\mathbb{Q})$ as desired. $\square$

9.4 Formality of Compact Kähler Manifolds

9.4.1 The 1-Minimal Model

Definition 9.20. Let A be a connected differential graded algebra. A **1-minimal model** for A is a pair $(\mathsf{M}_1(A), f_1)$ with $\mathsf{M}_1(A)$ a differential graded algebra which is an increasing union of degree 1 Hirsch extensions

$$k = \mathsf{M}_{1,0}(A) \subset \mathsf{M}_{1,1}(A) \subset \cdots \mathsf{M}_1(A)$$

and a morphism of differential graded algebras $f_1 : \mathsf{M}_1(A) \to A$ which induces an isomorphism on H^1 and an injection on H^2.

Before stating the existence and uniqueness result, we need to discuss base points. A choice of a base point $x \in X$ makes the De Rham and the Sullivan algebra into an augmented differential graded algebra. The model $k(t,dt)$ naturally is a k-augmented differential graded algebra and we say that a **homotopy** $h : A \to B \otimes k(t,dt)$ **preserves the augmentation** if the self-explanatory diagram

$$\begin{array}{ccc} A & \xrightarrow{h} & B \otimes k(t,dt) \\ \downarrow & & \downarrow \\ k & \hookrightarrow & k \otimes k(t,dt) \end{array}$$

is commutative.

Theorem 9.21 ([Sull]).

1) *Any connected differential graded algebra A has a 1-minimal model and any two 1-minimal models $(\mathsf{M}_1 A, f_1)$ and $(\mathsf{M}'_1 A, f'_1)$ are related by an isomorphism $\iota : \mathsf{M}_1 A \xrightarrow{\sim} \mathsf{M}'_1 A$ such that f_1 and $f'_1 \circ \iota$ are homotopic. If moreover A is augmented, there is a unique induced augmentation on $\mathsf{M}_1 A$. Moreover, f, ι and the homotopy between f and $f'_1 \circ \iota$ preserve augmentations.*
2) *Any homomorphism of k-augmented differential graded algebra's can be lifted in a functorial way to a homomorphism between the 1-minimal models. This homomorphism preserves the augmentation and is unique up to augmentation preserving homotopy.*

3) *If two (augmented) differential graded algebras are quasi-isomorphic, their 1-minimal models are isomorphic.*

We shall review the construction of the 1-minimal model. We refer to [Grif-Mo, Ch. XII] for details. It resembles the construction of the minimal model in the 1-connected case as outlined in the proof of Theorem 9.4. Here we let $V_1 = H^1(A)$ and set

$$\begin{aligned} \mathsf{M}_{1,0}A &= k \\ \mathsf{M}_{1,1}A &= k \otimes_{\varphi=0} \Lambda_1 V_1 \end{aligned}$$

where every element in V_1 has degree 1 and boundary zero. Choosing a k-linear section for the projection sending 1-cycles to their cohomology classes then defines

$$f_{1,1} : V_1 \to A.$$

For the second step we set

$$V_2 = \operatorname{Ker}\left(H^1(A) \wedge H^1(A) \xrightarrow{\ \wedge\ } H^2(A)\right).$$

We let

$$d : V_2 \to \mathsf{M}_{1,1}(A)^2 = H^1(A) \wedge H^1(A)$$

be the inclusion and we set

$$\mathsf{M}_{1,2} = \mathsf{M}_{1,1} \otimes_d \Lambda_1 V_2,$$

a Hirsch extension in degree 1. Then one defines

$$\begin{aligned} f_{1,2} : H^1(A) \wedge H^1(A) &\to A^2 \\ x \wedge y &\mapsto f_{1,1}(x) f_{1,1}(y). \end{aligned}$$

For the induction step, we assume that we have

$$f = f_{1,n} : \mathsf{M}_{1,n}A \to A$$

inducing an isomorphism on H^1 and we set

$$V_{n+1} := \operatorname{Ker}(f^* : H^2(\mathsf{M}_{1,n}A) \to H^2(A)).$$

Then we choose, linearly in $w \in V_{n+1}$, cocycle representatives $(m_w, a_w) \in \mathsf{M}_1(1;n)^2 \oplus A^1$ for the cone of f (see Definition A.7), i.e. $dm_w = 0$, $f(m_w) = da_w$ and $[m_w] = w$. Then we put

$$\mathsf{M}_{1,n+1}A := \mathsf{M}_{1,n}A \otimes_d \Lambda_1 V_{n+1}$$

$$d(w) := m_w, \ f_{1,n+1}(w) := da_w.$$

Doing this for the cohomology algebra $H^*(A)$ (with trivial differentials), we deduce:

Proposition 9.22. *The 1-minimal model of the cohomology algebra $H^*(A)$ is completely determined by $H^1(A)$ and the cup product $H^1(A) \wedge H^1(A) \to H^2(A)$.*

9.4.2 The De Rham Fundamental Group

In this subsection (X, x) is a path connected smooth pointed manifold with rational De Rham algebra A. Following Sullivan, we explain how the 1-minimal model $\mathsf{M}_1\mathsf{A}$ yields information about the fundamental group $\pi_1(X, x)$.

The starting observation is that a any exterior differential algebra $(\Lambda V, d)$ gives a Lie-algebra structure on $L = V^\vee$ by observing that $d : V \to V \wedge V$ dually gives a bracket $L \wedge L \to L$ for which the Jacobi identity holds since $d^2 = 0$. Next, consider an increasing set of differential graded algebras

$$\mathsf{M} = [k = \mathsf{M}_0 \subset \mathsf{M}_1 \subset \mathsf{M}_2 \subset \cdots]$$

such that each step is a degree 1 Hirsch extension, say by $\Lambda_1 V_1$, $\Lambda_1 V_2$ etc. Then the degree 1 elements in M_s form the vector space $V := V_1 \oplus V_2 \oplus \cdots \oplus V_s$ and since $d : V \to \mathsf{M}^2_{s-1} \subset \Lambda^2 V$, at any stage, dually we get a Lie-algebra $L_s(\mathsf{M}) := V^\vee$ and hence a tower $\cdots \to L_s(\mathsf{M}) \to L_{s-1}(\mathsf{M}) \to \cdots \to L_1(\mathsf{M})$. Applying this to the 1-minimal model $\mathsf{M}_1\mathsf{A}$ results in a tower

$$\cdots \to L_s(\mathsf{M}_1\mathsf{A}) \to \cdots \to L_1(\mathsf{M}_1\mathsf{A}) \to k. \tag{IX–1}$$

We explain how this tower is directly related to the fundamental group through its Malcev algebra. This concept makes sense for any finitely presented group π and any field k of characteristic 0, and is called the k-Malcev algebra $L(\pi; k)$. To construct it, first form the completion of the group ring $k[\pi]$ with respect to the augmentation ideal J:

$$\widehat{k[\pi]} = \lim_{\leftarrow} k[\pi]/J^m.$$

The diagonal $\Delta : k[\pi] \to k[\pi] \otimes k[\pi]$ extends to a continuous homomorphism $\hat{\Delta} : \widehat{k[\pi]} \to \widehat{k[\pi]} \hat{\otimes} \widehat{k[\pi]}$ and gives $\widehat{k[\pi]}$ the structure of a complete Hopf algebra containing the augmentation ideal $\hat{J}$.

Definition 9.23. The **Malcev algebra** $L(\pi, k)$ is the Lie algebra of **primitive elements** inside $\widehat{k[\pi]}$:

$$L(\pi; k) := \{x \in \hat{J} \mid \Delta x = x \hat{\otimes} 1 + 1 \hat{\otimes} x\}.$$

Equivalently, since it is filtered by its sub algebras $L(\pi; k)^{(s)} = L^{(s)}(\pi; k) \cap \widehat{J^s}$, setting $L_s(\pi; k) = L(\pi; k)/L(\pi; k)^{(s+1)}$ we can identify $L(\pi; k)$ with the inverse limit of the tower of nilpotent Lie-algebras

$$\cdots \to L_3(\pi; k) \to L_2(\pi; k) \to L_1(\pi; k).$$

This tower is related to the lower central series for π:

$$\pi \supset \pi^{(2)} \cdots \supset \pi^{(s)} \cdots, \quad \pi^{(s)} := [\pi, \pi^{(s-1)}] \tag{IX–2}$$

via the so called nilpotent completion of the tower of (finite) nilpotent groups

$$\cdots \longrightarrow \pi/\pi^{(s)} \longrightarrow \cdots \longrightarrow \pi/\pi^{(2)} \longrightarrow 1,$$

as we now explain. Let $\widehat{J^s}$ be the closure inside $\widehat{k[\pi]}$ of the s-th power of the augmentation ideal. The **set of group-like elements** inside $\widehat{k[\pi]}$ is defined to be

$$\pi(k) := \{a \in 1 + \hat{J} \mid \hat{\Delta}(x) = x \hat{\otimes} x\}.$$

These indeed form a group under multiplication and $\pi(k)$ is filtered by subgroups

$$\pi(k)^{(s)} := \pi(k) \cap (1 + \widehat{J^s}).$$

The map $g \mapsto 1 + (g-1)$, $g \in \pi$ induces a homomorphism $\pi \longrightarrow \pi(k)$ sending π^s to $\pi(k)^s$. If the quotient π/π^s is abelian, the quotient $\pi(k)/\pi(k)^s$ is just the usual tensor product $\pi/\pi^s \otimes k$. This motivates the tensor product notation

$$(\pi/\pi^{(s+1)}) \otimes k := \pi(k)/\pi(k)^{(s+1)}$$

so that there are natural maps $\pi/\pi^{(s+1)} \longrightarrow \pi/\pi^{(s+1)} \otimes k$. We further introduce

$$\pi \otimes k := \left[\cdots \longrightarrow \pi/\pi^{(s)} \otimes k \longrightarrow \cdots \longrightarrow \pi/\pi^{(2)} \otimes k \longrightarrow k \right]. \qquad \text{(IX–3)}$$

It follows in particular that for abelian groups π this yields the ordinary tensor product (over $\mathbb{Z}$). In the special case of the fundamental group $\pi_1(X, x)$, one calls $\pi_1(X, x) \otimes k$ the **De Rham fundamental group**. .

In the general situation, the Lie-algebra $L(\pi; k)$ and the group $\pi(k)$ are in one-two-one correspondence through the exponential map $\exp : L(\pi; k) \to \pi(k)$, since exp has a natural well-defined inverse log. The tower $\pi \otimes k$ and the tower for $L(\pi; k)$ then correspond to each other under the exponential map.

For all of the above, we refer to [Quil86, A.2.6, A.2.8], and [Chen79, (2.7.2)]. We can now state Sullivan's result:

Proposition 9.24 ([Sull]). *Let X be a path connected smooth manifold with finitely presented fundamental group. Let k be any field contained in $\mathbb{R}$. The inductive system* (IX–1) *defined by the* 1*-minimal model of the Sullivan algebra $\mathsf{A}(X) \otimes k$ is canonically isomorphic to the De Rham fundamental group $\pi_1(X, x) \otimes k$ or, equivalently, to the Malcev algebra $L(\pi_1(X); k)$.*

Remark 9.25. Hain's pro-mixed Hodge structure on the J-adic completion $\widehat{\mathbb{Q}\pi_1(X, x)}$ (see Cor. 8.32) induces a pro-mixed Hodge structure on the subalgebra $L(\pi; \mathbb{Q})$, the Malcev algebra. By the preceding discussion, this yields a pro-mixed Hodge structure on the De Rham fundamental group $\pi_1(X, x) \otimes \mathbb{Q}$. Hence, by Prop. 9.24 dually, the 1-minimal model itself gets an ind-mixed Hodge structure. This complements Theorem 9.10 valid in the simply connected situation.

A word of warning: this mixed Hodge structure in general differs from Morgan's mixed Hodge structure on the 1-minimal model, also to be found in [Mor] but which is not treated in this book.

9.4.3 Formality

Theorem 9.26 ([Del-G-M-S]). *Let X be a compact Kähler manifold. The minimal model, respectively the 1-minimal model of the real De Rham algebra $E_{\mathrm{DR}}(X)$ is isomorphic to the minimal model, the 1-minimal model of its cohomology algebra, respectively. One says that X is **formal**.*

Proof. The real De Rham complex admits also another operator, the operator $d^c = i(\partial - \bar{\partial})|\mathcal{E}(X)$ and the dd^c lemma says that an exact form is dd^c-exact, $\operatorname{Im} d = \operatorname{Im} d^c$ and a d-closed form is d^c-closed. This follows directly from Lemma 1.9. From it we deduce that the inclusion of differential graded algebras

$$(d^c\text{-closed forms inside } \mathcal{E}(X), d) \to (\mathcal{E}(X), d)$$

is a quasi-isomorphism. On the other hand, for the same reason d induces the zero map on the cohomology of the d^c-complex, $H^*((\mathcal{E}(X), d^c)$ which of course also computes the real cohomology of X. Using Lemma 1.9 again, we see that the map of differential graded algebras

$$(d^c\text{-closed forms inside } \mathcal{E}(X), d) \to (H^*((\mathcal{E}(X), d^c), d)$$

is a quasi-isomorphism. Combining all of the preceding facts, we see that the De Rham algebra is quasi-isomorphic to the De Rham cohomology algebra with $d = 0$. But quasi-isomorphic algebras have the same minimal and 1-minimal models. □

In the simply connected case the k-minimal model determines the k-homotopy type and by Prop. 9.24 the 1-minimal model determines $\pi_1(X) \otimes k$. This, together with Proposition 9.22 implies:

Corollary 9.27. *The real homotopy type of a simply connected compact Kähler manifold is completely determined by its cohomology algebra. The real De Rham fundamental group $\pi_1(X, x) \otimes \mathbb{R}$ of a connected compact Kähler manifold X is completely determined by $H^1(X; \mathbb{R})$ and the cup product $H^1(X; \mathbb{R}) \wedge H^1(X; \mathbb{R}) \to H^2(X; \mathbb{R})$.*

Remark 9.28. Although the previous results show that the rational homotopy groups of a simply connected algebraic variety X are completely determined by the cohomology algebra $H^*(X; \mathbb{Q})$, this is not true for the mixed Hodge structure. Counterexamples can be found in [C-C-M].

Let us next discuss Massey triple products. We start with a triple of real cohomology classes $[a] \in H^p(X)$, $[b] \in H^q(X)$, $[c] \in H^r(X)$ such that $[a] \wedge [b] = 0$, $[b] \wedge [c] = 0$. Select cochains f and g such that $df = a \wedge b$ and $dg = b \wedge c$. The **Massey triple product** $\langle [a], [b], [c] \rangle$ is by definition given by

$$\langle [a], [b], [c] \rangle = f \wedge c + (-1)^{p-1} a \wedge g \in \frac{H^{p+q+r-1}(X)}{[a] \wedge H^{q+r-1}(X) + [c] \wedge H^{p+q-1}(X)}.$$

Remark. Note that by definition, the Massey triple product of three elements is only well defined up to certain ambiguities and to say that it vanishes means that the triple product is zero modulo these ambiguities.

If A is any differential graded algebra, we can define Massey triple products for any triple of classes $[a], [b], [c] \in H^*(A)$ with $[a] \wedge [b] = 0 = [b] \wedge [c]$ and these depend functorially on A. Notice that triple products in $(H^*(A), d = 0)$ are always zero and so, if A is quasi-isomorphic to its cohomology algebra, the Massey triple products vanish. This holds in particular for compact Kähler manifolds:

Corollary 9.29. *The Massey triple products for a compact Kähler manifold vanish.*

This has consequences for the the group cohomology $H^*(\pi_1(X); \mathbb{R})$ of the fundamental group $\pi_1(X)$ of a compact Kähler manifold. Group cohomology for a group π is defined as the cohomology of the Eilenberg-Mac Lane space $K(\pi, 1)$. By successively attaching cells of dimensions 3, 4, etc. one can kill the higher homotopy groups of X and one obtains a continuous map $c : X \to K(\pi_1(X), 1)$ with the property that it induces an isomorphism in cohomology in degree 0 and 1 and an injection on cohomology in degree 2. So Massey triple products of degree 1 classes in $H^*(\pi_1(X); \mathbb{R})$ vanish, when considered as elements in $H^*(X; \mathbb{R})$. If we consider such a triple product, it is an element of a quotient of $H^2(\pi_1(X); \mathbb{R})$ and injectivity on H^2 then implies that such an element must be zero. We have shown:

Corollary 9.30. *Suppose that π is the fundamental group of a compact Kähler manifold. Then the Massey triple products of $H^1(\pi; \mathbb{R})$ must vanish.*

Example 9.31. Consider the Heisenberg group H_3 of upper triangular 3 by 3 integral matrices with 1 on the diagonal. We claim that it cannot be the fundamental group of any Kähler manifold. To see this, first note that the Malcev algebra of H_3 coincides with the Lie algebra of the real Heisenberg group $(H_3)_{\mathbb{R}}$. Indeed, the augmentation ideal of the group algebra $\mathbb{R}[H_3]$ as a real vector space is generated by the three matrices $X = X_{12}, Y = X_{23}, Z = X_{13}$, where X_{ij} is the matrix with 1 on the (i, j)-th entry and 0 elsewhere. The only non trivial commutation relation is $Z = [X, Y]$. These form the Lie-algebra of $(H_3)_{\mathbb{R}}$ and since $1 + X, 1 + Y, 1 + Z$ are group like this yields indeed the Malcev algebra for H_3 and the exponential map sends it isomorphically to $(H_3)_{\mathbb{R}} = H_3 \otimes \mathbb{R}$.

The dual of this Lie algebra is a free differential graded algebra M on three generators x, y, z with only one non-trivial derivation $dz = xy$. This differential graded algebra is the 1-minimal model for any topological space having H_3 as its fundamental group such as $K(H_3, 1)$ and by definition $H^1(M)$ is free of rank two with x and y as generators. Since $xx = xy = 0$ in $H^2(M)$, the Massey product $\langle x, x, y \rangle \in H^2(M)$ exists and equals xz. Since this non-zero in $H^2(M)$ the assertion follows.

Historical Remarks. The results in this chapter are due to Morgan (see [Mor]) and uses Sullivan's constructions from [Sull]. The extension to arbitrary algebraic varieties can be found in [Nav].

Part IV

Hodge Structures and Local Systems

10

Variations of Hodge Structure

The cohomology groups $H^k(X_t)$ of compact Kähler manifolds X_t which vary in a smooth family over a complex base manifold S define a local system over S and the varying Hodge flags form the prototype of a variation of Hodge structure. These satisfy certain axioms which have been verified by Griffiths ([Grif68]): the Hodge flags vary holomorphically and Griffiths' transversality holds: the natural flat connection shifts the index of the flags back by at most 1. The variations coming from families of compact Kähler manifolds are called *geometric variations*. In § 10.4 we discuss these and show that the local system defined by the cohomology of the fibres of such a family indeed underlies a variation of Hodge structure.

Flat connections are introduced in § 10.1 and in § 10.2 we briefly treat abstract variations of Hodge structures. We state some important results for these whose proofs depend on Schmid's asymptotic analysis [Sch73] which is beyond the scope of the present monograph. These results have have strong implications on the possible monodromy representations for local systems underlying an abstract variation of Hodge structure. We give two examples of such restrictions: implications for the Mumford-Tate groups and relation with big monodromy groups.

10.1 Preliminaries: Local Systems over Complex Manifolds

Let S be a complex manifold and let $\mathbb{V}$ be a locally constant sheaf of complex vector spaces. Then $\mathcal{V} := \mathbb{V} \otimes_{\mathbb{C}} \mathcal{O}_S$ is a holomorphic vector bundle on S. For v, f local sections of $\mathbb{V}$ and $\mathcal{O}_S$ respectively the assignment

$$\begin{aligned} \nabla : \mathcal{V} &\to \Omega^1_S \otimes_{\mathcal{O}_S} \mathcal{V} \\ v \otimes f &\mapsto df \otimes v, \end{aligned}$$

defines a $\underline{\mathbb{C}}_S$-linear map for which the Leibniz rule holds:

$$\begin{aligned} &\nabla(fs) = f\nabla(s) + df \otimes s, \\ &f \text{ a local section of } \mathcal{O}_S,\ s \text{ a local section of } \mathcal{V}. \end{aligned} \tag{X–1}$$

This is an example of a holomorphic connection:

Definition 10.1. Let $\mathcal{V}$ be an $\mathcal{O}_S$-module on an n-dimensional complex manifold S. A **holomorphic connection** on $\mathcal{V}$ is a $\mathbb{C}_S$-linear map $\nabla : \mathcal{V} \to \Omega^1_S \otimes_{\mathcal{O}_S} \mathcal{V}$ such that the Leibniz rule (X–1) holds. A local section m of $\mathcal{V}$ with $\nabla(m) = 0$ is called **horizontal**. We use the notation $\mathcal{V}^\nabla$ for $\mathrm{Ker}(\nabla)$.

Note that in this definition $\mathcal{V}$ is not supposed to be locally free. We just need a sheaf of $\mathcal{O}_S$-modules. If however $\mathcal{V}$ is locally free of finite rank, like for ordinary connections (see § B.3.3) we can describe the local structure of a holomorphic connection with respect to a frame $\{e_j\}$, say on a polycylinder $U \subset \mathbb{C}^n$ by the formula $\nabla(e_j) = \sum_{i=1}^m \omega_{ij} \otimes e_i$. The **connection matrix** is the matrix of holomorphic 1-forms on U given by

$$\omega_U = (\omega_{ij})$$

The derivative $\nabla(s)$ of an arbitrary holomorphic section $s = \sum_{j=1}^m g_j e_j$ is $\nabla(s) = \sum_{j=1}^m dg_j \otimes e_j + \sum_{i,j=1}^m g_j \omega_{ij} \otimes e_i$, which can be abbreviated as

$$\nabla|_U = d + \omega_U. \tag{X–2}$$

From Leibniz' rule (X–1) it follows that the difference between two holomorphic connections on the same $\mathcal{O}_S$-module $\mathcal{V}$ is an $\mathcal{O}_S$-linear endomorphism of $\mathcal{V}$. So the holomorphic connections on $\mathcal{V}$ form an affine space under the vector space $\mathrm{End}_{\mathcal{O}_S}(\mathcal{V})$ (unless no connection on $\mathcal{V}$ exists).

As with ordinary connections, we may extend ∇ to $\Omega^p(\mathcal{V})$ and use this to define the curvature:

Definition 10.2. Let $(\mathcal{V}, \nabla)$ be an $\mathcal{O}_S$-module with a connection. We let $\Omega^p_S(\mathcal{V}) := \Omega^p_S \otimes \mathcal{V}$. Cup product of holomorphic differential forms defines $\mathcal{O}_S$-linear maps $\wedge : \Omega^p_S \otimes \Omega^q_S(\mathcal{V}) \to \Omega^{p+q}_S(\mathcal{V})$ inducing

$$\begin{aligned} \nabla^{(p)} : \Omega^p_S(\mathcal{V}) &\to \Omega^{p+1}_S \otimes \mathcal{V} \\ \omega \otimes m &\mapsto d\omega \otimes m + (-1)^p \omega \wedge \nabla m \end{aligned}$$

1) The **curvature** of the connection is the map

$$F_\nabla := \nabla^{(1)} \circ \nabla : \mathcal{V} \to \Omega^2_S(\mathcal{V}) .$$

One easily checks that F_∇ is an $\mathcal{O}_S$-linear map. The connection ∇ is called **flat** or **integrable** if its curvature is zero.
2) Since for an integrable connection the composition $\nabla^{(p)} \circ \nabla^{(p-1)}$ is zero, putting $d_S = \dim S$, we can speak of its **De Rham complex**:

$$\Omega^\bullet_S(\mathcal{V}) := \left[0 \to \mathcal{V} \xrightarrow{\nabla} \Omega^1_S \otimes \mathcal{V} \xrightarrow{\nabla^{(1)}} \cdots \xrightarrow{\nabla^{(d_S)}} \Omega^{d_S}_S \otimes \mathcal{V}\right]. \tag{X–3}$$

Suppose that we have a vector bundle $\mathcal{V}$ of finite rank m. In terms of a local frame $e_1, \dots, e_m$ of sections of $\mathcal{V}|_U$, using (X–2) we find that the curvature is given by the $m \times m$ matrix

$$F_U = d\omega_U - \omega_U \wedge \omega_U$$

of holomorphic two-forms where we write $(\omega \wedge \omega)_{ij} = \sum_{k=1}^{s} \omega_{ik} \wedge \omega_{kj}$. Thus integrability of the connection is expressed as

$$\omega_U \wedge \omega_U = d\omega_U.$$

Clearly, if $(\mathcal{V}, \nabla)$ comes from a local system $\mathbb{V}$ on S, then ∇ is integrable. Indeed, locally such $(\mathcal{V}, \nabla)$ is isomorphic to a direct sum of a finite number of copies of $(\mathcal{O}_S, d)$ and the holomorphic Poincaré lemma shows that in that case, the de Rham complex $\Omega_S^\bullet(\mathcal{V})$ is a resolution of $\mathbb{V}$. In fact the converse holds:

Theorem 10.3. *Let $(\mathcal{V}, \nabla)$ be a holomorphic vector bundle on S with an integrable connection. Then*

$$\mathbb{V} := \mathcal{V}^\nabla$$

is a local system on S and $(\mathcal{V}, \nabla) \simeq \mathbb{V} \otimes_{\mathbb{C}} (\mathcal{O}_S, d)$. Moreover, $\nabla^{(p)} \circ \nabla^{(p-1)} = 0$ for all $p > 0$ and the de Rham complex (X–3) *is a resolution of* $\mathbb{V} = \mathrm{Ker}(\nabla)$.

Proof. The statement about the de Rham complex clearly follows from the other statements. The fact that $\mathbb{V}$ is a local system on S and $(\mathcal{V}, \nabla) \simeq \mathbb{V} \otimes_{\mathbb{C}} (\mathcal{O}_S, d)$ is classical: the integrability of the connection enables one to show that the solutions of $\nabla(s) = 0$ form locally on S a vector space of dimension equal to the rank of $\mathcal{V}$. For a proof see e. g. [Pham, p. 74]. □

Corollary 10.4. *For any complex manifold S one has an equivalence of categories between the category of complex local systems on S and the category of holomorphic vector bundles on S with an integrable connection.*

Remark 10.5. This result can be viewed as a prototype of the Riemann-Hilbert correspondence which will be treated later in generality (Theorem 11.7 and 13.64).

10.2 Abstract Variations of Hodge Structure

Definition 10.6. Let S be a complex manifold. A **variation of Hodge structure of weight** k on S consists of the following data:
1) a local system $\mathbb{V}_{\mathbb{Z}}$ of finitely generated abelian groups on S;
2) a finite decreasing filtration $\{\mathcal{F}^p\}$ of the holomorphic vector bundle $\mathcal{V} := \mathbb{V}_{\mathbb{Z}} \otimes_{\mathbb{Z}} \mathcal{O}_S$ by holomorphic subbundles (the **Hodge filtration**).

These data should satisfy the following conditions:

1) for each $s \in S$ the filtration $\{\mathcal{F}^p(s)\}$ of $\mathbb{V}(s) \simeq \mathbb{V}_{\mathbb{Z},s} \otimes_{\mathbb{Z}} \mathbb{C}$ defines a Hodge structure of weight k on the finitely generated abelian group $\mathbb{V}_{\mathbb{Z},s}$;
2) the connection $\nabla : \mathcal{V} \to \mathcal{V} \otimes_{\mathcal{O}_S} \Omega^1_S$ whose sheaf of horizontal sections is $\mathbb{V}_{\mathbb{C}}$ satisfies the **Griffiths' transversality condition**

$$\nabla(\mathcal{F}^p) \subset \mathcal{F}^{p-1} \otimes \Omega^1_S. \qquad \text{(X–4)}$$

The notion of a **morphism of variations of Hodge structure** is defined in the obvious way.

Given two variations $\mathbb{V}$, $\mathbb{V}'$ of Hodge structure over S of weights k and k', there is an obvious structure of variation of Hodge structure on the underlying local systems of $\mathbb{V} \otimes \mathbb{V}'$ and $\mathcal{H}om(\mathbb{V}, \mathbb{V}')$ of weights $k + k'$ and $k - k'$ respectively.

Examples 10.7. 1) Let V be a Hodge structure of weight k and $s_0 \in S$ a base point. Suppose that one has a representation $\rho : \pi_1(S, s_0) \to \mathrm{Aut}(V)$. Then the local system $\mathbb{V}(\rho)$ associated to ρ underlies a locally constant variation of Hodge structure. In this case the Hodge bundles $\mathcal{F}^p$ are even locally constant, so that $\nabla(\mathcal{F}^p) \subset \mathcal{F}^p \otimes \Omega^1_S$. This property characterizes the local systems of Hodge structures among the variations of Hodge structure. In case ρ is the trivial representation, we denote the corresponding variation by $\underline{V}_S$.
2) Let $f : X \to S$ be a proper and smooth morphism between complex algebraic manifolds. By Theorem C.10 such a morphism is locally differentiable trivial. Therefore the cohomology groups $H^k(X_s)$ of the fibres X_s fit together into a local system. By the fundamental results of Griffiths [Grif68] this local system underlies a variation of Hodge structure on S such that the Hodge structure at s is just the Hodge structure we have on $H^k(X_s)$. Such variations are called **geometric variations**. Below we sketch a proof of these results: Theorem 10.30 implies holomorphicity of the Hodge flag and Theorem 10.31 states the transversality property.

Definition 10.8. A **polarization** of a variation of Hodge structure $\mathbb{V}$ of weight k on S is a morphism of variations

$$Q : \mathbb{V} \otimes \mathbb{V} \to \underline{\mathbb{Z}(-k)}_S$$

which induces on each fibre a polarization of the corresponding Hodge structure of weight k.

Suppose that S is a complex manifold. Then the De Rham complex $\Omega^\bullet_S$ is a resolution of the constant sheaf $\underline{\mathbb{C}}_S$. If $\mathbb{V}$ is a local system underlying a variation of Hodge structure, and $(\mathcal{V}, \nabla)$ associated vector bundle equipped with its integrable connection, we can transport the Hodge filtration to the complex $\Omega^\bullet_S(\mathcal{V})$ by putting

$$F^p(\Omega^\bullet) := \left[0 \to F^p\mathcal{V} \xrightarrow{\nabla} \Omega^1 \otimes \mathcal{F}^{p-1}\mathcal{V} \to \dots\right] \qquad \text{(X–5)}$$

There also is a natural map

$$\alpha : \mathbb{V}_{\mathbb{Z}} \to \Omega^{\bullet}(\mathcal{V}) \tag{X–6}$$

which becomes a quasi-isomorphism after tensoring with $\mathbb{C}$. So we have all the ingredients for a Hodge complex of sheaves. Indeed, we have [Zuc79, Theorem 2.9]:

Proposition 10.9. *Suppose S is a compact Kähler manifold and let $(\mathbb{V}, \mathcal{F}^{\bullet})$ be a* polarizable *variation of Hodge structures of weight k on S. Then the above data* (X–5) *and* (X–6) *define a Hodge complex of sheaves $(\mathbb{V}_{\mathbb{Z}}, (\Omega^{\bullet}_S(\mathcal{V}), \alpha)$ of weight k on S.*

The existence of a polarization imposes strong restrictions on the underlying local system of a variation of Hodge structure.

Example 10.10. Let $\mathbb{V}$ be a polarized variation of Hodge structure on a connected complex manifold which is purely of type (p, p). The polarization being definite, the isometry group of the lattice is finite so that $\mathbb{V}$ has a finite monodromy group.

In the geometric setting of a smooth projective family $f : X \to S$ the Theorem of the Fixed Part 4.23 states that invariant classes are all restrictions of classes on a smooth compactification $\bar{X}$ of X. In terms of of Hodge structures this implies that the invariant classes inside $H^k(X_s; \mathbb{Q})$ form a Hodge substructure, since the restriction map $H^k(\bar{X}; \mathbb{Q}) \to H^k(X_s; \mathbb{Q})$ is a morphism of Hodge structures. In the abstract setting this remains true:

Theorem 10.11. *Let $\mathbb{V}$ be a variation of Hodge structure of weight k on a complex manifold S which is Zariski open in a compact complex manifold. Then $H^0(S, \mathbb{V})$ admits a Hodge structure of weight k. The evaluation map at a point $s \in S$ gives an isomorphism of $H^0(S, \mathbb{V})$ with the subspace of $\mathbb{V}_s$ left fixed by the action of $\pi_1(S, s)$. The inclusion of this subset into $\mathbb{V}_s$ is a morphism of Hodge structures. In other words, the variation of Hodge structure on $\mathbb{V}$ restricts to a constant variation of Hodge structure on its maximal constant local subsystem.*

This follows immediately from [Sch73, Theorem 7.22] stating that the (p, q)-components of a flat global section of $\mathbb{V}$ are themselves flat.

This theorem has the following obvious, but interesting consequence:

Corollary 10.12. *If $a \in H^0(S, \mathbb{V})$ has Hodge type (p, q) at some point $s \in S$, it has Hodge type (p, q) everywhere.*

We conclude:

Theorem 10.13. *The category of polarizable variations of $\mathbb{Q}$-Hodge structures on a given manifold, Zariski-open in a compact complex manifold is semi-simple.*

Proof. Suppose that $\mathbb{V}'$ is a subvariation of $\mathbb{V}$ and suppose that $\mathbb{V}$ is polarized. For every $s \in S$ the Hodge structure $\mathbb{V}_s$ is polarized, and if $\mathbb{V}'_s$ is a Hodge substructure of $\mathbb{V}_s$, this polarization induces an orthogonal projector in $\mathrm{End}(\mathbb{V}_s)$ with image equal to $\mathbb{V}'_s$. This projector commutes with the monodromy action since $\mathbb{V}'$ is a subsystem of $\mathbb{V}$ and so defines a projector $p \in \mathrm{End}(\mathbb{V})$. Since it is of type $(0,0)$ at the point s, it is everywhere of type $(0,0)$, i.e. $p \in \mathrm{End}_{\mathrm{VHS}}(\mathbb{V})$. It follows that $\mathbb{V} = \mathbb{V}' \oplus \mathbb{V}''$, $\mathbb{V}'' = \mathrm{Im}(1\!\!1 - p)$. □

Let us now consider a variation of Hodge structure of *even weight* $k = 2p$ over any smooth connected complex base S together with some section v of $\mathbb{V}_{\mathbb{Z}}$ on the universal cover of S. Let Y_v be the locus of all $s \in S$ where some determination $v(s')$, $s' \mapsto s$ is of type (p,p). This locus is a countable union of analytic subvarieties of S since the condition to belong to the Hodge bundle $\mathcal{F}^p$ is analytic and a local section v of $\mathbb{V}_Z$ is a Hodge vector in $\mathbb{V}_s$ precisely when $v(s) \in \mathcal{F}^p$. In case $Y_v \neq S$ we call v **special**. The union of all Y_v, with v special forms a thin subset of S. We call $s \in S$ **very general** with respect to $\mathbb{V}$ if it lies in the complement of this set. The very general points of S with respect to $\mathbb{V}$ form a dense subset. Now, if $s \in S$ is very general, by definition any Hodge vector in $\mathbb{V}_s$ extends to give a multivalued horizontal section of $\mathbb{V}$ everywhere of type (p,p).

We can now show how the monodromy group is related to the Mumford-Tate group of the Hodge structure at a very general $s \in S$ using the characterization (Theorem 2.15) of $\mathrm{MT}(\mathbb{V}_s)$ as the largest rationally defined algebraic subgroup of $\mathrm{GL}(\mathbb{V}_s) \times \mathbb{C}^*$ fixing the Hodge vectors in $\mathbb{V}_s^{m,n}(p)$, for all triples (m,n,p) with $(m-n)k - 2p = 0$. So we look at $s \in S$ which is very general for all local systems $\mathbb{V}(m,n)(p)$ with $(m-n)k - 2p = 0$. Then there is a local system $\mathbb{H}(m,n,p)$ on S whose stalk at s is $\mathrm{Hodge}(\mathbb{V}_s^{m,n}(p))$. Using this we deduce:

Proposition 10.14. *Let S be a smooth complex variety. For very general $s \in S$ a finite index subgroup of the monodromy group is contained in the Mumford-Tate group of the Hodge structure on $\mathbb{V}_s$.*

Proof. The Hodge structure on $\mathbb{H}(m,n,p)_s$ is polarizable and so there is a positive definite quadratic form on this space invariant under monodromy. Hence the monodromy acts on $\mathbb{H}(m,n,p)$ through a finite group. Since the Mumford-Tate group is algebraic, the Noetherian property then implies that finitely many triples (m,n,p) determine the Mumford-Tate group and so a finite index subgroup π' of the fundamental group has its image in the Mumford-Tate group. □

In case S is quasi-projective and $\mathbb{V}$ the local system $R^{2d} f_* \underline{\mathbb{Q}}_X$ of the rank $2d$-cohomology groups of the fibres of a smooth algebraic family $f : X \to S$, the validity of the Hodge conjecture would imply that analytic sets Y_v in fact are *algebraic*. Surprisingly this has been proved recently; it is a consequence of the following result due to Cattani, Deligne and Kaplan [C-D-K] which we quote without giving the proof.

Theorem 10.15. *Fix a natural number m. Suppose that we have a polarized variation $(\mathbb{V}, Q)$ of even weight $k = 2p$ on a compactifiable $S \subset \overline{S}$ whose compactifying divisor has normal crossings. Define*

$$S^{(m)} = \{(s,v) \mid s \in S,\ v \in (\mathbb{V}_{\mathbb{Z}})_s \text{ is a Hodge vector and } Q(v,v) \leq m\}.$$

Then the image of the projection of $S^{(m)}$ onto S is a finite disjoint union of analytic subspaces of S which, locally at every boundary point $p \in \overline{S} - S$, are traces of analytic subvarieties defined in an open neighbourhood of p inside $\overline{S}$.

As we noted before, the fact that the projection of $S^{(m)}$ onto S consists of a finite disjoint union of analytic subspaces of S is not hard; the difficult point is the assertion about the behaviour near the boundary.

This theorem can indeed be applied to the geometric setting of a smooth algebraic family $f : X \to S$ where X and S are quasi-projective. The vector bundle $\mathcal{V} = := \mathbb{V} \otimes_{\mathbb{Q}} \mathcal{O}_S$ where $\mathbb{V} = R^{2d} f_* \underline{\mathbb{Q}}_X$ is an algebraic vector bundle and the Hodge filtration gives algebraic subbundles. The **locus of Hodge classes** is the collection of vectors v_t in the fibre at $t \in S$ of $F^d\mathcal{V}$ which define Hodge classes in $H^{2d}(X_t; \mathbb{C})$. By Theorem 10.15 the locus of Hodge classes is algebraic in the sense that the component containing a given Hodge class $v \in H^{2d}(X_t; \mathbb{C})$ is an algebraic subvariety of $F^d\mathcal{V}$. It is called the **locus of the Hodge class** v. Its projection onto S is one of the components of $S^{(m)}$ figuring in the preceding theorem, where $m = Q(v,v)$.

Remark. Recently C. Voisin [Vois07] applied this theorem to *absolute Hodge classes* (Def. 2.37). To explain this, any projective manifold X is defined over a field k which is of finite transcendence degree over $\mathbb{Q}$, say $k = \mathbb{Q}(a)(t_1, \ldots, t_s)$ with a algebraic over $\mathbb{Q}$ and the t_j transcendent. So X can be considered as a fibre of a family $Y \to S$, defined over $\mathbb{Q}(a)$, where S is a Zariski-open in some affine s-space. Any irreducible cycle Z of X then can be viewed as a cycle $\tilde{Z}$ of Y finite over S and hence defined over a finite algebraic extension of $\mathbb{Q}$. So $\tilde{Z}$, the locus of the class of Z is defined over $\bar{\mathbb{Q}}$. More generally, it can be seen to be true for the Hodge locus of an absolute Hodge cycle: such a Hodge locus is also defined over $\bar{\mathbb{Q}}$ and the Galois conjugates of these loci are also Hodge loci. Voisin uses this remark to show that if the Hodge conjecture holds for absolute Hodge cycles on varieties defined over $\bar{\mathbb{Q}}$, then it hold for absolute Hodge classes in general. Moreover, under certain genericity assumptions on X a similar statement holds for all Hodge classes on X. So in a certain sense, the proof of the Hodge conjecture can be reduced to varieties over $\bar{\mathbb{Q}}$; this implies that one only has to test a *countable* number of cases.

10.3 Big Monodromy Groups, an Application

If $\mathbb{V}$ underlies a polarizable variation of Hodge structure, as we already saw (Theorem 10.13), the monodromy representations is fully reducible. Irreducible representations give **indecomposable local systems**. If we have a representation on a $\mathbb{Q}$-vector space which stays irreducible under field-extensions we say that the representation is **absolutely irreducible**. There is one particular type of such representations, namely representations with "big" monodromy group in the following sense.

Definition 10.16. Let $\mathbb{V}$ be a local system on a connected and locally 1-connected topological space S with monodromy representation

$$\rho : \pi_1(S, s) \to \mathrm{GL}(V), \quad V := V_s.$$

1) The **algebraic monodromy group** G^{mon} is the identity component of the smallest algebraic subgroup of $\mathrm{GL}(V)$ containing the monodromy group $\rho(\pi_1(S, s))$;
2) the monodromy group is said to be **big** if G^{mon} acts irreducibly on $V_{\mathbb{C}}$.

Remark 10.17. If $\rho : S' \to S$ is a finite unramified cover, the induced morphism between the fundamental groups $\rho_* : \pi_1(S', s') \to \pi_1(S, s)$, $\rho(s') = s$ identifies $\pi_1(S', s')$ with a normal subgroup of $\pi_1(S, s)$ of finite index. It acts on $\rho^*\mathbb{V}$ and the algebraic monodromy group for this action therefore is a connected normal subgroup of G^{mon} of finite index and hence equals G^{mon}. It follows that the property of having a big monodromy group is stable under finite unramified coverings.

To determine algebraic monodromy, we use the following criterion, due to Deligne [Del80]:

Criterion 10.18. *Let V be a finite dimensional complex vector space of dimension n equipped with a non-degenerate bilinear form Q which is either symmetric or anti-symmetric. Let $M \subset \mathrm{Aut}(V, Q)$ be an algebraic subgroup.*

1) *If Q is anti-symmetric we suppose that M contains the transvections $T_\delta : v \mapsto v + Q(v, \delta)\delta$, where δ runs over an M-orbit R which spans V. Then $M = \mathrm{Aut}(V, Q)(= \mathrm{Sp}(V))$.*
2) *If Q is symmetric, suppose that M contains the reflections $R_\delta : v \mapsto v - Q(v, \delta)\delta$ in "roots" δ, i.e. with $Q(\delta, \delta) = 2$ which form an M-orbit spanning V. Then either M is finite or $M = \mathrm{Aut}(V, Q)(= \mathrm{O}(V))$.*

Examples 10.19. 1) Let $V = V_{\mathbb{Z}} \otimes \mathbb{C}$, where $V_{\mathbb{Z}}$ is a free finite rank $\mathbb{Z}$-module equipped with a non-degenerate anti-symmetric bilinear form. The Zariski-closure inside $\mathrm{Sp}(V)$ of the group $\mathrm{Sp}(V_{\mathbb{Z}})$ of symplectic automorphisms of the lattice V_Z is the full group $\mathrm{Sp}(V)$. This follows from the fact that $\mathrm{Sp}(V_{\mathbb{Z}})$ contains all symplectic transvections T_v, $v \in V_{\mathbb{Z}}$ and for given non-zero $\delta \in V_{\mathbb{Z}}$, the elements $T_v\delta$, $v \in V_{\mathbb{Z}}$ span already V. It follows that the Zariski-closure of any subgroup of finite index in $\mathrm{Sp}(V_{\mathbb{Z}})$ is also the full symplectic group, hence is big.

2) Let $V = V_{\mathbb{Z}} \otimes \mathbb{C}$, where $V_{\mathbb{Z}}$ is a free finite $\mathbb{Z}$-module equipped with a non-degenerate symmetric bilinear form Q. If Q is definite, the orthogonal group preserving the lattice $V_{\mathbb{Z}}$ is of course finite and equals its Zariski-closure. Hence it is never big. In general it will contain reflections R_δ in all roots $\delta \in V_{\mathbb{Z}}$. Assuming that these roots contain at least one orbit which spans the lattice, we conclude in the indefinite case that the Zariski closure of $\mathrm{Aut}(V_{\mathbb{Z}}, Q)$ is the full orthogonal group.

The fact that $\mathbb{V}$ underlies a variation of Hodge structure imposes severe restrictions of Noether-Lefschetz type:

Theorem 10.20. *Let there be given a (rational) polarized weight k variation of Hodge structure over a smooth quasi-projective base S with big monodromy group. If $s \in S$ is very general with respect to* $\mathrm{Hom}(\mathbb{V}, \mathbb{V})$*, then $\mathbb{V}_s$ has no non-trivial rational Hodge substructures.*

Proof. Any projector $p : \mathbb{V}_s \to \mathbb{V}_s$ onto a Hodge substructure extends to a multivalued flat section of $\mathrm{Hom}(\mathbb{V}, \mathbb{V})$ everywhere of type $(0, 0)$. This flat section generates a sub Hodge structure of type $(0, 0)$ which by Example 10.10 has finite monodromy. By Remark 10.17 we may replace S by a finite unramified cover $q : S' \to S$. So, replacing S by S', we may assume that the flat section is uni-valued, i.e. invariant under the monodromy. This means that the projector p intertwines every element from the monodromy group and thus defines a sub system of $\mathbb{V}$. Since the latter is irreducible, this subsystem is either zero or all of $\mathbb{V}$. □

Remark 10.21. The proof from [Del72] asserting the truth of the theorem for certain variations related to K3-surfaces can be applied to our setting. The crucial result to use here is Prop. 10.14. Clearly the preceding proof is more elementary.

We now consider the tautological families of smooth complete intersections in projective space. We have:

Theorem 10.22. *The monodromy group for the tautological family of n-dimensional complete intersections in projective space is big except for quadrics, cubic surfaces or even-dimensional intersection of two quadrics.*

Proof. For simplicity we only consider hypersurfaces in $\mathbb{P}^{n+1}$ (to have the above set-up one views these as hyperplane sections of the Veronese embedded $\mathbb{P}^{n+1}$). By the Zariski-Van Kampen theorem [Kamp] (see also Prop. C.19) we may restrict to a Lefschetz pencil and then, applying Theorem C.23, the vanishing cocycles form (up to signs) one orbit under the monodromy group. So, by the remarks in Example 10.19 the Zariski closure of the monodromy group acting on the middle primitive cohomology group of Y is the full symplectic group if n is odd and either finite or the full orthogonal group if n is even. If n is even and M is finite, Q must be definite. For complete intersections in projective space the variable cohomology is just the primitive cohomology. Then the Hodge Riemann bilinear relations tell us that the signature is (a, b) with a, b respectively, the sum of the Hodge numbers $\dim H^{p,q}_{\text{prim}}$ with p even, odd respectively. Since $(2m)$-dimensional hypersurfaces always have $H^{m,m}_{\text{prim}} \neq 0$, one deduces that the form Q can only be definite if all Hodge numbers $h^{p,q}$ with $p \neq q$ are zero. It is easy to see that this only happens for $d = 2$ or for cubic surfaces. For the case of complete intersections, see [Del73]. □

Using Theorem 10.20 and Theorem 10.22 we deduce:

Corollary 10.23. *Except for quadric hypersurfaces, cubic surfaces and even-dimensional intersections of two quadrics, the generic stalk of the tautological variation of Hodge structures on primitive cohomology for smooth complete intersections in complex projective space does not contain non-trivial sub Hodge structures.*

10.4 Variations of Hodge Structures Coming From Smooth Families

The Griffiths transversality condition is inspired by the geometric case. See Theorem 10.31. We are now going to regard this more in detail. Let us consider a proper morphism $f : X \to S$ of complex manifolds of maximal rank and such such that X is bimeromorphic to a Kähler manifold. By the results of § 2.3 the cohomology groups of any compact complex submanifold of X admit a strong Hodge decomposition. In particular this applies to the fibres of f. We use the (standard) notation

$$\begin{aligned} &X_s = f^{-1}(s) \\ &\mathfrak{m}_s \subset \mathcal{O}_{S,s} \ \text{ the maximal ideal} \\ &\kappa(s) = \mathcal{O}_{S,s}/\mathfrak{m}_s \ \text{ the residue field.} \end{aligned}$$

By the topological proper base change theorem (cf. [Gode, p. 202]) applied to f, the stalks at $s \in S$ of the local systems $R^q f_* \underline{\mathbb{Z}}_X$ and $R^q f_* \underline{\mathbb{C}}_X$ can be written

$$R^q f_* \underline{\mathbb{Z}}_{X,s} \simeq H^q(X_s); \qquad R^q f_* \underline{\mathbb{C}}_{X,s} \simeq H^q(X_s; \mathbb{C}).$$

The sheaf of **relative De Rham cohomology group**

$$H^q_{\mathrm{DR}}(X/S) := R^k f_* \mathbb{C} \otimes_{\mathbb{C}} \mathcal{O}_S,$$

is locally free and

$$H^k_{\mathrm{DR}}(X/S)_s \otimes \kappa(s) \simeq R^k f_* \underline{\mathbb{C}}_{X,s} \simeq H^k(X_s; \mathbb{C}).$$

Proposition-Definition 10.24. *Let $f : X \to S$ be as above. The* ***Gauss-Manin connection*** ∇^{GM} *on $H^q_{\mathrm{DR}}(X/S)$ is the flat connection whose sheaf of locally constant sections is $R^q f_* \underline{\mathbb{C}}_X$. This is a locally constant sheaf whose fibre at $s \in S$ is $H^q(X_s;\mathbb{C})$. The Gauss-Manin connection is the natural connection on $R^q f_* f^{-1}\mathcal{O}_S$ coming from $d : f^{-1}\mathcal{O}_S \to f^{-1}\Omega^1_S$.*

Proof. Taking the q-th direct image of the exact sequence

$$0 \to f^{-1}\underline{\mathbb{C}}_S \to f^{-1}\mathcal{O}_S \xrightarrow{d} f^{-1}\Omega^1_S \xrightarrow{d} f^{-1}\Omega^2_S$$

one finds on S the exact sequence

$$0 \to R^q f_* f^{-1}\underline{\mathbb{C}}_S \to R^q f_* f^{-1}\mathcal{O}_S \xrightarrow{\nabla} R^q f_* f^{-1}\Omega^1_S \xrightarrow{\nabla} R^q f_* f^{-1}\Omega^2_S,$$

where we have abbreviated $\nabla^{\mathrm{GM}} = \nabla$. Hence $\nabla \circ \nabla = 0$, i.e. ∇ is a flat connection. Since its locally constant sections generate the sheaf $R^q f_* f^{-1}\underline{\mathbb{C}}_S$, this identifies ∇ with the unique flat connection on $R^q f_* f^{-1}\mathcal{O}_S$ whose sheaf of locally constant sections is $R^q f_* f^{-1}\underline{\mathbb{C}}_S$. □

We next introduce the relative De Rham complex. It comes from the exact sequence defining the bundle of **relative one-forms**

$$0 \to f^*\Omega^1_Y \to \Omega^1_X \to \Omega^1_{X/Y} \to 0. \qquad \text{(X–7)}$$

So, the bundle of relative 1-forms has rank

$$d_{X/Y} := \dim X - \dim Y$$

and we let

$$\Omega^p_{X/Y} := \bigwedge_{\mathcal{O}_X}^{p} \Omega^1_{X/Y}. \qquad \text{(X–8)}$$

These locally free sheaves form a complex, the relative de Rham complex. We give a slightly more general definition of this complex.

Definition 10.25. Let $f : X \to Y$ be a holomorphic map between complex spaces. The **relative de Rham complex** of f, denoted by $\Omega^\bullet_{X/Y}$, is the quotient of the Kähler de Rham complex $\Omega^\bullet_X$ (see (VII–6)) by the subcomplex generated locally by forms $f^*(\eta) \wedge \omega$ where η is a local section of Ω^1 and ω a local section of $\Omega^{\bullet-1}_X$:

$$\Omega^p_{X/Y} = \frac{\Omega^p_X}{f^*\Omega^1_Y \wedge \Omega^{p-1}_X}.$$

If X and Y are smooth and f is of maximal rank, then this is consistent with (X–7) and (X–8): in suitable local coordinates on X and Y the map f has the form

$$(z_1, \ldots, z_n) \mapsto (z_1, \ldots, z_k)$$

with $n = \dim(X)$, $k = \dim(Y)$ so that $\Omega^\bullet_{X/Y}$ is locally isomorphic to the exterior algebra over $\mathcal{O}_X$ on the generators $dz_{k+1}, \ldots, dz_n$.

We cite without proof the following alternative description of $H^k_{\mathrm{DR}}(X/S)$ in terms of the relative de Rham complex (see [Del70, Prop. I.2.28.]).

Theorem 10.26. *Let $f : X \to S$ be a proper and smooth holomorphic map between complex manifolds, let $q \in \mathbb{N}$ and let $\mathbb{V}$ be a local system of complex vector spaces on X. There is a natural isomorphism*

$$\mathcal{O}_S \otimes_{\mathbb{C}} R^q f_* \mathbb{V} \simeq R^q f_*(\Omega^\bullet_{X/S} \otimes_{\mathbb{C}} \mathbb{V}).$$

Corollary 10.27. *With notation as above, for each $q \in \mathbb{N}$ we have an isomorphism*

$$H^q_{\mathrm{DR}}(X/S) \simeq R^q f_*(\Omega^\bullet_{X/S}).$$

Let us give a description of the connection on $H^q_{\mathrm{DR}}(X/S)$ in these terms. First we observe, that for any sheaf F of $\mathbb{C}$-vector spaces on S and each $q \in \mathbb{N}$ we have a canonical isomorphism

$$R^q f_* \underline{\mathbb{C}}_X \otimes_{\underline{\mathbb{C}}_S} F \xrightarrow{\sim} R^q f_* f^{-1} F.$$

Indeed, this can be verified on the stalks. In particular, for $F = \mathcal{O}_S$, using that $\Omega^\bullet_{X/S}$ is a resolution of $f^{-1}\mathcal{O}_S$, one has

$$R^q f_* \underline{\mathbb{C}}_X \otimes_{\underline{\mathbb{C}}_S} \mathcal{O}_S \simeq R^q f_* f^{-1} \mathcal{O}_S \simeq R^q f_* \Omega^\bullet_{X/S}.$$

The Gauss-Manin connection can be described in terms of the relative De Rham complex as follows. Define the **Koszul filtration**

$$\mathrm{Koz}^q\, \Omega^\bullet_X = f^* \Omega^q_S \wedge \Omega^{\bullet - q}_X. \tag{X–9}$$

This is a subcomplex of $\Omega^\bullet_X$ for each q and one has

$$\mathrm{Gr}^q_{\mathrm{Koz}}\, \Omega^\bullet_X \simeq f^* \Omega^q_S \otimes_{\mathcal{O}_X} \Omega^\bullet_{X/S}[-q],$$

hence a short exact sequence

$$\begin{array}{ccccccc} 0 \longrightarrow & \mathrm{Gr}^1_{\mathrm{Koz}} & \longrightarrow \mathrm{Koz}^0 / \mathrm{Koz}^2 \longrightarrow & \mathrm{Gr}^0_{\mathrm{Koz}}\,. & \longrightarrow 0 \\ & \| & & \| & \\ & f^*(\Omega^1_S) \otimes \Omega^\bullet_{X/S}[-1] & & \Omega^\bullet_{X/S} & \end{array} \tag{X–10}$$

If $\dim S = 1$ this sequence reduces to the self-evident exact sequence

$$0 \to f^*(\Omega^1_S) \otimes \Omega^\bullet_{X/S}[-1] \to \Omega^\bullet_X \to \Omega^\bullet_{X/S} \to 0. \tag{X–11}$$

Theorem 10.28. *The Gauss-Manin connection ∇^{GM} is the connecting homomorphism*

$$\begin{array}{ccc} R^q f_* \,\mathrm{Gr}^0_{\mathrm{Koz}} & \xrightarrow{\partial} & R^{q+1} f_* \,\mathrm{Gr}^1_{\mathrm{Koz}} \\ \| & & \| \\ R^q f_* \Omega^\bullet_{X/S} & \xrightarrow{\nabla^{\mathrm{GM}}} & \Omega^1_S \otimes_{\mathcal{O}_S} R^q f_* \Omega^\bullet_{X/S} \end{array}$$

in the long exact sequence obtained by applying Rf_* to the exact sequence (X–10).

For a proof, we refer to [Katz-Oda].

We now study the Hodge filtration on $R^q f_* \Omega^\bullet_{X/S}$; it is obtained as follows. The relative de Rham complex inherits the trivial filtration σ from the absolute de Rham complex. We define

$$F^p R^q f_* \Omega^\bullet_{X/S} = \operatorname{Im}\left[R^q f_* \sigma^{\geq p} \Omega^\bullet_{X/S} \to R^q f_* \Omega^\bullet_{X/S}\right] \qquad \text{(X–12)}$$

We first remark that Proposition 2.22 has a relative analogue:

Proposition 10.29. *The spectral sequence*

$$'E_1^{p,q} = R^q f_* \Omega^p_{X/S} \Longrightarrow R^{p+q} f_* \Omega^\bullet_{X/S} = H^{p+q}_{\mathrm{DR}}(X/S)$$

degenerates at E_1.

To prove this, we use Grauert's base change theorem [Gr60]:

Theorem 10.30. *Let $f : X \to S$ be a proper holomorphic map with S reduced and connected, and $\mathcal{F}$ a coherent $\mathcal{O}_X$-module, flat over $f^{-1}\mathcal{O}_S$. Then for all $p \in \mathbb{Z}$ the function $s \mapsto \dim \mathbb{H}^p(X_s, \mathcal{F} \otimes \mathcal{O}_{X_s})$ is upper semicontinuous. Moreover the following are equivalent:*

1) *$s \mapsto \dim \mathbb{H}^p(X_s, \mathcal{F} \otimes \mathcal{O}_{X_s})$ is a constant function on S;*
2) *$R^p f_* \mathcal{F}$ is a locally free $\mathcal{O}_S$-module and for all $s \in S$ the natural map*

$$R^p f_* \mathcal{F} \otimes \kappa(s) \to H^p(X_s, \mathcal{F} \otimes \mathcal{O}_{X_s})$$

is an isomorphism.

Proof (of 10.29). For each $s \in S$ we have $\Omega^p_{X/S} \otimes_{\mathcal{O}_X} \mathcal{O}_{X_s} \simeq \Omega^p_{X_s}$. The functions $s \mapsto h^{p,q}(s) = \dim H^q(X_s, \Omega^p_{X_s})$ are upper semicontinuous on S and $b_n(X_s) = \sum_{p+q=n} h^{p,q}(s)$ is constant on S. Therefore $s \mapsto h^{p,q}(s)$ is constant on S, the sheaves $R^q f_* \Omega^p_{X/S}$ are locally free on S and the natural maps

$$R^q f_* \Omega^p_{X/S} \otimes \kappa(s) \to H^q(X_s, \Omega^p_{X_s})$$

are isomorphisms for all p, q. By decreasing induction on p one proves that the sheaves $F^p R^q f_* \Omega^\bullet_{X/S}$ are locally free on S and that their formation commutes with base change as well. Proposition 10.29 follows immediately from this. Moreover, the natural mappings $F^p R^q f_* \Omega^\bullet_{X/S} \to R^q f_* \Omega^\bullet_{X/S}$ are injective for all p. □

We now can prove :

Corollary 10.31 (Griffiths' transversality theorem). *The Gauss-Manin connection has the property*

$$\nabla^{\mathrm{GM}}(F^p H^k) \subseteq \Omega^1_S \otimes F^{p-1} H^k \ .$$

Proof. This follows from the filtered version of (X–10)

$$0 \to f^*\Omega^1_S \otimes \sigma^{\geq p-1}\Omega^\bullet_{X/S}[-1] \to \sigma^{\geq p}(\mathrm{Koz}_0 \,/\, \mathrm{Koz}_2) \to \sigma^{\geq p}\Omega^\bullet_{X/S} \to 0$$

after applying Rf_*:

$$\begin{array}{ccc} F^pR^kf_*(\Omega^\bullet_{X/S}) & \xrightarrow{\partial} & F^{p-1}R^kf_*(f^*\Omega^1_S \otimes \Omega^\bullet_{X/S}) \\ \| & & \| \\ F^pH^k & \xrightarrow{\nabla^{\mathrm{GM}}} & \Omega^1_S \otimes F^{p-1}H^k \end{array}$$

□

Corollary 10.32. *Let $f : X \to S$ be a proper and smooth holomorphic map between complex manifolds, let $q \in \mathbb{N}$. Suppose that X is bimeromorphic to a Kähler manifold. Then, with the Hodge filtration* (X–12)*, the local system $R^qf_*\underline{\mathbb{Z}}_X$ underlies a variation of Hodge structure.*

11

Degenerations of Hodge Structures

Usually, families come with singular fibres and it is very interesting to investigate what happens near such a fibre. In this monograph we mostly consider 1-parameter degenerations, where the base is a punctured disk and the family is smooth over the punctured disk $\Delta^* = \Delta - \{0\}$. The flat connection on any of the local systems coming from the cohomology of the smooth fibres over Δ^* acquires a logarithmic singularity and its residue is intimately related to the monodromy around the singular fibre. We explain this in the abstract setting in §11.1.1; this leads directly to a first version of the Riemann-Hilbert correspondence. A full version will be given in § 13.6.3.

What happens in the geometric setting is a concretization of results by Schmid who studied abstract degenerations of variations of Hodge structure in his fundamental study [Sch73]; we state his results in § 11.2.1 without giving proofs. The main result of this Chapter, the description of the limit mixed Hodge structure in the *geometric setting* is Theorem 11.22. As a consequence of the proof, in § 11.3.1 various other central results are derived: the local monodromy theorem and the local invariant cycle theorem. We also show that a variant of the Wang sequence is exact as a sequence of mixed Hodge structures, and we explain the Clemens-Schmid exact sequence. We close with a section containing some concrete examples of degenerations where the reader can appreciate the strength of these theorems.

11.1 Local Systems Acquiring Singularities

11.1.1 Connections with Logarithmic Poles

Let X be a complex manifold and let $D \subset X$ be a divisor which locally looks like the crossings of some coordinate hyperplanes. In § 4.1 we called this a normal crossing divisor. And if the irreducible components are smooth, it was called a simple normal crossing divisor.

Definition 11.1. Let $\mathcal{V}$ be a holomorphic vector bundle on X and let ∇ be a connection of $\mathcal{V}_{|U}$. Then ∇ is said to have **logarithmic poles along** D if it extends to a morphism

$$\nabla : \mathcal{V} \to \Omega^1_X(\log D) \otimes_{\mathcal{O}_X} \mathcal{V} \tag{XI–1}$$

which satisfies Leibniz' rule (X–1).

Suppose that D has *simple* normal crossings. For any irreducible component D_k of D the Poincaré residue map R_k along D_k is defined as follows. In a coordinate chart with coordinates $z_1, \ldots, z_n$ such that $z_1 = 0$ is an equation for D_k, writing $\omega \in \Omega^1(\log D)$ locally as $\omega = \eta \wedge (dz_1/z_1) + \eta'$ with η, η' not containing dz_1, the **Poincaré residue map** can be defined as

$$\begin{aligned} R_k : \Omega^1_X(\log D) &\to \mathcal{O}_{D_k} \\ \omega &\mapsto \eta\big|_{D_k}. \end{aligned}$$

In particular $R_k(dz_1) = 0$ and $R_k(z_1 \cdot \omega) = 0$, where ω is a local section of $\Omega^1(\log D)$. So for local sections f, m of $\mathcal{O}_X(-D_k)$ and $\mathcal{V}$ respectively one has $\nabla(fm) = df \otimes m + f\nabla m \in \operatorname{Ker}(R_k \otimes 1)$. This implies that the map $(R_k \otimes 1) \circ \nabla$ induces an $\mathcal{O}_{D_k}$-linear endomorphism

$$\operatorname{res}_{D_k}(\nabla) \in \operatorname{End}(\mathcal{V} \otimes \mathcal{O}_{D_k}), \tag{XI–2}$$

called the **residue of the connection** along D_k. If D_k is compact, the characteristic polynomial of $\operatorname{res}_{D_k}(\nabla)$ has constant coefficients (because these are global holomorphic functions on D_k).

Consider the special case where X is the unit disk Δ in the complex plane and D is the origin. We let $\Delta^* := \Delta - \{0\}$ and let T denote the monodromy automorphism of $(\mathcal{V}_{|\Delta^*})^\nabla$ determined by a counter-clockwise loop around 0. We let

$$\mathfrak{h} := \{u \in \mathbb{C} \mid \operatorname{Im}(u) > 0\}$$

be the upper half plane, which is the universal covering space of Δ^* via the map

$$\begin{aligned} \mathrm{e} : \mathfrak{h} &\to \Delta^* \\ u &\mapsto e^{2\pi i u}. \end{aligned}$$

Proposition 11.2. *T can be extended to an automorphism of $\mathcal{V}$ whose restriction T_0 to $\mathcal{V}(0)$ is given by*

$$T_0 = \exp(-2\pi \mathrm{i}\, \operatorname{res}_0(\nabla)).$$

This is classical. For proofs see [Del70, Thm II, 1.17] or [Ku, Prop. 8.7.1].

On the other hand, for every module $(\mathcal{V}, \nabla)$ on Δ^* equipped with an integrable connection there exist extensions to a logarithmic connection over Δ. Fix a section of the projection $\mathbb{C} \to \mathbb{C}/\mathbb{Z}$, say

$$\tau : \mathbb{C}/\mathbb{Z} \to \mathbb{C}. \tag{XI–3}$$

For instance, we may demand that the real part of $\tau(z)$ is in the interval $[0,1)$ or $[-1,0)$. Any choice of τ determines a branch of the logarithm as follows. The multivalued function $\frac{\log z}{2\pi \mathrm{i}}$ is univalent on $\mathbb{C}/\mathbb{Z}$ and so

$$\log_\tau(z) := 2\pi \mathrm{i} \tau \left(\frac{\log z}{2\pi \mathrm{i}}\right) \tag{XI–4}$$

is indeed univalent on $\mathbb{C}$.

Proposition 11.3. *Let $(\mathcal{V}, \nabla)$ be a holomorphic vector bundle on Δ^* equipped with an integrable connection. There exists a unique extension $\tilde{\mathcal{V}}(\tau)$ of $\mathcal{V}$ to a vector bundle on Δ such that ∇ extends to a logarithmic connection $\tilde{\nabla}(\tau)$ on $\tilde{\mathcal{V}}(\tau)$ whose residue at 0 has its eigenvalues in the image of τ. Moreover, we can choose a trivialization of $\tilde{\mathcal{V}}(\tau)$ by a frame such that the transition matrix of this frame to a multivalued locally constant frame on Δ^* is meromorphic on Δ, i.e without essential singularities at 0.*

This proposition is due to Manin [Ma]. Following [Del70, pp. 91-95], we give a brief
Sketch of the proof. Introduce the **canonical fibre**

$$\mathbb{V}_\infty := H^0(\mathfrak{h}, \mathrm{e}^*\mathcal{V})^\nabla, \tag{XI–5}$$

the $\mathbb{C}$-vector space of multivalued horizontal sections of $\mathcal{V}$.
i) Unipotent monodromy. We put

$$N = -\frac{1}{2\pi \mathrm{i}} \log T = \frac{1}{2\pi \mathrm{i}} \sum_{k>0} (I-T)^k/k.$$

For any holomorphic section v of $\mathrm{e}^*\mathcal{V}$ we define a new holomorphic section $\varphi(v)$ by the rule

$$\varphi(v)(u) = [\exp(2\pi \mathrm{i} u N)]v(u) = \sum \frac{(2\pi \mathrm{i})^k}{k!} u^k N^k v(u). \tag{XI–6}$$

If $v \in \mathbb{V}_\infty$ it transforms through the rule $v(u+1) = Tv(u)$, so $\varphi(s)$ is invariant under $u \mapsto u+1$ and hence descends to a section of $\mathcal{V}|_{\Delta^*}$. So, with $j : \Delta^* \hookrightarrow \Delta$ the inclusion, $\varphi(\mathbb{V}_\infty) \subset H^0(\Delta, j_*\mathcal{V}$ and we set

$$\tilde{\mathcal{V}} := \varphi(\mathbb{V}_\infty) \otimes_{\mathbb{C}} \mathcal{O}_\Delta \subset j_*\mathcal{V}.$$

We have

$$\nabla(\varphi(v)u) = 2\pi \mathrm{i} N[\varphi(v)] \otimes du = 2\pi \mathrm{i} N[\varphi(v)] \otimes \mathrm{e}^*\left(\frac{dt}{t}\right),$$

and so we obtain a logarithmic connection $\tilde{\nabla}$ on $\tilde{\mathcal{V}}$ with residue N at 0. The above extension corresponds to the special case where $\tau(0) = 0$. It has a special name:

Definition 11.4. The **canonical extension** $(\tilde{\mathcal{V}}, \tilde{\nabla})$ of $\mathcal{V}$ is the unique extension for which the residue of $\tilde{\nabla}$ at 0 has eigenvalues in the interval $[0, 1)$.

Other extensions corresponding to liftings τ are obtained replacing N by $N + \tau(0)I$. In t-coordinates on the unit disk, using the notation (XI–4), we then have

$$\varphi_\tau(v)(t) = t^{\tau(0)} \cdot \exp(2\pi \mathrm{i} \log_\tau t)[Nv(t)]. \qquad \text{(XI–7)}$$

For later reference we note that φ gives and explicit identification of the stalk of the canonical extension with the $\mathbb{C}$-vector space of multivalued horizontal sections:

$$\varphi : \mathbb{V}_\infty \xrightarrow{\sim} \tilde{\mathcal{V}}(0). \qquad \text{(XI–8)}$$

In particular a frame for the right hand side gives a holomorphic trivialization of $\tilde{\mathcal{V}}$ near 0.

ii) General case. The monodromy T acts on any fibre V of the vector bundle $\mathcal{V}|_{\Delta^*}$ and we have a decomposition $V = \bigoplus V_\lambda$ into generalized eigenspaces V_λ on which $T - \lambda I$ acts nilpotently. Hence $\lambda^{-1}T$ acts unipotently on V_λ. The vector spaces V_λ over the different points of Δ^* define a sub bundle to which the previous analysis applies leading to a decomposition

$$\mathcal{V} \simeq \bigoplus_\lambda (U_\lambda \mid \Delta^*) \otimes \mathcal{V}_\lambda$$

where now λ runs over all the eigenvalues of T. On the subbundle $\mathcal{V}_\lambda$ we have unipotent monodromy given by $\lambda^{-1}T$ and U_λ is the module $\mathcal{O}_\Delta$ trivialized by taking a flat frame for it and twisting with the function $u \mapsto \lambda^{-u}$ ensuring that the monodromy on it is multiplication by λ. On U_λ we also have a connection $f \mapsto df + \tau(\lambda) f \dfrac{dt}{t}$. Then we put $\tilde{\mathcal{V}}(\tau) = \bigoplus_\lambda U_\lambda \otimes \tilde{\mathcal{V}_\lambda}(\tau)$. □

Remark 11.5. Essentially the same proof can be applied in the situation where one is given a vector bundle equipped with an integrable connection on $X - D$, X a smooth complex manifold and D a normal crossing divisor. The result is that, provided all local monodromy-operators along the branches of D are quasi-unipotent, given τ, $\mathcal{V}$ extends to an essentially unique locally free $\mathcal{O}_X$-module $\tilde{\mathcal{V}}(\tau)$ equipped with a connection having logarithmic poles along D such that the eigenvalues of the residues in the image of τ. See [Malg79, Theorem 4.4].

11.1.2 The Riemann-Hilbert Correspondence (I)

We follow Malgrange's exposition [Bor87, IV]. Let us start with one variable; $\Delta \subset \mathbb{C}$ is a disk in the t-plane centred at 0, and $\mathbb{V}$ is a local system of $\mathbb{C}$-vector spaces on the punctured disk Δ^*, say of dimension m. Choose a multivalued flat frame $\{e_1, \dots, e_m\}$ for $\mathbb{V}$ over a possibly smaller punctured disk centred at 0, still denoted Δ^*. Then the connection matrix (X–2) of the integrable connection is holomorphic on Δ^*, but may have an essential singularity at 0. We have seen (Prop. 11.3) that $\mathcal{V} := \mathbb{V} \otimes_{\mathbb{C}} \mathcal{O}_{\Delta^*}$ extends to a holomorphic vector bundle $\tilde{\mathcal{V}}$ on Δ trivialized by a *holomorphic* frame in which the connection matrix has a simple pole. In particular $(\mathcal{V}, \nabla)$ extends to a logarithmic connection on $\tilde{\mathcal{V}}$. Observe that this also shows that the transformation matrix which transforms the new in the old frame in general has at most a pole at 0, i.e. the new frame is still a *meromorphic frame* for $\mathcal{V}$. We say therefore that the bundle $\tilde{\mathcal{V}}$ is a **meromorphic extension** of $\mathcal{V}$ and that ∇ is **regular** with respect to $\tilde{\mathcal{V}}$. So in dimension 1 "regular" with respect to a given extension just means that ∇ extends with logarithmic poles. Before discussing the several variables setting, we change our point of view a little making use of the sheaf

$$\mathcal{O}_\Delta(*0) = \mathcal{O}_\Delta[t^{-1}]$$

of meromorphic functions on Δ which are holomorphic on Δ^*. A choice of frame $\mathbf{v}$ for the space $\mathbb{V}_\infty$ of multivalued flat sections, by (XI–8) gives a holomorphic frame $\mathbf{v}(t)$ for $\tilde{\mathcal{V}}$ near the origin. In this frame, the connection matrix has entries in $\mathcal{O}_\Delta(*0)$. Regularity means that these entries have a pole of order at most 1. The extension $\tilde{\mathcal{V}}$ corresponds to the branch of the logarithm determined by (XI–4) after choosing the section τ with $\tau(0) = 0$. Any other choice of τ with $\tau(0) = k$ defines the extension $(\tilde{\mathcal{V}}^{(\tau)}, \tilde{\nabla})$ and (XI–7) shows that the trivializing frame $\mathbf{v}(t)$ must be replaced by $t^k T^{-k}\mathbf{v}(t)$. This shows that have $\tilde{\mathcal{V}} \otimes \mathcal{O}_\Delta(*0) = \bigcup_\tau \tilde{\mathcal{V}}^{(\tau)}$.

We now pass to several variables. Our point of departure is a complex manifold X, a (possibly reducible) hypersurface $D \subset X$ and a vector bundle $\mathcal{V}$ over $X - D$. Let $j : X - D \hookrightarrow X$ be the inclusion. Consider locally free $\mathcal{O}_X$-submodules of $j_*\mathcal{V}$ which restrict to $\mathcal{V}$ on $X - D$. Two such extensions $\tilde{\mathcal{V}}_1$ and $\tilde{\mathcal{V}}_2$ are called equivalent if locally on X there exist natural numbers a and b such that $\mathcal{I}_D^a\tilde{\mathcal{V}}_1 \subset \tilde{\mathcal{V}}_2$ and $\mathcal{I}_D^b\tilde{\mathcal{V}}_2 \subset \tilde{\mathcal{V}}_1$. An equivalence class of such modules is called a **meromorphic structure** on $\mathcal{V}$. As in the 1-variable case, let $\mathcal{O}_X(*D)$ be the sheaf of meromorphic functions on X whose restriction to $X - D$ is holomorphic. Then

$$\tilde{\mathcal{V}}(*D) := \tilde{\mathcal{V}} \otimes_{\mathcal{O}_X} \mathcal{O}_X(*D)$$

only depends on the meromorphic structure defined by $\tilde{\mathcal{V}}$.

Suppose now that D has normal crossings and that $(\mathcal{V}, \nabla)$ is an integrable connection. By Remark 11.5 there always exists a locally free extension $\tilde{\mathcal{V}}$ of $\mathcal{V}$ such that ∇ extends with logarithmic poles along D. Any section τ of $\mathbb{C} \to \mathbb{C}/\mathbb{Z}$ defines a locally free extension $\tilde{\mathcal{V}}^{(\tau)}$ of $\mathcal{V}$ and two sections τ, σ define two extensions $\tilde{\mathcal{V}}^{(\tau)}$ and $\tilde{\mathcal{V}}^{(\sigma)}$ which are equivalent. The sheaf $\tilde{\mathcal{V}}^{(\tau)}(*D)$ does not depend on the choice of τ; indeed, we have

$$\tilde{\mathcal{V}}(*D) := \bigcup_\tau \tilde{\mathcal{V}}^{(\tau)}. \tag{XI–9}$$

The resulting meromorphic structure on $\mathcal{V}$ is the unique meromorphic structure containing lattices like $\tilde{\mathcal{V}}^{(\tau)}$ such that ∇ extends to a meromorphic connection with a logarithmic pole of order one. The pair $(\tilde{\mathcal{V}}, \nabla)$ is called a **regular meromorphic extension** of $(\mathcal{V}, \nabla)$.

Remark 11.6. If X is an algebraic variety and $\mathcal{V}$ is an algebraic vector bundle on $X - D$, for all coherent $\tilde{\mathcal{V}}$ extending $\mathcal{V}$ one has $\tilde{\mathcal{V}}(*D) = j_*\mathcal{V}$. Hence $j_*\mathcal{V}$ determines a unique meromorphic structure; if ∇ is an algebraic connection on $\mathcal{V}$, it is regular if the analytic counterpart of $\mathcal{V}(*D)$ is the extension defined by (XI–9). Regularity does not depend on the chosen compactification of $X - D$. We can also test regularity using algebraic curves C mapping to X, say $u : C \to X$ such that $u(C) \not\subset D$. Then the connection is regular if $u^*\nabla$ is regular at all points $C \cap u^{-1}D$ where $u^*\nabla$ is defined as follows. Let $x_1, \ldots, x_n$ be local coordinates centred at $u(0) = y$ and let $\tilde{v}$ be a local section of $\tilde{\mathcal{V}}$. If $\nabla(\tilde{v}) = \sum_{i=1}^n dx_i \otimes v_i, \quad v_i \in \tilde{\mathcal{V}}(y)$. and $u_i(t) = x_i \circ u(t)$ set

$$(u^*\nabla)u^*\tilde{v} := \sum_{i=1}^n du_i \otimes u^*(v_i). \tag{XI–10}$$

It is not hard to verify that this definition does not depend on the choice of local coordinates, but only on $u^*\tilde{v}$ and that it indeed gives a connection on the vector bundle $u^*(\tilde{\mathcal{V}})|\Delta^*$.

We can now explain a first version of the Riemann-Hilbert correspondence as stated and proven in [Bor87, IV].

Theorem 11.7 (Riemann-Hilbert correspondence (first version))**.** *Let X be a complex manifold and D a divisor with normal crossings.*

1) *The assignment*

$$(\tilde{\mathcal{V}}, \nabla) \mapsto (\mathcal{V}, \nabla) = (\tilde{\mathcal{V}}, \nabla)\Big|_{X-D}$$

gives an equivalence

$$\left\{\begin{matrix}\text{regular meromorphic extensions to } X\\ \text{of vector bundles on } X-D \text{ equipped}\\ \text{with an integrable connection}\end{matrix}\right\} \longleftrightarrow \left\{\begin{matrix}\text{vector bundles on } X-D\\ \text{equipped with an inte-}\\ \text{grable connection}\end{matrix}\right\}.$$

2) *If U is a smooth complex algebraic variety, the assignment*

$$(\mathcal{V}, \nabla) \mapsto (\mathcal{V}^{\mathrm{an}}, \nabla)$$

gives an equivalence

$$\left\{\begin{matrix}\text{algebraic regular integrable connections}\\ \text{on algebraic vector bundles on } U\end{matrix}\right\} \longleftrightarrow \left\{\begin{matrix}\text{vector bundles on } U\\ \text{equipped with an inte-}\\ \text{grable connection}\end{matrix}\right\}.$$

11.2 The Limit Mixed Hodge Structure on Nearby Cycle Spaces

11.2.1 Asymptotics for Variations of Hodge Structure over a Punctured Disk

Let $\mathbb{V}$ be local system of finite rank $\mathbb{Z}$-modules on the punctured disk Δ^* underlying a polarized variation of Hodge structure of weight k, and let t be the standard coordinate on the disk. The monodromy is quasi-unipotent by the following theorem due to Borel [Sch73, Lemma 4.5, Thm 6.1]:

Theorem 11.8 (MONODROMY THEOREM). *Let $\mathbb{V}$ be a polarized variation of Hodge structure on the punctured disk Δ^*. Then the monodromy operator T is quasi-unipotent. More precisely: if $\ell = \max(\{p-q \mid \mathbb{V}^{p,q}_{\mathbb{C},t} \neq 0\})$ and $T = T_sT_u$ is the Jordan decomposition of T with T_u unipotent and T_s semisimple, then $(T_u - I)^{\ell+1} = 0$ and T_s has finite order.*

For the geometric case the proof that T is quasi-unipotent is rather straightforward. See Remark 11.20. The more subtle bound on the index of nilpotency is Theorem 11.42 below.

We set

$$N = \log T_u = \log(I + [T_u - I]) = \sum_{k\geq 1} \frac{(-1)^{k-1}[T_u - I]^k}{k}. \qquad \text{(XI–11)}$$

Observe that convergence is immediate since the right hand side is a finite sum.

Assume from now on that the monodromy is unipotent so that $T = \exp N$. As before, let $\mathrm{e} : \mathfrak{h} \to \Delta^*$, $\mathrm{e}(u) = \exp(2\pi\mathrm{i}u)$ be the universal cover of the punctured unit disk. Then

$$\mathbb{V}_\infty := H^0(\mathfrak{h}, \mathrm{e}^*\mathbb{V}_\mathbb{C})$$

is isomorphic to the canonical fibre at 0 of the canonical extension $\tilde{\mathcal{V}}$ of $\mathcal{V} = \mathbb{V} \otimes_\mathbb{Z} \mathcal{O}_{\Delta^*}$. It has a logarithmic connection extending ∇, which has a nilpotent residue R at 0 and we have $N = -2\pi\mathrm{i}R$. The identification (XI–8) depends on the choice of the holomorphic coordinate t and will be denoted

$$g_t : \mathbb{V}_\infty \xrightarrow{\sim} \tilde{\mathcal{V}}(0)$$

which we use to define an integral lattice in $\tilde{\mathcal{V}}(0)$:

$$\tilde{\mathcal{V}}(0)_\mathbb{Z} = g_t(\Gamma(\mathfrak{h}, \mathrm{e}^*\mathbb{V}_\mathbb{Z})).$$

The choice of a different local coordinate s on $(\Delta, 0)$ gives a new integral lattice $g_s(\Gamma(\mathfrak{h}, \mathrm{e}^*\mathbb{V}_\mathbb{Z})$ related to $g_t(\Gamma(\mathfrak{h}, \mathrm{e}^*\mathbb{V}_\mathbb{Z}))$ by

$$g_t(\Gamma(\mathfrak{h}, \mathrm{e}^*\mathbb{V}_\mathbb{Z})) = \exp(2\pi\mathrm{i}\alpha N) g_s(\Gamma(\mathfrak{h}, \mathrm{e}^*\mathbb{V}_\mathbb{Z}), \quad \alpha = \frac{dt}{ds}(0).$$

So a canonical object is the **nilpotent orbit** $\{\exp(2\pi i\beta)g_t(\Gamma(\mathfrak{h}, e^*\mathbb{V}_{\mathbb{Z}})) \mid \beta \in \mathbb{C}^*\}$.

Since N is nilpotent, by [Sch73, Lemma 6.4], we have a naturally defined weight filtration on the canonical fibre (XI–8) of $\mathbb{V}$:

Lemma-Definition 11.9. Given a nilpotent endomorphism N of a finite dimensional vector space V, the **weight filtration of N centred at k** is the unique increasing filtration $W = W(N,k)$ of V with the properties

1) $N(W_i) \subset W_{i-2}$, $i \geq 2$;
2) the map
$$N^\ell : \mathrm{Gr}^W_{k+l} V \to \mathrm{Gr}^W_{k-l} V$$
is an isomorphism for all $\ell \geq 0$.

Moreover, there is a Lefschetz-type decomposition

$$\mathrm{Gr}^W V = \bigoplus_{\ell=0}^{k} \bigoplus_{r=0}^{\ell} N^r [PV]_{k+\ell}$$

$$\text{with } PV_{k+\ell} := \mathrm{Ker}[N^{\ell+1} : \mathrm{Gr}^W_{k+\ell} V \to \mathrm{Gr}^W_{k-\ell-2} V].$$

and the endomorphism N has $\dim[PV]_{k+\ell}$ Jordan blocs of size $\ell+1$, $\ell = 0, \ldots, k$.

We now can formulate Schmid's result [Sch73, Theorem 6.16]:

Theorem 11.10. *The Hodge bundles $\mathcal{F}^p$ of $\mathcal{V}$ extend to holomorphic subbundles $\tilde{\mathcal{F}}^p$ of $\tilde{\mathcal{V}}$, and the triple*

$$\mathbb{V}^{\mathrm{Hdg}}_\infty := (\tilde{\mathcal{V}}(0)_{\mathbb{Z}}, W_\bullet(N,k), \tilde{\mathcal{F}}^\bullet(0))$$

is a mixed Hodge structure. Taking into account the ambiguity of the integral structure, we obtain a "nilpotent orbit" of mixed Hodge structures.

The proof for the cohomology of 1-parameter degenerations is given below in § 11.2.7.

Remark. In Schmid's work the notion of nilpotent orbit has a more precise meaning reflecting the asymptotic properties of the period map.

11.2.2 Geometric Set-Up and Preliminary Reductions

We let X be a complex manifold, $\Delta \subset \mathbb{C}$ the unit disk and $f : X \to \Delta$ a holomorphic map smooth over the punctured disk Δ^*. We say that f is a **one-parameter degeneration**. In general $X_0 = f^{-1}(0)$ can have arbitrarily bad singularities, but after suitable blowings up, X_0 can be assumed to have only simple normal crossings on X. Let μ be the least common multiple of the multiplicities of the components of the divisor X_0 and consider the map $m : t \mapsto t^\mu$ sending Δ to itself. For the moment, let us denote by Δ' the source of the map m and let W be the normalisation of the fibre product $X \times_{\Delta} \Delta'$. In general W is a V-manifold. Blowing up the singularities we obtain a manifold X' and a morphism $f' : X' \to \Delta'$. We call $f' : X' \to \Delta'$ the **μ-th root fibration of f**. The fibre X'_0 has simple normal crossings, but unless $\dim X = 3$, the components introduced in the last blowing up might not be reduced. The semistable reduction theorem [K-K-M-S] states that repeating the above procedure a finite number of times we can achieve this:

Theorem 11.11. *Let $f : X \to \Delta$ be as above. Then there exists $m \in \mathbb{N}$ such that for the m-th root $f' : X' \to \Delta'$ of f the special fibre has simple normal crossings and such that all its components are reduced.*

Unless further notice we shall henceforth assume that $f : X \to \Delta$ is smooth over Δ^* and that $E := f^{-1}(0)$ is a simple normal crossing divisor all of whose components are reduced.

Notation. We let E_i be the components of E and we let

$$E_J = \bigcap_{i \in J} E_j, \; E(m) = \coprod_{|J|=m} E_J. \tag{XI–12}$$

Introduce the universal cover

$$\mathrm{e} : \mathfrak{h} \to \Delta^*, \quad \mathrm{e}(u) = \exp(2\pi \mathrm{i} u)$$

of the punctured disk Δ^* and the **canonical fibre** X_∞ of $f : X \to \Delta$ as the fibre product

$$X_\infty := X \times_{\Delta^*} \mathfrak{h}$$

leading to the **specialization diagram**

$$\begin{array}{ccccc} X_\infty & \xrightarrow{k} & X & \xleftarrow{i} & E \\ \downarrow{\scriptstyle f_\infty} & & \downarrow{\scriptstyle f} & & \downarrow \\ \mathfrak{h} & \xrightarrow{\mathrm{e}} & \Delta & \leftarrow & \{0\}. \end{array} \tag{XI–13}$$

We also need a special set of neighbourhoods at any given point $x \in E$. Choose a system $(z_0, \ldots, z_n)$ of local coordinates on a neighbourhood U of Q in X centred at x, such that $f(z_0, \ldots, z_n) = z_0 \cdots z_k$. Put

$$V_{r,\eta} = \{z \in U \mid \|z\| < r \text{ and } |f(z)| < \eta\} \tag{XI–14}$$

for $0 < \eta \ll r \ll 1$. These form a fundamental system of neighbourhoods of x in X.

11.2.3 The Nearby and Vanishing Cycle Functor

The canonical fibre X_∞ is homotopic to any fibre X_t of f since f_∞ is differentiably a product. The total space X is homotopy equivalent to E by a fibre preserving retraction $r : X \to E$. So the inclusion $i_t : X_t \hookrightarrow X$ followed by the retraction can be seen as the specialization map $r_t : X_t \to E$. The complex $(Rr_t)_* i_t^* \underline{\mathbb{Z}}_X = \psi_f \underline{\mathbb{Z}}_X$ is the complex of nearby cocycles. Note that using (B–21) we have

$$\mathbb{H}^q(\psi_f \underline{\mathbb{Z}}_X) = H^q(X_t) \tag{XI–15}$$

and so it gives a rather elaborate way to calculate the cohomology of the smooth fibre. The point is that there is an alternative complex-analytic description of this complex which allows us to enlarge it into a mixed Hodge complex of sheaves:

Lemma 11.12. *The complex of vanishing cocycles has the same cohomology sheaves as $i^* Rk_*(k^* \underline{\mathbb{Z}}_X)$.*

Proof. Using the notation (XI–14), the Milnor fibre $M_{f,x}$ is the intersection of X_t with $V_{r,\eta}$ for t small but non-zero. For t real $M_{f,x}$ embeds in $k^{-1}V_{r,\eta}$ through $z \mapsto (z, \frac{\log t}{2\pi \mathrm{i}})$ and it can be seen that this is a homotopy equivalence. Hence the inclusion induces

$$H^q(M_{f,x}) \simeq \lim_{r,\eta} H^q(k^{-1}(V_{r,\eta})) = \left[R^q k_* \underline{\mathbb{Z}}_{X_\infty}\right]_x = \left[H^q(i^* Rk_*(k^* \underline{\mathbb{Z}}_X))\right]_x .$$

By (C–7), we have $H^q(M_{f,x}) = \left[H^q(\psi_f \underline{\mathbb{Z}}_X)\right]_x$, and this concludes the proof. □

Motivated by Lemma 11.12, we now consider any bounded below complex of sheaves of R-modules $\mathcal{K}^\bullet$ on X and set

$$\psi_f \mathcal{K}^\bullet := i^* Rk_* k^* \mathcal{K}^\bullet.$$

This is a bounded below complex of sheaves of R-modules (note that we defined Rk_* of a complex which is bounded from below as k_* of its Godement resolution). One has a natural morphism of complexes $\mathcal{K}^\bullet \to Rk_* k^* \mathcal{K}^\bullet$ (see (B–24)) and hence a morphism of complexes

$$\mathrm{sp} : i^* \mathcal{K}^\bullet \to \psi_f \mathcal{K}^\bullet \quad \text{(the } \textbf{specialization} \text{ of } \mathcal{K}^\bullet\text{)}. \tag{XI–16}$$

As before, setting

$$\phi_f \mathcal{K}^\bullet := \mathrm{Cone}^\bullet(\mathrm{sp}),$$

projection on the second factor induces the **canonical map**

$$\mathrm{can} : \psi_f \mathcal{K}^\bullet \to \phi_f \mathcal{K}^\bullet \tag{XI–17}$$

which occurs in the triangle for the cone (A–15), which in this setting is called the **specialization triangle**

$$\begin{array}{ccc} i^*\mathcal{K}^\bullet & \xrightarrow{\text{sp}} & \psi_f\mathcal{K}^\bullet \\ & {\scriptstyle [1]}\nwarrow \quad \swarrow{\scriptstyle \text{can}} & \\ & \phi_f\mathcal{K}^\bullet & \end{array}$$

We call ψ_f and ϕ_f the **nearby** and **vanishing cycle functors** associated to f. They map bounded complexes of sheaves of R-modules on X to similar objects on E, and induce functors with the same names on the level of the derived categories.

Note that in diagram (XI–13) one may as well restrict f to a small disk around 0 and replace the map $\mathrm{e} : \mathfrak{h} \to \Delta$ by its restriction to a subset $\{u \in \mathfrak{h} \mid \mathrm{Im}(u) > K\}$ for some $K \in \mathbb{R}$, $K > 0$. The map $h : X_\infty \to X_\infty$ given by $h(x, u) = (x, u + 1)$ satisfies $k \circ h = k$, hence we have an automorphism h^* of $Rk_*k^*\mathcal{K}^\bullet$ and of $\psi_f\mathcal{K}^\bullet$. The formula for the monodromy analogous to (C–5) becomes

$$T := (h^*)^{-1} : \psi_f\mathcal{K}^\bullet \to \psi_f\mathcal{K}^\bullet \qquad \left(\begin{array}{c}\textbf{monodromy trans-}\\ \textbf{formation} \text{ for } f.\end{array}\right) \qquad \text{(XI–18)}$$

The inverse is put here because T follows a counterclockwise loop. We have an induced monodromy action on $\phi_f\mathcal{K}^\bullet$, also denoted by T. Note that $(T - I)\circ\mathrm{sp} = 0$, and the analogue of the variation map (C–6) is the map

$$\mathrm{var} : \phi_f\mathcal{K}^\bullet \to \psi_f\mathcal{K}^\bullet \qquad (\textbf{variation} \text{ for } f), \qquad \text{(XI–19)}$$

defined by $\mathrm{var}(x, y) = Ty - y$ for local sections x, y of $i^*\mathcal{K}^\bullet[1]$ and $\psi_f\mathcal{K}^\bullet$ respectively. It is a morphism of complexes such that

$$T - I = \mathrm{var}\circ\mathrm{can} \ \text{ on } \psi_f\mathcal{K}^\bullet\ .$$

Then also $\mathrm{can}\circ\mathrm{var} = T - I$ on $\phi_f\mathcal{K}^\bullet$.

Under the assumption that all components of E are Kähler we are going to construct mixed Hodge complexes of sheaves ψ_f^{Hdg} and ϕ_f^{Hdg} on E such that in the derived category of bounded above complexes of sheaves of $\mathbb{Q}$-vector spaces on E we have $[\psi_f^{\mathrm{Hdg}}]_{\mathbb{Q}} \simeq \psi_f\underline{\mathbb{Q}}_X$ and $[\phi_f^{\mathrm{Hdg}}]_{\mathbb{Q}} \simeq \phi_f\underline{\mathbb{Q}}_X$; moreover we'll show that there is an exact sequence of mixed Hodge complexes of sheaves

$$0 \to \mathcal{H}dg^\bullet(E) \to [\psi_f^{\mathrm{Hdg}}]_{\mathbb{Q}} \xrightarrow{\mathrm{can}} [\phi_f^{\mathrm{Hdg}}]_{\mathbb{Q}} \to 0\ .$$

By Lemma 11.12 this will put a mixed Hodge structure on the cohomology of X_∞ which by definition is **the limit mixed Hodge structure**.

11.2.4 The Relative Logarithmic de Rham Complex and Quasi-unipotency of the Monodromy

We first construct a complex of sheaves on E which is quasi-isomorphic to $\psi_f\underline{\mathbb{C}}_X$ but which is closer in nature to the complexes occurring in mixed

Hodge theory, the **relative de Rham complex on X with logarithmic poles along E:**

$$\Omega^\bullet_{X/\Delta}(\log E) := \Omega^\bullet_X(\log E)/f^*\Omega^1_\Delta(\log 0) \wedge \Omega^{\bullet-1}_X(\log E).$$

Outside of E this complex coincides with the relative de Rham complex from Definition 10.25. Its cohomology sheaves are given by:

Theorem 11.13. *Let $X = \mathbb{C}^{n+1}$ with coordinates $(z_0, \dots, z_n)$, and let $f : X \to \Delta$ be given by $t = f(z_0, \dots, z_n) = z_0 \cdots z_k$ for some $k \in \mathbb{N}$ with $0 \le k \le n$. Let E be the zero set of t. Put $\xi_i = dz_i/z_i$ for $i = 0, \dots, k$. Then*

1) *$H^0(\Omega^\bullet_{X/\Delta}(\log E))_0 \simeq \mathbb{C}\{t\}$;*
2) *$H^1(\Omega^\bullet_{X/\Delta}(\log E))_0$ is the $\mathbb{C}\{t\}$-module with generators $\xi_0, \dots, \xi_k$ and the single relation $\sum_{i=0}^k \xi_i = 0$;*
3) *$H^q(\Omega^\bullet_{X/\Delta}(\log E))_0 \simeq \bigwedge^q_{\mathbb{C}\{t\}} H^1(\Omega^\bullet_{X/\Delta}(\log E))_0$ for $q > 1$.*

Proof. The complex $\Omega^\bullet_{X/\Delta}(\log E)_0$ can be considered as a double complex where the differential d is written as $d_1 + d_2$ such that d_1 involves differentiation with respect to the first $k+1$ variables and d_2 differentiation with respect to the last $n-k$ variables. The complex $(\Omega^\bullet_{X/\Delta}(\log E)_0, d_2)$ is acyclic in positive degrees by the relative Poincaré Lemma, so it is quasi-isomorphic to $(\mathrm{Ker}(d_2), d_1)$. Hence we may reduce to the case $n = k$.

Since $\xi_0 = -\sum_{j\ge 1}\xi_j$ we see that for $i = 1, \dots, n$ we have for all $g \in \mathcal{O}_{X,0}$

$$dg = \sum_{j=1}^n D_j(g)\xi_j, \quad D_j := z_j\partial/\partial z_j - z_0\partial/\partial z_0.$$

So in this case the complex is isomorphic to the Koszul complex on $R := \mathbb{C}\{z_0, \dots, z_n\}$ with operators D_i:

$$R \xrightarrow{d_0} V \otimes R \xrightarrow{d_1} \Lambda^2 V \otimes R \to \cdots \to \Lambda^n V \otimes R,$$

where $V = \mathbb{C}\xi_1 \oplus \cdots \oplus \mathbb{C}\xi_n$ and

$$d_\ell(\xi_{i_1} \wedge \cdots \wedge \xi_{i_\ell} \otimes g) = \sum_j \xi_j \wedge \xi_{i_1} \wedge \cdots \wedge \xi_{i_\ell} \otimes D_j g.$$

The operator D_i multiplies a monomial $z_0^{a_0} \cdots z_k^{a_k}$ by the integer $a_i - a_0$. Hence the cohomology of the complex may be computed monomial by monomial. One only gets a non-zero contribution for those monomials on which all D_i are zero, i.e. for the powers of t. □

Corollary 11.14. 1) *$H^0(\Omega^\bullet_{X/\Delta}(\log E) \otimes \mathcal{O}_E) \simeq \underline{\mathbb{C}}_E$;*
2) *$H^1(\Omega^\bullet_{X/\Delta}(\log E) \otimes \mathcal{O}_E)_0$ is the $\mathbb{C}$-vector space with generators $\xi_0, \dots, \xi_k$ and the single relation $\sum_{i=0}^k \xi_i = 0$;*

3) $H^q(\Omega^\bullet_{X/\Delta}(\log E) \otimes \mathcal{O}_E) \simeq \bigwedge^q_{\underline{\mathbb{C}}_E} H^1(\Omega^\bullet_{X/\Delta}(\log E) \otimes \mathcal{O}_E)$ *for* $q > 1$.

Let us next consider the logarithmic extension of the Gauss-Manin connection. We start with the extension of (X–11):

$$0 \to f^*(\Omega^1_\Delta(\log 0)) \otimes \Omega^\bullet_{X/\Delta}[-1] \to \Omega^\bullet_X(\log E) \to \Omega^\bullet_{X/\Delta}(\log E) \to 0. \quad \text{(XI–20)}$$

We then have:

Proposition 11.15. *The connecting homomorphism*

$$\delta^q_0 : R^q f_* \Omega^\bullet_{X/\Delta}(\log E)_0 \to R^q f_* \Omega^\bullet_{X/\Delta}(\log E)_0$$

for the long exact sequence associated to (XI–21) *is residue at* 0 *of the of the logarithmic extension of the Gauss-Manin connection.*

Proof. Since outside E this is just the sequence (X–11), Theorem 10.28 shows that the connecting homomorphism

$$\delta : R^q f_* \Omega^\bullet_{X/\Delta}(\log E) \to \Omega^1_\Delta \otimes_{\mathcal{O}_\Delta} R^q f_* \Omega^\bullet_{X/\Delta}(\log E)$$

in the long exact sequence for the derived image sheaves restricts to the Gauss-Manin connection on the punctured disk. On the origin we get its logarithmic extension (XI–1). This notion has been explained in § 11.1.1 where we also explained the residue (XI–2) for such a logarithmic extension. Taking residues, transforms the sequence (XI–20) into the following exact sequence on E:

$$\left.\begin{aligned} 0 \to \Omega^\bullet_{X/\Delta}(\log E) \otimes \mathcal{O}_E[-1] \xrightarrow{\wedge(dt/t)} &\ \Omega^\bullet_X(\log E) \otimes \mathcal{O}_E \\ \longrightarrow &\ \Omega^\bullet_{X/\Delta}(\log E) \otimes \mathcal{O}_E \to 0. \end{aligned}\right\} \quad \text{(XI–21)}$$

Using the above calculations, the result follows. □

We are going to relate the complex $\Omega^\bullet_{X/\Delta}(\log E) \otimes \mathcal{O}_E$ to $\psi_f \underline{\mathbb{C}}_{X^*}$ by a chain of quasi-isomorphisms as follows.

1) On $X^* := X - E$ we have the quasi-isomorphism $\underline{\mathbb{C}}_{X^*} \hookrightarrow \Omega^\bullet_{X^*}$ which induces a quasi-isomorphism

$$\psi_f \underline{\mathbb{C}}_{X^*} \to \psi_f \Omega^\bullet_{X^*}.$$

2) The inclusion

$$k_* \Omega^\bullet_{X_\infty} \to k_* \mathcal{C}_{\mathrm{Gdm}} \Omega^\bullet_{X_\infty}$$

induces a quasi-isomorphism

$$i^* k_* \Omega^\bullet_{X_\infty} \to \psi_f \Omega^\bullet_{X_\infty}.$$

The reason for this is that $k : X_\infty \to X$ is a Stein map hence $R^q k_* \mathcal{F} = 0$ for $q > 0$ if $\mathcal{F}$ is a coherent sheaf on X_∞; apply this to each $\Omega^p_{X_\infty}$.

3) Consider the inclusion $\alpha : i^*\Omega^\bullet_X(\log E)[\log t] \hookrightarrow i^*k_*\Omega^\bullet_{X^*}$. Here $d\log t = dt/t$ as one would expect, and local sections of $i^*\Omega^\bullet_X(\log E)[\log t]$ have the form $\sum_{i=0}^m \omega_i(\log t)^i$. We will show below that the inclusion map α of this subcomplex is a quasi-isomorphism.
4) We also have the map $\beta : i^*\Omega^\bullet_X(\log E)[\log t] \to \Omega^\bullet_{X/\Delta}(\log E) \otimes O_E$ given by $\sum_{i=0}^m \omega_i(\log t)^i \mapsto \omega_0$. We will show below that β is a quasi-isomorphism.

Modulo the two claims, this chain of quasi-isomorphism thus shows:

Theorem 11.16. *If X is a complex manifold and $f : X \to \Delta$ is holomorphic such that $E = f^{-1}(0)$ is a reduced divisor with normal crossings on X, then*

$$\psi_f(\underline{\mathbb{C}}_X) \simeq \Omega^\bullet_{X/\Delta}(\log E) \otimes \mathcal{O}_E$$

in the derived category D^+(sheaves of $\mathbb{C}$-vector spaces on E).

Proof. We need to show that the maps α and β are quasi-isomorphisms. Fix a point $x \in E$. Using the special neighbourhoods (XI–14) we have

$$\left[H^q(i^*k_*\Omega^\bullet_{X_\infty})\right]_x = \lim_{r,\eta} H^q(\Gamma(k^{-1}(V_{r,\eta}), \Omega^\bullet_{X_\infty})).$$

The natural inclusion

$$k^{-1}(V_{r,\eta}) \hookrightarrow \{(z,u) \in \mathbb{C}^{n+1} \times \mathbb{C} \mid \prod_{i=0}^k z_i = \mathrm{e}(u)\}$$

can be seen to be a homotopy equivalence. The restriction map

$$H^q((\mathbb{C}^*)^{k+1} \times \mathbb{C}^{n-k} \times \mathbb{C}; \mathbb{C}) \to H^q(k^{-1}(V_{r,\eta}); \mathbb{C})$$

is surjective. The former is the q-th exterior power of the $\mathbb{C}$-vector space with basis $\xi_0, \ldots, \xi_k$ and the latter is obtained by dividing out the relation $dt/t = 0$. This computes the stalk of the cohomology sheaf of $i^*k_*\Omega^\bullet_{X_\infty}$ at x and shows that α is a quasi-isomorphism.

The stalk at x of $H^q(\Omega^\bullet_X(\log E))$ has already been computed (see Proposition 4.3). The result is that H^0 has stalk $\mathbb{C}$, the stalk of H^1 at x is the vector space spanned by the classes of $\xi_0, \ldots, \xi_k$, and $H^q = \bigwedge^q H^1$. Let H^1 denote the subspace of $\Omega^1_X(\log E)_x$ spanned by $\xi_0, \ldots, \xi_k$ and $H^q = \bigwedge^q H^1 \subset \Omega^q_X(\log E)_x$. Let $H^q[\log t]$ denote the subspace of $\Omega^q_X(\log E)_x$ consisting of elements of the form $\sum_{i=0}^s \omega_i(\log t)^i$ with $\omega_i \in H^q$. Then one can easily check that $H^\bullet[\log t]$ is a subcomplex of $\Omega^\bullet_X(\log E)[\log t]_x$ which is quasi-isomorphic to it, and that the inclusion $H^\bullet \hookrightarrow H^\bullet[\log t]$ induces surjective maps $H^q \to H^q(H^\bullet[\log t])$ with as kernel the elements $(dt/t) \wedge \eta$ with $\eta \in H^{q-1}$. This implies that β is also a quasi-isomorphism. □

Using the description of the Gauss-Manin connection given in Prop. 11.15 and its relation to the monodromy (Prop. 11.2) we deduce:

Corollary 11.17. *The monodromy T on $\mathbb{H}^q(E, \psi_f(\underline{\mathbb{C}}_X))$ is related to the connecting homomorphism*

$$\mathrm{res}_0(\nabla) : \mathbb{H}^q(E, \Omega^\bullet_{X/\Delta}(\log E) \otimes \mathcal{O}_E) \to \mathbb{H}^q(E, \Omega^\bullet_{X/\Delta}(\log E) \otimes \mathcal{O}_E)$$

in the long exact hypercohomology sequence of the exact sequence (XI–21) *by the formula*

$$T = \exp(-2\pi i \mathrm{res}_0(\nabla)).$$

Corollary 11.18. *If X is a complex manifold, S a Riemann surface, and $f : X \to S$ a proper holomorphic map the union of whose singular fibres E is a reduced divisor with normal crossings on X, then the sheaves $R^m f_* \Omega^\bullet_{X/S}(\log E)$ are locally free on S and*

$$R^m f_* \Omega^\bullet_{X/S}(\log E)_t \otimes_{\mathcal{O}_{S,t}} \kappa(t) \simeq \mathbb{H}^m(X_t, \Omega^\bullet_{X/S}(\log E) \otimes \mathcal{O}_{X_t}) \text{ for all } t \in S.$$

Proof. The map f cannot be constant, hence it is flat. Moreover there is a discrete subset Σ of S of critical values, because f is proper. Then $E = f^{-1}(\Sigma)$.

It follows from Theorem 11.16 that the function

$$t \mapsto \dim_{\mathbb{C}} \mathbb{H}^m(X_t, \Omega^\bullet_{X/S}(\log E) \otimes \mathcal{O}_{X_t})$$

is locally constant on S. The result follows from a trivial generalization of [Gr-Rie, Theorem on p. 211]. □

Corollary 11.19. *The eigenvalues of the monodromy operator T are all* 1.

Proof. By Cor. 11.17 it suffices to show that the eigenvalues of $R = \mathrm{res}_0(\nabla)$ are integers. Now R acts on the terms of the spectral sequence of hypercohomology

$$E_2^{p,q} = H^p(E, H^q(\Omega^\bullet_{X/S}(\log E) \otimes_{\mathcal{O}_X} \mathcal{O}_E)) \implies \mathbb{H}^{p+q}(E, \Omega^\bullet_{X/S}(\log E) \otimes_{\mathcal{O}_X} \mathcal{O}_E).$$

It follows that the eigenvalues of R must occur as eigenvalues of its action on the $E_2^{p,q}$-terms. We saw (Prop. 11.15) that the action on the sheaf $H^q(\Omega^\bullet_{X/S}(\log E)) \otimes \mathcal{O}_E)$ is nothing but the connecting homomorphism δ_0^q for the long exact cohomology sequence (XI–21). A computation in local coordinates (notation as in Theorem 11.13) shows that on generators for $\Omega^\bullet_{X/S}(\log E)$ we have $\delta_0^q(t^a \xi_{k_1} \wedge \cdots \wedge \xi_{i_q}) = a t^a \xi_{k_1} \wedge \cdots \wedge \xi_{i_q}$. After tensoring with $\mathcal{O}_E$ only the generators with $a = 0$ survive and so the eigenvalues of R are zero indeed.

Remark 11.20. In the situation where X_0 has simple normal crossings but the least common multiple e of the multiplicities of E is possibly ≥ 2 the preceding computation has to be modified and shows that the residue of R has eigenvalues of the form a/e and so are rational. It follows that in this more general situation T_0 (and hence also T) is quasi-unipotent.

11.2.5 The Complex Monodromy Weight Filtration and the Hodge Filtration

We define a filtration W on $\Omega^\bullet_{X/\Delta}(\log E) \otimes \mathcal{O}_E$ by

$$W_k\Omega^\bullet_{X/\Delta}(\log E) \otimes \mathcal{O}_E := \text{Image of } W_k\Omega^\bullet_X(\log E) \text{ in } \Omega^\bullet_{X/\Delta}(\log E) \otimes \mathcal{O}_E.$$

It is clear that this is a filtration by subcomplexes, and at first sight it might be a good ingredient for a mixed Hodge complex of sheaves. Let us however calculate the cohomology sheaves of $\mathrm{Gr}^W_m\, \Omega^\bullet_{X/\Delta}(\log E) \otimes \mathcal{O}_E$.

The sheaf $\Omega^\bullet_{X/\Delta}(\log E) \otimes \mathcal{O}_E$ is the cokernel of the map

$$\theta : \Omega^{\bullet-1}_X(\log E) \to \Omega^\bullet_X(\log E); \quad \theta(\omega) = \frac{dt}{t} \wedge \omega$$

so that we need to investigate θ. Note that dt/t induces a global section of $\Omega^1_X(\log E) \otimes \mathcal{O}_E$ and hence that θ maps $W_k\Omega^p_X(\log E) \otimes \mathcal{O}_E$ to $W_{k+1}\Omega^{p+1}_X(\log E) \otimes \mathcal{O}_E$ and induces maps

$$\theta : \mathrm{Gr}^W_k\, \Omega^{p+k}_X(\log E) \otimes \mathcal{O}_E \to \mathrm{Gr}^W_{k+1}\, \Omega^{p+k+1}_X(\log E) \otimes \mathcal{O}_E.$$

Recalling that $\mathrm{Gr}^W_0\, \Omega^p_X(\log E) \otimes \mathcal{O}_E \simeq \tilde{\Omega}^p_E$ and applying the residue maps (IV–2) to the terms of the sequence

$$\begin{aligned} 0 \to \mathrm{Gr}^W_0\, \Omega^p_X(\log E) \otimes \mathcal{O}_E \xrightarrow{\theta} \ & \mathrm{Gr}^W_1\, \Omega^{p+1}_X(\log E) \otimes \mathcal{O}_E \\ & \xrightarrow{\theta} \mathrm{Gr}^W_2\, \Omega^{p+2}_X(\log E) \otimes \mathcal{O}_E \ldots \end{aligned}$$

one obtains the sequence

$$0 \to \tilde{\Omega}^p_E \to a_{1*}\Omega^p_{E(1)} \to a_{2*}\Omega^p_{E(2)} \to \cdots$$

in which the maps are nothing but the alternating sums of restriction maps. Hence the sequences are exact.

We find $\mathrm{Gr}^W_m(\Omega^\bullet_{X/\Delta}(\log E) \otimes \mathcal{O}_E)$ as the cokernel of the map

$$\theta : \mathrm{Gr}^W_{m-1}(\Omega^{\bullet-1}_X(\log E) \otimes \mathcal{O}_E) \to \mathrm{Gr}^W_m\, \Omega^\bullet_X(\log E) \otimes \mathcal{O}_E,$$

but the exactness just proved entails exactness of the sequence

$$\begin{aligned} 0 \to \mathrm{Gr}^W_m(\Omega^\bullet_{X/\Delta}(\log E) \otimes \mathcal{O}_E) \xrightarrow{\theta} \ & \mathrm{Gr}^W_{m+1}(\Omega^{\bullet+1}_X(\log E) \otimes \mathcal{O}_E) \\ & \xrightarrow{\theta} \mathrm{Gr}^W_{m+2}(\Omega^{\bullet+2}_X(\log E) \otimes \mathcal{O}_E). \end{aligned}$$

Again, by the residue maps this sequence transforms into

$$0 \to \mathrm{Gr}^W_m\, \Omega^\bullet_{X/\Delta}(\log E) \otimes \mathcal{O}_{E[m]} \to (a_{m+1})_*\Omega^\bullet_{E(m+1)} \to (a_{m+2})_*\Omega^\bullet_{E(m+2)}$$

so that

$$
\begin{aligned}
H^m \operatorname{Gr}_m^W \Omega^\bullet_{X/\Delta}(\log E) \otimes \mathcal{O}_E &\simeq \operatorname{Ker}((a_{m+1})_* \underline{\mathbb{C}}_{E(m+1)} \to (a_{m+2})_* \underline{\mathbb{C}}_{E(m+2)}) \\
&\simeq \underline{\mathbb{C}}_{E[m+1]}
\end{aligned}
$$

where $E[m+1]$ is the set of points of E of multiplicity at least $m+1$. Moreover, $H^q \operatorname{Gr}_m^W \Omega^\bullet_{X/\Delta}(\log E) \otimes \mathcal{O}_E = 0$ whenever $q \neq m$. In particular, $\operatorname{Gr}_0^W \Omega^\bullet_{X/\Delta}(\log E) \otimes \mathcal{O}_E$ is a resolution of $\underline{\mathbb{C}}_E$ and there is no hope that we get a pure Hodge structure out of it!

However, this computation provides us with a resolution of $\Omega^\bullet_{X/\Delta}(\log E) \otimes \mathcal{O}_E$ as follows. Define a bi-filtered double complex of sheaves

$$
(A^{\bullet,\bullet}, d', d'', W(M), W)
$$

on E by

$$
A^{p,q} = \Omega_X^{p+q+1}(\log E)/W_p \Omega_X^{p+q+1}(\log E) \text{ for } p,q \geq 0
$$

with differentials

$$
\begin{aligned}
d' &: A^{p,q} \to A^{p+1,q}, \\
d'' &: A^{p,q} \to A^{p,q+1},
\end{aligned}
$$

defined by

$$
\begin{aligned}
d'(\omega) &= (dt/t) \wedge \omega, \\
d''(\omega) &= d\omega
\end{aligned}
$$

and two filtrations, the **weight filtration**, respectively the **monodromy weight filtration**

$$
\begin{aligned}
W_r A^{p,q} &= \text{image of } W_{r+p+1} \Omega_X^{p+q+1}(\log E) \text{ in } A^{p,q}, \\
W(M)_r A^{p,q} &= \text{image of } W_{r+2p+1} \Omega_X^{p+q+1}(\log E) \text{ in } A^{p,q}. \qquad \text{(XI–22)}
\end{aligned}
$$

We have maps

$$
\mu : \Omega^q_{X/\Delta}(\log E) \otimes \mathcal{O}_E \to A^{0,q}, \ \omega \mapsto (-1)^q (dt/t) \wedge \omega \bmod W_0
$$

defining a morphism of complexes

$$
\mu : \Omega^\bullet_{X/\Delta}(\log E) \otimes \mathcal{O}_E \to s(A^{\bullet,\bullet}).
$$

Here $s(A^{\bullet,\bullet})$ is the associated single complex. The exactness of the sequences above shows that

$$
\mu : (\Omega^\bullet_{X/\Delta}(\log E) \otimes \mathcal{O}_E, W, F) \to (s(A^{\bullet,\bullet}), W, F)
$$

is a bi-filtered quasi-isomorphism, if we equip $s(A^{\bullet,\bullet})$ with the filtration F given by

$$
F^r s(A^{\bullet,\bullet}) = \bigoplus_p \bigoplus_{q \geq r} A^{p,q}. \qquad \text{(XI–23)}
$$

Hence this map is a quasi-isomorphism, and

$$s(A^{\bullet,\bullet}) \simeq \psi_f \underline{\mathbb{C}}_X.$$

Moreover, because $d'W(M)_r \subset W(M)_{r-1}$ we find that

$$\begin{aligned} Gr_r^{W(M)} s(A^{\bullet,\bullet}) &\simeq \bigoplus\nolimits_{k\geq 0,-r} Gr^W_{r+2k+1} \Omega^\bullet_X(\log E)[1] \\ &\simeq \bigoplus\nolimits_{k\geq 0,-r} (a_{r+2k+1})_* \Omega^\bullet_{E(r+2k+1)}[-r-2k] \end{aligned} \tag{XI–24}$$

where the last isomorphism is defined by res_{r+2k+1} (see § 4.2).

Introduce the morphism

$$\left.\begin{array}{ccc} A^{pq} & \xrightarrow{\nu} & A^{p+1,q-1} \\ \| & & \| \\ \Omega_X^{p+q+1}(\log E)/W_p\Omega_X^{p+q+1} & \longrightarrow & \Omega_X^{p+q+1}(\log E)/W_{p+1}\Omega_X^{p+q+1}, \\ \omega & \longmapsto & \omega \bmod W_{p+1}. \end{array}\right\} \tag{XI–25}$$

It commutes with d' and d'', so it induces an endomorphism of the associated simple complex $s(A^{\bullet,\bullet})$ which we also denote by ν. It maps $W(M)_r$ to $W(M)_{r-2}$ and F^p to F^{p-1}.

Theorem 11.21. *The following diagram is commutative:*

$$\begin{array}{ccc} \mathbb{H}^q(E, \Omega^\bullet_{X/\Delta}(\log E)\otimes \mathcal{O}_E) & \xrightarrow{\mu} & \mathbb{H}^q(E, s(A^{\bullet\bullet})) \\ \downarrow{\scriptstyle \mathrm{res}_0(\nabla)} & & \downarrow{\scriptstyle -\nu} \\ \mathbb{H}^q(E, \Omega^\bullet_{X/\Delta}(\log E)\otimes \mathcal{O}_E) & \xrightarrow{\mu} & \mathbb{H}^q(E, s(A^{\bullet\bullet})) \end{array}$$

Proof. Let $B^\bullet = \mathrm{Cone}^\bullet(\nu)[-1]$. It becomes a double complex with $B^{pq} = A^{pq} \oplus A^{p,q-1}$. Define maps $\tilde{\mu} : \Omega^q_X(\log E)\otimes \mathcal{O}_E \to B^{0q} = A^{0q}\oplus A^{0,q-1}$ by

$$\tilde{\mu}(\omega) = \big(\omega \wedge (dt/t) \mathrm{mod}\ W_0, (-1)^{q-1}\omega \bmod W_0\big)\,.$$

We obtain the commutative diagram with exact columns, in which the horizontal maps are quasi-isomorphisms

$$\begin{array}{ccc} 0 & & 0 \\ \downarrow & & \downarrow \\ \Omega^\bullet_{X/\Delta}(\log E)\otimes\mathcal{O}_E[-1] & \xrightarrow{\theta} & s(A^{\bullet,\bullet})[-1] \\ \downarrow & & \downarrow \\ \Omega^\bullet_X(\log E)\otimes\mathcal{O}_E & \xrightarrow{\eta} & B^\bullet \\ \downarrow & & \downarrow \\ \Omega^\bullet_{X/\Delta}(\log E)\otimes\mathcal{O}_E & \xrightarrow{\theta} & s(A^{\bullet,\bullet}) \\ \downarrow & & \downarrow \\ 0 & & 0 \end{array}$$

We can conclude the theorem using Corollary 11.17 and the fact that the connecting homomorphism in the long exact hypercohomology sequence of the right hand column is the map induced by $-\nu$ (see Example A.11). □

Note that a suitable Tate twist turns $(a_{r+2k+1})_* \Omega^\bullet_{E(r+2k+1)}[-r-2k]$ into the complex part of a pure Hodge complex of sheaves of weight r. What we need is a similar construction over $\mathbb{Q}$ involving only the weight filtration.

11.2.6 The Rational Structure

To define the rational component of our mixed Hodge complex of sheaves ψ_f^{Hdg}, we imitate the construction of the double complex $A^{\bullet,\bullet}$ on the rational level. Note that the ingredients of the construction of $A^{\bullet,\bullet}$ are:

1) the logarithmic de Rham complex $\Omega^\bullet_X(\log E)$ with its weight filtration W and its multiplication;
2) the global section $d\log t$ of $\Omega^\bullet_X(\log E)$;

In § 4.4 we have defined a rational analogue of the logarithmic de Rham complex with its weight filtration: using the logarithmic structure we constructed the complexes

$$K^\bullet_m \subset \cdots \subset K^\bullet_\infty \subset \cdots$$

where

$$K^q_p = i^* \mathrm{Sym}^{p-q}_{\mathbb{Q}}(\mathcal{O}_X) \otimes \bigwedge^q_{\mathbb{Q}} (\mathcal{M}^{\mathrm{gp}}_{X,D} \otimes_{\mathbb{Z}} \mathbb{Q}),$$

One has multiplications

$$K^q_p \otimes K^s_r \to K^{q+s}_{p+r}\,;$$
$$K^q_\infty \otimes K^s_\infty \to K^{q+s}_\infty\,.$$

We have the global section $\tilde{\theta} = 1 \otimes t \otimes 2\pi\mathrm{i}$ of $K^1_1(1)$, which under $\phi_1 : K_1(1) \to \Omega^1_X(\log E)$ maps to dt/t . This motivates to define the filtered double complex $(C^{\bullet,\bullet}, W(M))$ by

$$C^{p,q} = \left(i^* K^{p+q+1}_\infty / i^* K^{p+q+1}_p \right)(p+1) \text{ for } p \geq 0 \text{ and } p+q \geq -1 \qquad \text{(XI–26)}$$

with differential $\mathrm{d} = d' + d''$ where

$$\left.\begin{array}{ll} d' : \ C^{p,q} \to C^{p+1,q}, & d'(x \otimes y) = x \otimes (t \wedge y) \\ d'' : \ C^{p,q} \to C^{p,q+1}, & d''(x \otimes y) = d_K(x \otimes y), \end{array}\right\} \qquad \text{(XI–27)}$$

where d_K is the differential in the complex $i^* K_\infty$. It carries the filtration

$$W(M)_r C^{p,q} = \ \text{image of } i^* K^{p+q+1}_{r+2p+1}(p+1) \text{ in } C^{p,q}.$$

Referring to the complex monodromy weight filtration (XI–22) and the Hodge filtration (XI–23) the map ϕ induces a filtered quasi-isomorphism

$$(s(C^{\bullet,\bullet}), W(M)) \otimes_{\mathbb{Q}} \mathbb{C} \to (s(A^{\bullet,\bullet}), W(M)) . \tag{XI–28}$$

Note that

$$(1 \otimes t) \cdot W(M)_r C^{p,q} \subset \operatorname{Im}\left[K^{p+q+2}_{r+2p+2}(p+2) \to C^{p+1,q} = W(M)_{r-1} C^{p+1,q}\right]$$

so

$$\operatorname{Gr}_r^{W(M)} s(C^{\bullet,\bullet}) \simeq \bigoplus \left(\operatorname{Gr}^W_{r+2p+1} K^{\bullet}_{n+p+1}\right)(p+1)[1]$$

which underlies a Hodge complex of sheaves of weight $r+2p+1-2(p+1)+1 = r$.

We still have to show that the rational structure on $\psi_f \underline{\mathbb{C}}_X$ defined by the quasi-isomorphism (XI–28) is the same as the one given by $\psi_f \underline{\mathbb{Q}}_X$. To this end we construct a sequence of quasi-isomorphisms similar to that in the proof of Theorem 11.16.

Consider the map $k : X_\infty \to X$. Note that we have a natural quasi-isomorphism

$$\underline{\mathbb{Q}}_{X_\infty} \to k^* K_\infty$$

which induces a quasi-isomorphism

$$\psi_f \underline{\mathbb{Q}}_X \to i^* R k_* k^* K_\infty .$$

The latter complex contains $i^* K_\infty$ and we have the diagram

$$\begin{array}{ccccc} Rk_*k^*K_\infty & \leftarrow & i^*K_\infty & \to & C^{\bullet,\bullet} \\ \downarrow & & \downarrow & & \downarrow \\ \psi_f \underline{\mathbb{C}}_X & \leftarrow & i^*\Omega^\bullet_X(\log E)[u] & \to & A^{\bullet,\bullet} \end{array}$$

in which the vertical arrows become quasi-isomorphisms after tensoring with $\mathbb{C}$ and the bottom row consists of quasi-isomorphisms. Hence also the top row consists of quasi-isomorphisms.

11.2.7 The Mixed Hodge Structure on the Limit

We thus have shown:

Theorem 11.22. *Let $f : X \to \Delta$ be a proper holomorphic map from a complex manifold X to the unit disk in $\mathbb{C}$, smooth over the punctured disk. Suppose that $E = f^{-1}(0)$ is a reduced divisor with simple normal crossings on X and that the irreducible components of E are Kähler. Referring to* (XI–22), (XI–23) *and* (XI–28) *the data*

$$\psi_f^{\mathrm{Hdg}} := (\psi_f \underline{\mathbb{Z}}_X, (s(C^{\bullet,\bullet}), W(M)), (s(A^{\bullet,\bullet}), W(M), F))$$

together with the quasi-isomorphisms constructed above, constitute a marked mixed Hodge complex of sheaves on E.

We write $H^k(X_\infty)$ for the mixed Hodge structure which this complex puts on the hypercohomology group $\mathbb{H}^k(E, \psi_f \underline{\mathbb{Z}}_X)$.

Consider the E_1-term of the weight spectral sequence

$$ {}_{W(M)}E_1^{-r,q+r} = \bigoplus_{k\geq 0,r} H^{q-r-2k}(E(2k+r+1);\mathbb{Q})(-r-k). \tag{XI–29}$$

The restrictions $k \geq 0, -r$ come from (XI–24).

Notation. Following [G-N] we introduce

$$\left.\begin{aligned} K_{\mathbb{Q}}^{ijk} &= H^{i+j-2k+n}(E(2k-i+1);\mathbb{Q})(i-k) \quad \text{if } k\geq 0, i \\ &= 0 \text{ else,} \\ K_{\mathbb{Q}}^{ij} &= \textstyle\bigoplus_k K_{\mathbb{Q}}^{ijk}. \end{aligned}\right\} \tag{XI–30}$$

Observe that the E_1-term of the weight spectral sequence then reads

$$ {}_{W(M)}E_1^{-r,q+r} = \bigoplus_k H^{q-r-2k}(E(2k+r+1);\mathbb{Q})(-r-k) = \bigoplus_k K_{\mathbb{Q}}^{-r,q-n,k}.$$

Corollary 11.23. *Under the same hypotheses we have*

1) *The* ***monodromy weight spectral sequence***

$$ {}_{W(M)}E_1^{-r,q+r} = \bigoplus_k K_{\mathbb{Q}}^{-r,q-n,k} \Rightarrow H^q(X_\infty;\mathbb{Q})$$

degenerates at E_2.

2) *The* ***Hodge spectral sequence***

$$ {}_F E_1^{pq} = H^q(E, \Omega^p_{X/\Delta}(\log E)\otimes \mathcal{O}_E) \Rightarrow H^{p+q}(X_\infty;\mathbb{C})$$

degenerates at E_1.

Corollary 11.24. *Under the same hypotheses, if $\epsilon > 0$ is sufficiently small, then for all $t \in \Delta^*$ with $|t| < \epsilon$ the Hodge spectral sequence*

$$ {}_F E_1^{pq} = H^q(X_t, \Omega^p_{X_t}) \Rightarrow H^{p+q}(X_t;\mathbb{C})$$

degenerates at E_1.

Proof. Define $h^{pq}(t) := \dim H^q(X_t, \Omega^p_{X/\Delta}(\log E)\otimes \mathcal{O}_{X_t})$ for $t\in\Delta$. Take $\epsilon > 0$ so small that $h^{pq}(t) \leq h^{pq}(0)$ for all $t \in \Delta$ with $|t| < \epsilon$ and all $p, q \geq 0$. For $t \neq 0$ one has $\Omega^p_{X/\Delta}(\log E)\otimes\mathcal{O}_{X_t} \simeq \Omega^p_{X_t}$ so by the Hodge spectral sequence we have

$$\sum_{p,q} h^{pq}(t) \geq \sum_k \dim H^k(X_t;\mathbb{C}) \text{ for } t\neq 0$$

with equality if and only if the Hodge spectral sequence for X_t degenerates at E_1. We have

$$\sum_k \dim H^k(X_\infty;\mathbb{C}) = \sum_{p,q} h^{pq}(0) \geq \sum_{p,q} h^{pq}(t) \geq \sum_k \dim H^k(X_t;\mathbb{C})$$

so equality must hold everywhere as $\dim H^k(X_\infty;\mathbb{C}) = \dim H^k(X_t;\mathbb{C})$ by Theorem 11.18. □

Corollary 11.25. *We have* $\dim F^p H^k(X_\infty) = \dim F^p H^k(X_t)$, $t \in \Delta^*$.

Recall (see Def. 3.13) that any mixed Hodge complex of sheaves has a Hodge-Grothendieck class in the Grothendieck ring $K_0(\mathfrak{hs})$ which, when composed with the Hodge number polynomial, yields the Hodge-Euler polynomial. Recall also that $\mathbb{L} = [\mathbb{Q}(-1)] = H^2(\mathbb{P}^1) \in K_0(\mathfrak{hs})$.

Corollary 11.26. *For the Hodge-Grothendieck class we have*

$$\left.\begin{aligned} \chi_{\mathrm{Hdg}}(\psi_f^{\mathrm{Hdg}}) &= \textstyle\sum_{b\geq 1}(-1)^{b-1}\chi_{\mathrm{Hdg}}(E(b))\cdot\left[\sum_{a=0}^{b-1}\mathbb{L}^a\right] \\ &= \textstyle\sum_{b\geq 1}(-1)^{b-1}\chi_{\mathrm{Hdg}}(E(b)\times\mathbb{P}^{b-1}). \end{aligned}\right\} \tag{XI–31}$$

and the Hodge-Euler polynomial is given by

$$e_{\mathrm{Hdg}}(\psi_f^{\mathrm{Hdg}}) = \sum_{b\geq 1}(-1)^{b+1} e_{\mathrm{Hdg}}(E(b))\frac{u^b v^b - 1}{uv - 1}.$$

Proof. We use the monodromy weight spectral sequence (XI–29) to calculate the Hodge-Grothendieck class. We put $r+k = a$, $2k+r+1 = b$, $q-r-2k = c$. Then $a, c \geq 0$, $b \geq 1$, and since $k = b-a-1$, we have the restriction $0 \leq a \leq b-1$. We find:

$$\begin{aligned} \chi_{\mathrm{Hdg}}(\psi_f^{\mathrm{Hdg}}) &= \sum_{b\geq 1, c\geq 0}\sum_{a=0}^{b-1}(-1)^{c+b+1}[H^c(E(b))(-a)] \\ &= \sum_{b\geq 1}(-1)^{b+1}\chi_{\mathrm{Hdg}}(E(b))\cdot\left[\sum_{a=0}^{b-1}\mathbb{L}^a\right]. \end{aligned}$$

The formula for the Hodge-Euler polynomial then follows. □

Remark 11.27. The motivic interpretation of the Euler-Hodge character (Remark 5.56) suggest to introduce the **motivic nearby fibre**

$$\psi_f^{\mathrm{mot}} := \sum_{b\geq 1}(1)^{b-1}E(b)\times\mathbb{P}^{b-1} \in K_0(\mathrm{Var})$$

so that $\chi_{\mathrm{Hdg}}(\psi_f^{\mathrm{mot}}) = \chi_{\mathrm{Hdg}}(\psi_f^{\mathrm{Hdg}})$.

11.3 Geometric Consequences for Degenerations

11.3.1 Monodromy, Specialization and Wang Sequence

We keep the notations and hypotheses from the preceding sections. Recall the map

$$\mathrm{sp} : \underline{\mathbb{Z}}_E = i^*\underline{\mathbb{Z}}_X \to \psi_f\underline{\mathbb{Z}}_X.$$

We will lift this map to the level of mixed Hodge complexes of sheaves.

Note that the endomorphism ν of $s(A^{\bullet,\bullet})$ (XI–25) has its companion on $s(C^{\bullet,\bullet})$:

$$\begin{array}{ccc} \left(i^*K_\infty^{p+q+1}/i^*K_p^{p+q+1}\right)(p+1) & \to & \left(i^*K_\infty^{p+q+1}/i^*K_{p+1}^{p+q+1}\right)(p+1) \\ \| & & \| \\ C^{p,q} & \xrightarrow{\nu} & C^{p+1,q-1}(-1). \end{array} \tag{XI–32}$$

The two companions define a morphism of (marked) mixed $\mathbb{Q}$-Hodge complexes of sheaves $\nu : [\psi_f^{\mathrm{Hdg}}]_{\mathbb{Q}} \to [\psi_f^{\mathrm{Hdg}}]_{\mathbb{Q}}(-1)$. Moreover it follows from Theorem 11.21 and Prop. 11.2 that $N = \log T$ is the map on cohomology induced by $2\pi \mathrm{i}\nu$. We conclude:

Theorem 11.28. *The map*

$$\frac{\log T}{2\pi \mathrm{i}} : H^k(X_\infty;\mathbb{Q}) \to H^k(X_\infty;\mathbb{Q})(-1)$$

is a morphism of mixed $\mathbb{Q}$-Hodge structures.

Consider the mixed $\mathbb{Q}$-Hodge complex of sheaves $\mathrm{Ker}(\nu) \subset [\psi_f^{\mathrm{Hdg}}]_{\mathbb{Q}}$. We extend it to a mixed Hodge complex of sheaves on E by adding $\underline{\mathbb{Z}}_E$ as the integral component, together with the identifications

$$\underline{\mathbb{Z}}_E \otimes \mathbb{Q} = \underline{\mathbb{Q}}_E \simeq i^*\underline{\mathbb{Q}}_X \simeq \mathrm{Ker}(\nu_{\mathbb{Q}}).$$

The $\mathbb{C}$-component of $\mathcal{H}dg^\bullet(E)$ is the double complex $\bigoplus_{p,q\geq 0} \Omega^q_{E(p+1)}$ and $\Omega^q_{E(p+1)}$ is identified with $\mathrm{Ker}(\nu_{\mathbb{C}})^{p,q} = Gr^W_{p+1}\Omega_X^{p+q+1}(\log E)$ by means of the Poincaré residue map. We conclude that we have a morphism of mixed Hodge complexes of sheaves $\mathrm{Ker}(\nu) \to \mathcal{H}dg^\bullet(E)$ which is a quasi-isomorphism on all levels. So the inclusion

$$\mathrm{sp}^{\mathrm{Hdg}} : \mathrm{Ker}(\nu) \to \psi_f^{\mathrm{Hdg}}$$

is a lifting of the specialization map sp to the level of mixed Hodge complexes of sheaves. Consequently we have:

Theorem 11.29. *The* ***specialization map***

$$\mathrm{sp} : H^*(E) \to H^*(X_\infty)$$

is a morphism of mixed Hodge structures.

We extract one interesting consequence of the proof:

Corollary-Definition 11.30. *Define*

$$\phi_f^{\mathrm{Hdg}} := \mathrm{Cone}^\bullet\left(\mathrm{Ker}(\nu) \xrightarrow{\mathrm{sp}^{\mathrm{Hdg}}} \psi_f^{\mathrm{Hdg}}\right).$$

The ***vanishing cohomology*** $\mathbb{H}^*(E, \phi_f^{\mathrm{Hdg}})$ *carries a natural mixed Hodge structure. Its Hodge-Grothendieck class is*

$$\begin{aligned}\chi_{\mathrm{Hdg}}(\phi_f^{\mathrm{Hdg}}) &= \chi_{\mathrm{Hdg}}(\psi_f^{\mathrm{Hdg}}) - \chi_{\mathrm{Hdg}}(E) \\ &= \sum_{b\geq 1}(-1)^{b+1}\chi_{\mathrm{Hdg}}(E(b))\cdot[\mathbb{L}+\cdots+\mathbb{L}^{b-1}]\end{aligned}$$

and its Hodge-Euler polynomial equals

$$e_{\mathrm{Hdg}}(\phi_f^{\mathrm{Hdg}}) = \sum_{b\geq 1}(-1)^{b+1}e_{\mathrm{Hdg}}(E(b))\left[\frac{ub-u^bv^b}{1-uv}\right].$$

Proof. By (III–13) the Hodge-Grothendieck class is the difference $\chi_{\mathrm{Hdg}}(\psi_f^{\mathrm{Hdg}})-\chi_{\mathrm{Hdg}}(E)$. Using (V–16) and (XI–31), the result about the Hodge-Grothendieck class follows. □

Remark 11.31. Continuing remark 11.27, the **motivic vanishing cycle** can be defined as $\phi_f^{\mathrm{mot}} = \psi_f^{\mathrm{mot}} - [E] \in K_0(\mathrm{Var})$. Explicitly

$$\phi_f^{\mathrm{mot}} := \sum_{b\geq 2}(-1)^{b-1}[E(b)\times\mathbb{P}^{b-2}\times\mathbb{A}^1].$$

We continue the Hodge theoretic discussion for the vanishing and nearby cycle functors. The specialization triangle lifts as follows

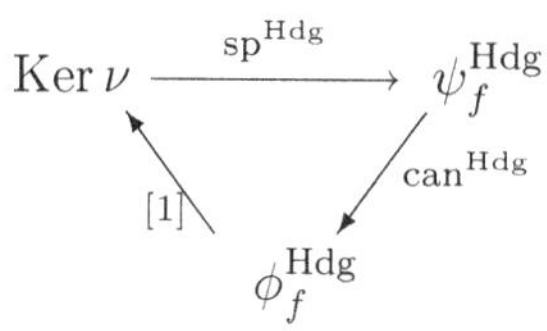

Take local sections x of $\mathrm{Ker}(\nu)[1]$ and y of $[\psi_f^{\mathrm{Hdg}}]_{\mathbb{Q}}$ respectively and define the morphism of rational mixed Hodge complexes

$$\mathrm{V}: [\phi_f^{\mathrm{Hdg}}]_{\mathbb{Q}} \to [\psi_f^{\mathrm{Hdg}}]_{\mathbb{Q}}(-1), \quad \mathrm{V}(x,y) = \nu(y). \tag{XI–33}$$

By construction it satisfies $\mathrm{V}\circ\mathrm{can}^{\mathrm{Hdg}} = \nu$. So it induces morphisms of mixed Hodge structures

$$V: \mathbb{H}^k(E, \phi_f\underline{\mathbb{Q}}_X) \to H^k(X_\infty;\mathbb{Q})(-1)$$

for which $V\circ\mathrm{can} = \nu = N/2\pi\mathrm{i}$. This map fits into the following useful exact sequence:

Theorem 11.32. *We have the long exact sequence of mixed Hodge structures*

$$\to H^k_E(X;\mathbb{Q}) \to \mathbb{H}^k(E,\phi_f\underline{\mathbb{Q}}_X) \xrightarrow{\mathrm{V}} H^k(X_\infty;\mathbb{Q})(-1) \to H^{k+1}_E(X;\mathbb{Q}) \to$$

Proof. Consider the mixed $\mathbb{Q}$-Hodge complex of sheaves $\mathrm{Coker}(\nu)$. We add the integral component $i^*Rj_*\underline{\mathbb{Z}}_{X^*}(1)/\underline{\mathbb{Z}}_X(1)[1]$ to it. We claim that this yields a mixed Hodge complexes of sheaves quasi-isomorphic to $\mathcal{H}dg^\bullet(X, X-E)[1]$. Indeed, $\nu : C^{p,q} \to C^{p+1,q-1}(-1)$ is surjective for $p > 0$ so that

$$\mathrm{Coker}(\nu) = (C^{0,\bullet}(-1), \theta)$$

and $(i^*K_\infty/i^*K_0)[1] \to \mathrm{Coker}(\nu)_{\mathbb{Q}}$ is a quasi-isomorphism. In a similar way we find that $(\Omega^\bullet_X(\log E)/W_0)[1] \to \mathrm{Coker}(\nu)_{\mathbb{C}}$ is a bi-filtered quasi-isomorphism. Note that $\mathrm{Ker}(V)$ is the acyclic complex $\mathrm{Cone}^\bullet(\mathrm{Ker}(\nu) \xrightarrow{\mathrm{id}} \mathrm{Ker}(\nu))$, hence

$$[\phi_f^{\mathrm{Hdg}}]_{\mathbb{Q}} \to \widetilde{[\phi_f^{\mathrm{Hdg}}]}_{\mathbb{Q}} := [\phi_f^{\mathrm{Hdg}}]_{\mathbb{Q}}/\,\mathrm{Ker}(V)$$

is a weak equivalence. By construction we have

$$0 \to \widetilde{[\phi_f^{\mathrm{Hdg}}]}_{\mathbb{Q}} \xrightarrow{\mathrm{V}} [\psi_f^{\mathrm{Hdg}}]_{\mathbb{Q}}(-1) \to \mathrm{Coker}(\nu) \to 0,$$

an exact sequence of mixed $\mathbb{Q}$-Hodge complexes of sheaves. Take the long exact hypercohomology sequence of this and use the calculation of $\mathrm{Coker}(\nu)$ which we just have made. □

The **Wang sequence** is the exact sequence (see [Wang] or [Mil68, Sect. 8])

$$\cdots \to H^k(X^*) \to H^k(X_\infty) \xrightarrow{T-I} H^k(X_\infty) \to H^{k+1}(X^*) \to \cdots$$

In order to obtain an exact sequence of mixed Hodge structures of this kind, we have to modify it, because $T - I$ is not a morphism of mixed Hodge structures. However, N and $T - I$ have the same kernel and cokernel, and $N/2\pi\mathrm{i}$ is a morphism of mixed Hodge structures, induced by the map ν. From ν, (suitably defined over the *rationals*) we shall obtain a long exact sequence of mixed Hodge structures, which is the natural analogue of the Wang sequence.

We first give an alternative for the rational component of $\mathcal{H}dg^\bullet(X^*)$ in terms of logarithmic structures (§ 4.4), analogous to the case of a smooth variety, using the fact that E is the reduced zero set of the holomorphic function $t : X \to \mathbb{C}$. As in § 4.4 the inclusion $i : E \hookrightarrow X$ gives rise to the complex $K^\bullet_\infty = \lim_p K^\bullet_p$ and which is an incarnation of $Rj_*\underline{\mathbb{Q}}_{X^*}$, such that the subcomplex $K^\bullet_p$ is quasi-isomorphic to $\tau_{\le p}Rj_*\underline{\mathbb{Q}}_{X^*}$. Moreover, t is a global section of K^1_1 and as we have seen before induces multiplication maps $\tilde{\theta} : K^r_p \to K^{r+1}_{p+1}$ and $\tilde{\theta} : K^r_\infty \to K^{r+1}_\infty$ by the formula

$$\tilde{\theta}(x \otimes y) := x \otimes (t \wedge y).$$

Moreover

$$K^\bullet_p/K^\bullet_{p-1} \xrightarrow[\sim]{\mathrm{qis}} R^p j_*\underline{\mathbb{Q}}_{T^*}[-p] \simeq (a_p)_*\underline{\mathbb{Q}}_{E(p)}(-p)[-p], \quad p \ge 1.$$

so in defining $\left[\mathcal{H}dg^\bullet(X^*)\right]_{\mathbb{Q}}$ we can replace $(a_p)_*\underline{\mathbb{Q}}_{E(p)}$ by

$$L_p^\bullet := \left(K_p^\bullet / K_{p-1}^\bullet\right)(p)[p].$$

The mapping

$$(a_p)_* \underline{\mathbb{Q}}_{E(p)} \to (a_{p+1})_* \underline{\mathbb{Q}}_{E(p+1)}$$

then corresponds to

$$2\pi \mathrm{i} \tilde{\theta} : L_p^\bullet \to L_{p+1}^\bullet$$

and we take the quotient of the resulting complex

$$\bigoplus_{p \geq 0} i^* L_p^\bullet \oplus i^* K_\infty^\bullet$$

by the image of $i^* K_0^\bullet$ under the map $(-\tilde{\alpha}, \tilde{\beta})$ where $\tilde{\alpha}$ the map induced by $\tilde{\theta}$ and β is the inclusion of $i^* K_0^\bullet$ into $i^* K_\infty^\bullet$. This complex we call $\left[\widetilde{\mathcal{H}dg^\bullet(X^*)}\right]_{\mathbb{Q}}$. It maps naturally to the previously defined $\left[\mathcal{H}dg^\bullet(X^*)\right]_{\mathbb{C}}$, preserving the weight filtration W, and defines hence a mixed $\mathbb{Q}$-Hodge complex of sheaves $\widetilde{\mathcal{H}dg^\bullet(X^*)}$ computing the rational cohomology of X^*.

Also note that we dispose of a commutative diagram

$$\begin{array}{ccc} \left[\widetilde{\mathcal{H}dg^\bullet(X^*)}\right]_{\mathbb{Q}} & \xrightarrow{\rho_{\mathbb{Q}}} & s(C^{\bullet,\bullet}) \\ \downarrow & & \downarrow \\ \left[\mathcal{H}dg^\bullet(X^*)\right]_{\mathbb{C}} & \xrightarrow{\rho_{\mathbb{C}}} & s(A^{\bullet,\bullet}) \end{array}$$

where the map $\rho_{\mathbb{Q}}$ is defined as follows:

- it maps a section ω of $i^* K_\infty^q$ to

$$(-1)^q \tilde{\theta}(\omega) \in i^* K_\infty^{q+1} / i^* K_0^{q+1} = C^{p,0}$$

- it restricts to the inclusion

$$i^* K_{p+1}^{p+q+1} / i^* K_p^{p+q+1} \to C^{p,q}$$

  on the remaining summands.
- these maps add up to zero on the image of K_0;

The map $\rho_{\mathbb{C}}$ is defined analogously. It follows that we get a morphism of marked mixed $\mathbb{Q}$-Hodge complexes of sheaves

$$\rho : \widetilde{\mathcal{H}dg^\bullet(X^*)} \to [\psi_f^{\mathrm{Hdg}}]_{\mathbb{Q}}$$

which induces a morphism of mixed $\mathbb{Q}$-Hodge structures

$$\rho' : H^k(X^*; \mathbb{Q}) \to H^k(X_\infty; \mathbb{Q})$$

which is part of our desired Wang sequence.

Let us compare $\mathrm{Cone}^\bullet(\rho)$ with $[\psi_f^{\mathrm{Hdg}}](-1)$. For simplicity we restrict the treatment to $\mathbb{C}$-coefficients. Define a map $\tilde{\nu} : \mathrm{Cone}^\bullet(\rho_{\mathbb{C}}) \to s(A^{\bullet,\bullet})$ as follows: let

$$((x_{pq}), (y_{rs}), z_{m+1})$$

represent a section of $\mathrm{Cone}^\bullet(\rho_{\mathbb{C}})$, where $p+q=m$, $r+s=m+1$ and

$$x_{pq} \in A^{p,q}, \quad y_{rs} \in W_0 A^{r,s}, \quad z_m \in \Omega_X^{m+1}(\log E) \ .$$

We define

$$\tilde{\nu}(x_{pq}) = \nu(x_{pq}), \quad \tilde{\nu}(y_{rs}) = 0, \quad \tilde{\nu}(z_{m+1}) = (-1)^{m+1} z_{m+1} \bmod W_0 \in A^{0,m}$$

Then clearly, the composition of $\tilde{\nu}$ with the inclusion $s(C^{\bullet,\bullet}) \hookrightarrow \mathrm{Cone}^\bullet(\rho_{\mathbb{C}})$ coincides with ν. Moreover, $\tilde{\nu}$ is a morphism of complexes, compatible with filtrations W and F, whose kernel is acyclic. This shows that $\tilde{\nu}$ is a quasi-isomorphism, and the usual sequence for the cone of ρ then yields the modified Wang sequence we are after. It is left to the reader to check that in fact $\tilde{\nu}$ can also be defined over $\mathbb{Q}$ in a compatible way. This shows:

Theorem 11.33. *The modified Wang sequence*

$$\cdots H^k(X^*;\mathbb{Q}) \to H^k(X_\infty;\mathbb{Q}) \xrightarrow{\nu} H^k(X_\infty;\mathbb{Q})(-1) \to H^{k+1}(X^*;\mathbb{Q}) \cdots \text{(XI–34)}$$

is an exact sequence of mixed $\mathbb{Q}$-Hodge structures.

11.3.2 The Monodromy and Local Invariant Cycle Theorems

We investigate the monodromy weight spectral sequence whose E_1-term, after tensoring with $\mathbb{R}$ can be identified (as real Hodge structures) as follows

$${}_{W(M)}E_1^{-r,q+r} \otimes \mathbb{R} \simeq K_{\mathbb{Q}}^{-r,q-n} \otimes \mathbb{R},$$

with the $K_{\mathbb{Q}}^{i,j}$ as in (XI–30). For simplicity of notation we set

$$K^{ij} = K_{\mathbb{Q}}^{ij} \otimes \mathbb{R}, \quad K^{i,j,k} = K_{\mathbb{Q}}^{i,j,k} \otimes \mathbb{R}$$

so that for $k \geq 0, i$

$$K^{ij} = H^{i+j-2k+n}(E(2k-i+1);\mathbb{R})(i-k)$$

and else $K^{ij}=0$. These K^{ij} come from the single complex associated to the double complex $C^{\bullet,\bullet}$ defined in (XI–26) with differentials (XI–27). It follows that the differential d_1 of the monodromy weight spectral sequence is a sum $d_1 = d_1' + d_1''$ of two morphisms of real Hodge structures induced by the two differentials d', d'':

$$d' : K^{i,j,k} \to K^{i+1,j+1,k+1},$$
$$d'' : K^{i,j,k} \to K^{i+1,j+1,k}.$$

From Proposition 4.10, we see that d' is the restriction map and that $d'' = -\gamma$ with γ the Gysin map. We also recall (XI–32) that the double complex $C^{\bullet,\bullet}$ admits a map ν essentially induced by the identity on the logarithmic De Rham complex. On the level of hypercohomology it induces a morphism of mixed Hodge structures

$$\nu : K^{i,j,k} \to K^{i+2,j,k+1}(-1)$$

which is the identity whenever its source and target are both nonzero (i.e. whenever $k \geq 0$, i). We deduce from this:

Proposition 11.34. *With notations as above, if the components of E are Kähler, then*

1) *for all $i \geq 0$, ν induces an isomorphism $\nu^i : K^{-i,j} \xrightarrow{\sim} K^{i,j}(-i)$;*
2) $\mathrm{Ker}(\nu^{i+1}) \cap K^{-i,j} = K^{-i,j,0}$.

We now suppose in addition that there exists a class $\mu \in H^2(E;\mathbb{R})(1)$ which restricts to a Kähler class on each component of E. We speak of a **one-parameter Kähler degeneration**. Cup product with the restriction of μ to the appropriate intersection of components of E defines mappings

$$\mu : K^{i,j,k} \to K^{i,j+2,k}(1)$$

for all $k \geq 0, i$. Because of our hypothesis, the mappings μ commute with d', d'' and ν. By the hard Lefschetz theorem 1.30, iteration of μ induces isomorphisms

$$\mu^j : K^{i,-j} \simeq K^{i,j}(j)$$

for all $j \geq 0$. Hence, if we define

$$K_0^{-i,-j} := \mathrm{Ker}(\mu^{j+1}) \cap \mathrm{Ker}(\nu^{i+1}) \cap K^{-i,-j}$$

then

$$K_0^{-i,-j} = \mathrm{Ker}(\mu^{j+1}) \cap K^{-i,-j,0} = H^{n-i-j}_{\mathrm{prim}}(E(i+1);\mathbb{R})(-i)$$

and we have a double primitive decomposition

$$K^{r,s} = \sum_{i,j \geq 0} \nu^i \mu^j K_0^{r-2i,s-2j}(i-j).$$

The linear mapping

$$\psi : K^{\bullet,\bullet} \otimes_{\mathbb{R}} K^{\bullet,\bullet} \to \mathbb{R}(-n)$$

defined by (recall that $\varepsilon(a) = (-1)^{a(a-1)/2}$)

$$\psi(x,y) = \begin{cases} \varepsilon(i+j-n)\left(\dfrac{1}{2\pi \mathrm{i}}\right)^{n-2k-i} \displaystyle\int_{E(2k+i+1)} x \wedge y & \text{if } x \in K^{-i,-j,k},\ y \in K^{i,j,k+i} \\ 0 & \text{else.} \end{cases}$$

is a pairing on $K^{\bullet,\bullet}$ which as we shall see, induces a polarization on the primitive parts, using μ and ν:

Theorem 11.35. i) ψ *is a morphism of bigraded real Hodge structures (where* $\mathbb{R}(-n)$ *has bidegree* (n,n)*);*
ii) $\psi(y,x) = (-1)^n \psi(x,y)$;
iii) $\psi(\nu x, y) + \psi(x, \nu y) = 0$;
iv) $\psi(\mu x, y) + \psi(x, \mu y) = 0$;
v) $\psi(d'x, y) = \psi(x, d''y)$;
vi) $\psi(d''x, y) = \psi(x, d'y)$.

Proof. i) Recall (II–2) that for a smooth d-dimensional compact Kähler manifold Z the **trace map** $\mathrm{tr} : H^{2d}(Z;\mathbb{R}) \to \mathbb{R}(-d)$ is given on de Rham representatives by

$$\mathrm{tr}(\omega) = \left(\frac{1}{2\pi \mathrm{i}}\right)^d \int_Z \omega$$

This is an isomorphism of real Hodge structures. As

$$\psi(x,y) = \varepsilon(i+j-n)\,\mathrm{tr}(x \wedge y)$$

and cup product is a morphism of real Hodge structures, ψ is a morphism of real Hodge structures.
ii) For $x \in K^{-i,-j,k}$, $y \in K^{i,j,k+i}$, one has

$$\begin{aligned}\psi(y,x) &= \varepsilon(-i-j-n)\,\mathrm{tr}(y \wedge x) = \varepsilon(-i-j-n)(-1)^{n+i+j}\,\mathrm{tr}(x \wedge y) \\ &= (-1)^n \varepsilon(i+j-n)\,\mathrm{tr}(x \wedge y) = (-1)^n \psi(x,y),\end{aligned}$$

as $y \wedge x = (-1)^{i+j+n} x \wedge y$.
iii) Let $x \in K^{-i,-j-2,k}$, $y \in K^{i,j,k+i}$. Then

$$\begin{aligned}\psi(\mu x, y) &= \varepsilon(i+j-n)\,\mathrm{tr}(\mu x \wedge y) \\ &= -\varepsilon(i+j-n+2)\,\mathrm{tr}(x \wedge \mu y) = -\psi(x, \mu y)\end{aligned}$$

because $x \otimes \mu y \in K^{-i,-j-2,k} \otimes K^{i,j+2,k+i}$ and $\varepsilon(a+2) = -\varepsilon(a)$. This proves iii) and the proof of iv) is similar.
v) For $x \in K^{-i-1,-j-1,k-1}$, $y \in K^{i,j,k+i}$ we have

$$\begin{aligned}\psi(d'x, y) &= \varepsilon(i+j-n)\,\mathrm{tr}(d'x \wedge y) = \varepsilon(i+j-n)\,\mathrm{tr}(x \wedge \gamma y) \\ &= -\varepsilon(i+j-n)\,\mathrm{tr}(x \wedge d''y) = \psi(x, d''y)\end{aligned}$$

because $d'' = -\gamma$ with γ the Gysin map (see Prop. 4.7) which is the transpose of the restriction map d'.
vi) is a consequence of v) and ii).

Proposition 11.36. *The form* $Q : K_0^{-i,-j} \otimes K_0^{-i,-j} \to \mathbb{R}$ *defined by*

$$Q(x,y) = (2\pi \mathrm{i})^{n+i-j} \psi(x, \nu^i \mu^j C y)$$

is symmetric positive definite.

Proof. We can write

$$Q(x,y) = \varepsilon(i+j-n) \int_{E(i+1)} ((2\pi \mathrm{i})^i x \wedge (\mu/(2\pi \mathrm{i})^j C(2\pi \mathrm{i})^i y).$$

Note that $\xi = (2\pi \mathrm{i})^i x, \eta = (2\pi \mathrm{i})^i y \in H^{n-i-j}_{\mathrm{prim}}(E(i+1);\mathbb{R})$, $L = \mu/(2\pi \mathrm{i})$ is the Lefschetz operator on $E = E(i+1)$ and and that the form $(\xi,\eta) \mapsto \varepsilon(n-i-j)\int_E (C\xi \wedge L^j \eta) = \varepsilon(i+j-n)\int_E(\xi \wedge L^j C\eta)$ is positive definite by the classical Hodge-Riemann bilinear relations (Thm. 1.33). □

In section 1.2.2 we have seen that the Lie algebra $\mathfrak{sl}(2,\mathbb{R})$ has the following generators:

$$\ell = \begin{pmatrix} 0 & 0 \\ 1 & 0 \end{pmatrix}, \quad \lambda = \begin{pmatrix} 0 & 1 \\ 0 & 0 \end{pmatrix}, \quad b = \begin{pmatrix} -1 & 0 \\ 0 & 1 \end{pmatrix}.$$

We have two commuting endomorphisms ν and μ on $K^{\bullet,\bullet}$ for which

$$\nu^i : K^{-i,j} \xrightarrow{\cong} K^{i,j} \text{ and } \mu^j : K^{i,-j} \xrightarrow{\cong} K^{i,j}.$$

Therefore $K^{\bullet,\bullet}$ admits a unique representation ρ of $\mathrm{SL}(2,\mathbb{R}) \times \mathrm{SL}(2,\mathbb{R})$ with

$$\begin{aligned} L_1 &:= d\rho(\ell,0) = \nu & d\rho(b,0) &= \text{multiplication with } i \text{ on } K^{i,j}; \\ L_2 &:= d\rho(0,\ell) = \mu & d\rho(0,b) &= \text{multiplication with } j \text{ on } K^{i,j} \end{aligned}$$

We define

$$\begin{aligned} &\Lambda_1 := d\rho(\lambda,0), \\ &\Lambda_2 := d\rho(0,\lambda), \\ &\mathbf{w} := \rho(w,w), \quad \text{where } w = \begin{pmatrix} 0 & 1 \\ -1 & 0 \end{pmatrix}. \end{aligned}$$

Lemma 11.37. *The bilinear form on* $K^{\bullet,\bullet}$ *given by*

$$\phi(x,y) = (2\pi \mathrm{i})^n \psi(x, C\mathbf{w}y)$$

is symmetric and positive definite.

Proof. For $(r,s) \neq (i,j)$, the spaces $K_0^{-i,-j}$ and $K_0^{-r,-s}$ are perpendicular with respect to ϕ. So writing $x = \sum \nu^p \mu^q x_{i-2p,j-2q} \in K^{i,j}$ and $y = \sum \nu^r \mu^s y_{i-2r,j-2s}$ with the $x_{k,\ell} K_0^{k,\ell}$ and $y_{m,n} \in K_0^{m,n}$, using Prop. 1.26 we have

$$\begin{aligned} \psi(x, C\mathbf{w}y) &= \sum (-1)^{p+q} \psi(x_{i-2p,j-2q}, \nu^p \mu^q C\mathbf{w} \nu^r \mu^s y_{i-2r,j-2s}) \\ &= \sum (-1)^{p+q+r+s} \frac{r!s!}{(r-i)!(s-j)!} \psi(x_{i-2p,j-2q}, C\nu^{p+r-i} \mu^{q+s-j} y_{i-2r,j-2s}) \\ &= \sum \frac{p!q!}{(p-i)!(q-j)!} \psi(x_{i-2p,j-2q}, C\nu^{2p-i} \mu^{2q-j} y_{i-2r,j-2s}). \end{aligned}$$

But the left hand side is just the form Q from Prop. 11.36 which states that it is symmetric positive definite on $K_0^{i,j}$ □

Recall that $K^{\bullet,\bullet}$ carries a differential d of bidegree $(1,1)$ which commutes with ν and ℓ. We let d^* denote its adjoint with respect to ϕ.

Lemma 11.38.

$$d^* = \boldsymbol{w}^{-1} d \boldsymbol{w}.$$

Proof.

$$\begin{aligned}\phi(dx, y) &= (2\pi \mathrm{i})^n \psi(dx, C\mathbf{w}y)\\ &= (2\pi \mathrm{i})^n \psi(x, dC\mathbf{w}y)\\ &= \phi(x, (C\mathbf{w})^{-1} dC\mathbf{w}y)\\ &= \phi(x, \mathbf{w}^{-1}d\mathbf{w}y). \qquad \square\end{aligned}$$

We define

$$\Delta : K^{\bullet,\bullet} \to K^{\bullet,\bullet}$$

by

$$\Delta = dd^* + d^* d.$$

Then the inclusion of $\mathrm{Ker}(\Delta)$ into $K^{\bullet,\bullet}$ induces an isomorphism of $\mathrm{Ker}(\Delta)$ with $H(K^{\bullet,\bullet}, d)$.

Theorem 11.39. $\mathrm{Ker}(\Delta)$ *is an invariant subspace of* $K^{\bullet,\bullet}$.

Proof. (See [G-N]). Consider $\mathrm{End}(K^{\bullet,\bullet})$ as a representation space for the group $\mathrm{SL}(2,\mathbb{R}) \times \mathrm{SL}(2,\mathbb{R})$. For representations σ, τ of $\mathrm{SL}(2,\mathbb{R})$ in vector spaces V_1, V_2 we have the representation $\sigma \boxtimes \tau$ in $V_1 \otimes V_2$ given by

$$\sigma \boxtimes \tau(g_1, g_2) = \sigma(g_1) \otimes \tau(g_2).$$

Because $d \in \mathrm{End}(K^{\bullet,\bullet})^{1,1}$ and $L_1(d) = L_2(d) = 0$ (as d commutes with L_1, L_2), the sub representation W of $\mathrm{End}(K^{\bullet,\bullet})$ generated by d is isomorphic to $\rho_1 \boxtimes \rho_1$, and d is a dominant vector. Hence $d^* = \mathbf{w}^{-1}(d) = \Lambda_1 \Lambda_2(d)$.

Consider the composition map

$$c : W \otimes W \to \mathrm{End}(K^{\bullet,\bullet}),\ f \otimes g \mapsto f \circ g.$$

Then c is equivariant with respect to the action of $\mathrm{SL}(2,\mathbb{R}) \times \mathrm{SL}(2,\mathbb{R})$. Therefore, to show that $\Delta = c(d^* \otimes d + d \otimes d^*)$ is invariant, it suffices to show that it is the image under c of an invariant element of $W \otimes W$. Note that $d^2 = 0$ so $d \otimes d \in \mathrm{Ker}(c)$. Hence $\mathrm{Ker}(c)$ contains the whole $\mathfrak{sl}(2,\mathbb{R}) \times \mathfrak{sl}(2,\mathbb{R})$-orbit of $d \otimes d$. Note that

$$\begin{aligned}\Lambda_1 \Lambda_2(d \otimes d) &= \Lambda_1(\Lambda_2(d) \otimes d + d \otimes \Lambda_2(d)\\ &= d \otimes d^* + d^* \otimes d + \Lambda_1(d) \otimes \Lambda_2(d) + \Lambda_2(d) \otimes \Lambda_1(d)\end{aligned}$$

lies in the kernel of c and the element

$$d \otimes d^* + d^* \otimes d - \Lambda_1(d) \otimes \Lambda_2(d) - \Lambda_2(d) \otimes \Lambda_1(d)$$

is invariant. Hence $d \otimes d^* + d^* \otimes d$ maps to an invariant element under c. $\square$

As the cohomology of the complex $(K^{\bullet,\bullet}, d)$ is just the E_2-term of the weight spectral sequence, which degenerates at the E_2-term, we conclude:

Theorem 11.40. *The map ν induces isomorphisms*

$$\nu^r : \mathrm{Gr}^{W(M)}_{k+r} H^k(X_\infty;\mathbb{Q}) \xrightarrow{\cong} \mathrm{Gr}^{W(M)}_{k-r} H^k(X_\infty;\mathbb{Q})(-r)$$

and the weight filtration $W(M)$ on $H^k(X_\infty;\mathbb{Q})$ coincides with the weight filtration of $N = \log T$ centred at k.

Corollary 11.41. *The filtration induced by $W(M)$ on $H^k(X_\infty, \mathbb{Q})$ coincides with the weight filtration $W(N,k)$ of $N = \log(T_u)$ centred at k (see Lemma-Definition 11.9).*

Corollary 11.42 (Monodromy Theorem)**.** *Suppose that for integers k, ℓ one has $H^{p,k-p}(X_t;\mathbb{C}) = 0$ for all $p > k/2+\ell$. Then $N^{\ell+1} = 0$ on $H^k(X_\infty;\mathbb{C})$.*

Proof. As N is a morphism of mixed Hodge structures of type $(-1,-1)$, its powers are strictly compatible with the Hodge and weight filtrations. Therefore it suffices to show that

$$N^{\ell+1} : \mathrm{Gr}^r_F H^k(X_\infty;\mathbb{C}) \to \mathrm{Gr}^{r-\ell-1}_F H^k(X_\infty;\mathbb{C})$$

is the zero map for all r. This is clearly the case, because the conditions on k and ℓ imply that either $\mathrm{Gr}^r_F H^k(X_\infty;\mathbb{C}) = 0$ or $\mathrm{Gr}^{r-\ell-1}_F H^k(X_\infty;\mathbb{C}) = 0$ (or both).

Theorem 11.43 (Local invariant cycle theorem)**.** *Let $X \to \Delta$ be a Kähler degeneration over a disk Δ centred at 0. For all $k \geq 0$ the sequence*

$$H^k(E;\mathbb{Q}) \xrightarrow{\mathrm{sp}} H^k(X_\infty;\mathbb{Q}) \xrightarrow{T-I} H^k(X_\infty;\mathbb{Q})$$

is exact. Concretely: the invariant classes in the generic fibre X_∞ are the classes in the image of the specialization map, i.e. the classes which are the restrictions from classes on the total space (provided Δ is small enough).

Proof (Deligne). As $N = 2\pi \mathrm{i}\nu$ we have $\mathrm{Ker}(N) = \mathrm{Ker}(T-I) = \mathrm{Ker}(\nu)$. So it suffices to show that the sequence

$$H^k(E;\mathbb{Q}) \xrightarrow{\mathrm{sp}} H^k(X_\infty;\mathbb{Q}) \xrightarrow{\nu} H^k(X_\infty;\mathbb{Q})(-1)$$

is an exact sequence of mixed Hodge structures. The specialization map is induced by the retraction $r : X \to X_0$, i.e. $\mathrm{sp} : H^k(E) \simeq H^k(X) \to H^k(X_\infty)$. Hence the following commutative diagram with an exact row

$$\begin{array}{ccccc} H^k(E;\mathbb{Q}) & \xrightarrow{\alpha} & H^k(X^*;\mathbb{Q}) & \to & H^{k+1}_E(X;\mathbb{Q}) \\ & \searrow^{\mathrm{sp}} & \downarrow{\scriptstyle\beta} & & \\ & & \mathrm{Ker}(\nu) & & \end{array}$$

where X^* is considered as a deleted neighbourhood of E. By the Wang sequence (11.33) we see that β is surjective. Moreover, by Theorem 11.40 $\operatorname{Ker}(\nu) = W_k \operatorname{Ker}(\nu)$. So the restriction of β to $W_k H^k(X - E;\mathbb{Q})$ is also surjective. But $H^k(E;\mathbb{Q})$ maps surjectively to $W_k H^k(X^*;\mathbb{Q})$, as Corr. 6.28 tells us that $H_E^{k+1}(X;\mathbb{Q})$ has weights $\geq k+1$. Hence $\beta \circ \alpha$ is surjective. □

A rather involved diagram chase can be used to show the following consequence.

Corollary 11.44 (The Clemens-Schmid exact sequence). *Combine the exact sequence for the pair (X, X^*) and the modified Wang sequence* (XI–34) *into the following commutative diagram of mixed Hodge structures where the dashed arrows are the compositions of the two obvious maps. Then the two horizontal sequences are exact.*

$$\begin{array}{ccccccccc}
\cdots \to H^k(X) & \dashrightarrow & H^k(X_\infty) & \xrightarrow{\nu} & H^k(X_\infty)(-1) & \dashrightarrow & H^{k+2}(X,X^*) \to \cdots \\
\searrow & & \nearrow & & \searrow & & \nearrow \\
& H^k(X^*) & & & & H^{k+1}(X^*) & \\
\nearrow & & \searrow & & \nearrow & & \searrow \\
\cdots \xrightarrow{\nu} H^{k-1}(X_\infty) & \dashrightarrow & H^{k+1}(X,X^*) & \to & H^{k+1}(X) & \dashrightarrow & H^{k+1}(X_\infty) \xrightarrow{\nu} \cdots
\end{array}$$

11.4 Examples

We shall give some examples which show how knowledge of the central fibre and the pure Hodge structure on the cohomology of a smooth fibre X_t of f can be used to figure out the Hodge numbers of the limiting mixed Hodge structure.

As noted before (Lemma-Def. 11.9), the weight filtration on $H^k(X_\infty;\mathbb{Q})$ is uniquely determined by the monodromy, and the length of a Jordan bloc of $N = \log T$ of size ℓ is the dimension of the primitive space weight ℓ subspace of $H^k(X_\infty;\mathbb{Q})$.

We also know (Lemma 11.25) that the Hodge filtration spaces of the limiting mixed Hodge structure and the ordinary pure Hodge structure have the same dimension, i.e. $\dim F^m H^k(X_t) = \dim F^m H^k(X_\infty)$ and hence

$$e_{\mathrm{Hdg}}(\psi_f)_{|v=1} = e_{\mathrm{Hdg}}(X_t)_{|v=1}. \qquad \text{(XI–35)}$$

In the examples we have $H^{2k}(X_\infty) = \mathbb{L}^k$ for $k \neq n = \dim X_t$. On the middle cohomology we have (see Def. 11.9):

$$\operatorname{Gr}^W H^n(X_\infty) = \bigoplus_{\ell=0}^{n} \bigoplus_{r=0}^{\ell} N^r [PH^n(X_\infty)]_{n+\ell}$$

where $[P^n((X_\infty)]_\ell$ is pure of weight ℓ. The number $p_\ell = \dim[P^n((X_\infty)]_\ell$ of Jordan blocs of size $\ell+1$ appears in the dimension formula

$$\dim \operatorname{Gr}_\ell^W = \sum_{k\geq 0} p_{\ell+2k}. \tag{XI–36}$$

The latter can be determined from the formula

$$\chi_{\mathrm{Hdg}}(\psi_f)|_{u=v=t} = (-1)^n \sum_{\ell=n}^{2n} \dim \operatorname{Gr}_\ell^W t^\ell + (1+t^2+\cdots t^{2n}) + \gamma_n t^n, \tag{XI–37}$$

where $\gamma_n = \frac{1}{2}(-1+(-1)^{n+1})$. The last term is non-zero only if n is even and corrects the fact that in this case t^n should not be present in the terms taking care of the cohomology in degrees $\neq n$.

Examples 11.45. 1) Let $F, L_1, \ldots, L_d \in \mathbb{C}[X_0, X_1, X_2]$ be homogeneous forms with $\deg F = d$ and $\deg L_i = 1$ for $i = 1, \ldots, d$, such that $F \cdot L_1 \cdots L_d = 0$ defines a reduced divisor with normal crossings on $\mathbb{P}^2(\mathbb{C})$. We consider the space

$$X = \{([x_0, x_1, x_2], t) \in \mathbb{P}^2 \times \Delta \mid \prod_{i=1}^d L_i(x_0, x_1, x_2) + tF(x_0, x_1, x_2) = 0\}$$

where Δ is a small disk around $0 \in \mathbb{C}$. Then X is smooth and the map $f : X \to \Delta$ given by the projection to the second factor has as its zero fibre the union $E_1 \cup \cdots \cup E_d$ of the lines $E_i : L_i = 0$. These lines are in general position. The formula (XI–31) gives us

$$e_{\mathrm{Hdg}}(\psi_f) = d(1+uv) - \binom{d}{2}(1+uv) = (1 - \binom{d-1}{2})(1+uv).$$

The general fibre X_t is a smooth projective curve of degree d. Substituting $v = 1$ in the preceding formula gives indeed $g = \binom{d-1}{2}$, the genus of a smooth plane curve of degree d. Setting $u = v = t$ we see that there are only even weight terms for $H^1(X_\infty)$ and its only primitive subspace has weight 2 and dimension g (since $\dim H^1(X_\infty) = 2p_2 = 2g$) and in particular N has g Jordan blocs of size 2, i.e. is "maximally unipotent".

2) If we consider the same example, but replace $\mathbb{P}^2$ by $\mathbb{P}^3$ and curves by surfaces, lines by planes, then the space X will not be smooth but has ordinary double points at the points of the zero fibre where two of the planes meet the surface $F = 0$. There are $d\binom{d}{2}$ of such points, d on each line of intersection. If we blow these up, we obtain a family $f : X_\infty \to \Delta$ whose zero fibre $D = E \cup F$ is the union of components E_i, $i = 1, \ldots, d$ which are copies of $\mathbb{P}^2$ blown up in $d(d-1)$ points, and components $F_j, j = 1, \ldots, d\binom{d}{2}$ which are copies of $\mathbb{P}^1 \times \mathbb{P}^1$. Thus

$$e_{\mathrm{Hdg}}(D(1)) = d(1 + (d^2 - d + 1)uv + u^2v^2) + d\binom{d}{2}(1+uv)^2.$$

The double point locus $D(2)$ consists of the $\binom{d}{2}$ lines of intersections of the E_i together with the $d^2(d-1)$ exceptional lines in the E_i. So

$$e_{\mathrm{Hdg}}(D(2)) = d(d-1)(d+\frac{1}{2})(1+uv).$$

Finally $D(3)$ consists of the $\binom{d}{3}$ intersection points of the E_i together with one point on each component F_j, so

$$\chi_{\mathrm{Hdg}}(D(3)) = \binom{d}{3} + d\binom{d}{2} = \frac{1}{3}d(d-1)(2d-1).$$

We get

$$e_{\mathrm{Hdg}}(\psi_f) = \left(\binom{d-1}{3}+1\right)(1+u^2v^2) + \frac{1}{3}d(2d^2-6d+7)uv.$$

in accordance with the Hodge numbers for a smooth degree d surface:

$$h^{2,0} = h^{0,2} = \binom{d-1}{3}, \quad h^{1,1} = \frac{1}{3}d(2d^2-6d+7).$$

The formula (XI–37) shows that there are only weight 4 and 2 primitive spaces, that $\dim \mathrm{Gr}^2_W = \frac{1}{3}d(2d^2-6d+7)$ and $\dim W_4 = \binom{d-1}{3}$. It follows that the monodromy on $H^2(X_\infty)$ has $\binom{d-1}{3}$ Jordan blocs of size 3 and $\frac{1}{2}d^3 - d^2 + \frac{1}{2}d + 1$ blocks of size 1.

3) Consider a similar smoothing of the union of two transverse quadrics in $\mathbb{P}^3$: $Q_1Q_2 + tF_4 = 0$. The generic fibre is a smooth K3-surface and after blowing up the 16 double points $\Sigma = \{t = Q_1 = Q_2 = F_4 = 0\}$ of the total space of the family (a hypersurface inside $\Delta \times \mathbb{P}^3$); the special fibre consists of eighteen smooth components which intersect transversally according to the following pattern:

i) $E(1)$ has two components which are blowings up of Q_i in Σ, $i = 1, 2$, and 16 exceptional divisors isomorphic to $\mathbb{P}^1 \times \mathbb{P}^1$; hence $e_{\mathrm{Hdg}}(E(1)) = 18(1+uv)^2 + 32uv$.

ii) $E(2)$ consists of the 32 exceptional lines together with the strict transform of the intersection of the two quadrics, which is an elliptic curve; hence $e_{\mathrm{Hdg}}(E(2)) = 33(1+uv) - u - v$;

iii) $E(3)$ consists of 16 points: one point on each exceptional $\mathbb{P}^1 \times \mathbb{P}^1$, so $e_{\mathrm{Hdg}}(E(3)) = 16$.

We get

$$e_{\mathrm{Hdg}}(\psi_f) = 1 + u + v + 18uv + uv^2 + uv^2 + u^2v^2$$

Putting $v = 1$ we get $2 + 20u + 2u^2$, in agreement with the Hodge numbers $(1, 20, 1)$ on the H^2 of a K3-surface. Setting $u = v = t$, formula (XI–31) shows that $\dim \mathrm{Gr}^1_W = \dim \mathrm{Gr}^3_W = 2$, $\dim \mathrm{Gr}^2_W = 18$, $\dim \mathrm{Gr}^4_W = \dim W_0 = 0$. Hence the weight filtration is $W_4 = W_3 \supset W_2 \supset W_1 \supset W_0 = 0$ which implies that $N^2 = 0$, and since the non-trivial gradeds have dimensions $2, 18, 2$, the monodromy has two Jordan blocs of size 2 and 18 blocs of size 1.

12

Applications of Asymptotic Hodge theory

Mixed Hodge theory has many applications to singularity theory. For instance, we show in § 12.1 that the cohomology of the Milnor fibre admits a mixed Hodge structure and we give applications to the spectrum. Here the full force of our geometric approach becomes clear since, in contrast to Schmid's theory, we can localize around a singularity.

In § 12.2 we explain Grothendieck's induction principle, the proof of which uses variations of Hodge structures.

12.1 Applications to Singularities

12.1.1 Localizing Nearby Cycles

The main difference between the techniques used by W. Schmid and those in Chapter 11 is that we have a local Hodge-theoretic description of the *sheaf* of nearby cycles. This enables one to construct mixed Hodge structures on cohomology groups of certain subsets of the general fibre which "are near" to certain subsets of the special fibre. This is particularly interesting for unions of components of the special fibre.

Let $f : X \to S$ be a proper holomorphic map with X a complex manifold of dimension $n+1$ and S the unit disk in $\mathbb{C}$, such that $E = f^{-1}(0)$ is a reduced divisor with strict normal crossings. Write E as the union of its irreducible components:

$$E = \bigcup_{i \in I} E_i.$$

For $J \subset I$ we define

$$E(J) = \bigcup_{i \in J} E_i \xrightarrow{v_J} E; \; E'(J) = E - E(I-J) \xrightarrow{u_J} E.$$

Proposition 12.1. *Let $U(J)$ be a tubular neighbourhood of $E(J)$ in X. For $t \in S - \{0\}$ sufficiently small we have*

$$\mathbb{H}^k(E(J), (v_J)^* \psi_f \underline{\mathbb{Q}}_X) \simeq H^k(U(J) \cap X_t; \mathbb{Q});$$
$$\mathbb{H}^k(E(J), (u_J)_! \psi_f \underline{\mathbb{Q}}_X) \simeq H^k_c(U(J) \cap X_t; \mathbb{Q}).$$

Proof. As $U(J)$ is a tubular neighbourhood of $E(J)$ in X, it has a retraction onto $E(J)$, and the inclusion of $E(J)$ into $E \cap U(J)$ is a homotopy equivalence. □

We want to define a mixed Hodge complex of sheaves whose rational component is quasi-isomorphic to $(v_J)^* \psi_f \underline{\mathbb{Q}}_X$. This problem is analogous to restricting the logarithmic de Rham complex to a union of components of the divisor, which we did in § 6.3.2 and which produced the mixed Hodge complex of sheaves $\mathcal{K}_{\mathrm{DR}}(C \log D)$. Here the construction is similar.

For any (smooth) intersection Z of components of E we have the inclusion $\iota_Z : Z \to E$ and we define

$$\iota_Z^* A^{p,q} = \Omega_X^{p+q+1}(\log E) \otimes \mathcal{O}_Z / W_p.$$

Then $\iota_Z^* A^{\bullet,\bullet}$ is quasi-isomorphic to $\Omega^\bullet_{X/S}(\log E) \otimes \mathcal{O}_Z$. It inherits the filtrations F and $W(M)$.

On the rational level we have a similar construction. Instead of $\Omega^\bullet_X(\log E)$ we consider K_∞ (see § 4.4) with its filtration W, and restrict it (topologically) to Z. In this way we obtain

$$\iota_Z^* C^{p,q} = \begin{cases} \iota_Z^* K_\infty^{p+q+1}(p+1)/W_{p-1} & \text{if } \; p \geq 0, \; p+q+1 \geq 0; \\ 0 & \text{else} \end{cases} .$$

The complex $\iota_Z^* C^{\bullet,\bullet}$ inherits the filtration $W(M)$. Let us write

$$E(k) \cap Z := \bigsqcup_{J \subset I, \sharp(J) = k} E(J) \cap Z.$$

Then

$$Gr_r^{W(M)} \iota_Z^* C^\bullet \simeq \bigoplus_k (a_{r+2k+1})_* \underline{\mathbb{Q}}_{E(r+2k+1) \cap Z}(-r-k)[-r-2k].$$

We have a similar relation between $s(\iota_Z^* A^{\bullet,\bullet}, W(M))$ and $s(\iota_Z^* C^{\bullet,\bullet}, W(M)) \otimes \mathbb{C}$ as in the global case and so we obtain for each Z a mixed Hodge complex of sheaves $\iota_Z^* \psi_F^H \underline{\mathbb{Z}}_X$.

For an inclusion $Z' \subset Z$ of closed strata of E we have a restriction map $\iota_Z^* \psi_F^H \underline{\mathbb{Z}}_X \to \iota_{Z'}^* \psi_F^H \underline{\mathbb{Z}}_X$. If Z runs over the smooth intersections of $E(J)$, we obtain a J-cocubical system of mixed Hodge complexes of sheaves, the total complex of which we denote by $v_J^* \psi_f \underline{\mathbb{Q}}_X^H$. Its underlying rational complex is quasi-isomorphic to $v_J^* \psi_f \underline{\mathbb{Q}}_X$. Hence we obtain a mixed Hodge structure on $H^k(X_\infty \cap U(J); \mathbb{Q})$.

Proposition 12.2.

$$\mathrm{Gr}_F^p H^k(X_\infty \cap U(J); \mathbb{C}) \simeq H^{k-p}(E(J), \Omega^p_{X/S}(\log E) \otimes \mathcal{O}_{E(J)}).$$

Proof. We have a filtered quasi-isomorphism

$$(\Omega^p_{X/S}(\log E) \otimes \mathcal{O}_{E(J)}, F) \to (v_J^* A^{\bullet,\bullet}, F)$$

and the Hodge spectral sequence degenerates at E_1. □

12.1.2 A Mixed Hodge Structure on the Cohomology of Milnor Fibres

In order to put a mixed Hodge structure on the cohomology of the Milnor fibre we apply the localisation procedure of § 12.1.1 to the exceptional divisor coming from blowing up a singular point on the special fibre of a degeneration. More generally, consider $g : Y \to \Delta$, a flat projective map of relative dimension n (so Y may be singular), and let C be a closed subvariety of $g^{-1}(0)$. We shall suppose however that g is as close to a genuine degeneration as possible: 0 is the only critical value of g and $Y - g^{-1}(0)$ is smooth. Indeed, blowing up along C, extracting roots and doing possibly some further blowings up brings us in the situation of § 12.1.1. To be precise, we have the following variant of Thm. 11.11:

Theorem 12.3. *There exist a positive integer m and a commutative diagram*

$$\begin{array}{ccc} X & \xrightarrow{\rho} & Y \\ \downarrow f & & \downarrow g \\ \Delta' & \xrightarrow{\mu} & \Delta \end{array}$$

with Δ' is a disk in $\mathbb{C}$ centred at 0, $\mu(t) = t^m$, X is smooth, f is projective with special fibre $E = f^{-1}(0) = \bigcup_{i \in I} E_i$, a reduced divisor with strict normal crossings on X. Moreover, the inverse image of C under ρ is a union of components of E and the automorphism $t \mapsto \exp(2\pi \mathrm{i}/m)t$ of Δ' lifts to X in such a way that it preserves fibres of ρ.

Proof. First blow up the subvariety C inside Y to reduce to the case that C is a divisor. Then blow up Y in order to reduce to the case when Y is smooth and apply Thm. 11.11. Inspection of the proof of this theorem given in [K-K-M-S] shows that we can extract roots in such a way that the components of the special fibre become nonsingular and that the required lift exists. □

Remark 12.4. In [Ste77a] the semistable reduction theorem has been avoided, at the cost of admitting quotient singularities in the special fibre. In that approach, first Y is resolved to obtain $\tilde{Y} \to S$ such that the special fibre $\tilde{Y}_0$ becomes a (non-reduced) divisor with (strict) normal crossings, one chooses m which is a common multiple of all multiplicities of the components of the special fibre, and considers the normalization of the fibre product $\tilde{Y} \times_S S'$.

Notation. With notations as in the previous theorem, we let

$$\iota_C^* \psi_g^H \underline{\mathbb{Q}}_Y := R\rho_* v_J^* \psi_f^H \underline{\mathbb{Q}}_X.$$

It is a mixed Hodge complex of sheaves on C. In the special case that C is a one-point set $\{y\}$ we just write $\psi_{g,y}^H \underline{\mathbb{Q}}_Y$.

Observe that $\psi_{g,y}\underline{\mathbb{Q}}_Y$ is nothing but the stalk at y of the complex of sheaves $\psi_g \underline{\mathbb{Q}}_Y$. Using (C–7), we deduce, and this is the crucial observation, that $H^k(\psi_{g,y}^H \underline{\mathbb{Q}}_Y) = H^k(\mathrm{Mil}_{g,y}; \mathbb{Q})$. This implies:

Lemma-Definition 12.5. *Let $g : Y \to \Delta$ be a holomorphic map of complex manifolds, smooth away from the origin. Let $y \in g^{-1}(0)$ and let $\mathrm{Mil}_{g,y}$ the corresponding Milnor fibre. The complex $\psi_{g,y}^H \underline{\mathbb{Q}}_Y$ is a mixed Hodge complex of sheaves on $\mathrm{Mil}_{g,y}$ endowing $H^k(\mathrm{Mil}_{g,y}; \mathbb{Q})$ with a mixed Hodge structure.*

In fact, every smoothing of an isolated singularity can be globalized i.e. be put in a family as above (see [Ste95bis]). It follows from the local nature of the constructions that the resulting mixed Hodge structures on the cohomology groups depend only on the local smoothing, i.e. on the germ of the holomorphic map $g : (Y, y) \to (S, 0)$, and not on the choice of globalization. In this way one obtains invariants of the smoothing from the mixed Hodge structure on the cohomology of its Milnor fibre.

Example 12.6 (ordinary multiple point). Suppose $g : Y \to S$ is a projective family of relative dimension n over the disk S with Y smooth, g smooth outside a point $y_0 \in Y$ where g has an isolated singularity and such that $g(y_0) = 0$. We assume that y_0 is an ordinary multiple point of g. This means that the projectivized tangent cone of $g^{-1}(0)$ at y_0 is nonsingular. Let m denote the multiplicity of $g^{-1}(0)$ at y_0. We form the fibre product $Y' = Y \times_S S'$ where $S' \to S$ is the m-fold ramified cover $t \mapsto t^m$. Then Y' has the unique ordinary multiple point $(y_0, 0)$. Let X denote the blowing up of Y' with centre $(y_0, 0)$, and let $f : X \to S'$, $\rho : X \to Y$ denote the induced maps. The fibre $f^{-1}(0)$ has two components: the strict transform E_1 of the special fibre of g and the exceptional component $E_2 = \rho^{-1}(y_0)$, which is isomorphic to a smooth hypersurface of degree m in $\mathbb{P}^{n+1}(\mathbb{C})$ and which contains $E_1 \cap E_2$ as a nonsingular hyperplane section.

The E_1-term of the weight spectral sequence in this case only has the contributions

$$H^{q-2}(E_1 \cap E_2)(-1) \xrightarrow{d_1^{-1,q}} H^q(E_1) \oplus H^q(E_2) \xrightarrow{d_1^{0,q}} H^q(E_1 \cap E_2).$$

One has $\mathrm{Gr}_W^{q+1} H^q(X_\infty) = \mathrm{Ker}(d_1^{-1,q+1})$, $\mathrm{Gr}_W^{q-1} H^q(X_\infty) = \mathrm{Coker}(d_1^{0,q-1})$ and $\mathrm{Gr}_W^q H^q(X_\infty) = \mathrm{Ker}(d_1^{0,q}) / \mathrm{Im}(d_1^{-1,q})$. The Gysin map $H^{q-2}(E_1 \cap E_2)(-1) \to H^q(E_2)$ is injective unless $q = n+1$ and the restriction map $H^q(E_2) \to H^q(E_1 \cap E_2)$ is surjective unless $q = n-1$ by the Lefschetz hyperplane theorem. Hence the only possibly non-zero E_2-terms are the $E_2^{0,q}$ for $0 \le q \le 2n$, $E_2^{-1,n+1}$, and $E_2^{1,n-1}$. We see from this that $H^q(X_\infty)$ is pure of weight q unless $q = n$, in which case at most the weights $n-1, n, n+1$ occur.

12.1.3 The Spectrum of Singularities

Let (V, F, γ) be a triple consisting of a finite-dimensional complex vectorspace V provided with a decreasing filtration F and an automorphism γ of finite order preserving F. There is a decomposition

$$(V, F) = \bigoplus_{-1<a\leq 0} (V_a, F) \quad V_a = \mathrm{Ker}(\gamma - \exp(-2\pi i a)).$$

Observe that the numbers a which occur are rational numbers, so that for a given integer n we can introduce the following invariant in the group ring of the rational numbers:

$$\mathrm{Sp}_n(V, F, \gamma) := \sum_a \sum_p [\dim \mathrm{Gr}_F^p V_a] \cdot (n - p + a) \in \mathbb{Z}[\mathbb{Q}].$$

A word of warning: in the group ring $\mathbb{Z}[\mathbb{Q}]$ we have the rule: $(a) \cdot (b) = (a + b)$ whenever $a, b \in \mathbb{Q}$. Also (0) is the unit in this ring.

Apply this in the following setting. Let $g : Y \to \mathbb{C}$ be a regular function on an n-dimensional algebraic variety Y; for $x \in Y$ with $g(x) = 0$ we take for V the i-th cohomology at x of the complex of sheaves $\psi_g^H \underline{\mathbb{Q}}_Y$, for F the Hodge filtration and for $\gamma = T_s$, the semi-simple part of the monodromy. In other words, letting $i_x : x \hookrightarrow Y$ be the inclusion, we put

$$\mathrm{Sp}^i(g, x) := \mathrm{Sp}_n(\mathbb{H}^i(i_x^* \psi_g^H \underline{\mathbb{Q}}_Y), F, T_s)$$

which, after taking a suitable alternating sum yields the invariant

$$\mathrm{Sp}(g, x) := (-1)^n [\sum_i (-1)^i \mathrm{Sp}^i(g, x) - (0)],$$

the **spectrum** of the singularity (g, x).

Example 12.7. Let $g : \mathbb{C} \to \mathbb{C}$ be given by $g(z) = z^{n+1}$. Then X is the normalization of the fibre product $Y \times_S S'$ with $S' = \mathbb{C} \to S = \mathbb{C}$ given by $t \mapsto t^{n+1}$. So X is the disjoint union of $n + 1$ copies of $\mathbb{C}$ each of which is mapped isomorphically to S' by f, and γ permutes these components cyclically. Hence

$$H^0(i_x^* \psi_g^H \underline{\mathbb{Q}}_Y) = \mathbb{Q}^{n+1}$$

and γ is the cyclic permutation of its basis vectors. Hence

$$\mathrm{Sp}(g, x) = \sum_{i=1}^{n} \left(-\frac{i}{n+1} \right).$$

We list a number of interesting properties of the spectrum.

1) **Symmetry**: if Y is smooth of dimension $n + 1$ and $x \in Y$ is an isolated critical point of $f : Y \to \mathbb{C}$ with $f(x) = 0$, then $\mathrm{Sp}(f, x)$ is invariant under the automorphism of the group ring $\mathbb{Z}[\mathbb{Q}]$ given by $(b) \mapsto (n - 1 - b)$.

2) **The Thom-Sebastiani property**: for function germs $g : (Y, y) \to (\mathbb{C}, 0)$ and $h : (Z, z) \to (\mathbb{C}, 0)$ we have the germ $g \oplus h : (Y \times Z, (y, z)) \to (\mathbb{C}, 0)$ given by $(g \oplus h)(u, v) = g(u) + h(v)$. With this notation one has

$$\mathrm{Sp}(g \oplus h, (y, z)) = (1) \cdot \mathrm{Sp}(g, y) \cdot \mathrm{Sp}(h, z).$$

See [Var81, Theorem 7.3] (or also [SchS85]) for the isolated hypersurface case and [DL99] for the general case. (M. Saito proved an even more general theorem for mixed Hodge modules).

3) **Semi-continuity**. This deals exclusively with case of isolated hypersurfaces. For any subset $A \subset \mathbb{Q}$ we have a group homomorphism

$$\deg_A : \mathbb{Z}[\mathbb{Q}] \to \mathbb{Z}, \quad \sum_b n_b \cdot (b) \mapsto \sum_{b \in A} n_b.$$

By a result of M. Saito [Sa83] the geometric genus of $(X, x) = f^{-1}(0)$ is equal to $\deg_{(-1,0]} \mathrm{Sp}(f, x)$. This invariant is upper semicontinuous under deformation: if the function f is deformed such that the singularity x splits into several singular points $x_1, \ldots, x_k$ all occurring in one fibre, then the sum of the geometric genera of the singular points $x_1, \ldots, x_k$ is not more than the geometric genus of x. This is a special case of

Theorem 12.8. *If the function f is deformed to a function f_t such that the singularity x splits into several singular points $x_1, \ldots, x_k$ all occurring in one fibre $f_t^{-1}(s)$, then for each half open interval I of length one:*

$$\sum_{i=1}^{k} \deg_I \mathrm{Sp}(f_t - s, x_i) \leq \deg_I \mathrm{Sp}(f, x).$$

This was proved in [Var83] for the case of low weight deformations of weighted homogeneous isolated hypersurface singularities (in which case the theorem even holds for an open interval of length one) and in [Ste85] in general. When applied to a homogeneous singularity, Varchenko's result yields an upper bound for the number of isolated singular points of given types which may occur on a projective hypersurface of given degree and dimension (the *spectrum bound*).

4) **Hertling's conjecture**. Let $a_1 \leq a_2 \leq \cdots \leq a_\mu$ denote the sequence of spectrum numbers of an isolated hypersurface singularity $f : (\mathbb{C}^{n+1}, 0) \to (\mathbb{C}, 0)$. The symmetry property show that the mean of the spectrum numbers is equal to $\frac{n-1}{2}$. But what about its variance? Claus Hertling [Hert01] conjectured

Conjecture 12.9.

$$\frac{1}{\mu} \sum_{i=1}^{\mu} \left(a_i - \frac{n-1}{2} \right)^2 \leq \frac{a_\mu - a_1}{12}$$

with equality if and only if f is a deformation with constant Milnor number of a weighted homogeneous singularity.

He proved himself, using the theory of Frobenius manifolds, that equality holds in the weighted homogeneous case. The conjecture was verified for functions of two variables by Brélivet [Bre].

For a generalization of the notion of spectrum, including symmetry, Thom-Sebastiani and semi-continuity, to the case of isolated complete intersection singularities, see [ES98].

12.2 An Application to Cycles: Grothendieck's Induction Principle

We shall discuss Grothendieck's suggested approach to the generalized Hodge conjecture by means of induction and give some examples in which this idea can be applied successfully.

We first recall the statement of the conjecture for a smooth projective variety X

$$GHC(X,m,c): \begin{cases} \forall H' \text{ a } \mathbb{Q}\text{-Hodge substructure of } H^m(X;\mathbb{Q}) \\ \qquad\qquad \text{of level} \le m-2c, \\ \exists Z \subset X \text{ a subvariety of codimension} \ge c \\ \qquad\qquad \text{such that } H' \text{ is supported on } Z. \end{cases}$$

An important source of examples arises as follows.

Lemma 12.10. *Let X and Y be smooth projective variety and let $Z \subset X \times Y$ be a degree c correspondence from Y to X, i.e. Z is an equidimensional subvariety of $X \times Y$ of dimension* $\dim X - c$. *Let $p: Z \to X$, $q: Z \to Y$ be induced by the two projections. Suppose that there is a surjection*

$$p_! \circ q^* : H^{m-2c}(Y;\mathbb{Q}) \twoheadrightarrow H^m(X;\mathbb{Q}).$$

Then $H^m(X)$ has level $\le m-2c$ and $GHC(X,m,c)$ holds.

Proof. Let $\nu : \tilde{Z} \to Z$ be a resolution of singularities. We have a factorisation of $p_! \circ q^*$:

$$\begin{array}{ccccc} & & & & H^{m-2c}(\tilde{Z};\mathbb{Q}) \to H^m(X;\mathbb{Q}), \\ & & & & \uparrow \\ H^{m-2c}(Y;\mathbb{Q}) & \to & H^{m-2c}(X\times Y) & \to & H^{m-2c}(Z) \end{array}$$

where the first three maps are induced by the projection $X \times Y \to Y$, the inclusion $Z \to X \times Y$ and ν respectively. The last map is the Gysin map for $\tilde{Z} \to X$. By assumption this last map is surjective and since it is a morphism of type (c,c), the image has level $\le m-2c$. □

Examples 12.11. Known examples come from surjective Abel-Jacobi maps. The situation is as follows. We have a smooth projective variety X of dimension $2m-1$ and a relative family $Z \to S$ of $(m-1)$-cycles on X over a smooth variety S which are cohomologically trivial. By the universal property of the Albanese variety, the Abel-Jacobi map $S \to J^m(X)$ factors over $\mathrm{Alb}(S) = J^s(S) \to J^m(X)$, $s = \dim S$ and the basic assumption is that the last map is *surjective.* Its tangent map $F^sH^{2s-1}(S) \to F^mH^{2m-1}(X)$ is induced by the homomorphism $p_! \circ q^*$ on the level of cohomology. The assumption implies that this homomorphism is also surjective. To apply the Lemma we need to have a degree $m-1$ correspondence, but this is only the case if S is a curve. So in this case $GHC(X, 2m-1, m-1)$ follows. In the general situation, we can take repeated hyperplane sections until we get a curve C. Restricting $Z \to S$ to this curve gives a Gysin map $H^1(C) \to H^{2s-1}(S) \to H^{2m-1}(X)$, where the first map comes from the inclusion $C \hookrightarrow S$. The Lefschetz hyperplane theorem (C.15) tells us that this map is a surjection. The lemma now shows that the entire cohomology $H^{2m-1}(X;\mathbb{Q})$ is supported on the image $C \to X$ of this curve and the result follows.
Geometric examples include :

1) The lines on a cubic threefold X form a surface, the Fano surface F and the Abel-Jacobi map $J^1(F) \to J^2(X)$ is surjective. Hence $GHC(X, 3, 1)$ holds. This result is due to Clemens and Griffiths [C-G].
2) The same holds for the Abel-Jacobi map for the Fano variety of lines on the quartic threefold. See [Tj72], [B-M] and [Let].
3) Let X be a complete intersection of two or three quadrics and $\dim X = 2m-1$. There is a family of codimension m-cycles whose Abel-Jacobi map is surjective and so $GHC(X, 2m-1, m-1)$ holds. This is due to Tjurin [Tj75].
4) Let X be a cubic 5-fold. Collino has shown [Coll] that there is a family of planes on X with surjective surjective Abel-Jacobi map so that $GHC(X, 5, 2)$ holds.

Theorem 12.12 (Main result). *Let X be a smooth projective variety of dimension $n+1$ and let Y be a generic hyperplane section of X. Suppose either that n is even or that the restriction map $H^n(X) \to H^n(Y)$ is not an isomorphism. Let us assume moreover that*

1) $GHC(Y, n-1, p-1)$ holds;
2) The variable middle cohomology $H^n_{\mathrm{var}}(Y;\mathbb{Q})$ of Y is supported on a codimension p subvariety of Y.

Then $GHC(X, n+1, p)$ holds.

Proof. Assume that Y is the generic fibre of some Lefschetz pencil. We recall the set-up of § 4.5.4. So, if B is the base locus of the pencil, the blow up $\tilde{X} = \mathrm{Bl}_B X$ fibres as

$$f : \tilde{X} \to \mathbb{P}^1.$$

The non-critical values of f form a Zariski open $j : U \hookrightarrow \mathbb{P}^1$ over which we have the local system $\mathbb{V}$ of classes of the vanishing cocycles. The assumptions ensure that Theorem 4.26 applies.

The middle cohomology of X is a direct summand of the middle cohomology of $\tilde{X}$. It is part of the three-step Leray filtration $L^\bullet$ on $H^{n+1}(\tilde{X};\mathbb{Q})$. Indeed, by Remark 4.28) the Leray filtration satisfies

$$\begin{aligned} L^1H^{n+1}(\tilde{X};\mathbb{Q}) &= H^{n+1}(X;\mathbb{Q}), \\ L^2H^{n+1}(\tilde{X};\mathbb{Q}) &= H^{n-1}(Y;\mathbb{Q})(-1) \end{aligned}$$

in accord with splitting of Hodge structures

$$H^{n+1}(X;\mathbb{Q}) = H^{n-1}(Y;\mathbb{Q})(-1) \oplus H^1(j_*\mathbb{V}).$$

so that $H^1(j_*\mathbb{V}) = L^1/L^2 \cong (L^2)^\perp \cap L^1$. All these isomorphism are isomorphisms of Hodge structures.

Assumption 1) implies that the generalized Hodge conjecture is true for L^2 and so it needs only to be verified for $(L^2)^\perp \cap L^1$ for which we use the second assumption. Since the stalk at $u \in U'$ is the variable cohomology of the fibre Y_u, this assumption says that provided u is generic, there is a codimension p cycle Z_u on Y_u such that

$$\operatorname{Im}(H^n_{Z_u}(Y_u) \to H^n(Y_u)) \supset \mathbb{V}_u. \tag{XII–1}$$

The cycles Z_u do not necessarily fit together to a global cycle on $\tilde{X}$, but, as will be shown shortly, there exists a codimension p cycle Z on $\tilde{X}$ which is flat over $\mathbb{P}^1$ such that for generic u the intersection of Z with Y_u contains Z_u as a component. Replacing Z_u by this (possibly larger) cycle, the assertion (XII–1) is still true and we are going to show that the cycle Z in fact supports $(L^2)^\perp \cap L^1$. But let us first show how to construct the cycle Z. Consider the family over $\mathbb{P}^1$ consisting of the Chow varieties of effective codimension p cycles in the fibres Y_u of $f : \tilde{X} \to \mathbb{P}^1$. Over the complement of a countable set these form an algebraic fibre bundle. So this bundle has a multi-section passing through the point corresponding to Z_u. This multi-section sweeps out the desired cycle Z, fitting into the commutative diagram

$$\begin{array}{ccc} Z & \xrightarrow{\quad h \quad} & \tilde{X} \\ & {\scriptstyle g}\searrow \quad \swarrow{\scriptstyle f} & \\ & \mathbb{P}^1 & \end{array}$$

Let $\sigma : \tilde{Z} \to Z$ be a resolution of singularities and we put $\tilde{h} := h\circ\sigma$. We are going to show that the image of Gysin map for $\tilde{h}$,

$$H^{n+2-2p}(\tilde{Z}) \xrightarrow{\tilde{h}_!} H^{n+1}(\tilde{X}),$$

contains $(L^2)^\perp \cap L^1$. We do this by showing that this Gysin map induces a surjection on the relevant part of the Leray spectral sequences:

$$\tilde{h}_! : H^1(\mathbb{P}^1, R^{n+1-2p}\tilde{g} * \mathbb{Q}) \twoheadrightarrow H^1(\mathbb{P}^1, R^n f_*\mathbb{Q}) = H^1(\mathbb{P}^1, j_*\mathbb{V}). \qquad \text{(XII–2)}$$

Indeed, the induced map on the Leray-filtrations induces $L^1(\tilde{g}) \to L^1(f)$ and if this becomes a surjection on Gr^L_1, the image of $\tilde{h}_!$ contains $(L^2 f)^\perp \cap L^1 f$ as wanted.

To show (XII–2), recall (formula VII–2) that the image of the Gysin map $H^{n+1-2p}(\tilde{Z}_u) \to H^n_{Z_u}(Y_u) \to H^n(Y_u)$, u generic, is the same as the image of $H^n_{Z_u}(Y_u) \to H^n(Y_u)$. Our assumption (XII–1) can thus be formulated in terms of the image of this Gysin map. Indeed, letting $j' : U' \hookrightarrow \mathbb{P}^1$ be the inclusion of the Zariski open subset of $\mathbb{P}^1$ over which both f and g are smooth, the Gysin maps fit together to give a morphism of local systems

$$\tilde{h}_! : \mathbb{W} := j'^* R^{n+1-2p}\tilde{g}_*\mathbb{Q} \to j'^* R^n f_*\mathbb{Q}.$$

Now assumption (XII–1) means that the image of this map contains $\mathbb{V}|U'$. But more is true: since $\tilde{h}_!$ is a morphism of polarized variations of Hodge structures we can use complete reducibility (Theorem 10.13) and hence the local system $\mathbb{V}|U'$ is a *direct factor* of $j'^* R^n f_*\mathbb{Q}$. If we project onto this factor we obtain a surjective morphism of (polarizable) variations of Hodge structures $\varphi : \mathbb{W} \to \mathbb{V}|U'$. Again by complete reducibility, we obtain a direct sum decomposition $\mathbb{W} \cong \mathrm{Ker}\,\varphi \oplus \mathbb{V}|U'$. So $\tilde{h} : \tilde{Z} \to \tilde{X}$ induces a surjection

$$H^1(j'_*\mathbb{W}) \twoheadrightarrow H^1(j'_*(\mathbb{V}|U')) = H^1(j_*\mathbb{V}). \qquad \text{(XII–3)}$$

Secondly, observe that the adjunction morphism

$$R^{n+1-2p}\tilde{g}_*\mathbb{Q} \to j'_* j'^* R^{n+1-2p}\tilde{g}_*\mathbb{Q} = j'_*\mathbb{W}$$

is an isomorphism away from the critical values of $\tilde{g}$. Hence the induced map

$$H^1(R^{n+1-2p}\tilde{g}_*\mathbb{Q}) \twoheadrightarrow H^1(j'_*\mathbb{W})$$

must be a surjection. Combining this and the surjection (XII–3) yields the desired surjection (XII–2). □

Examples 12.13. The first remark one should make is that for X a complete intersection of dimension $(n+1)$ in projective space, the Lefschetz hyperplane theorem C.15 implies that for $k \neq (n+1)$ the cohomology groups $H^k(X)$ are either zero or induced by a linear section and so the Hodge conjecture is only interesting in the middle dimension. For the same reason, for complete intersections the first hypotheses in Grothendieck's principle is always verified.

1) The generalized Hodge conjecture for curves on a cubic threefold in $\mathbb{P}^4$ is true. Indeed hypothesis (ii) is verified, since a general hyperplane section Y is a cubic surface so that $H^2_{\mathrm{var}}(Y) = H^2_{\mathrm{prim}}(Y)$ has pure type $(1,1)$ and so is supported on a divisor.
The generalized Hodge conjecture in this case means that all of $H^3(X;\mathbb{Q})$ (which itself is of level 1) is supported on a divisor.

2) We use the previous example to prove the generalized Hodge conjecture for curves on a cubic fourfold X in $\mathbb{P}^5$. So $n = 3$, $p = 1$.
Indeed, all of $H^3(Y) = H^3_{\text{var}}(Y)$, Y a smooth hyperplane section of X, is supported on a divisor inside Y by what we just proved.
Applying Remark 7.7 it also follows that the Hodge conjecture is true for $(2,2)$-classes.
3) Using Example 12.11 2), $GHC(4,1)$ and the Hodge conjecture for $(2,2)$-classes are true for X a quartic fourfold in $\mathbb{P}^5$ $(n = 3, p = 1)$.
4) Let X be a cubic 6-fold in $\mathbb{P}^7$ so that $n = 5$, $p = 2$. Note that $GHC(Y, 6, 2)$ is true (Example 12.11 4)) so that $GHC(X, 6, 2)$ is true as well as the classical Hodge conjecture for $(3,3)$-classes.
5) Using Example 12.11 3), we can inductively show $GHC(X, 2m, m-1)$ and the classical Hodge conjecture on (m,n)-classes for X an even-dimensional intersection of 2 or 3 quadrics.

13

Perverse Sheaves and D-Modules

In this chapter we discuss the general version of the Riemann-Hilbert correspondence. We had a first glimpse before: see Remark 10.5 and also § 11.1.2. Here we extend it to an equivalence (in the derived category) between constructible complexes and certain types of D-modules. A crucial point here is a duality operator which generalizes Poincaré duality on the constructible side.

In § 13.1 we introduce this duality operator, Verdier duality. Ordinary cohomology is not preserved under this duality operator. Instead, one needs to work with intersection cohomology. In the topological context this is explained in § 13.2.1. The intersection complexes become self-dual *after a shift in the degrees.* After this shift we obtain the perverse complexes treated in § 13.2.2.

The Riemann-Hilbert correspondence links perverse complexes to D-modules. We give a brief introduction to these in § 13.3. An important example comes from variations of Hodge structures. These D-modules are in addition filtered, and we treat filtered D-modules in § 13.5. For the Riemann-Hilbert correspondence holonomic D-modules play a central role, discussed in § 13.6. In that section we state (without proof) the full Riemann-Hilbert correspondence.

13.1 Verdier Duality

In the context of general topological spaces one cannot hope for a good duality theory generalizing Poincaré duality. However, as soon as the space is locally compact and finite dimensional, there does exist such a generalization, but only if one works in a suitable derived category; this is Verdier duality. We explain this in this section, and we also explain how Verdier duality implies the classical Poincaré duality theorems.

13.1.1 Dimension

For this subsection we follow [Bor84, V.1.B].

Definition 13.1. Let X be a locally compact space. Its **dimension** $\dim(X)$ is the smallest $n \in \mathbb{N} \cup \infty$ for which $H_c^{n+1}(X, \mathcal{F}) = 0$ for all sheaves $\mathcal{F}$ on X.

Examples 13.2. The dimension can be calculated using only constant sheaves on open sets. A topological n-dimensional manifold has dimension n. More generally, an n-dimensional pseudomanifold (see §C.1.1) is an n-dimensional space. Any complex space of complex dimension n has dimension $2n$.

On such spaces instead of the possibly unbounded Godement resolution we can take its n-truncated complex

$$_c\mathcal{C}^\bullet_{\mathrm{Gdm}}(\mathcal{F}) := \tau_{\leq n}\mathcal{C}^\bullet_{\mathrm{Gdm}}(\mathcal{F}),$$

by definition the **c-Godement resolution** of $\mathcal{F}$. As we show now, the sheaves in this resolution are **c-soft** in the sense that sections on compact subsets of X extend to the entire space:

Lemma 13.3. *The c-Godement resolution of $\mathcal{F}$ is a c-soft resolution.*

Proof. Godement sheaves being flabby are c-soft so we only need to see that the last sheaf in the resolution, say $\mathcal{B}$ is a soft sheaf. From the exact sequence (B–15) it follows that it is sufficient to show that for all open sets U we have $H_c^1(U, \mathcal{B}) = 0$. First note that, setting $\mathcal{Z}^k = \mathrm{Ker}(d : \mathcal{C}^k_{\mathrm{Gdm}}(\mathcal{F}) \to \mathcal{C}^{k+1}_{\mathrm{Gdm}}(\mathcal{F}))$, we have an exact sequence $0 \to \mathcal{Z}^k(\mathcal{F}) \to \mathcal{C}^k_{\mathrm{Gdm}}(\mathcal{F}) \to \mathcal{Z}^{k+1} \to 0$ and hence (B–14) shows that $H_c^1(U, \mathcal{Z}_k) \simeq H_c^2(U, \mathcal{Z}_{k-1}) \cdots \simeq H_c^k(U, \mathcal{Z}_1) \simeq H_c^{k+1}(U, \mathcal{F})$. In particular, since $\mathcal{Z}^n = \mathcal{B}$, we have $H_c^1(U, \mathcal{B}) = H_c^{n+1}(U, \mathcal{F})$ which vanishes by definition of the dimension. □

In this calculation we used the crucial fact that c-soft sheaves are acyclic for the functor Γ_c of global sections with compact support ([Iver, II.Theorem 2.7]). From the above Lemma it then follows that the c-Godement resolution is an acyclic resolution and hence, by the abstract De Rham theorem B.18 we have

$$H_c^q(X, \mathcal{F}) = H^q\left(\Gamma_c(X, {_c\mathcal{C}^\bullet_{\mathrm{Gdm}}}(\mathcal{F}))\right).$$

A similar statement is true for complexes $\mathcal{F}^\bullet$ of sheaves of R-modules on X:

$$\mathbb{H}_c^q(X, \mathcal{F}^\bullet) = H^q\left(s\Gamma_c(X, {_c\mathcal{C}^\bullet_{\mathrm{Gdm}}}(\mathcal{F}^\bullet))\right). \tag{XIII–1}$$

13.1.2 The Dualizing Complex

Details and proofs of the results in this subsection can be found in [Iver, V, VI] or [Bor84, V, 7].

We first assume that R is a *field* and explain how to dualize a complex of R-modules $C^\bullet$. Referring to the definition of the Hom-complexes (A–10), we set

$$DC^\bullet = \mathrm{Hom}^\bullet_R(C^\bullet, R),$$

i.e. $DC^q = \mathrm{Hom}_R(C^{-q}, R)$ and $d : DC^q \to DC^{q+1}$ is $(-1)^{q+1}$ times the transpose of $d : C^{-q-1} \to C^{-q}$. Let X be a topological space and $U \subset X$ any open subset and apply this construction to the complex of compactly supported sections of the c-Godement resolution of the constant sheaf $\underline{R}_X|U$, yielding a complex, say $D^\bullet(U)$. For an inclusion $j : V \hookrightarrow U$ of open sets the maps $D^q(U) \to D^q(V)$ induced by "extension by zero"-maps $\Gamma_c(V, {}_c\mathcal{C}^{-q}_{\mathrm{Gdm}}(\underline{R}_X)) \to \Gamma_c(U, {}_c\mathcal{C}^{-q}_{\mathrm{Gdm}}(\underline{R}_X))$ define a presheaf-structure. Here one uses that we are working with c-soft sheaves. In fact we even get a complex of sheaves, the (topological or Verdier) R-**dualizing complex** ${}^{\mathrm{Ve}}\mathbb{D}\underline{R}_X$:

$$ {}^{\mathrm{Ve}}\mathbb{D}\underline{R}_X(U) := DC^\bullet, \quad C^\bullet = \Gamma_c(U, {}_c\mathcal{C}^\bullet_{\mathrm{Gdm}}(\underline{R}_X)). \tag{XIII–2} $$

If R is no longer a field, one has to assume that R has finite cohomological dimension and choosing an injective resolution $R^\bullet$ of R one defines $DC^\bullet = \mathrm{Hom}^\bullet_R(C^\bullet, R^\bullet)$. The same construction as before then goes through.

Example 13.4. Suppose that X is an n-dimensional topological manifold and $R = \mathbb{Z}$. Use the standard injective resolution $0 \to \mathbb{Z} \to \mathbb{Q} \to \mathbb{Q}/\mathbb{Z} \to 0$ to define ${}^{\mathrm{Ve}}\mathbb{D}\underline{R}_X$. This complex turns out to be quasi-isomorphic to the **orientation sheaf** $\mathcal{O}r_X$, viewed as a complex in degree $-n$:

$$ {}^{\mathrm{Ve}}\mathbb{D}\underline{\mathbb{Z}}_X \xrightarrow[\sim]{\text{qis}} \mathcal{O}r_X[n]. $$

In fact, by [Bor84, V, 7.10] ${}^{\mathrm{Ve}}\mathbb{D}\underline{\mathbb{Z}}_X[-n]$ is an injective resolution of the orientation sheaf. Since X is **orientable** if the orientation sheaf is isomorphic to a constant sheaf, for an orientable manifold we have

$$ {}^{\mathrm{Ve}}\mathbb{D}\underline{\mathbb{Z}}_X[-n] \xrightarrow[\sim]{\text{qis}} \underline{\mathbb{Z}}_X. \tag{XIII–3} $$

We want to see the effect of Verdier duality on cohomology. Using (A–21), we find $\mathbb{H}^{-q}(X, {}^{\mathrm{Ve}}\mathbb{D}\underline{R}_X) = \mathrm{Ext}^{-q}(\Gamma_c(X, \underline{R}^\bullet_X), R)$. For R a field the right hand side equals $\mathrm{Hom}_R(H^q_c(X; R), R)$, the R-dual for $H^q_c(X; R)$ which can be identified with $H^{BM}_q(X; R)$. For $R = \mathbb{Z}$ this is conjecturally the case. It is certainly the case when X is compact, since then Borel-Moore homology is the same as singular homology. More generally [Bor-M, Thm. 3.8 and §5]:

Proposition 13.5. *Let X be paracompact having a one-point compactification which is a CW-complex, then there is a canonical identification (with $\mathbb{Z}$-coefficients)*

$$ \mathbb{H}^{-q}(X, {}^{\mathrm{Ve}}\mathbb{D}\underline{\mathbb{Z}}_X) = H^{BM}_q(X) $$
$$ \mathbb{H}^{-q}_c(X, {}^{\mathrm{Ve}}\mathbb{D}\underline{\mathbb{Z}}_X) = H_q(X). $$

Remark 13.6. 1) This theorem applies to compact manifolds, or more generally, to complements of a closed submanifold of a compact manifold. In particular we can take for X a complex algebraic manifold.

2) For an orientable manifold X of dimension n the last identity can be rewritten as $H^q_c(X; \mathbb{Z}) = H^{q-n}(X, {}^{\mathrm{Ve}}\mathbb{D}\underline{\mathbb{Z}}_X) = H_{n-q}(X, \mathbb{Z})$ which is just the ordinary Poincaré duality (Theorem B.24). Similarly, the first identity can be rephrased as Poincaré duality for Borel-Moore homology (loc. cit.).

13.1.3 Statement of Verdier Duality

For this subsection we refer to [Iver, Chap. VI] and also to [Bor84, V, 7].

The dualizing complex is a special case of a more general construction. To explain this, again, for simplicity we assume that R is a field, and we leave the modifications in the general case to the reader. First we define the R-dual for a sheaf $\mathcal{F}$ of R-modules. Note that the complex ${}_c\mathcal{C}^\bullet_{\mathrm{Gdm}}(\mathcal{F})$ is a c-soft resolution for $\mathcal{F}$. Since c-softness was all that was needed to define good restriction maps to get a presheaf, we may use in (XIII–2) the complex ${}_c\mathcal{C}^\bullet_{\mathrm{Gdm}}(\mathcal{F})$ instead of ${}_c\mathcal{C}^\bullet_{\mathrm{Gdm}}(\underline{R}_X)$ and still obtain a complex of sheaves of R-modules on X, the Verdier dual of $\mathcal{F}$, denoted ${}^{\mathrm{Ve}}\mathbb{D}_X(\mathcal{F})$. In particular, applying this to the constant sheaf we find back the dual complex.

More generally, we can define the Verdier dual ${}^{\mathrm{Ve}}\mathbb{D}_X(\mathcal{F}^\bullet)$ for a complex $\mathcal{F}^\bullet$ of sheaves as we now explain. Recall the construction of the tensor product $\mathcal{F}^\bullet \otimes \mathcal{G}^\bullet$ given by formula (A–2) at the end of § A.1.1. Then the dual complex for $\mathcal{F}^\bullet$ is the complex of sheaves given by

$$U \mapsto {}^{\mathrm{Ve}}\mathbb{D}_X(\mathcal{F}^\bullet; R)(U) := DC^\bullet, \quad C^\bullet = \Gamma_c(U, {}_c\mathcal{C}^\bullet_{\mathrm{Gdm}}(\mathcal{F}^\bullet))).$$

Again, as in the special cases we treated before, "extension by zero" can be used to define restriction maps giving a presheaf structure to each of the constituents of this complex.

From the definition it follows that whenever $\mathcal{F}^\bullet$ is bounded, bounded above, respectively bounded below, its dual is bounded, bounded below, respectively bounded above. The Verdier duality functor thus yields:

$${}^{\mathrm{Ve}}\mathbb{D}_X : D^- \begin{pmatrix} \text{sheaves of } R\text{-} \\ \text{modules on } X \end{pmatrix} \longrightarrow D^+ \begin{pmatrix} \text{sheaves of } R\text{-} \\ \text{modules on } X \end{pmatrix}. \tag{XIII–4}$$

The complex ${}^{\mathrm{Ve}}\mathbb{D}_X\mathcal{F}^\bullet$ turns out to be isomorphic to the Hom-complex of sheaves $\mathcal{H}om^\bullet(\mathcal{F}^\bullet, {}^{\mathrm{Ve}}\mathbb{D}_X\underline{R}_X)$. At this point we recall that for any two sheaves $\mathcal{F}$, $\mathcal{G}$ of R-modules on X the sheaf $\mathcal{H}om(\mathcal{F},\mathcal{G})$ is defined by means of the presheaf $U \mapsto \mathrm{Hom}_R(\mathcal{F}(U),\mathcal{G}(U))$. Moreover, the construction of Hom-complexes can also be done on the level of sheaves: for any two complexes $\mathcal{F}^\bullet$ and $\mathcal{G}^\bullet$ of sheaves on X, the graded presheaf

$$U \mapsto \mathcal{H}om^\bullet(\mathcal{F}^\bullet,\mathcal{G}^\bullet)(U) := \mathrm{Hom}^\bullet_R(\mathcal{F}^\bullet(U),\mathcal{G}^\bullet(U))$$

is made into a complex by using for d the same formula as for the usual Hom-complexes (cf. Appendix A.2.1). The definition of the Verdier duality isomorphism uses the following construction. Let $U \subset X$ be an open set, $\mathcal{F}$ and $\mathcal{G}$ two sheaves of R-modules on X, and S an R-module. For any $s \in \Gamma(U,\mathcal{F})$, $t \in \Gamma_c(U,\mathcal{G})$ the tensor product $s \otimes t$ belongs to $\Gamma_c(U,\mathcal{F}\otimes\mathcal{G})$ and we define the natural map

$$\left.\begin{array}{ccc} \mathrm{Hom}_R\,(\Gamma_c(U,\mathcal{F}\otimes\mathcal{G}),S) & \to & \mathrm{Hom}_R\,(\mathcal{F}(U),\mathrm{Hom}_R(\Gamma_c(U,\mathcal{G}),S)) \\ h & \mapsto & \{s \longmapsto \{t \mapsto h(s\otimes t\}\}. \end{array}\right\} \tag{XIII–5}$$

Now we can state:

Proposition 13.7 (Verdier duality). *Let X be a locally compact n-dimensional space and let $\mathcal{F}^\bullet$ a bounded* above *complex of sheaves of R-modules. The maps* (XIII–5) *for $\mathcal{F} = \mathcal{F}^p$, $\mathcal{G} = {}_{\mathrm{c}}\mathcal{C}^q_{\mathrm{Gdm}}(\underline{R}_X)$ and $S = R$ induce a natural isomorphism of bounded below complexes of sheaves of R-modules on X*

$$ {}^{\mathrm{Ve}}\mathbb{D}_X(\mathcal{F}^\bullet) \xrightarrow{\sim} \mathcal{H}om^\bullet(\mathcal{F}^\bullet, {}^{\mathrm{Ve}}\mathbb{D}\underline{R}_X). $$

13.1.4 Extraordinary Pull Back

For more details on the results in this subsection, we refer to [Iver, Chap. VI] and [Bor84, V, 7].

If X and Y are locally compact and $f : X \to Y$ any continuous map a proper direct image functor $f_!$ can be defined which transforms sheaves of R-modules on X to sheaves of R-modules on Y; if f is proper, $f_! = f_*$, and if f is the inclusion of an open subset, it has a right adjoint. See § B.2.5. The adjoint $f^!$ cannot be defined in general; some finiteness assumption is needed. Here we assume that X is finite dimensional. Then, for a given bounded below complex $\mathcal{G}^\bullet$ of sheaves of R-modules on Y, using the bounded c-Godement resolution of the constant sheaf $\underline{R}_X$ the assignment

$$ U \mapsto f^!\mathcal{G}^\bullet(U) := \mathrm{Hom}^\bullet_R(f_! \circ j_!\, {}_{\mathrm{c}}\mathcal{C}^\bullet_{\mathrm{Gdm}}(\underline{R}_U), \mathcal{G}^\bullet), \quad j : U \subset X \text{ the inclusion,} $$

defines a bounded below complex $f^!\mathcal{G}^\bullet$ of sheaves of R-modules on X. It determines a functor between derived categories

$$ f^! : D^+\begin{pmatrix} \text{sheaves of} \\ R\text{-modules on } Y \end{pmatrix} \longrightarrow D^+\begin{pmatrix} \text{sheaves of} \\ R\text{-modules on } X \end{pmatrix}. \tag{XIII–6} $$

It sends $D^{\mathrm{b}}\{\text{sheaves of } R\text{-modules on } Y\}$ to $D^{\mathrm{b}}\{\text{sheaves of } R\text{-modules on } X\}$ and is called the **extra-ordinary pull back**.

Examples 13.8. 1) For the inclusion $j : X \hookrightarrow Y$ of an open subset this functor $j^!$ coincides with j^{-1} of Appendix B (cf. (B–22)). For a closed embedding $i : Z \hookrightarrow X$ functor $i^!$ as defined above, coincides with the functor (B–33) of Appendix B.

2) Applying this to the case where Y is a point and $\mathcal{G}^\bullet = R^\bullet$, observing that $R^\bullet = R$ in the derived category of bounded R-modules, we get:

$$ {}^{\mathrm{Ve}}\mathbb{D}\underline{R}_X = a_X^! R, \quad a_X : X \to \text{point}. \tag{XIII–7} $$

This can be used as a *definition* of the dualizing complex. Similarly, Prop. 13.7 can be turned around and can be viewed as a definition for the Verdier duality operation:

$$ {}^{\mathrm{Ve}}\mathbb{D}_X(\mathcal{F}^\bullet) := R\mathcal{H}om^\bullet(\mathcal{F}^\bullet, {}^{\mathrm{Ve}}\mathbb{D}\underline{R}_X). \tag{XIII–8} $$

We can now formulate the most general form of Verdier duality:

Theorem 13.9 (Verdier duality II). *Let $f : X \to Y$ be a continuous map between locally compact spaces of finite dimension and let R be a commutative ring of finite cohomological dimension. Let $\mathcal{F}^\bullet$ be a bounded complex of sheaves of R-modules on X, and let $\mathcal{G}^\bullet$ be a bounded above complex of sheaves of R-modules on Y. Then there is a canonical isomorphism*

$$R\mathcal{H}om^\bullet(Rf_!\mathcal{F}^\bullet, \mathcal{G}^\bullet) \xrightarrow{\sim} Rf_*\left(R\mathcal{H}om^\bullet(\mathcal{F}^\bullet, f^!\mathcal{G}^\bullet)\right)$$

in $D^+\{$sheaves of R-modules on $Y\}$, i.e. $f^!$ is a right adjoint for $Rf_!$.

Since $f^!({}^{Ve}\mathbb{D}\underline{R}_X) = {}^{Ve}\mathbb{D}\underline{R}_Y$ one deduces from Theorem 13.9 that under duality the pair of adjoint functors (f^{-1}, Rf_*) is interchanged against $(f^!, Rf_!)$:

Corollary 13.10. *Let $f : X \to Y$ be a continuous map between locally compact spaces of finite dimension. We have a diagram*

$$\begin{array}{ccc} D^{\mathrm{b}}\begin{pmatrix}\text{sheaves of}\\ R\text{-modules on } X\end{pmatrix} & \underset{Rf_!}{\overset{f^{-1}}{\rightleftarrows}} & D^{\mathrm{b}}\begin{pmatrix}\text{sheaves of}\\ R\text{-modules on } Y\end{pmatrix} \\ \updownarrow {}^{Ve}\mathbb{D}_X & & {}^{Ve}\mathbb{D}_Y \updownarrow \\ D^{\mathrm{b}}\begin{pmatrix}\text{sheaves of}\\ R\text{-modules on } X\end{pmatrix} & \underset{Rf_*}{\overset{f^{!}}{\rightleftarrows}} & D^{\mathrm{b}}\begin{pmatrix}\text{sheaves of}\\ R\text{-modules on } Y\end{pmatrix} \end{array} \tag{XIII–9}$$

which incorporates two commutativity relations $Rf_*\circ{}^{Ve}\mathbb{D}_X = {}^{Ve}\mathbb{D}_Y\circ Rf_!$ and ${}^{Ve}\mathbb{D}_X\circ f^{-1} = f^!\circ{}^{Ve}\mathbb{D}_Y$.

Remark 13.11. i) As a special case, when $j : U \hookrightarrow X$ is an open inclusion, we have ${}^{Ve}\mathbb{D}_U j^{-1}\mathcal{F}^\bullet = j^{-1}{}^{Ve}\mathbb{D}_X\mathcal{F}^\bullet$, and if $i : Z \hookrightarrow X$ is the inclusion of a closed subset, ${}^{Ve}\mathbb{D}_X i_*\mathcal{F}^\bullet = i_*{}^{Ve}\mathbb{D}_Z\mathcal{F}^\bullet$.

ii) The theorem generalizes to the situation where $f_!$ has finite cohomological dimension, i.e. the fibres of f are uniformly bounded in dimension.

iii) Applying the duality isomorphism to the identity morphism $f^!\mathcal{G}^\bullet \to f^!\mathcal{G}^\bullet$ we get a morphism $Rf_!\circ f^!\mathcal{G}^\bullet \to \mathcal{G}^\bullet$. Taking $\mathcal{G}^\bullet = {}^{Ve}\mathbb{D}\underline{R}_Y$, since by (XIII–7) $f^!{}^{Ve}\mathbb{D}\underline{R}_Y = {}^{Ve}\mathbb{D}\underline{R}_X$, this morphism becomes

$$182007Rf_!{}^{Ve}\mathbb{D}\underline{R}_X \to {}^{Ve}\mathbb{D}\underline{R}_Y \tag{XIII–10}$$

which induces $\mathbb{H}_c^{-q}(Y, Rf_!{}^{Ve}\mathbb{D}\underline{R}_X) \to \mathbb{H}_c^{-q}(Y, {}^{Ve}\mathbb{D}\underline{R}_Y)$. Using (B–28) and Prop. 13.5, in integral homology this gives back the usual induced map in homology

$$f_* : H_q(X) \to H_q(Y) \tag{XIII–11}$$

13.2 Perverse Complexes

13.2.1 Intersection Homology and Cohomology

In this subsection we fix a commutative ring R with 1.

Complex algebraic varieties, and more generally, complex analytic spaces embeddable in a compact analytic space as the complement of a closed subspace admit Whitney stratifications (see Property C.7, 2–3). By Cor. C.6 any such analytic space X has the structure of an oriented pseudomanifold of dimension $2n$ where n is the complex dimension. More precisely, the strata of the Whitney stratification combine to give a filtration $X = X_n \supset X_{n-1} \supset \cdots X_0$, where $X_k - X_{k-1}$ is an oriented $2k$-dimensional topological manifold. Its connected components are the $2k$-dimensional strata of the Whitney stratification. We shall only consider such stratifications.

If one considers usual homology and cohomology there is no straightforward generalization of Poincaré-duality due to the presence of singularities. Instead it is better to start with the constant sheaf on a dense smooth open subset of X and then extend it successively on the singular locus as a complex of sheaves, the intersection complex as we now explain. We consider more generally a local system $\mathbb{V}$ of R-modules on the dense open smooth subset $U_1 = X - X_{n-1}$ of X. We view X as the increasing union of open sets $U_1 \subset U_2 \cdots \subset U_{n+1} = X$ where

$$U_k = X - X_{n-k}, \quad j_k^0 : U_k \hookrightarrow U_{k+1}. \qquad \text{(XIII–12)}$$

Definition 13.12. Let $\mathbb{V}$ be a local system of R-modules on U_1. The associated **intersection complex** $\mathcal{IC}_X^\bullet\, \mathbb{V}$ is inductively defined as follows. Set $\mathcal{IC}_X^\bullet\, \mathbb{V} = \mathbb{V}$ on U_1 and, assuming $\mathcal{IC}_X^\bullet\, \mathbb{V}$ defined on U_k set

$$\mathcal{IC}_X^\bullet\, \mathbb{V}|U_{k+1} = \tau_{\le k-1} Rj_{k*}^0 (\mathcal{IC}_X^\bullet\, \mathbb{V}|U_k)$$

Its hypercohomology groups are the intersection (co)homology groups with coefficients $\mathbb{V}$

$$IH_k^{\mathrm{BM}}(X;\mathbb{V}) = IH^{2n-k}(X;\mathbb{V}) := \mathbb{H}^{2n-k}(X, \mathcal{IC}_X^\bullet\, \mathbb{V}).$$

Similarly, we define intersection (co)homology with compact support with coefficients $\mathbb{V}$

$$IH_k(X;\mathbb{V}) = IH_c^{2n-k}(X;\mathbb{V}) := \mathbb{H}_c^{2n-k}(X, \mathcal{IC}_X^\bullet\, \mathbb{V}).$$

If $\mathbb{V} = \underline{R}_{U_1}$ one conventionally just writes $IH_k^{\mathrm{BM}}(X;R) = IH^{2n-k}(X;R)$ respectively $IH_k(X;R) = IH_c^{2n-k}(X;R)$.

Remark 13.13. 1) Intersection (co)homology seemingly depends on the stratification. We shall explain below (see Theorem 13.19) that this is not the case.

2) What we explained above is the so-called "middle perversity" intersection complex. There are other perversities as well, but the middle perversity is the only one behaving well under Verdier duality. See Cor. 13.20.

3) Intersection complexes (for any perversity) can also be defined for any pseudomanifold.

13.2.2 Constructible and Perverse Complexes

In this subsection R denotes a field of characteristic 0.

As in the previous subsection, let $X = X_n \supset X_{n-1} \supset \cdots X_0$ be an n-dimensional Whitney-stratified complex space. For any given local system $\mathbb{V}$ on $X - X_{n-1}$ the intersection complex $\mathcal{IC}^\bullet_X \mathbb{V}$ is an example of a (weakly) constructible complex.

Definition 13.14. i) A sheaf $\mathcal{F}$ of R-modules on X is **weakly constructible** if there exists a stratification such that the restriction of $\mathcal{F}$ to each stratum is a locally constant sheaf; it is **constructible** if, moreover, its stalks are finitely generated R-modules;

ii) a bounded complex $\mathcal{F}^\bullet$ of sheaves on X is **(cohomologically) constructible** the sheaves $H^q(\mathcal{F}^\bullet), q \in \mathbb{Z}$ are constructible with respect to some stratification. We set

$$D^{\mathrm{b}}_{\mathrm{cs}}(X;R) := \left\{ \begin{array}{l} \text{derived category of} \\ \text{constructible complexes of} \\ \text{sheaves of } R\text{-modules on } X \end{array} \right\}. \tag{XIII–13}$$

iii) A cohomologically constructible complex $\mathcal{F}^\bullet$ of sheaves on X is called **perverse** if, setting $S_k = X_k - X_{k-1}$, the following two conditions hold

$$\left. \begin{array}{l} \text{For all } x \in S_k \text{ one has } H^j(\mathcal{F}^\bullet)_x = 0,\ j > -k; \\ \text{For all } x \in S_k \text{ one has } H^j_c(\mathcal{F}^\bullet)_x = 0,\ -j > -k. \end{array} \right\} \tag{XIII–14}$$

Here we have set

$$H^j_c(\mathcal{F}^\bullet)_x := \varprojlim_U \mathbb{H}^j_c(U, \mathcal{F}^\bullet),$$

where U runs over the open neighbourhoods of x in X .

We next summarize the behaviour of constructible and perverse sheaves under duality and under morphisms. For proofs see for instance [Kash-S, § 8.4, § 8.5] or [Bor84, Chap. V, 8 –10]:

Proposition 13.15. i) *On complex analytic spaces the Verdier duality operator preserves cohomological constructibility.*

ii) *Let $f : X \to Y$ be a morphism between complex analytic varieties. The functors f^{-1} and $f^!$ preserve cohomological constructible complexes. If f is proper or a morphism between complex algebraic varieties, Rf_* and $Rf_!$ preserve (cohomological) constructible complexes.*

Remark 13.16. In general the functors f^{-1}, $f^!$, Rf_* and $Rf_!$ do not preserve perversity and we need to work with constructible complexes. For instance, given a constructible complex $\mathcal{F}^\bullet$ of R-modules on a complex analytic variety X and a closed subvariety $i : Z \hookrightarrow X$, the **adjunction triangle** relating the morphisms induced by i and $j : X - Z \hookrightarrow X$ is a distinguished triangle in $D^{\mathrm{b}}_{\mathrm{cs}}(X;R)$:

$$
\begin{array}{ccc}
Ri_*i^!\mathcal{F}^\bullet & \longrightarrow & \mathcal{F}^\bullet \\
 {\scriptstyle [1]}\nwarrow & & \swarrow {\scriptstyle \alpha_{Z,X}(\mathcal{F}^\bullet)} \\
 & Rj_*j^{-1}\mathcal{F}^\bullet &
\end{array}
\qquad \text{(XIII–15)}
$$

Note that the above definition of perversity uses the stratification. However, if for some $x \in S_k$ we have $H^j(\mathcal{F}^\bullet)_x \neq 0$, we must have $j \leq -k$. Recall that the closure of the set $\{x \in X \mid H^j(\mathcal{F}^\bullet)_x \neq 0\}$ is the *support* of the sheaf $H^j(\mathcal{F}^\bullet)$. By the very definition of constructibility it is the closure of a stratum and so contained in X_k for some k. So $\dim(\operatorname{Supp} H^j(\mathcal{F}^\bullet)) = k \leq -j$. If R is a field, we can rephrase the second condition using Verdier duality. Verdier duality preserves constructibility by Prop. 13.15. Also, by [Bor84, V.9.7] we have $H^j_c(\mathcal{F}^\bullet)^\vee_x = H^{-j}({}^{\mathrm{Ve}}\mathbb{D}_X\mathcal{F}^\bullet)_x$. It follows that the closure of $\{x \mid H^j_c(\mathcal{F}^\bullet)_x \neq 0\}$ coincides with the support of $H^{-j}({}^{\mathrm{Ve}}\mathbb{D}_X\mathcal{F}^\bullet)$. So, replacing j by $-j$ we get the same condition for the support of $H^j({}^{\mathrm{Ve}}\mathbb{D}_X\mathcal{F}^\bullet)$. Summarizing, we have:

Lemma 13.17. *Assume that R is a field of characteristic zero. Let $\mathcal{F}^\bullet$ be a bounded complex of sheaves of R-modules which is cohomologically constructible. Then it is perverse if and only if the following two conditions hold:*

$$
\begin{aligned}
\dim \operatorname{Supp} H^j(\mathcal{F}^\bullet) &\leq -j \quad \text{(the support condition);}\\
\dim \operatorname{Supp} H^j({}^{\mathrm{Ve}}\mathbb{D}_X\mathcal{F}^\bullet) &\leq -j \quad \text{(the cosupport condition).}
\end{aligned}
$$

In particular $\mathcal{F}^\bullet$ is perverse if and only if its Verdier-dual is perverse and the notion of perversity does not depend on the chosen stratification.

Example 13.18. 1) If $i : Z \hookrightarrow X$ is the inclusion of a closed subvariety and $\mathcal{F}^\bullet$ perverse on Z, then by Remark 13.11.i) its extension $i_*\mathcal{F}^\bullet$ to X is perverse. By the same remark, of $j : U \hookrightarrow X$ is the inclusion of an open set and $\mathcal{G}^\bullet$ perverse on X, then $j^{-1}\mathcal{G}^\bullet$ is perverse on U.
2) The constant sheaf $\underline{\mathbb{Q}}_X$ on a complex manifold is not perverse, but $\underline{\mathbb{Q}}_X[n]$ is. More generally, the complex $\underline{\mathbb{Q}}_X[n]$ is perverse on a locally complete intersection variety X. The support condition for $\mathbb{Q}[n]$ is empty. The cosupport condition states that the complex dimension of the locus of points x such that $H^{n+k}_c(U_x, \mathbb{Q}) \neq 0$ for all sufficiently small neighbourhoods U_x of x should at most be k. Assume that X has been stratified. By Prop. C.8 for x on a stratum of dimension d, for all sufficiently small neighbourhoods U of x, we have $H^k_c(U, \mathbb{Q}) = 0$ unless $k = n+d, n+d+1, 2n$. The cosupport condition then follows.

Motivated by this last example, we consider any local system $\mathbb{V}$ over a dense open subset of the regular locus of a complex variety X of pure dimension d_X and introduce its **perverse extension** to X of $\mathbb{V}$

$$
{}^\pi\mathbb{V}_X := \mathcal{IC}^\bullet_X\, \mathbb{V}[d_X].
$$

It is indeed a perverse complex and hence has an intrinsic meaning. In fact we have

Theorem 13.19 ([Bor84, Chap.V, 4]). *Let X be an n-dimensional stratifyable complex space and let U be a dense smooth Zariski open subset of X on which we have a a local system of R-modules of finite rank $\mathbb{V}$. The complex ${}^{\pi}\mathbb{V}_X$ is up to quasi-isomorphism the unique perverse complex of sheaves of R-modules on X which restricts over U to $\mathbb{V}[n]$ and which has no non-trivial perverse sub- or quotient complexes with support on $X - U$.*

Remark. A word of warning. In the topological setting of n-dimensional pseudo-manifolds of [Bor84] the perversity is an increasing function $[2, n] \to \mathbb{Z}_{\geq 0}$ and the intersection complex itself is called perverse. In the analytic setting, perversities are decreasing functions $2\mathbb{N} \to \mathbb{Z}$ and they are normalized by setting $p(0) = 0$. We stick to the latter convention. The middle perversity π then is given by $\pi(2k) = -k$ (instead of by $\pi(2k) = \pi(2k-1) = k - 1$). We leave it to the reader to make the translation.

Corollary 13.20. *We have* ${}^{\mathrm{Ve}}\mathbb{D}_X({}^{\pi}\mathbb{V}_X) = {}^{\pi}(\mathbb{V}_X^{\vee})$.

Proof. By Theorem 13.19 ${}^{\pi}\mathbb{V}_X$ is perverse and hence, by Lemma 13.17 so is its Verdier dual. It has no trivial perverse sub- or quotient complexes with support on $X - U$. So, again by Theorem 13.19 it is a shifted intersection complex, i.e. of the form ${}^{\pi}\mathbb{W}_X$ for a local system on a suitable open subset of X. Since restriction to open sets commutes with the Verdier dual (Remark 13.11 i.) we have $\mathbb{W}_X = \mathbb{V}_X^{\vee}$. □

By Corollary 13.20 the dual of an intersection complex is again an intersection complex (up to a shift); this implies:

Proposition 13.21. *There is a natural isomorphism*

$$IH_k(X; \mathbb{V}^{\vee}) \xrightarrow{\sim} IH_{2n-k}^{\mathrm{BM}}(X; \mathbb{V})^{\vee}.$$

Remark. a) In addition, [Bor84, V, 9.16] states that the above isomorphism is induced by the natural pairing

$$IH_k(X; \mathbb{V}^{\vee}) \otimes IH_{2n-k}^{\mathrm{BM}}(X; \mathbb{V}) \longrightarrow R$$

which therefore is non-degenerate.

b) We have sketched a proof for R a field; the assertion remains true for a commutative ring with unit. See [Bor84, Ch. V].

We shall now explain why the R-perverse complexes on X form an abelian category which we shall denote $\mathrm{Perv}(X; R)$.

Lemma 13.22. *The category* $\mathrm{Perv}(X; R)$ *is an abelian category.*

Proof. The category $D^{\mathrm{b}}_{\mathrm{cs}}(X; R)$ inherits the triangulated structure from the category of bounded complexes of R-sheaves on X. Let $D^{\mathrm{b}}_{\mathrm{cs}}(X; R)_{\leq 0}$ respectively $D^{\mathrm{b}}_{\mathrm{cs}}(X; R)_{\geq 0}$ be the subcategory whose objects are the complexes satisfying the support condition, respectively the co-support condition. By [B-B-D, Ch. 2] this defines a t-structure (Def. A.21) whose core consists of the perverse complexes. Since a core is an abelian subcategory this shows that $\mathrm{Perv}(X; R)$ is abelian. □

The truncation functors for the t-structure in the preceding proof define the **perverse truncation functors** ${}^{\pi}\tau$. The **perverse cohomology** of any complex $\mathcal{F}^\bullet$ in $D^{\mathrm{b}}_{\mathrm{cs}}(X;R)$ are by definition the perverse objects obtained by applying the cohomology functor (A–18):

$$ {}^{\pi}H^k(\mathcal{F}^\bullet) := {}^{\pi}\tau_{\leq 0}\, {}^{\pi}\tau_{\geq 0}(\mathcal{F}^\bullet[k]). \tag{XIII–16} $$

Examples 13.23. i) If $\mathcal{F}^\bullet$ is itself perverse, then ${}^{\pi}H^k(\mathcal{F}^\bullet) = 0$ unless $k = 0$ and then ${}^{\pi}H^0(\mathcal{F}^\bullet) = \mathcal{F}^\bullet$.

ii) Let $f : X \to Y$ be a smooth proper morphism between complex manifolds and let $\mathbb{V}$ be a local system on X. The direct image $Rf_*\mathbb{V}$ is a complex of sheaves whose cohomology sheaves $R^k f_*\mathbb{V}$ are locally constant and we have ${}^{\pi}H^k(Rf_*\mathbb{V}) = Rf_*^{k-m}\mathbb{V}[m]$, where $m = \dim Y$. The terms in the Leray spectral sequence then read

$$ \mathbb{H}^p(X, {}^{\pi}H^q(Rf_*\mathbb{V})) = \mathbb{H}^p(X, R^{q-m} f_*\mathbb{V}[m]) = H^{p+m}(X, R^{q-m} f_*\mathbb{V}) \tag{XIII–17} $$

We next consider functors on the category of perverse complexes: Every additive functor T between triangulated categories equipped with t-structures induces a functor ${}^tT$ (see formula (A–20)) between the cores of the categories. If T is a t-exact functor one of course has ${}^tT = T$. As remarked in Examples 13.18 the two functors f^{-1} for an open embedding and f_* for a closed embedding are t-exact in the perverse setting. In this setting we will rather use ${}^{\pi}T$ to denote the functor between the perverse complexes associated to a functor T between complexes of sheaves. As an example, let X be a complex variety and let $j : U \hookrightarrow X$ the inclusion of some dense open smooth subset, then

$$ {}^{\pi}j_!\underline{\mathbb{Q}}_U = j_!(\underline{\mathbb{Q}}_U[d_X]) $$

and similarly for ${}^{\pi}j_*\underline{\mathbb{Q}}_U$. Since we have a natural morphism $j_! \to j_*$ of functors we can define the perverse intermediate direct image functor

$$ {}^{\pi}j_{!*}\mathcal{F}^\bullet = \mathrm{Im}[{}^{\pi}j_!\mathcal{F}^\bullet \to {}^{\pi}j_*\mathcal{F}^\bullet], \quad \mathcal{F}^\bullet \in \mathrm{Perv}(U;R) \tag{XIII–18} $$

which produces a perverse complex out of a perverse complex $\mathcal{F}^\bullet$. Using these notions, we have the following description of the intersection complex which does not make use of stratifications [B-B-D, § 0]:

Proposition 13.24. *Let X be a complex variety, $Z \subset X$ an equidimensional subvariety of X, and $\mathbb{V}$ a local system of finite dimensional R-vector spaces defined over an open dense subset U of Z consisting of smooth points of Z. With $i : Z \hookrightarrow X$ and $j : U \hookrightarrow Z$ the inclusions the intersection complex $\mathcal{IC}^\bullet_X\,\mathbb{V}[d_Z]$ is equal to*

$$ {}^{\pi}\mathbb{V}_Z := i_*({}^{\pi}j_{!*}\mathbb{V}) = i_*(j_{!*}\mathbb{V}[d_Z]), \quad \dim Z = d_Z. \tag{XIII–19} $$

Motivated by this proposition we introduce:

Definition 13.25. Let X, $Z \subset X$, and $\mathbb{V}$ as before. Then, using the above notation (XIII–19), ${}^{\pi}\mathbb{V}_Z$ is called the **perverse extension** of the local system $\mathbb{V}$ to X.

Remark. Strictly speaking, the perverse extension of a local system is *not* an extension of the system, but only of the system viewed as a complex put in a suitable degree.

In case Z is irreducible and $\mathbb{V}$ comes from an irreducible representation of $\pi_1(U)$, its perverse extension to X is a simple object in the category $\mathrm{Perv}(X;R)$. Here we recall that an object in an abelian category is **simple** if it has no non-trivial sub objects and quotient objects. An abelian category is called **artinian** if every object admits a Jordan-Hölder sequence, i.e. a finite filtration whose successive gradeds are simple. To see that ${}^{\pi}\mathbb{V}_Z$ is indeed simple, note that it is clearly simple over the largest open subset $U \subset Z$ on which $\mathbb{V}$ is locally constant, and by the characterization of the intersection complex (Prop. 13.18), it has no perverse sub-objects or quotient-objects supported on the complement of U in Z.

Lemma 13.26 ([B-B-D]). *If $\mathcal{F}^\bullet \in$* $\mathrm{Perv}(X;R)$ *and $\mathcal{F}^\bullet$ is cohomologically constructible with respect to a finite stratification of X, then $\mathcal{F}^\bullet$ admits a Jordan-Hölder sequence. In particular, if X is complex algebraic,* $\mathrm{Perv}(X;R)$ *is artinian. The simple objects $\mathcal{F}^\bullet$ are the intersection complexes ${}^{\pi}\mathbb{V}_Z$ supported on an irreducible subspace $Z \subset X$ and where $\mathbb{V}$ is associated to an irreducible representation of $\pi_1(U)$, $U \subset Z$ the largest open subset of Z over which $\mathcal{F}^\bullet$ is locally constant. They are perverse on Z as well as on X.*

So, by Prop. 13.24, if X is a complex algebraic variety, an R-perverse complex admits a finite filtration such that the successive quotients are exactly the intersection complexes ${}^{\pi}\mathbb{V}_Z$ supported on some irreducible subvariety Z of X which are associated to an irreducible local system $\mathbb{V}$ of R-vector spaces on a dense open subset of Z contained in the regular locus of Z. Since $\mathrm{Perv}(X;R)$ is abelian, we can consider its Grothendieck group (Def. A.4):

Corollary 13.27. *The Grothendieck group $K_0(\mathrm{Perv}(X;R))$ is generated by the classes of ${}^{\pi}\mathbb{V}_Z$, where $Z \subset X$ is an irreducible algebraic subvariety and $\mathbb{V}$ an irreducible local system defined over a dense open subset of Z contained in the regular locus of Z.*

13.2.3 An Example: Nearby and Vanishing Cycles

We recall briefly the set-up from § 11.2.3. Let X be a complex manifold and consider a one-parameter degeneration $f : X \to \Delta$, i.e. over the punctured disk $\Delta^* = \Delta - \{0\}$ the map f has maximal rank. Fix a ring R. With reference to the diagram (XI–13) for any bounded below complex $\mathcal{K}^\bullet$ of R-modules on X we have introduced the complex of nearby and vanishing cycles, defined by

$$\psi_f \mathcal{K}^\bullet = i^* Rk_* k^* \mathcal{K}^\bullet$$
$$\phi_f \mathcal{K}^\bullet = \mathrm{Cone}^\bullet(\mathrm{sp}).$$

Let us recall (see (XI–16)) that the specialization morphism $\mathrm{sp} : i^*\mathcal{K}^\bullet \to \psi_f \mathcal{K}^\bullet$ is the morphism induced by the natural map $\mathcal{K}^\bullet \to Rk_* k^* \mathcal{K}^\bullet$. We fix a Whitney stratification of X adapted to the map f in the sense that the open stratum is $X - X_0$. Suppose from now on that $\mathcal{K}^\bullet$ is (analytically) cohomologically constructible (Def. 13.14). From the description of the retraction map as given in § 11.2.3, it is clear that if the Whitney stratification is such that the cohomology sheaves of $\mathcal{K}^\bullet$ are constant along strata, also its nearby cohomology sheaves will be constant along the same strata. Since a given stratification can be refined to Whitney stratification, we can always assume this (property C.7.2). Moreover, the preceding description of the specialization map shows that then also its vanishing cohomology is constant along the strata. In other words, we have shown:

Proposition 13.28. *Let X be an analytic space, $f : X \to \Delta$ a surjective analytic map, $\mathcal{K}^\bullet$ a bounded below complex of sheaves of R-modules on X with constructible cohomology. Then its nearby and vanishing cohomology sheaves are constructible as well, i.e. the cohomology of $\psi_f \mathcal{K}^\bullet$ and $\phi_f \mathcal{K}^\bullet$ is constructible.*

Perversity is a much deeper property. In fact, there are shifts, and we cite a result, due to O. Gabber [Bry]:

Proposition 13.29. *In the setting of the previous proposition, if $\mathcal{K}^\bullet$ is a perverse complex, the complexes $\phi_f \mathcal{K}^\bullet[-1]$ and $\psi_f \mathcal{K}^\bullet[-1]$ are perverse on X_0.*

13.3 Introduction to D-Modules

13.3.1 Integrable Connections and D-Modules

Let X be a complex manifold. Germs of vector fields on X are standard examples of differential operators of order ≤ 1. Multiplication with germs of holomorphic functions give 0-th order differential operators. Together, as a sheaf of $\mathcal{O}_X$-algebras they generate the sheaf $\mathcal{D}_X$ of differential operators on X. More precisely, we can recursively define this sheaf as follows. A local section P of $\mathrm{End}_{\mathbb{C}}(\mathcal{O}_X)$ is said to be a **differential operator**

- of order 0 if it is multiplication with a germ $g \in \mathcal{O}_X$, i.e. $Pf = gf$ for all $f \in \mathcal{O}_X$;
- of order $\leq m$ (for m a positive integer) if for all germs $g \in \mathcal{O}_X$ the operator

$$f \mapsto P(gf) - gP(f)$$

  has order $\leq (m-1)$.

We let $F_m^{\mathrm{ord}}\mathcal{D}_X$ denote the sheaf of differential operators of order $\leq m$ and

$$\mathcal{D}_X = \bigcup_{m\geq 0} F_m^{\mathrm{ord}}\mathcal{D}_X \subset \mathrm{End}_{\mathbb{C}}(\mathcal{O}_X).$$

Using multi-index notation $I = \{i_1, \dots, i_n\}$, $|I| = \sum_k i_k$, in local coordinates $(U, z_1, \dots, z_n)$ an m-th order differential operator can be uniquely written as

$$P = \sum_{|I|\leq m} P_I \partial^I, \quad P_I \in \mathcal{O}_X(U), \, \partial^I = \partial_1^{i_1} \cdots \partial_n^{i_n}.$$

We define its m**-th order symbol** as

$$\sigma_m(P) := = \sum_{|I|=m} \xi^I \in \mathrm{Sym}^m(T_X)(U), \quad \xi_j = [\partial/\partial z_j].$$

Clearly σ_m induces isomorphisms

$$[\sigma_m] : \mathrm{Gr}_m^{F^{\mathrm{ord}}} \mathcal{D}_X \xrightarrow{\sim} \mathrm{Sym}^m_{\mathcal{O}_X}(T_X). \tag{XIII–20}$$

and

$$[\sigma] : \mathrm{Gr}\, \mathcal{D}_X \xrightarrow{\sim} \mathrm{Sym}(T_X). \tag{XIII–21}$$

We observe:

Lemma 13.30. *Let $\pi : T^\vee X \to X$ be the holomorphic cotangent bundle of X. Then the sheaf $\mathrm{Gr}^{F^{\mathrm{ord}}} \mathcal{D}_X$ can be identified with the subsheaf $\pi_* \mathcal{O}_{T^\vee X}$ which which restrict to a polynomial on the cotangent space $T_x^\vee X$ for each $x \in X$.*

A sheaf $\mathcal{M}$ of left $\mathcal{D}_X$-modules is called a $\mathcal{D}_X$**-module**, or, if no confusion is possible, a D-module. So $\mathcal{M}$ admits a left multiplication with germs of vector fields, or, in other words, we obtain a Lie-algebra *representation*. Conversely, there is a unique way of extending a Lie-algebra representation to a D-module:

Lemma 13.31. *Let X be a complex manifold and $\mathcal{M}$ an $\mathcal{O}_X$-module. Then there is a 1-1 correspondence between*

a) *(left) $\mathcal{D}_X$-module structures on $\mathcal{M}$,*
b) *$\mathcal{O}_X$-linear homomorphisms of sheaves of Lie algebras*

$$\rho : T_X \to \mathcal{H}om_{\mathbb{C}_X}(\mathcal{M}, \mathcal{M}). \tag{XIII–22}$$

satisfying Leibniz' rule

$$\rho(\theta)(fm) = \theta(f)m + f\rho(\theta)m$$

for local sections f, θ, m of $\mathcal{O}_X$, T_X, $\mathcal{M}$ respectively.

For $\mathcal{M}$ a locally free $\mathcal{O}_X$-module, these structures correspond to an integrable connection on $\mathcal{M}$.

Example 13.32. a) The structure sheaf $\mathcal{O}_X$ is a trivial example of a $\mathcal{D}_X$-module: if ξ is a local holomorphic vector field on X then $\xi(f) = d_\xi f$, the directional derivative of f in the direction of ξ.

b) There is an action of T_X on k-forms given by the **Lie-derivative**. Given a local holomorphic k-form α on X Lie derivation along a local vector field ξ on X produces a k-form $L_\xi(\alpha)$.

$$L_\xi(\alpha) = di_\xi(\alpha) + i_\xi d\alpha,$$

where i_ξ stands for contraction. It is well known that L_ξ is indeed a derivation, i.e. satisfies the Leibniz rule. Likewise, one has $L_{[\xi,\eta]} = [L_\xi, L_\eta]$, i.e. this gives a Lie-algebra homomorphism. However, for all $f \in \mathcal{O}_X$, one has

$$\begin{aligned} L_{f\xi}(\alpha) - fL_\xi(\alpha) &= df \wedge i_\xi\alpha \\ L_{f\xi}(\alpha) - L_\xi(f\alpha) &= -i_\xi(df \wedge \alpha). \end{aligned}$$

The first formula shows that the Lie-algebra homomorphism is *not* $\mathcal{O}_X$-linear and we don't have a connection. The last formula shows that for top-degree differential forms ω the *right action* $\omega\xi = L_\xi\omega$ becomes $\mathcal{O}_X$-linear. In order to obtain a Lie-algebra action of T_X we need to change signs by letting ξ act as $-L_\xi$ since then $-L_{[\xi,\eta]} = [-L_\eta, -L_\xi]$. We reserve a special notation for this right $\mathcal{D}_X$-module:

$$\omega_X := \Omega_X^{d_X}, \quad d_X = \dim(X). \qquad \text{(XIII–23)}$$

13.3.2 From Left to Right and Vice Versa

Many examples of $\mathcal{D}_X$-modules come naturally as right D-modules and so it is useful to be able to pass from right $\mathcal{D}_X$-modules to left $\mathcal{D}_X$-modules and vice-versa. Locally, one can do this by inverting the order of differential operators, but this does not work well globally. To achieve this, we use the right $\mathcal{D}_X$-module ω_X from (XIII–23). It can indeed be used to turn left $\mathcal{D}_X$-modules $\mathcal{M}$ into right $\mathcal{D}_X$-modules by tensoring on the left by ω_X. The action

$$(\omega \otimes m)\xi := \omega\xi \otimes m - \omega \otimes \xi m$$

extends to a right $\mathcal{D}_X$-module structure on

$$\mathcal{M}^{\text{rgt}} := \omega_X \otimes_{\mathcal{O}_X} \mathcal{M}.$$

Similarly for any right $\mathcal{D}_X$-module $\mathcal{N}$, we obtain a left $\mathcal{D}_X$-module structure on $\mathcal{N}^{\text{left}} := \mathcal{H}om_{\mathcal{O}_X}(\omega_X, \mathcal{N})$ by putting

$$(\xi\phi)(\omega) := \phi(\omega\xi) - \phi(\omega)\xi, \qquad \begin{array}{l} \xi, \omega \text{ germs of a section of } T(X), \text{resp. } \omega_X \\ \phi \text{ germ of a section of } \mathcal{N}^{\text{left}}. \end{array}$$

The last two constructions give an equivalence between the categories of left and right $\mathcal{D}_X$-modules. The results of these two operations are also called **the left-right and right-left transforms**.

For tensor product and homomorphisms the following rules apply. The proofs are left to the reader.

Observation 13.33. Given two left D-modules, $\mathcal{M}$ and $\mathcal{N}$ the Leibniz rule makes the tensor product $\mathcal{M} \otimes_{\mathcal{O}_X} \mathcal{N}$ *over the structure sheaf* into a D-module. Similarly, the module $\mathcal{H}om_{\mathcal{O}_X}(\mathcal{M}, \mathcal{N})$ is a left $\mathcal{D}_X$-module by the rule

$$(\xi\phi)(m) = \xi(\phi(m)) - \phi(\xi m),$$

where ϕ is a germ of an $\mathcal{O}_X$-homomorphism from $\mathcal{M}$ to $\mathcal{N}$, ξ is a local vector field and m a local section of $\mathcal{M}$. For tensor products over $\mathcal{D}_X$ some care must be taken. For example, consider $\mathcal{R} \otimes_{\mathcal{D}_X} \mathcal{L}$, where $\mathcal{R}$ is a right $\mathcal{D}_X$-module, and $\mathcal{L}$ has a left action. Then the result in general is only an $\mathcal{O}_X$-module structure; the vector fields no longer act.

13.3.3 Derived Categories of D-modules

Let X be a complex manifold of dimension n and let $\mathcal{M}$ be a (left) $\mathcal{D}_X$-module.. The category of $\mathcal{D}_X$-modules has enough injectives (see Example A.22.1)) so that we can derive all left exact functors. For instance, we can derive the Hom-functor. To do this, first recall that $\mathfrak{A}^\circ$ stands for the category opposite to $\mathfrak{A}$, i. e having the same objects as $\mathfrak{A}$ put with all arrows reversed. We now put

$$\begin{aligned} D^-(\mathcal{D}_X\text{-modules})^\circ \times D^+(\mathcal{D}_X\text{-modules}) &\to D^+(\text{sheaves of } \mathbb{C}_X\text{-modules}) \\ (\mathcal{N}^\bullet \ , \ \mathcal{M}^\bullet) &\longmapsto R\mathcal{H}om_{\mathcal{D}_X}(\mathcal{N}^\bullet, \mathcal{M}^\bullet), \end{aligned}$$

which can be represented by the complex $\mathcal{H}om^\bullet(\mathcal{N}^\bullet, I^\bullet(\mathcal{M}^\bullet))$ with $I^\bullet(\mathcal{M}^\bullet)$ an injective resolution of $\mathcal{M}^\bullet$.

We also need to derive the left tensor product. We have seen (Observation. 13.33) that tensoring right by left D-modules in general only gives $\mathcal{O}_X$-modules. A boundedness property (see [Kash-S, §2.6]) then guarantees that there is a well defined bi-functor

$$\begin{aligned} D^{\mathrm{b}}(\mathcal{D}_X^{\mathrm{rgt}}\text{-modules}) \times D^+(\mathcal{D}_X\text{-modules}) &\to D^+(\text{sheaves of } \mathcal{O}_X\text{-modules}) \\ (\mathcal{N}^\bullet \ , \ \mathcal{M}^\bullet) &\longmapsto \mathcal{N}^\bullet \overset{\mathrm{L}}{\otimes}_{\mathcal{D}_X} \mathcal{M}^\bullet. \end{aligned}$$

In examples we often have explicit finite resolutions coming from the De Rham complex (X–3) of some D-module $\mathcal{M}$:

$$\mathrm{DR}_X(\mathcal{M}) = \left[\left(\mathcal{M} \to \Omega^1_X(\mathcal{M}) \to \cdots \to \Omega^{d_X}_X(\mathcal{M})\right)\right][d_X], \qquad \text{(XIII–24)}$$

where the derivatives in the complex come from the integrable connection associated to the D-module structure. For example, by [Bor87, § VII.3.5], the

Rham complex of $\mathcal{D}_X$ is a locally free left resolution of ω_X (XIII–23) by right $\mathcal{D}_X$-modules. So, if $\mathcal{M}$ is a (left) $\mathcal{D}_X$-module, we have

$$\mathrm{DR}_X(\mathcal{M}) = \omega_X \overset{\mathrm{L}}{\otimes}_{\mathcal{D}_X} \mathcal{M}$$

which motivates to define for any bounded below complex $\mathcal{M}^\bullet$ of $\mathcal{D}_X$-modules

$$\mathrm{DR}_X(\mathcal{M}^\bullet) := \omega_X \overset{\mathrm{L}}{\otimes}_{\mathcal{D}_X} \mathcal{M}^\bullet. \tag{XIII–25}$$

The shift d_X is the same as the shift needed to turn a local system into a perverse complex (Theorem 13.19), tying in with the Riemann-Hilbert correspondence, as we see later (Theorem 13.61). This (modified) **De Rham complex** is a bounded complex of $\mathcal{O}_X$-modules whose maps are only $\mathbb{C}_X$-linear.

Definition 13.34. The **dualizing left D-module** is the module

$$\mathcal{D}_X^* = \mathcal{H}om_{\mathcal{O}_X}(\omega_X, \mathcal{D}_X^{\mathrm{rgt}}) = \mathcal{D}_X^{\mathrm{rgt}} \otimes_{\mathcal{O}_X} \omega_X^{-1}.$$

where we view it as a left-module via right-left transform of the right $\mathcal{D}_X$-module structure on $\mathcal{D}_X$. Call this the **natural left-structure**. The left-structure on $\mathcal{D}_X$ persists on the dualizing module and is called the **supplementary left-structure**.

Using the supplementary left structure on this module, given a left $\mathcal{D}_X$-module $\mathcal{M}$ the **dual module**

$$\mathcal{M}^* := \mathcal{H}om_{\mathcal{D}_X}(\mathcal{M}, \mathcal{D}_X^*) \tag{XIII–26}$$

gets the structure of a left $\mathcal{D}_X$-module. When working in the derived category it is conventionally put in degree $-d_X$; in fact the definition can be extended to bounded complexes of $\mathcal{D}_X$-modules as follows

$$\mathbb{D}_X(\mathcal{M}^\bullet) := R\mathcal{H}om_{\mathcal{D}_X}(\mathcal{M}^\bullet, \mathcal{D}_X^*)[d_X]. \tag{XIII–27}$$

This **dual complex** indeed turns out to be a bounded complex of (left) $\mathcal{D}_X$-modules. Applying the De Rham functor to this construction gives [Bor87, VI.9.7]:

$$\mathrm{DR}_X(\mathbb{D}_X(\mathcal{M}^\bullet)) = R\mathcal{H}om_{\mathcal{D}_X}(\mathcal{M}^\bullet, \mathcal{O}_X)[d_X]. \tag{XIII–28}$$

13.3.4 Inverse and Direct Images

Let $f : X \to Y$ be a holomorphic map between complex manifolds. Set

$$d_X = \dim X, \quad d_Y = \dim Y, \quad d_{X/Y} = \dim X - \dim Y.$$

Recall that for any $\mathcal{O}_Y$-module $\mathcal{N}$, the **analytic inverse image** is defined as

$$f^*\mathcal{N} = \mathcal{O}_X \otimes_{f^{-1}\mathcal{O}_Y} f^{-1}\mathcal{N}.$$

When $\mathcal{N}$ is a left $\mathcal{D}_Y$-module we want to make this $\mathcal{O}_X$-module into a left $\mathcal{D}_X$-module. So let ξ be a local holomorphic vector field on X. The image $f_*\xi$ can be viewed as a local section of the inverse image $f^{-1}T(Y)$ and it acts on $f^*\mathcal{N}$ by the chain rule $\xi(u \otimes n) = \xi(u) \otimes n + u \otimes f_*\xi(n)$, where u is a germ of a holomorphic function on X and n a local section of $f^{-1}\mathcal{N}$.

We need an alternative procedure which is better suited for the derived category. It uses the **transfer module**

$$\mathcal{D}_{X\to Y} := f^*\mathcal{D}_Y \tag{XIII–29}$$

This is not only a left $\mathcal{D}_X$-module, but also has a natural right $f^{-1}\mathcal{D}_Y$-module structure; we say that $\mathcal{D}_{X\to Y}$ has a $(\mathcal{D}_X, f^{-1}\mathcal{D}_Y)$**-bimodule structure**. It can be tensored on the right by any left $f^{-1}\mathcal{D}_Y$-module such as $f^{-1}\mathcal{N}$ and produces a left $\mathcal{D}_X$-module. This gives an alternative definition of the inverse image

$$f^*\mathcal{N} = \mathcal{D}_{X\to Y} \otimes_{f^{-1}\mathcal{D}_Y} f^{-1}\mathcal{N} \tag{XIII–30}$$

which extends to the derived category as

$$\begin{array}{rcl} f^! : D^+(\mathcal{D}_Y\text{-modules}) & \to & D^+(\mathcal{D}_X\text{-modules}) \\ \mathcal{N}^\bullet & \mapsto & \mathcal{D}_{X\to Y} \overset{\mathbf{L}}{\otimes}_{f^{-1}\mathcal{D}_Y} f^{-1}(\mathcal{N}^\bullet)[d_{X/Y}]. \end{array} \tag{XIII–31}$$

Example 13.35. Let $u : \Delta \to X$ be a holomorphic map of the unit disk to X and let $\mathcal{M}$ be a $\mathcal{D}_X$-module seen as an $\mathcal{O}_X$-module with a connection ∇. We have defined the pull back connection $u^*\nabla$ using formula (XI–10). The corresponding $\mathcal{D}_\Delta$-module is $u^*\mathcal{M}$, and, when viewed as a complex in degree $-d_{X/Y}$ is $u^!\mathcal{M}$. By Remark 11.6, this holds more generally for any smooth curve $u : C \to X$ mapping to X.

Using the $(\mathcal{D}_X, f^{-1}\mathcal{D}_Y)$-bimodule $\mathcal{D}_{X\to Y}$, the **direct image** for right D-modules $\mathcal{R}$ is easy to define:

$$f_+\mathcal{R} := f_*\mathcal{R} \otimes_{\mathcal{D}_X} \mathcal{D}_{X\to Y}.$$

We have seen that the tensor product in general admits no D-module structure. Here however the right module structure on the transfer module persists so that the direct image is a right $\mathcal{D}_X$-module.

To define the direct image for left D-modules, we need the transfer module $\mathcal{D}_{Y\leftarrow X}$ which is a $(f^{-1}\mathcal{D}_Y, \mathcal{D}_X)$-bimodule. It is defined as follows. First recall (Def. 13.34) that the dualizing left D-module $\mathcal{D}_X^*$ has two left-module structures, the natural left-structure and the supplementary left-structure. So the inverse image $f^*\mathcal{D}_Y^*$ of the dualizing D-module has a left $\mathcal{D}_X$-structure, coming from the natural left-structure on $\mathcal{D}_Y^*$ but retains the left $f^{-1}\mathcal{D}_Y$-structure coming from the supplementary left-structure on $\mathcal{D}_Y^*$. Using the **relative canonical bundle**

$$\omega_{X/Y} := (f^*\omega_Y)^{-1} \otimes \omega_X,$$

we can transfer the left $\mathcal{D}_X$-structure to the right, obtaining the sought after $(f^{-1}\mathcal{D}_Y, \mathcal{D}_X)$-bimodule

$$\mathcal{D}_{Y \leftarrow X} := f^*(\mathcal{D}_Y^*) \otimes_{\mathcal{O}_X} \omega_X = f^*\mathcal{D}_Y \otimes_{\mathcal{O}_Y} \omega_{X/Y}. \tag{XIII–32}$$

Tensoring on the right with a left $\mathcal{D}_X$-module $\mathcal{M}$ produces the left $f^{-1}\mathcal{D}_Y$-module

$$\mathrm{DR}_{X/Y}(\mathcal{M}) := \mathcal{D}_{Y \leftarrow X} \otimes_{\mathcal{D}_X} \mathcal{M} \tag{XIII–33}$$

Next we consider the adjoint mapping (B–23) (as a map of sheaves of rings)

$$\mathcal{D}_Y \to f_* f^{-1}\mathcal{D}_Y.$$

Here f_* is the topological direct image. The adjoint mapping makes the topological direct image $f_*\mathcal{A}$ of an $f^{-1}\mathcal{D}_Y$-module $\mathcal{A}$ on X into a left $\mathcal{D}_Y$-module on Y which we denote by $\tilde{f}_*(\mathcal{A})$. In particular, applying this to $\mathcal{A} = \mathrm{DR}_{X/Y}(\mathcal{M})$ defines the **direct image**

$$f_+\mathcal{M} := \tilde{f}_* \left(\mathcal{D}_{Y \leftarrow X} \otimes_{\mathcal{D}_X} \mathcal{M}\right),$$

again a left $\mathcal{D}_Y$-module. This extends to the derived category as follows

$$\begin{array}{rclc} f_+ : D^+(\mathcal{D}_X\text{-modules}) & \to & D^+(\mathcal{D}_Y\text{-modules}) & \\ \mathcal{M}^\bullet & \mapsto & R\tilde{f}_* \left(\mathcal{D}_{Y \leftarrow X} \overset{\mathrm{L}}{\otimes}_{\mathcal{D}_X} \mathcal{M}^\bullet\right). & \end{array} \tag{XIII–34}$$

The cohomology sheaves of these complexes (the higher direct images) are often denoted as follows

$$\int_f^q \mathcal{M} := R^q f_+ \mathcal{M} = H^q(f_+\mathcal{M}).$$

We use the duality operator to define the two remaining functors f^+, $f_!$ associated to f, i.e.

$$f^+ := \mathbb{D}_X \circ f^! \circ \mathbb{D}_Y \quad f_! := \mathbb{D}_Y \circ f_+ \circ \mathbb{D}_X.$$

In other words, we get a diagram

$$\begin{array}{ccc} D^+(\mathcal{D}_X\text{-modules}) & \underset{f_!}{\overset{f^+}{\rightleftarrows}} & D^+(\mathcal{D}_Y\text{-modules}) \\ \updownarrow \mathbb{D}_X & & \mathbb{D}_Y \updownarrow \\ D^+(\mathcal{D}_X\text{-modules}) & \underset{f_+}{\overset{f^!}{\rightleftarrows}} & D^+(\mathcal{D}_Y\text{-modules}) \end{array} \tag{XIII–35}$$

analogous to the topological version (XIII–9).

13.3.5 An Example: the Gauss-Manin System

Let $f : X \to S$ be a proper submersion between complex manifolds. Consider the p-th direct image of the $\mathcal{D}_X$-module $\mathcal{O}_X$, which is the **Gauss-Manin system** $R^p f_+ \mathcal{O}_X$. Since f is a submersion, one can easily show that the shifted relative De Rham complex (Def. 10.25)

$$\mathrm{DR}^\bullet_{X/S}(\mathcal{O}_X) := \Omega^\bullet_{X/S}[d_{X/S}]$$

is a locally free resolution of the transfer module $\mathcal{D}_{S \leftarrow X}$ ([Pham, Lemme 14.3.5]) and hence, by the definition of the direct image we conclude that the Gauss-Manin system equals $R^p f_*(\mathrm{DR}^\bullet_{X/S}(\mathcal{O}_X))$ as an $\mathcal{O}_S$-module. It carries a D-module structure in the guise of a flat connection uniquely determined by its sheaf of locally constant sections, the local system $R^{p+d_{X/S}} f_* \underline{\mathbb{C}}_X$. In other words, up to a shift in degree, this is exactly the Gauss-Manin connection.

More generally, consider direct images of any (left) $\mathcal{D}_X$-module $\mathcal{M}$ given by a flat connection $\nabla : \mathcal{M} \to \Omega^1_X \otimes_{\mathcal{O}_X} \mathcal{M}$. The Koszul filtration $\mathrm{Koz}^q\, \Omega^\bullet_X$ induces on $\Omega^\bullet_X \otimes_{\mathcal{O}_X} \mathcal{M}$ the filtration $\mathrm{Koz}^q\, \Omega^\bullet_X \otimes_{\mathcal{O}_X} \mathcal{M}$. We have

$$\mathrm{Gr}^q_{\mathrm{Koz}}(\Omega^\bullet_X \otimes \mathcal{M}) = f^* \Omega^q_S \otimes \Omega^{\bullet - q}_{X/S} \mathcal{M}.$$

As in Thm. 10.28, the connecting morphism in the long exact sequence relating $\mathrm{Gr}^0_{\mathrm{Koz}}$ and $\mathrm{Gr}^1_{\mathrm{Koz}}$ yields a flat connection which is again called **Gauss-Manin connection**

$$\nabla^{\mathrm{GM}} : R^p f_* \Omega^\bullet_{X/S} \mathcal{M} \to \Omega^1_S \otimes_{\mathcal{O}_S} R^p f_* \Omega^\bullet_{X/S} \mathcal{M}.$$

This connection gives $R^p f_* \Omega^\bullet_{X/S} \mathcal{M}$ the structure of a left $\mathcal{D}_S$-module, which in the derived category of bounded complexes of $\mathcal{D}_S$-modules computes the $(p + d_{X/S})$-th direct image sheaf $R^{p+d_{X/S}} f_+ \mathcal{M}$. To get rid of this shift, one uses the shifted **relative De Rham complex**

$$\mathrm{DR}^\bullet_{X/S}(\mathcal{M}) := \Omega^\bullet_{X/S} \otimes_{\mathcal{O}_X} \mathcal{M}\,[d_{X/S}],$$

a complex of left $f^{-1}(\mathcal{D}_S)$-modules placed in degrees $-d_{X/S}, \ldots, 0$. Using it, we thus find

$$\int_f^p \mathcal{M} = R^p(f_+ \mathcal{M}) = R^p f_*(\mathrm{DR}^\bullet_{X/S}\, \mathcal{M}).$$

13.4 Coherent D-Modules

We shall speak of $\mathcal{O}_X$-modules and (left or right) $\mathcal{D}_X$-modules if we mean a sheaf of $\mathcal{O}_X$-modules or a sheaf of (left or right) $\mathcal{D}_X$-modules.

13.4.1 Basic Definitions

Let us first recall that $\mathcal{O}_X$ is a sheaf of noetherian rings. This is by no means trivial. See for instance [Gu-Ro, Chap. II.C]. In particular, if $\mathcal{F}$ is an $\mathcal{O}_X$-module, its stalk $\mathcal{F}_x$ at $x \in X$ is Noetherian (i.e. finitely generated) if and only if it is finitely presented in the sense that there is an exact sequence.

$$\mathcal{O}^q_{X,x} \to \mathcal{O}^p_{X,x} \to \mathcal{F}_x \to 0.$$

However, if $\mathcal{F}$ is locally finitely generated, it need not be finitely presented over any smaller set. Coherence is a property that guarantees that this.

Definition 13.36. An $\mathcal{O}_X$-module $\mathcal{F}$ is **coherent** if it is first of all locally finitely generated, i.e. every point has a neighbourhood U over which there exists a surjection

$$\mathcal{O}^p_U \to \mathcal{F}|U \to 0,$$

and secondly if every homomorphism $\mathcal{O}^q_U \to \mathcal{F}|U$ has a kernel which is locally finitely generated.

It is a deep theorem due to Oka that $\mathcal{O}_X$ is coherent [Gu-Ro, Chap. IV.C]. This implies (as can be seen easily) that $\mathcal{F}$ is a coherent $\mathcal{O}_X$-module if and only if it is locally finitely presented.

Let us now pass to $\mathcal{D}_X$ and left $\mathcal{D}_X$-modules. We can and do use the same definition for coherence as before:

Definition 13.37. A $\mathcal{D}_X$-module $\mathcal{M}$ is **coherent** if it is first of all locally finitely generated, i.e. every point has a neighbourhood U over which there exists a surjection

$$\mathcal{D}^p_U \to \mathcal{M}|U \to 0,$$

and secondly if every homomorphism $\mathcal{D}^q_U \to \mathcal{M}|U$ has a kernel which is locally finitely generated.

From the fact that $\mathcal{O}_X$ is coherent it is not hard to see [Bor87, II.§3] that $\mathcal{D}_X$ is coherent (as a left-$\mathcal{D}_X$-module) and from this one deduces the following lemma.

Lemma 13.38. *A D-module is coherent if and only if it is locally finitely presented: locally over an open subset $U \subset X$ we have an exact sequence of $\mathcal{D}(U)$-modules*

$$\mathcal{D}(U)^q \xrightarrow{\alpha} \mathcal{D}(U)^p \to \mathcal{M}(U) \to 0.$$

This presentation is related to the following **differential system** on U:

$$\sum_{j=1}^q A_{ij} u_j = 0 \qquad (A_{ij}) = \text{ matrix of } \alpha \text{ with respect to the standard bases.} \qquad \text{(XIII–36)}$$

If one associates to $\mathbf{u} = (u_1, \ldots, u_q) \in \mathcal{O}(U)^q$ the homomorphism

$$\begin{aligned}\varphi_{\mathbf{u}} : \mathcal{D}(U)^q &\to \mathcal{O}(U)\\ (P_1, \ldots, P_q) &\mapsto \sum P_j(u_j)\end{aligned}$$

the map $\varphi_{\mathbf{u}}$ factors over $\mathcal{M}(U)$ if and only if $\mathbf{u}$ is a solution of the system (XIII–36). In this way we get a 1–1 correspondence

$$\text{Solutions of (XIII–36)} \overset{\text{1-1}}{\longleftrightarrow} \text{Sol}(\mathcal{M}(U)) := \text{Hom}_{\mathcal{D}(U)}(\mathcal{M}(U), \mathcal{O}(U))$$

This example motivates the following definition.

Definition 13.39. Let $\mathcal{M}$ be a coherent $\mathcal{D}_X$-module. Its **solution module** is the $\mathcal{D}_X$-module

$$\text{Sol}(\mathcal{M}) := \mathcal{H}om_{\mathcal{D}_X}(\mathcal{M}, \mathcal{O}_X).$$

In the derived category $D^{\mathrm{b}}_{\mathrm{coh}}(\mathcal{D}_X)$, this leads to the **solution complex**

$$\text{Sol}(\mathcal{M}^\bullet) := R\mathcal{H}om_{\mathcal{D}_X}(\mathcal{M}^\bullet, \mathcal{O}_X)[d_X].$$

Recalling (XIII–28)) we see that we have

$$\text{Sol}(\mathcal{M}^\bullet) = \text{DR}_X(\mathbb{D}_X\mathcal{M}^\bullet). \tag{XIII–37}$$

Examples 13.40. 1) The structure sheaf $\mathcal{O}_X$ is a left $\mathcal{D}_X$-module, generated globally by the section 1. In local coordinates $(z_1, \ldots, z_n)$ on an open set $U \subset X$ the kernel of the sheaf homomorphism ev : $\mathcal{D}_X \to \mathcal{O}_X$ given by $P \mapsto P(1)$ is generated by the vector fields $\partial_1, \ldots, \partial_n$. Hence $\mathcal{O}_X$ is a $\mathcal{D}_X$-module locally of finite presentation, and therefore a coherent $\mathcal{D}_X$-module. A coordinate invariant description of Ker(ev) can be given as follows. The sheaf T_X of germs of holomorphic tangent vectors is locally free of rank n over $\mathcal{O}_X$. Hence the tensor product $\mathcal{D}_X \otimes_{\mathcal{O}_X} T_X$ is a locally free left $\mathcal{D}_X$–module. The map $P \otimes \theta \mapsto P\theta$ defines a homomorphism of left $\mathcal{D}_X$-modules $\mathcal{D}_X \otimes T_X \to \mathcal{D}_X$ and it represents Ker(ev). This shows that $\mathcal{O}_X$ is a coherent $\mathcal{D}_X$-module.
2) Every locally free $\mathcal{D}_X$-module of finite rank is coherent.
3) Every $\mathcal{O}_X$-coherent $\mathcal{D}_X$-module $\mathcal{M}$ is locally free as an $\mathcal{O}_X$-module. To see this, it suffices to show that $\mathcal{M}_x$ is a free $\mathcal{O}_{X,x}$-module for any $x \in X$. Let $\mathfrak{m}_x$ denote the maximal ideal of $\mathcal{O}_{X,x}$ and choose elements $e_1, \ldots, e_r$ in $\mathcal{M}_x$ which map to a $\mathbb{C}$-basis of the fibre

$$\mathcal{M}(x) := \mathcal{M}_x / \mathfrak{m}_x \mathcal{M}_x.$$

By Nakayama's lemma, $\mathcal{M}_x$ is generated by $e_1, \ldots, e_r$. These generators form a free basis. We prove this by contradiction. Indeed, if not there would be a relation $\sum_{i=1}^r f_i e_i = 0$ such that not all the f_i are zero. Let k to be the minimum of the orders of vanishing at x of f_i. We call this minimum the *order of the relation.* For simplicity, assume that f_1 realizes this minimum. We arrive at a contradiction as follows. We cannot have $k = 0$, since in that case the classes of the e_i in $\mathcal{M}(x)$ become dependent. On the other hand, if $k > 0$, we can reduce order of the relation: choose i such that in local coordinates, $\partial_i f_1$ vanishes to order $k-1$ at x (this is possible since the relation is of order k). Then, writing $\partial_i e_j = \sum_k b_{jk} e_k$, we find

$$0 = \partial_i \Big(\sum_{j=1}^{s} f_j e_j\Big) = \Big(\partial_i f_1 + \sum_{k=1}^{s} f_k b_{k1}\Big) e_1 + \sum_{j=2}^{s} \Big(\partial_i f_j + \sum_{k=1}^{s} f_l b_{kj}\Big) e_j$$

which is a relation of lower order. This contradiction indeed shows that $\mathcal{M}$ is locally free as an $\mathcal{O}_X$-module.

4) Let $\mathcal{M}_X$ denote the sheaf of germs of meromorphic functions on X. This is a $\mathcal{D}_X$-module which is not locally of finite type.

13.4.2 Good Filtrations and Characteristic Varieties

Let X be a complex manifold of dimension n and $\mathcal{M}$ be a $\mathcal{D}_X$-module.

Definition 13.41. A **filtration** on $\mathcal{M}$ is an increasing and exhaustive ($\mathcal{M} = \bigcup_p F_p\mathcal{M}$) sequence of $\mathcal{O}_X$-submodules $(F_p\mathcal{M})_{p\in\mathbb{Z}}$ such that $F_r^{\text{ord}}\mathcal{D}_X[F_p\mathcal{M}] \subset F_{p+r}\mathcal{M}$ $\forall r, s \in \mathbb{Z}$. It is called a **good filtration** if moreover

1) every point has a neigbourhood on which $F_p\mathcal{M} = 0$ for all $p \ll 0$, and such that for all $r \gg 0$ one has $F_r^{\text{ord}}\mathcal{D}_X[F_p\mathcal{M}] = F_{p+r}\mathcal{M}$;
2) each $F_p\mathcal{M}$ is a coherent $\mathcal{O}_X$-module.

A **filtered D-module** is a D-module equipped with a good filtration.

Basic examples of good filtrations are given by variations of Hodge structure (Def. 10.6):

Example 13.42. Let $\mathbb{V}$ be a local system on S of real vector spaces of finite dimension and suppose that $\mathcal{V} := \mathbb{V} \otimes \mathcal{O}_S$ underlies a real variation of Hodge structures with Hodge filtration $\mathcal{F}^\bullet$. The flat connection on $\mathcal{V}$ makes $\mathcal{V}$ into a $\mathcal{D}_S$-module and the filtration given by $\mathcal{F}_\bullet$ is good, since the transversality condition ensures that $F_1^{\text{ord}}\mathcal{D}_X\mathcal{F}_k \subset \mathcal{F}_{k+1}$.

For every coherent $\mathcal{D}_X$-module $\mathcal{M}$ a good filtration exists locally on X: starting from a local presentation $\bigoplus^a \mathcal{D}_X \xrightarrow{v} \bigoplus^b \mathcal{D}_X \xrightarrow{u} \mathcal{M} \to 0$ one can put $F_p\mathcal{M} := u(\oplus^b F_p^{\text{ord}}\mathcal{D}_X)$ for $p \geq 0$ while $F_pM = 0$ for $p < 0$. Then $F_r^{\text{ord}}\mathcal{D}_X[F_p\mathcal{M}] = F_{p+r}\mathcal{M}$ for all $r, p \in \mathbb{Z}$. Conversely, if $\mathcal{M}$ locally possesses a good filtration, $\mathcal{M}$ is coherent.

To test if a given filtration is good, the following Lemma is useful (see [Bor87, II.4]. To state it, recall (XIII–21) that $\operatorname{Gr}\mathcal{D}_X$ is the graded sheaf of rings associated to $\mathcal{D}_X$ with respect to the order filtration. Similarly, we set

$$\operatorname{Gr}_F \mathcal{M} = \bigoplus_k \operatorname{Gr}_F^k \mathcal{M}.$$

Lemma 13.43. *Let $(\mathcal{M}, F)$ be a $\mathcal{D}_X$-module equipped with a filtration. Then F is good precisely when $\mathrm{Gr}_F \mathcal{M}$ is coherent as a $\mathrm{Gr}\, \mathcal{D}_X$-module.*

It is also important (and easy to show) that any two good filtrations F and G on a given $\mathcal{D}_X$-module $\mathcal{M}$ are locally commensurable in the sense that locally there exist two integers a and b such for all $p \in \mathbb{Z}$ we have $F_{p-a}\mathcal{M} \subset G_p\mathcal{M} \subset F_{p+b}\mathcal{M}$. Using this, one proves

Proposition 13.44. *Let $\mathcal{I}(\mathcal{M}, F)$ be the annihilator of $\mathrm{Gr}_F \mathcal{M}$. Then $\sqrt{\mathcal{I}(\mathcal{M}, F)}$ does not depend on the choice of the good filtration F on $\mathcal{M}$.*

Proof. Let $P \in \mathcal{D}_X$ be of order m and let $\alpha = \sigma_m(P)$ be its symbol. For some integers a, b we have $F_{p-a} \subset G_p \subset F_{p+b}$. Suppose that $\alpha^s \in \mathcal{I}(M, F)$. Then $P^{ks} F_r \subset F_{r+ksm-1}$ for all $r \in \mathbb{Z}$ and hence

$$P^{ks} G_p \subset P^{ks} F_{p+b} \subset F_{p+b+ksm-k} \subset G_{p-1+ksm}$$

as soon as $k \geq a+b+1$ and then $\alpha^k \in \mathcal{I}(\mathcal{M}, G)$. □

Since locally good filtrations F exist, we deduce from this that there exists a globally defined sheaf of ideals $\sqrt{\mathcal{I}(\mathcal{M})} \subset \mathrm{Gr}(\mathcal{D}_X)$ which locally coincides with $\sqrt{\mathcal{I}(\mathcal{M}, F)}$. Recall (Lemma 13.30) that $\mathrm{Gr}(\mathcal{D}_X)$ consists of the sheaf of functions on the total space of the holomorphic cotangent bundle of X which are polynomial on each fibre. The zeros of the ideal $\sqrt{\mathcal{I}(\mathcal{M})}$ thus define a subvariety of the cotangentbundle, the **characteristic variety** of $\mathcal{M}$, which in each fibre is a cone. It will be denoted

$$\mathrm{Char}(\mathcal{M}) := \bigcup_{x \in X} V(\sqrt{\mathcal{I}_x}) \subset T(X)^\vee. \qquad \text{(XIII–38)}$$

We finally remark that if we have a good filtration F on $\mathcal{M}$, the characteristic variety can also be seen as the support of the sheaf $\mathrm{Gr}^F(\mathcal{M}) \subset \mathrm{Gr}(\mathcal{D}_X)$ (inside the cotangent bundle).

Examples 13.45. 1) Let $\mathcal{M} = \mathcal{O}_X$. Then a good filtration is given by $F_p\mathcal{M} = 0$ for $p < 0$ and $F_p\mathcal{M} = \mathcal{M}$ for $p \geq 0$. The same procedure holds if $\mathcal{M}$ is a $\mathcal{D}_X$-module which is coherent as an $\mathcal{O}_X$-module. The characteristic variety of such a $\mathcal{D}_X$-module is the zero section of the cotangent bundle. Conversely, suppose that the characteristic variety of $\mathcal{M}$ consists of the zero section. Then for local coordinates $(z_1, \ldots, z_n)$ on $U \subset X$, considering the differentials dz_j as local functions w_j on the total space of the cotangent bundle, $(z_1, \ldots, z_n, w_1, \ldots, w_n)$ give a set of local coordinates on $T^\vee(U) \simeq U \times \mathbb{C}^n$. Then $\sqrt{\mathcal{I}(\mathcal{M})}$ is generated by $(w_1, \ldots, w_n)$. This means that $\mathrm{Gr}_F \mathcal{M}$ is killed by a power of the ideal $(w_1, \ldots, w_n)$ and hence is a finitely generated $\mathcal{O}_U$-module. Hence $\mathcal{M}$ is itself a finitely generated $\mathcal{O}_X$-module i.e. $\mathcal{M}$ is a coherent $\mathcal{O}_X$-module and hence locally free by Ex. 13.40. 3).

2) Let $D \subset X$ be a submanifold of codimension one. Recall

$$\mathcal{O}_X(*D) := \bigcup_m \mathcal{O}_X(mD),$$

the sheaf of meromorphic functions on X, holomorphic on $X - D$ and having a pole along D. Let $\mathcal{M} = \mathcal{O}_X(*D)/\mathcal{O}_X$ and put $F_p\mathcal{M} = 0$ for $p < 0$ and $F_p\mathcal{M} = \mathcal{O}_X(pD)/\mathcal{O}_X$ if $p \geq 0$. This defines a good filtration on $\mathcal{M}$. If $\mathcal{N}_{D|X} = \mathcal{O}_X(D)/\mathcal{O}_X$ is the normal bundle of D in X, then $\mathrm{Gr}_p^F(\mathcal{M}) = 0$ for $p \leq 0$ and $\mathrm{Gr}_p^F(\mathcal{M}) \simeq \mathcal{N}_{D|X}^{\otimes p}$ for $p > 0$. Let $(z_1, \ldots, z_n)$ be local coordinates on X such that D is given by $z_1 = 0$. Let $\delta(z_1)$ be the class of z_1^{-1} modulo $\mathcal{O}_X$. Then $\delta(z_1)$ locally generates $\mathrm{Gr}^F(\mathcal{M})$ over $\mathrm{Gr}(\mathcal{D}_X) \simeq \mathcal{O}_X[w_1, \ldots, w_n]$. The annihilator ideal of this generator is generated by $z_1, w_2, \ldots, w_n$. Hence $\mathrm{Char}(\mathcal{M})$ is the **conormal bundle** of D in X, i.e. the subspace of $T^\vee(X)$ consisting of pairs (x, α) such that the covector α vanishes on tangent vectors to D.

3) Let

$$0 \to \mathcal{M}' \to \mathcal{M} \to \mathcal{M}'' \to 0$$

be an exact sequence of $\mathcal{D}_X$-modules. If two of these are coherent, the third one is coherent too. In that case, we have

$$\mathrm{Char}(\mathcal{M}) = \mathrm{Char}(\mathcal{M}') \cup \mathrm{Char}(\mathcal{M}'').$$

Applying this to the defining sequence for $\mathcal{O}_X(*D)/\mathcal{O}_X$ it follows that the characteristic variety of $\mathcal{O}_X(*D)$ is the union of the zero section and the conormal bundle of D.

4) The order filtration on $\mathcal{D}_X$ is a good filtration. We see that $I(\mathcal{D}_X, F^{\mathrm{ord}})$ is the zero ideal, so the characteristic variety of $\mathcal{D}_X$ is the whole cotangent bundle.

13.4.3 Behaviour under Direct and Inverse Images

There is no reason for the inverse image or the direct image of a coherent D-module to be coherent. We quote

Proposition 13.46. *Let $f : X \to Y$ be a* proper *holomorphic map between complex manifolds and let $\mathcal{M}$ be a coherent $\mathcal{D}_X$-module admitting a global good filtration (Def. 13.41). Then the cohomology sheaves of the complex $f_+\mathcal{M}$ are coherent.*

The proof uses Grauert's coherence theorem [Gr60] and can be found in [Ma-Sa, Lect. 3]. In the special case of a closed embedding we have [Bor87, § VI.7.11]:

Theorem 13.47 (Kashiwara's equivalence). *Let $i : X \hookrightarrow Y$ be the embedding of a smooth closed subvariety X into Y. Then i_+ induces an equivalence between coherent $\mathcal{D}_X$-modules and coherent $\mathcal{D}_Y$-modules with support on X.*

In the codimension 1 case, in local coordinates $(t, x_1, \dots)$ for which the submanifold X is given by the equation $t = 0$, and if $\mathcal{M}$ has support on X we put $\mathcal{M}_0 = \{m \in \mathcal{M} \mid t \cdot m = 0\}$, a $\mathcal{D}_X$-module. Then

$$\mathcal{M} = i_+(\mathcal{M}_0) = \bigoplus_{k \geq 0} \mathcal{M}_k, \qquad \mathcal{M}_k = (\partial/\partial t)^k \mathcal{M}_0,$$

and $t^k : \mathcal{M}_k \xrightarrow{\sim} \mathcal{M}_0$.

The behaviour under inverse images is more problematic as shown by the following example: $Y = \mathbb{C}^n$, $X = \mathbb{C}^{n-1}$ and f is the closed embedding given by identifying $\mathbb{C}^{n-1} \subset \mathbb{C}^n$ as the hyperplane $z_n = 0$. Then the inverse image of the coherent $\mathcal{D}_Y$-module $\mathcal{D}_Y$ itself is given by $\mathcal{O}_X \otimes_{\mathcal{O}_Y} \mathcal{D}_Y = \mathcal{D}_X[\partial_{z_n}]$ which is not coherent as a $\mathcal{D}_X$-module. We need the notion of a non-characteristic map which we introduce now.

Let $f : X \to Y$ be any holomorphic map and let $df_* : T(X) \to f^*T(Y)$ the natural map. Dually we have the map df^* and the base change map φ:

$$\left.\begin{aligned} df^* : f^*T^\vee(Y) &\to T^\vee(X) \\ \varphi : f^*T^\vee(Y) &\to T^\vee(Y) \end{aligned}\right\}. \tag{XIII–39}$$

Introduce the **conormal space to f**

$$N^\vee(X/Y) := \operatorname{Ker}(df^*) = (df^*)^{-1}(\text{0-section of } T^\vee(Y)) \subset f^*T^\vee(Y).$$

Note that this is not always a *vector bundle*, but it is if f has constant rank. It intervenes in the following crucial notion.

Definition 13.48. One says that f is **non-characteristic** with respect to a coherent $\mathcal{D}_Y$-module $\mathcal{N}$ if f is **transverse** to the characteristic variety for $\mathcal{N}$ in the sense that

$$N^\vee(X/Y) \cap \varphi^{-1}(\operatorname{Char}(\mathcal{N})) \subset (\text{0-section of } f^*T^\vee(Y)).$$

Equivalently

$$df^* : \varphi^{-1}\operatorname{Char}(\mathcal{N}) \to T^\vee(X) \quad \text{is proper and finite onto its image.}$$

Example 13.49. If $f : X \to Y$ is everywhere submersive, f is non-characteristic with respect to all coherent $\mathcal{D}_Y$-modules.

If we have any good filtration F on $\mathcal{N}$, we may consider the induced filtration

$$F_\bullet(f^*\mathcal{N}) = (\mathcal{D}_{X\to Y}, F^{\mathrm{ord}}) \otimes_{f^{-1}\mathcal{D}_Y} f^{-1}(F_\bullet \mathcal{N}), \tag{XIII–40}$$

where $(\mathcal{D}_{X\to Y}, F^{\mathrm{ord}}) = f^*(\mathcal{D}_Y, F^{\mathrm{ord}})$. This should be interpreted as follows:

$$F_m(f^*\mathcal{N}) = \operatorname{Image}\left[\sum_{p+q} F_p^{\mathrm{ord}}(\mathcal{D}_{X\to Y}, F^{\mathrm{ord}}) \otimes_{f^{-1}\mathcal{O}_Y} f^{-1}(F_q \mathcal{N})\right] \text{ in } f^*\mathcal{N}.$$

We have the following result by Kashiwara [Kash80]:

Lemma 13.50. *Let $f : X \to Y$ be holomorphic and suppose that f is non-characteristic with respect to the coherent $\mathcal{D}_Y$-module $\mathcal{N}$. Then $f^*\mathcal{N}$ is coherent. Moreover, if $\mathcal{N}$ admits a global good $\mathcal{D}_Y$-filtration, the induced filtration on $f^*\mathcal{N}$ as defined above* (XIII–40) *is a good $\mathcal{D}_X$-filtration*

13.5 Filtered D-modules

Up to now we assumed existence of a good filtration but we did not incorporate it in our data. Now we consider pairs $(\mathcal{M}, F)$ of a $\mathcal{D}_X$-module together with a good filtration and study the corresponding category $F\mathcal{D}_X$. The morphisms are those $\mathcal{D}_X$-linear maps which respect the filtration.

We first introduce the De Rham complex in the context of filtered modules. Loosely speaking, we consider Ω_X^k as having filtering degree $-k$. More formally, we introduce

$$\left.\begin{aligned} &F_\ell(\mathrm{DR}_X^\bullet\,\mathcal{M}) = \\ &= \Big[F_\ell\mathcal{M} \to \Omega_X^1(F_{\ell+1}\mathcal{M}) \to \cdots \to \Omega_X^{d_X}(F_{\ell+d_X}\mathcal{M})\Big]\,[-d_X]. \end{aligned}\right\} \quad \text{(XIII–41)}$$

This defines a filtration of $\mathrm{DR}_X^\bullet\,\mathcal{M}$ by subcomplexes, the **filtered De Rham complex** $\mathrm{DR}_X^\bullet(\mathcal{M}, F)$. It is a filtered complex of $\mathcal{O}_X$-modules whose morphisms are only $\mathbb{C}_X$-linear.

Examples 13.51. 1) Let $\mathcal{M} = \mathcal{O}_X$ with the trivial one-step filtration i.e. $F_{-1}\mathcal{O}_X = 0$, $F_0\mathcal{O}_X = \mathcal{O}_X$. Then the filtration of the De Rham complex is the usual trivial filtration σ (Example A.34.1) re-indexed to make it increasing:

$$\mathrm{DR}_X^\bullet(\mathcal{O}_X, \sigma) = (\Omega_X^\bullet, \sigma)[d_X]$$

2) Consider ω_X (XIII–23) as a right $\mathcal{D}_X$-module with one-step filtration $F_{d_X-1}\omega_X = 0$, $F_{d_X}\omega_X = \omega_X$.

13.5.1 Derived Categories

The category $F\mathcal{D}_X$ of filtered $\mathcal{D}_X$-modules is an additive category in which every morphism has a kernel, coimage, image and cokernel. However, the category is not abelian (see Example A.5.1) but in Appendix A.3.1 it is explained how to circumvent this problem so that one can still define the corresponding derived category (see Def. A.36). Observe that the existence of a good filtration F on a $\mathcal{D}_X$-module $\mathcal{M}$ implies that $\mathcal{M}$ is a coherent $\mathcal{D}_X$-module and that the $F_k\mathcal{M}$ are $\mathcal{O}_X$-coherent submodules; hence the cohomology sheaves of complexes of filtered $\mathcal{D}_X$-modules are automatically coherent $\mathcal{O}_X$-modules. This explains the notation

$$D^{\mathrm{b}}_{\mathrm{coh}}(F\mathcal{D}_X) := \left\{\begin{array}{l}\text{derived category of bounded}\\ \text{complexes of filtered } \mathcal{D}_X\text{-modules}\end{array}\right\}.$$

13.5.2 Duality

For details and proofs in this section we refer to [Sa88, §2.4].

The dualizing left $\mathcal{D}_X$-module $\mathcal{D}_X^* = \mathcal{H}om_{\mathcal{D}_X}(\omega_X, \mathcal{D}_X^{\mathrm{rgt}})$ is naturally filtered by the order filtration on $\mathcal{D}^{\mathrm{rgt}}$ shifted by $-d_X$ because of the sheaf ω_X^{-1}:

$$F_p^{\mathrm{ord}}\mathcal{D}_X^* = \mathcal{H}om_{\mathcal{D}_X}(\omega_X, F_{p-d_X}^{\mathrm{ord}}\mathcal{D}_X) = F_{p-d_X}^{\mathrm{ord}}\mathcal{D}_X \otimes \omega_X^{-1}.$$

We view this at the same time as a complex in degree $-d_X$ and we shall write this succinctly as

$$(\mathcal{D}, F)_X^* = (\mathcal{D}_X, F_{\mathrm{ord}}) \otimes \omega_X^{-1}[d_X].$$

If $\mathcal{M}$ is a filtered $\mathcal{D}_X$-module, its dual module $\mathcal{M}^* = \mathcal{H}om_{\mathcal{D}_X}(\mathcal{M}, \mathcal{D}_X^*)$, as a Hom sheaf, is filtered in a natural way as well:

$$F_p(\mathcal{M}^*) := \{\phi \in \mathcal{H}om_{\mathcal{D}_X}(\mathcal{M}, \mathcal{D}_X^*) \mid \forall i,\ \phi(F_i(\mathcal{M}) \subset F_{p+i}^{\mathrm{ord}}\mathcal{D}_X^*\}.$$

Assuming (as we do) that F is good, then it turns out that this dual filtration is good and we let $(\mathcal{M}, F)^*$ be the resulting filtered module. To have a duality operator compatible with the non-filtered case (XIII–27) we place the module $(\mathcal{M}, F)^*$ in degree $-d_X$ so that we get a complex:

$$\mathbb{D}_X(\mathcal{M}, F) := (\mathcal{M}, F)^*[d_X] = [\mathcal{H}om_{\mathcal{D}_X}((\mathcal{M}, F), (\mathcal{D}, F)_X^*)]\ [d_X].$$

This extends as an involution $\mathbb{D}_X$ to the filtered derived category $D^{\mathrm{b}}_{\mathrm{coh}}(F\mathcal{D}_X)$.

13.5.3 Functoriality

Let $X \to Y$ be a holomorphic map between complex manifolds. We have already defined (§13.3.4) the operators f^*, $\tilde{f}_*$ on the level of D-modules as well as their extensions $f^!$ and f_+ to the derived categories of bounded complexes of D-modules. Let us now extend these definitions to the filtered setting. We start with the functor $\tilde{f}_*$. In the non-filtered setting we defined $f_+\mathcal{M}$ using the transfer module $\mathcal{D}_{Y\leftarrow X} = f^*\mathcal{D}_Y \otimes_{\mathcal{O}_Y} \omega_{X/Y}$. Remembering the rule that Ω^k has filtering degree $-k$, the filtration by order on $\mathcal{D}_Y$ induces a natural filtration

$$F_m^{\mathrm{ord}}\mathcal{D}_{Y\leftarrow X} := f^*\left(F_{m+d_{X/Y}}^{\mathrm{ord}}\mathcal{D}_X\right) \otimes_{\mathcal{O}_X} \omega_{X/Y}.$$

We then put the tensor product filtration on the relative De Rham module (XIII–33):

$$\mathrm{DR}^\bullet_{X/Y}(\mathcal{M}, F) := (\mathcal{D}_{Y\leftarrow X}, F^{\mathrm{ord}}) \otimes_{\mathcal{D}_X} (\mathcal{M}, F)$$

which we consider as a complex in the derived category of bounded complexes of filtered $f^{-1}\mathcal{D}_Y$-modules. Finally we apply $\tilde{f}_*$:

$$f_+(\mathcal{M}, F) := [\tilde{f}_*\, \mathrm{DR}^\bullet_{X/Y}\, \mathcal{M}, \tilde{f}_* F]$$

However, this direct image filtration need not be exhaustive, i.e. the union of subsheaves $\tilde{f}_* F_k$ need not be $\tilde{f}_*\mathcal{M}$ and thus is not good. But, if f is proper the filtration is exhaustive and good.

Example 13.52. Suppose that $f : X \to Y$ is of maximal rank, but not necessarily proper. In § 13.3.5 we showed that in the derived category the transfer module can be replaced by the "true" relative De Rham complex $\mathrm{DR}^\bullet_{X/Y}(\mathcal{O}_X)$, properly shifted to the left. As such it has a natural filtration just as the ordinary De Rham complex: in (XIII–41) just replace $\Omega^\bullet_X$ by $\Omega^\bullet_{X/Y}$. Since $F_m(\mathrm{DR}^\bullet_{X/Y}(\mathcal{O}_X))$ is a resolution of $F_m^{\mathrm{ord}}\mathcal{D}_{Y\leftarrow X}$ this shows that the two descriptions are compatible in the derived category:

$$(\mathcal{D}_{Y\leftarrow X}, F_{\mathrm{ord}}) = (\Omega^\bullet_{X/Y}, \sigma)[d_{X/Y}] \text{ in } D^{\mathrm{b}}(Ff^{-1}\mathcal{D}_Y).$$

If moreover f is proper, we have

$$f_+(\mathcal{O}_X, \sigma) = (f_+\Omega^\bullet_{X/Y}, \sigma)[d_{X/Y}] \text{ in } D^{\mathrm{b}}(F\mathcal{D}_Y). \tag{XIII–42}$$

Example 13.53. Let $a_X : X \to \mathrm{pt}$ be the constant map. By (XIII–42) the filtered direct image $(a_X)_+(\mathcal{O}_X, \sigma)$ gives the cohomology of X together with the Hodge filtration (made increasing).

We end this section with a result concerning the De Rham functor. We have seen that the morphisms of the De Rham complex of any $\mathcal{D}_X$-module are only $\underline{\mathbb{C}}_X$-linear. However, for a *filtered* $\mathcal{D}_X$-module, the morphisms in any of the associated graded complexes are still $\mathcal{O}_X$-linear. This implies the following result.

Lemma 13.54. *Let $K_0(X)$ be the Grothendieck group for the category of coherent $\mathcal{O}_X$-modules. The De Rham functor induces the* ***De Rham characteristic***

$$\left.\begin{array}{lll} \chi_{\mathrm{DR}}(X) : & K_0(F\mathcal{D}_X) & \to K_0(X)[u, u^{-1}] \\ & [(\mathcal{V}, F)] & \mapsto \sum(-1)^j[\mathrm{Gr}_j^F \mathrm{DR}^\bullet_X(\mathcal{V}, F)]u^j. \end{array}\right\} \tag{XIII–43}$$

This is compatible with proper pushforwards, in the sense that for $f : X \to Y$ a proper map between algebraic manifolds, we have a commutative diagram

$$\begin{array}{ccc} K_0(F\mathcal{D}_X) & \to & K_0(X)[u, u^{-1}] \\ \downarrow{\scriptstyle f_+} & & \downarrow{\scriptstyle f_*} \\ K_0(F\mathcal{D}_Y) & \to & K_0(Y)[u, u^{-1}]. \end{array}$$

13.6 Holonomic D-Modules

13.6.1 Symplectic Geometry

The total space of the cotangent bundle $\pi : T^\vee X \to X$ of a complex manifold X carries a natural 1-form Θ: if η is a tangent vector at a point $(x, \xi) \in T^\vee X$ we put $\Theta(\eta) = \xi[d\pi(\eta)]$. The 2-form $\omega = d\Theta$ turns out to be a non-degenerate skew-symmetric form: in local coordinates $(z_1, \ldots, z_n)$ on X, the differentials

dz_j define local functions w_j on $T^\vee X$ and $(z_1, \dots, z_n, w_1, \dots, w_n)$ then give local coordinates on the cotangent bundle in which $\omega = \sum_j dz_j \wedge dw_j$.

A complex **symplectic manifold** is a complex manifold U equipped with a non-degenerate skew-symmetric 2-form ω. It has even dimension $2n$. For any subspace $V \subset T_u U$ we put

$$V^\perp := \{a \in T_u U \mid \omega(a, V) = 0\}.$$

and we say that V is **isotropic**, respectively **involutive** if $V \subset V^\perp$, respectively $V \supset V^\perp$. Any isotropic subspace has dimension at most n and any involutive subspace has dimension at least n. A maximal isotropic subspace is called **Lagrangian**. A subvariety S of U is called involutive, respectively Lagrangian, if $T_u S \subset T_u U$ is involutive, respectively Lagrangian for all regular points $u \in S$. An involutive submanifold of U is Lagrangian if and only if $\dim U = n$.

Returning to the cotangent bundle $T^\vee X$, we can identify its tangent space at any point $(x, \xi) \in T^\vee X$ with $T_x X \times T_x^\vee X$. The symplectic form restricts on it as

$$[(t, \tau), (t', \tau')] \mapsto \tau'(t) - \tau(t').$$

Then a subspace of the form $A \times B$ with A and B subspaces of $T_x X$ and $T_x^\vee X$ respectively, is Lagrangian precisely when B is the annihilator of A. If A happens to be the tangent space to a submanifold $Z \subset X$, then B is exactly the conormal space to Z at that point.

More generally, the conormal space of an irreducible subvariety $Z \subset X$ by definition is the closure inside $T^\vee(X)$ of the conormal bundle of its smooth locus, and denoted

$$N^\vee(Z/X) \quad \text{(the conormal space to } Z \text{ in } X\text{)}.$$

It is Lagrangian since this is Lagrangian on the dense open subset of smooth points of Z.

Conversely, any irreducible *conical* Lagrangian subvariety V of the cotangent bundle $T^\vee(X)$ of some complex manifold X is the conormal space of the analytic variety $\pi(V)$, where $\pi : T^\vee(X) \to X$. To see this, we may assume that V differs from the 0-section of the cotangent bundle. Then we can pass to the associated projective bundle $\mathbb{P}T^\vee(X)$ in which V defines a subvariety V'. The projection $\pi' : \mathbb{P}T^\vee(X) \to X$ being proper, the proper mapping theorem of Remmert and Grauert [Gr60] then implies that $\pi'(V') = Z = \pi(V)$ is an analytic subvariety of X. We need to see that its conormal space coincides with V. By irreducibility, it suffices to show this over a suitable dense open subset of Z, for example the set of smooth points over which π has maximal rank. We may thus assume that V and Z are smooth and that $\pi : V \to Z$ has maximal rank. Identifying the tangent space at a point (x, ξ) of V with $T_x X \times T_x^\vee X$, the tangent space to V is a Lagrangian subspace of the form $T_x Z \times B$. Since V is smooth and conical, B is the restriction of V to $T_x^\vee X \subset T^\vee X$. Since V

is Lagrangian, B is the annihilator of $T_x Z$. But this means exactly that V is the conormal bundle to Z.

Note also that if $f : X \to Y$ is any holomorphic map between complex manifolds and $Z = f^{-1}W$ is the inverse image of a complex subvariety $W \subset Y$, the map $df : T^\vee Y \to T^\vee X$ induces $(df)^* : f^* N^\vee(W/Y) \to N^\vee(Z/X)$ (see (XIII–39)) which is an isomorphism over the points where f is has maximal rank.

13.6.2 Basics on Holonomic D-Modules

It is a deep theorem that for a coherent $\mathcal{D}_X$-module $\mathcal{M}$ the characteristic variety $\mathrm{Char}(\mathcal{M}) \subset T^\vee(X)$ is involutive. See [Malg79]. Hence the following concept is natural.

Lemma-Definition 13.55. A coherent $\mathcal{D}_X$-module is **holonomic** if its characteristic variety is Lagrangian, or, equivalently if

$$\dim \mathrm{Char}(\mathcal{M}) = d_X.$$

In that case, from what we said in § 13.6.1, the characteristic subvariety $\mathrm{Char}(\mathcal{M})$ consists of the union of conormal spaces to irreducible subvarieties of X.

The corresponding derived category is

$$D^{\mathrm{b}}_{\mathrm{h}}(\mathcal{D}_X) := \left\{ \begin{array}{l} \text{derived category of bounded complexes} \\ \text{of coherent } \mathcal{D}_X\text{-modules with} \\ \text{holonomic cohomology sheaves.} \end{array} \right\} \qquad \text{(XIII–44)}$$

An object of this category is called **a holonomic complex**. So a holonomic complex of (coherent) D-modules need not consist of holonomic modules; only its cohomology sheaves should be holonomic.

Lemma 13.56 ([Bor87, VI.3.7]). *A coherent $\mathcal{D}_X$-module $\mathcal{M}$ is holonomic if and only if the complex $\mathbb{D}_X\mathcal{M}$ has only cohomology in degree 0; the resulting map*

$$\mathcal{M} \mapsto \mathcal{M}^* := H^0(\mathbb{D}_X\mathcal{M})$$

is an involution on the category of holonomic $\mathcal{D}_X$-modules. It extends to an involution $\mathbb{D}_X$ on the category $D^{\mathrm{b}}_{\mathrm{h}}(\mathcal{D}_X)$.

Remark. A word of warning at this point: in [Bor87] the D-modules are allowed to be quasi-coherent instead of coherent. We apply the theory only to filtered D-modules which, as we have seen, are automatically coherent. In any case, Borel shows that the derived category D^{b}(coherent $\mathcal{D}_X$-modules) is the same as the category of bounded complexes of quasi-coherent $\mathcal{D}_X$-modules whose cohomology is coherent. So there is no loss of generality in restricting to coherent $\mathcal{D}_X$-modules from the start.

For proper direct images we have [Kash80]:

Proposition 13.57. *Let $f : X \to Y$ be a proper holomorphic map. Then f_+ preserves complexes with coherent cohomology and sends holonomic complexes on X to holonomic complexes on Y:*

$$f_+ : D^{\mathrm{b}}_{\mathrm{h}}(\mathcal{D}_X) \to D^{\mathrm{b}}_{\mathrm{h}}(\mathcal{D}_Y).$$

Moreover, we have $\mathbb{D}_Y \circ f_+ = f_+ \circ \mathbb{D}_X$, i.e. the duality functor commutes with direct image.

The functor f_+ preserves the full subcategory of bounded holonomic complexes of filtered *holonomic D-modules.*

For inverse images the basic result reads:

Proposition 13.58. *Let $f : X \to Y$ be a holomorphic map. Then f^* preserves holonomicity. On complexes of coherent $\mathcal{D}_Y$-modules with holonomic cohomology sheaves the functor f^* extends to*

$$f^! : D^{\mathrm{b}}_{\mathrm{h}}(\mathcal{D}_Y) \to D^{\mathrm{b}}_{\mathrm{h}}(\mathcal{D}_X).$$

The inverse image f^* behaves well on filtered modules provided f is non-characteristic for $\mathcal{N}$. Indeed, if this is the case, by Lemma 13.50 the inverse $f^*(\mathcal{N}, F)$ is a filtered module. If moreover $\mathcal{N}$ is holonomic, essentially by the previous Proposition 13.58, the module $f^*\mathcal{N}$ is holonomic. For filtered holonomic *complexes* a further condition is needed on the filtration. It is subsumed in the following definition.

Definition 13.59. *Let $(\mathcal{N}, F)$ be a filtered $\mathcal{D}_Y$-module. We say that $(\mathcal{N}, F)$ is* ***non-characteristic*** *with respect to $f : X \to Y$ if*

1) *f is non-characteristic with respect to $\mathcal{N}$ (see Def. 13.48);*
2) *$\mathcal{T}or_k^{f^{-1}\mathcal{O}_Y}(f^{-1}\operatorname{Gr}^F \mathcal{N}, \mathcal{O}_X) = 0$, $k \neq 0$.*

A complex of filtered $\mathcal{D}_Y$-modules is non-characteristic with respect to f if its constituents are.

The second condition ensures that filtrations behave well under tensor products; we can in fact show ([Sa88, 3.5.2]):

Lemma 13.60. *Let $(\mathcal{N}, F)^\bullet$ be a bounded complex of filtered holonomic $\mathcal{D}_Y$-modules. Suppose that $(\mathcal{N}, F)^\bullet$ is non-characteristic with respect to $f : X \to Y$. Then $f^!(\mathcal{N}, F)^\bullet$ is a holonomic complex of filtered $\mathcal{D}_X$-modules.*

13.6.3 The Riemann-Hilbert Correspondence (II)

The link between D-modules and perverse complexes is given by Kashiwara's theorem, one of the central ingredients of the Riemann-Hilbert Correspondence: [Kash74]

Theorem 13.61. *Let X be a complex analytic manifold. The De Rham complex of a holonomic $\mathcal{D}_X$-module is a perverse complex.*

The version we are going to discuss is only valid in the *algebraic setting* of algebraic $\mathcal{D}$-modules on smooth (but not necessarily compact) complex algebraic varieties. This guarantees finiteness properties of the inverse and direct images of $\mathcal{D}$-modules.

The central notion which generalizes regular connections is that of regular holonomic complexes. We have seen that a connection ∇ on a smooth algebraic variety U with good compactification X is regular if and only if for all *compact* curves $u : C \to X$ mapping to X and not contained in the boundary $D = X - U$ the pull back connection $u^*\nabla$ is regular at all points mapping to the boundary. Equivalently, by Remark 11.6 and Example 13.35, a $\mathcal{D}_U$-module $\mathcal{M}$ is regular if and only if for all such maps $u : C \to X$, the complex $u^!\mathcal{M}$ is regular holonomic on a Zariski-open subset of C. This allows us to consider directly maps of non-compact algebraic curves C into U so that we need not mention any good compactification in what follows. Extending this to complexes leads to the following definition.

Definition 13.62. Let $\mathcal{M}^\bullet$ be a holonomic complex on U, i. e. $\mathcal{M}^\bullet \in D^{\mathrm{b}}_{\mathrm{h}}(\mathcal{D}_U)$. We say that $\mathcal{M}^\bullet$ is **regular holonomic** if for all holomorphic maps $u : C \to U$ of a smooth irreducible curve C into U, the complex $u^!\mathcal{M}^\bullet$ is regular holonomic on C.

Remark. If $(\mathcal{M}, F)$ is (regular) filtered holonomic, $\tilde{f}_*(\mathcal{M}, F)$ is (regular) filtered holonomic as well.

We already saw (Lemma 13.56) that duality preserves holonomicity. It can be seen to respect regularity as well. Also, by Prop. 13.58, if $f : X \to Y$ is a morphism between smooth complex algebraic varieties and $\mathcal{N}^\bullet$ is holonomic, then also $f^!\mathcal{N}^\bullet$ is holonomic. It is not hard to see that $f^!$ preserves regular holonomic complexes as well. As to f_+, by Prop. 13.58 the direct image $f_+\mathcal{M}^\bullet$ of a holonomic complex under a proper map f is holonomic, and again regularity is preserved in this case. See [Bor87, VII.§12]. Summarizing, we thus have:

Proposition 13.63. *Regular holonomicity is preserved by $f^!$, by duality, and, for proper morphisms, by f_+.*

Extending the notation (XIII–44) to the regular situation, we set

$$D^{\mathrm{b}}_{\mathrm{rh}}(\mathcal{D}_X) := \left\{ \begin{array}{l} \text{derived category of bounded complexes} \\ \text{of coherent } \mathcal{D}_X\text{-modules whose cohomology} \\ \text{sheaves are regular holonomic.} \end{array} \right\} \qquad \text{(XIII–45)}$$

The De Rham functor (XIII–25) produces out of any complex of $\mathcal{D}_X$-modules a bounded complex of sheaves of complex vector spaces (though we deal with algebraic D-modules, for the De Rham functor we first pass to the associated *analytic* D-modules).

By Kashiwara's theorem 13.61 these complexes are certainly cohomologically constructible if the $\mathcal{D}_X$-modules are holonomic, and even perverse if we have a single module in degree 0. Surprisingly, the Riemann-Hilbert correspondence says that such a module must be quasi-isomorphic to a unique regular holonomic complex of $\mathcal{D}_X$-modules. In fact, we even have:

Theorem 13.64 (RIEMANN-HILBERT CORRESPONDENCE (II)). *Let X be a complex algebraic manifold.*

1) *Recalling the notation* (XIII–45) *and* (XIII–13), *the De Rham functor*

$$\mathrm{DR}_X : D^{\mathrm{b}}_{\mathrm{rh}}(\mathcal{D}_X) \longrightarrow D^{\mathrm{b}}_{\mathrm{cs}}(\mathbb{C}_X)$$

is an equivalence of categories. Under this equivalence the regular holonomic $\mathcal{D}_X$-modules correspond exactly to the category of $\mathbb{C}$-perverse complexes. I.e. the cohomology sheaves of a regular holonomic complex $\mathcal{M}^\bullet$ is concentrated in degree 0 if and only if $\mathrm{DR}_X(\mathcal{M}^\bullet)$ is perverse.

2) *Under the De Rham equivalence the Verdier duality operator ${}^{\mathrm{Ve}}\mathbb{D}_X$ (§ 13.1.3) and the duality operator $\mathbb{D}_X$* (XIII–27) *correspond in the sense that $\mathrm{DR}_X \circ \mathbb{D}_X = {}^{\mathrm{Ve}}\mathbb{D}_X \circ \mathrm{DR}_X$. In particular, the solution complex of a holonomic complex* (XIII–37) *is Verdier dual to its De Rham complex.*

Let $f : X \to Y$ be a morphism between smooth complex algebraic varieties.

1) *We have a commutative diagram*

$$\begin{array}{ccc} D^{\mathrm{b}}_{\mathrm{rh}}(\mathcal{D}_Y) & \xrightarrow{\mathrm{DR}_Y} & D^{\mathrm{b}}_{\mathrm{cs}}(\mathbb{C}_Y) \\ \downarrow f^! & & \downarrow f^! \\ D^{\mathrm{b}}_{\mathrm{rh}}(\mathcal{D}_X) & \xrightarrow{\mathrm{DR}_X} & D^{\mathrm{b}}_{\mathrm{cs}}(\mathbb{C}_X), \end{array}$$

where $f^!$ on sheaves of complex vector spaces stands for the extra-ordinary pull back defined in § 13.1.4.

2) *We have a commutative diagram*

$$\begin{array}{ccc} D^{\mathrm{b}}_{\mathrm{rh}}(\mathcal{D}_X) & \xrightarrow{\mathrm{DR}_X} & D^{\mathrm{b}}_{\mathrm{cs}}(\mathbb{C}_X) \\ \downarrow f_+ & & \downarrow Rf_* \\ D^{\mathrm{b}}_{\mathrm{rh}}(\mathcal{D}_Y) & \xrightarrow{\mathrm{DR}_Y} & D^{\mathrm{b}}_{\mathrm{cs}}(\mathbb{C}_Y), \end{array}$$

where Rf_* is the usual (derived) direct image for sheaves of complex vector spaces defined in § B.2.5. In particular, Rf_* preserves cohomologically constructible complexes, even if f is not proper (compare with Prop. 13.15).

A full proof can be found in [Bor87, VIII].

Example 13.65. Let X be a compact complex manifold, let $Y \subset X$ be a divisor with normal crossings and let $\mathbb{V}$ be a complex local system on $U = X - Y$. Under the Riemann-Hilbert correspondence the perverse complex ${}^{\pi}\mathbb{V}$ corresponds to a unique regular holonomic $\mathcal{D}_X$-module ${}^{\pi}\mathcal{V}$, and

$$ {}^{\pi}\mathcal{V}\,|_U = \mathcal{V} = \mathbb{V} \otimes_{\mathbb{C}} \mathcal{O}_U . $$

Let $(\tilde{\mathcal{V}}, \tilde{\nabla})$ be the unique extension of $(\mathcal{V}, \nabla)$ to a holomorphic vector bundle equipped with a holomorphic connection whose residues along the components of D have eigenvalues in $[0, 1)$ (see § 11.1.2). With $j : U \hookrightarrow X$ the inclusion, let ${}^{\pi}\mathcal{V}$ be the $\mathcal{D}_X$-submodule of $j_*\mathcal{V}$ generated by $\tilde{\mathcal{V}}$. Then by [Kash-Ka87a, Theorem 1.1]

$$ \mathrm{DR}_X({}^{\pi}\mathcal{V}) \simeq {}^{\pi}\mathbb{V}. $$

Remark 13.66. 1) Regular holonomic $\mathcal{D}_X$-modules form an abelian category, whose Grothendieck group is the same as for $D^{\mathrm{b}}_{\mathrm{rh}}(\mathcal{D}_X)$. The De Rham functor induces an isomorphism

$$ K_0(\text{Regular holonomic } \mathcal{D}_X\text{-modules}) \xrightarrow[\mathrm{DR}_X]{\sim} K_0(\mathrm{Perv}(X; \mathbb{C})) \qquad \text{(XIII–46)} $$

2) It follows from this theorem and the definitions that the De Rham-functor (on the level of regular holonomic modules) transforms the diagram induced by (XIII–35)

$$ \begin{array}{ccc} D^{\mathrm{b}}_{\mathrm{rh}}(\mathcal{D}_X) & \underset{f_+}{\overset{f^+}{\leftrightarrows}} & D^{\mathrm{b}}_{\mathrm{rh}}(\mathcal{D}_Y) \\ \updownarrow{\scriptstyle \mathbb{D}_X} & & {\scriptstyle \mathbb{D}_Y}\updownarrow \\ D^{\mathrm{b}}_{\mathrm{rh}}(\mathcal{D}_X) & \underset{f_!}{\overset{f^!}{\leftrightarrows}} & D^{\mathrm{b}}_{\mathrm{rh}}(\mathcal{D}_Y) \end{array} $$

into the diagram (XIII–9).

Historical Remarks.

Historically, perverse complexes have been called *perverse sheaves* and have two origins. Firstly, from the geometric side there is Deligne's sheaf-theoretic treatment of intersection homology. Secondly, the notion of a core of a triangulated category as explained in the thesis of Verdier leads to constructible complexes of sheaves. The latter point of view has been formalized in the basic work [B-B-D]. This point of view explains why a perverse complex of sheaves as an object in an abelian category behaves very much like a single sheaf, explaining the terminology perverse sheaf.

D-module theory started around 1970 with M. Sato's introduction of *algebraic analysis* and J. Bernstein's work on the Bernstein-Sato polynomial. Kashiwara at around 1980 founded the algebraic theory of micro-local analysis. The use of sheaf theory turned out to be crucial in globalizing the theory; it has been especially effective with regard to involutive systems. The Kashiwara-Malgrange filtration stems from this period. It is important in studying monodromy around singularities by analytic means and is a crucial ingredient in M. Saito's work on Hodge modules. These topics will be treated in the next chapter.

The two are related through the Riemann-Hilbert correspondence. Our final version of it is due to Z. Mebkhout, and this can be seen as the crowning achievement in the theory of D-modules as it has been treated by the Grothendieck school.

14

Mixed Hodge Modules

The definition of mixed Hodge modules is very involved. For this reason it is more suitable to start with an axiomatic introduction. This makes it possible to deduce important results rather painlessly, such as the existence of pure Hodge structures on the intersection cohomology groups.

Hodge modules are generalizations of variations of Hodge structures. In fact, a variation of Hodge structures is a basic example of a Hodge module. In the world of Hodge modules the underlying (complexified) local system with its integrable connection and filtration is to be considered as a filtered D-module. The local system itself must be replaced by its perverse intersection complex. In general, a Hodge module consists of a filtered D-module and a perverse complex of $\mathbb{Q}$-vector spaces such that the De Rham complex of the D-module is isomorphic to the complexification of the perverse complex. In addition a whole lot of extra properties have to be satisfied inductively which makes it hard to verify them in concrete situations. The definition of a Hodge module as well as some of its properties are given in § 14.3. Essential for Hodge modules is the way they behave under the vanishing and nearby cycle functors with respect to any locally defined holomorphic function. On the level of D-modules this necessitates to introduce the so-called V-filtration which encodes the action of the monodromy around the hypersurface $(t = 0)$. This is explained in § 14.2.

Mixed Hodge modules generalize Hodge modules in the same way as mixed Hodge structures generalize Hodge structures. For the same reason mixed modules have a better functorial behaviour. Since mixed Hodge modules over a point are exactly the graded polarizable mixed Hodge structures, applying the direct image functor to the constant map, by functoriality one obtains a mixed Hodge structure on the cohomology groups. All of the mixed Hodge structures on cohomology discussed in this book, and many more, can be obtained in this way. We explain this briefly in § 14.1.

14.1 An Axiomatic Introduction

14.1.1 The Axioms

We recall (XIII–13) that for any complex algebraic variety X the derived category of bounded cohomologically constructible complexes of sheaves of $\mathbb{Q}$-vector spaces on X is denoted $D^{\mathrm{b}}_{\mathrm{cs}}(X;\mathbb{Q})$ and that it contains as a full subcategory the category $\mathrm{Perv}(X;\mathbb{Q})$ of perverse $\mathbb{Q}$-complexes. The Verdier duality operator ${}^{\mathrm{Ve}}\mathbb{D}_X$ is an involution on $D^{\mathrm{b}}_{\mathrm{cs}}(X;\mathbb{Q})$ preserving $\mathrm{Perv}(X;\mathbb{Q})$. We also recall that by Cor. 13.10, associated to a morphism $f: X \to Y$ between complex algebraic varieties, there are pairs of adjoint functors (f^{-1}, Rf_*) and $(f^!, Rf_!)$ between the respective derived categories of cohomologically constructible complexes which are interchanged by Verdier duality. Let us now state the axioms:

A) For each complex algebraic variety X there exists an abelian category $\mathsf{MHM}(X)$, the category of **mixed Hodge modules** on X with the following properties:
 - There is a faithful functor

$$\mathrm{rat}_X : D^{\mathrm{b}}\mathsf{MHM}(X) \to D^{\mathrm{b}}_{\mathrm{cs}}(X;\mathbb{Q}). \tag{XIV–1}$$

   such that $\mathsf{MHM}(X)$ corresponds to $\mathrm{Perv}(X;\mathbb{Q})$. We say that $\mathrm{rat}_X M$ is the underlying rational perverse sheaf of M. Moreover, we say that

   $M \in \mathsf{MHM}(X)$ **is supported on** $Z \iff \mathrm{rat}_X M$ is supported on Z.
 - There is a faithful functor

$$\mathrm{Dmod}_X : D^{\mathrm{b}}\mathsf{MHM}(X) \to D^{\mathrm{b}}_{\mathrm{coh}}(\mathcal{D}_X). \tag{XIV–2}$$

   We say that $\mathrm{Dmod}_X(M)$ is the underlying $\mathcal{D}_X$-module.
 - We demand that the triangle

$$\begin{array}{ccc} D^{\mathrm{b}}\mathsf{MHM}(X) & \xrightarrow{\mathrm{rat}_X \otimes \mathbb{C}} & D^{\mathrm{b}}_{\mathrm{cs}}(X;\mathbb{C}) \\ & {\searrow}^{\mathrm{Dmod}_X} \quad {\nearrow}_{\mathrm{DR}_X} & \\ & D^{\mathrm{b}}_{\mathrm{coh}}(\mathcal{D}_X) & \end{array}$$

   is commutative up to isomorphisms. More precisely, for each mixed Hodge module M there is an isomorphism

$$\alpha : \mathrm{rat}_X(M) \otimes \mathbb{C} \xrightarrow{\sim} \mathrm{DR}_X\left[\mathrm{Dmod}_X(M)\right].$$

   This isomorphism is called the comparison isomorphism.

B) The category of mixed Hodge modules supported on a point is the category of graded polarizable rational mixed Hodge structures; the functor "rat" associates to the mixed Hodge structure the underlying rational vector space.

C) Each object M in $\mathsf{MHM}(X)$ admits a **weight filtration** W such that
- morphisms preserve the weight filtration strictly;
- the object $\mathrm{Gr}_k^W M$ is semisimple in $\mathsf{MHM}(X)$;
- if X is a point the W-filtration is the usual weight filtration for the mixed Hodge structure.

Since $\mathsf{MHM}(X)$ is an abelian category, the cohomology groups of any complex of mixed Hodge modules on X are again mixed Hodge modules on X. With this in mind, we say that for complex $M^\bullet \in D^{\mathrm{b}}\mathsf{MHM}(X)$ the **weight** satisfies

$$\mathrm{weight}[M^\bullet]\begin{cases}\leq n,\\ \geq n\end{cases} \iff \mathrm{Gr}_i^W H^j(M^\bullet)=0 \begin{cases}\text{for } i>j+n\\ \text{for } i<j+n.\end{cases}$$

We observe that if we consider the weight filtration on the mixed Hodge modules which constitute a complex $M^\bullet \in D^{\mathrm{b}}\mathsf{MHM}(X)$ of mixed Hodge modules we get a filtered complex in this category. Moreover, by assumption this filtration is strict in the sense of Lemma-Def. A.35 and so the two functors H and Gr^W can be interchanged:

$$\mathrm{Gr}_i^W H^j(M^\bullet) = H^j(\mathrm{Gr}_i^W M^\bullet), \quad \forall i,j\in\mathbb{Z}; \tag{XIV–3}$$

There is a subtle point here: a priori the filtration W on the left is the induced filtration coming from the weight spectral sequence, and need not coincide with the intrinsic weight filtration on $H^j(M^\bullet)$ as a mixed Hodge module. However, the E_1-terms (the terms on the right) are pure mixed Hodge modules and hence the left hand side is pure as well. So, by induction the filtration W on the left coincides with the intrinsic weight filtration.

D) The duality functor ${}^{\mathrm{Ve}}\mathbb{D}_X$ of Verdier lifts to $\mathsf{MHM}(X)$ as an involution, also denoted ${}^{\mathrm{Ve}}\mathbb{D}_X$, in the sense that ${}^{\mathrm{Ve}}\mathbb{D}_X \circ \mathrm{rat}_X = \mathrm{rat}_X \circ {}^{\mathrm{Ve}}\mathbb{D}_X$.

E) For each morphism $f: X\to Y$ between complex algebraic varieties, there are induced functors $f_*, f_!: D^{\mathrm{b}}\mathsf{MHM}(X)\to D^{\mathrm{b}}\mathsf{MHM}(Y)$, $f^*, f^!: D^{\mathrm{b}}\mathsf{MHM}(Y)\to D^{\mathrm{b}}\mathsf{MHM}(X)$ interchanged under ${}^{\mathrm{Ve}}\mathbb{D}_X$ and which lift the analogous functors on the level of constructible complexes.

F) The functors $f_!, f^*$ do not increase weights in the sense that if $M^\bullet$ has weights $\leq n$, the same is true for $f_! M^\bullet$ and $f^* M^\bullet$.

G) The functors $f^!, f_*$ do not decrease weights in the sense that if $M^\bullet$ has weights $\geq n$, the same is true for $f^! M^\bullet$ and $f_* M^\bullet$.

By way of terminology, we say that $M^\bullet \in D^{\mathrm{b}}\mathsf{MHM}(X)$ is **pure of weight** n if it has weight $\geq n$ and weight $\leq n$, i.e. for all $j\in\mathbb{Z}$ the cohomology sheaf

$H^j(M^\bullet)$ has pure weight $j+n$: only $\mathrm{Gr}_W^{j+n} H^j(M^\bullet)$ might be non-zero. We say that a morphism **preserves weights**, if it neither decreases or increases weights. Since for a proper map $f_* = f_!$ axioms F) and G) entail:

H) For proper maps between complex algebraic varieties f_* preserves weights.

14.1.2 First Consequences of the Axioms

We begin the study of complexes of negative or positive weight:

Lemma 14.1. a) *Suppose that* $K^\bullet \in D^{\mathrm{b}}\mathsf{MHM}(X)$ *has weight* ≥ 0. *There is a surjective quasi-isomorphism* $K^\bullet \twoheadrightarrow \bar{K}^\bullet$ *such that* $\mathrm{Gr}_p^W \bar{K}^q = 0$ *for* $q > p$.
b) *Suppose that* $K^\bullet \in D^{\mathrm{b}}\mathsf{MHM}(X)$ *has weight* ≤ 0. *There is an injective quasi-isomorphism* $\bar{K}^\bullet \hookrightarrow K^\bullet$ *such that* $\mathrm{Gr}_p^W \bar{K}^q = 0$ *for* $q < p$.

Proof. We shall only prove a); the proof of b) is similar. From (XIV–3) we see that for every p the complex $\mathrm{Gr}_p^W K^\bullet$ has cohomology in degrees $\leq p$ only. We now make the following

Observation 14.2. *Let* $L^\bullet$ *be a bounded complex in an abelian category and suppose that* $H^q(L^\bullet) = 0$ *for* $q > p$. *If* L^p *is semi-simple, there is a surjective quasi-isomorphism* $L^\bullet \twoheadrightarrow \bar{L}^\bullet$ *with* $\bar{L}^q = 0$ *for* $q > p$. *In fact, there exists an acyclic subcomplex* $C^\bullet \subset L^\bullet$ *such that the quotient complex is zero in degrees* $> p$.

Proof (of the observation). Since L^p is semi-simple, there is a direct sum-decomposition $L^p = Z^p \oplus C^p$ where $Z^p = \mathrm{Ker}(d^p)$. By assumption $d : C^p \xrightarrow{\sim} \mathrm{Im}(d^p) = \mathrm{Ker}(d^{p+1})$ and so the complex $C^\bullet$ which equals $L^\bullet$ in degrees $> p$, is equal to C^p in degree p and zero in degrees $< p$ is acyclic. The quotient complex by construction is zero in degrees $> p$. □

Now apply the above to $L^\bullet = \mathrm{Gr}_p^W K^\bullet$ starting with the smallest $p = p_0$ for which $W_p K^\bullet \neq 0$. Then the subcomplex $C^\bullet$ in Observation 14.2 is a subcomplex of $K^\bullet$ itself. Replace $K^\bullet$ by the quotient so that the new complex which we continue to denote $K^\bullet$ has the property $W_{p_0} K^q = 0$ for $q > p_0$. In the next step divide out the corresponding subcomplex $C^\bullet$ of $\mathrm{Gr}_{p_0+1}^W K^\bullet$. By assumption, this subcomplex only lives in degrees $\geq p_0 + 1$ and in these degrees $\mathrm{Gr}_{p_0+1}^W K^\bullet = W_{p_0+1} K^\bullet$ so that it is actually a subcomplex of $K^\bullet$. The quotient complex, still denoted $K^\bullet$ has the property that $\mathrm{Gr}_{p_0+1}^W K^q = 0$ for $q > p_0 + 1$. Moreover, since we did not change $K^\bullet$ in degrees $< p_0 + 1$, we still have that $W_{p_0} K^q = 0$ for $q > p_0$ Continuing in this way, we obtain our surjective quasi-isomorphism $K^\bullet \to \bar{K}^\bullet$ such that $\mathrm{Gr}_p^W \bar{K}^q = 0$ for $q > p$. □

We say that $M^\bullet$ **has smaller weight than** $N^\bullet$ if for some n the complex $M^\bullet$ has weights $\leq n$ while $N^\bullet$ has weights $\geq n + 1$.

Lemma 14.3. *Suppose that $M^\bullet$ has smaller weights than $N^\bullet[p]$. Then*

$$\mathrm{Ext}^p_{D^{\mathrm{b}}\mathsf{MHM}(X)}(M^\bullet, N^\bullet) = 0.$$

Proof. For simplicity of notation we omit the dots for complexes and we abbreviate $D = D^{\mathrm{b}}\mathsf{MHM}(X)$. One may assume $p = 0$, since $\mathrm{Ext}^p_D(M, N) = \mathrm{Hom}_D(M, N[p])$ (see (A–23)). Now we need to show that the assumption on weights implies $\mathrm{Hom}_D(M, N) = 0$. By shifting the degrees we may assume that M has weights ≤ -1 and N has weights ≥ 0. Apply Lemma 14.1 to $M[1]$ and N, obtaining quasi-isomorphic complexes $\bar{M}[1]$ and $\bar{N}$ respectively. A morphism $f : M \to N$ induces a morphism $\bar{f} : \bar{M} \to \bar{N}$ which in the derived category replaces f. For a fixed q we have $W_p\bar{M}^q = \bar{M}^q$ for $p > q$, while $W_p\bar{N}^q = 0$ for $p \leq q$. It follows from strictness that such a morphism must be zero. □

Corollary 14.4. *If $M^\bullet$ is pure of weight n, we have a (non-canonical) isomorphism*

$$M^\bullet \simeq \bigoplus_p H^p M^\bullet[-p].$$

Proof. If $M^\bullet$ has weight n, $\tau_{\leq p}M^\bullet$ and $Gr^\tau_p M^\bullet$ have weight n as well, because $W_k(\tau_{\leq p}M^\bullet) = \tau_{\leq p}W_k(M^\bullet)$, and similarly for Gr^τ_p. Since $\tau_{\leq p-1}M^\bullet[1]$ has weight $n+1$, we have $\mathrm{Ext}^1(\mathrm{Gr}^\tau_p M^\bullet, \tau_{\leq p-1}M^\bullet) = \mathrm{Hom}(\mathrm{Gr}^\tau_p M^\bullet, \tau_{\leq p-1}M^\bullet[1]) = 0$, and hence, there is a non-canonical splitting

$$\tau_{\leq p}M^\bullet \simeq \tau_{\leq p-1}M^\bullet \oplus \mathrm{Gr}^\tau_p M^\bullet.$$

Since by (A–27) the complex $\mathrm{Gr}^\tau_p M^\bullet$is quasi-isomorphic to $H^p(M^\bullet)[-p]$ the result follows by induction. □

If $M^\bullet$ is a complex of mixed Hodge modules on X its cohomology $H^qM^\bullet$ is a mixed Hodge module on X. A consequence of Axiom A) then is:

Lemma 14.5. *The cohomology functors $H^q : D^{\mathrm{b}}\mathsf{MHM}(X) \to \mathsf{MHM}(X)$ are compatible with the functor rat_X in the sense that for any bounded complex $M^\bullet$ of mixed Hodge modules we have*

$$\mathrm{rat}_X[H^qM^\bullet] = {}^{\pi}H^q[\mathrm{rat}_X M^\bullet],$$

i.e. we need to work with the perverse *cohomology functor* (XIII–16).

Axiom E) and B) imply:

Lemma 14.6. *Let $a_X : X \to \mathrm{pt}$ be the constant map to the point. For any complex $M^\bullet$ of mixed Hodge modules on X*

$$\mathbb{H}^p(X, M^\bullet) := H^p((a_X)_*M^\bullet) \qquad \text{(XIV–4)}$$

is a mixed Hodge structure.

Applying Lemma 14.5 (for a complex of sheaves over a point perverse cohomology is ordinary cohomology) we find:

Corollary 14.7. *Let M be mixed Hodge module whose rational component is the perverse complex $\mathcal{F}^\bullet$. Then the hypercohomology group $\mathbb{H}^p(X, \mathcal{F}^\bullet)$ is the rational vector space underlying $\mathbb{H}^p(X, M)$ and hence the former gets a rational mixed Hodge structure.*

We now want to explain how this leads to rational mixed Hodge structures on ordinary and compactly supported cohomology by taking for M a suitable Hodge module. To start with, from axiom A) and B) we see that there is a unique element

$$\mathbb{Q}^{\mathrm{Hdg}} \in \mathsf{MHM}(\mathrm{pt}\ ), \quad \mathrm{rat}(\mathbb{Q}^{\mathrm{Hdg}}) = \mathbb{Q}(0), \qquad \text{(XIV–5)}$$

the unique Hodge structure on $\mathbb{Q}$ of type $(0,0)$. The next lemma explains how the various cohomology groups can be expressed using direct and inverse image functors.

To simplify notation, for any morphism f of complex varieties we shall we shall from now on write f_* and $f_!$ instead of Rf_* and $Rf_!$.

Lemma 14.8. *We have the following identifications*

$$\begin{aligned} H^k(X;\mathbb{Q}) &= H^k(\mathrm{pt}, (a_X)_* a_X^* \mathbb{Q}) \\ H_{-k}(X;\mathbb{Q}) &= H^k(\mathrm{pt}, (a_X)_! a_X^! \mathbb{Q}) \\ H_c^k(X;\mathbb{Q}) &= H^k(\mathrm{pt}, (a_X)_! a_X^* \mathbb{Q}) \\ H_{-k}^{\mathrm{BM}}(X;\mathbb{Q}) &= H^k(\mathrm{pt}, (a_X)_* a_X^! \mathbb{Q})). \end{aligned}$$

Moreover, if $i : Z \hookrightarrow X$ is a closed embedding, we have

$$H_Z^k(X;\mathbb{Q}) = H^k(\mathrm{pt}, (a_X)_* i_* i^! a_X^* \mathbb{Q}) = H^k(\mathrm{pt}, (a_Z)_* i^! a_X^* \mathbb{Q}).$$

Proof. The first assertion assertion is the special case (B–21) for $f = a_X$ and $\mathcal{F}^\bullet = \underline{\mathbb{Q}}_X$, while the second assertion is (B–28). The third and fourth are Verdier dual to these. Indeed, Prop. 13.5 for $\mathbb{Q}$-coefficients states that $\mathbb{H}^{-q}(X, a_X^! \mathbb{Q}) = H_q^{BM}(X;\mathbb{Q})$ and $\mathbb{H}_c^{-q}(X, a_X^! \mathbb{Q}) = H_q(X;\mathbb{Q})$; now apply again (B–21), respectively (B–28). For the last assertion we observe that $\mathbb{H}^k(X, i_* i^! \mathbb{Q}) = \mathbb{H}^k(Z, i^! \mathbb{Q}) = H_Z^k(X;\mathbb{Q})$ by (B–34). □

Motivated by Lemma 14.8, using axiom D) and E) we do the same for the complex of mixed Hodge modules $\mathbb{Q}^{\mathrm{Hdg}}$ from (XIV–5).

$$\left.\begin{aligned} \underline{\mathbb{Q}}_X^{\mathrm{Hdg}} &:= a_X^* \mathbb{Q}^{\mathrm{Hdg}} \in D^{\mathrm{b}}\mathsf{MHM}(X) \\ \mathbb{D}\underline{\mathbb{Q}}_X^{\mathrm{Hdg}} &:= a_X^! \mathbb{Q}^{\mathrm{Hdg}} \in D^{\mathrm{b}}\mathsf{MHM}(X). \end{aligned}\right\} \qquad \text{(XIV–6)}$$

Since the axioms guarantee that the underlying rational component of these two complexes is equal to $\underline{\mathbb{Q}}_X$, respectively $\mathbb{D}(\underline{\mathbb{Q}}_X)$, Cor. 14.7 implies:

Corollary 14.9. *Let X be a complex algebraic variety and $i : Z \hookrightarrow X$ a subvariety. The complexes of mixed Hodge modules* $(a_X)_* \underline{\mathbb{Q}}_X^{\mathrm{Hdg}}$, $(a_X)_! \mathbb{D}\underline{\mathbb{Q}}_X^{\mathrm{Hdg}}$, $(a_X)_! \underline{\mathbb{Q}}_X^{\mathrm{Hdg}}$, $(a_X)_* \mathbb{D}\underline{\mathbb{Q}}_X^{\mathrm{Hdg}}$, *respectively* $i_* i^! \underline{\mathbb{Q}}_X^{\mathrm{Hdg}}$ *put mixed Hodge structures on cohomology, homology, cohomology with compact support, Borel-Moore homology, and cohomology with support in Z respectively.*

Remark 14.10. These mixed Hodge structures coincide with the ones constructed by Deligne and were discussed in Chap. 4 and 5. This is not hard to prove if X is a smooth algebraic variety, or if X can be embedded in a smooth algebraic variety. See the remark at end of [Sa90, § 4.5]. It is true in general, but highly non-trivial since Saito's approach does not work well in the setting of cubical or simplicial spaces. See [Sa00, Cor. 4.3].

As to functoriality, let $f : X \to Y$ be a morphism of complex algebraic varieties. Let M and N be bounded complexes with constructible cohomology on X respectively Y. Consider the adjoint relations

$$\mathrm{Hom}(f^* M^\bullet, N^\bullet) = \mathrm{Hom}(M^\bullet, f_* N^\bullet), \quad \mathrm{Hom}(f_! M^\bullet, N^\bullet) = \mathrm{Hom}(M^\bullet, f^! N^\bullet).$$

Apply the first relation with $M^\bullet = \underline{\mathbb{Q}}_Y$, $N^\bullet = \underline{\mathbb{Q}}_X$ and the identity morphism. Observing that $f^* M^\bullet = N^\bullet$ we get a morphism $\underline{\mathbb{Q}}_Y \to f_* \underline{\mathbb{Q}}_X$. Now apply the maps $(a_Y)_*$, $(a_X)_*$ to source, respectively target. This yields an induced map between complexes of vector spaces over a point, and in cohomology this gives $f^* : H^*(Y;\mathbb{Q}) \to H^*(X;\mathbb{Q})$. The dual argument yields a map between complexes over a point which in cohomology gives the morphisms $f_* : H_*(X;\mathbb{Q}) \to H_*(Y;\mathbb{Q})$, and, if X and Y are compact and smooth, at the same time give the Gysin morphisms.

Using in this construction the mixed Hodge modules $M = \underline{\mathbb{Q}}_Y^{\mathrm{Hdg}}$ and $N = \underline{\mathbb{Q}}_X^{\mathrm{Hdg}}$, we deduce that these morphisms are indeed morphisms of mixed Hodge structures. In a moment we need in particular the induced morphism

$$f^\sharp : (a_Y)_* \underline{\mathbb{Q}}_Y^{\mathrm{Hdg}} \to (a_X)_* \underline{\mathbb{Q}}_X^{\mathrm{Hdg}} \qquad \text{(XIV–7)}$$

14.1.3 Spectral Sequences

Since the category of mixed Hodge modules is abelian, the canonical filtration τ preserves complexes of mixed Hodge modules. Therefore, the second spectral sequence for any functor T sending mixed Hodge modules to mixed Hodge modules is a spectral sequences of mixed Hodge modules:

$$E_2^{p,q} = H^p T(H^q(M^\bullet) \Longrightarrow H^{p+q} T(M^\bullet).$$

Let $h : X \to Z$ be a morphism between complex algebraic varieties and let $M^\bullet$ be any complex of mixed Hodge modules on X. Suppose that $h = g \circ f$. By axiom E) there is a complex of mixed Hodge modules $f_* M^\bullet$ on Y and a complex of mixed Hodge modules $g_*(f_* M^\bullet)$ quasi-isomorphic to $h_* M^\bullet$ on

Z. Apply the above remark to the spectral sequence for the direct image functor g_* for complexes of the form $f_*M^\bullet$. We arrive at the following result, which states how the cohomology modules $H^q h_* M^\bullet$, which are mixed Hodge modules, behave when h decomposes:

Theorem 14.11. *Let $f : X \to Y$ and $g : Y \to Z$ be morphisms between complex algebraic varieties and let $M^\bullet$ be a bounded complex of mixed Hodge modules on X. Then the spectral sequence*

$$E_2^{pq} = H^p g_* \circ H^q f_* M^\bullet \implies H^{p+q}(g \circ f)_* M^\bullet$$

is a spectral sequence of mixed Hodge modules on Z.

Here we recall that $f_*M^\bullet$ is a complex of objects in the *abelian* category of mixed Hodge modules on Y. Hence its cohomology $H^q f_* M^\bullet$ is a mixed Hodge module on Y and likewise $H^p g_* \circ H^q f_* M^\bullet$ and $H^{p+q}(g \circ f)_* M^\bullet$ are mixed Hodge modules on Z.

Let us now take for Z a point, $g = a_Y$ so that $g \circ f = a_X$. The mixed Hodge modules from the preceding theorem are thus all mixed Hodge modules over a point, i.e. they are mixed Hodge structures. Using the notation (XIV–4) we thus have:

Corollary 14.12. *Let $f : X \to Y$ be a morphism between complex algebraic varieties and let $M^\bullet$ be a bounded complex of mixed Hodge modules on X. Then the* perverse *Leray spectral sequence*

$$\mathbb{H}^p(Y, H^q[f_* M^\bullet]) \implies \mathbb{H}^{p+q}(X, M^\bullet)$$

is a spectral sequence of mixed Hodge structures.

In particular we may take $M^\bullet = \underline{\mathbb{Q}}_X^{\mathrm{Hdg}}$. The direct image $f_*\underline{\mathbb{Q}}_X^{\mathrm{Hdg}}$ is a complex of mixed Hodge modules on Y, whose rational component is the complex $Rf_*\underline{\mathbb{Q}}_X$. Then the rational component of its cohomology sheaf $H^q f_* \underline{\mathbb{Q}}_X^{\mathrm{Hdg}}$ is in general *not* equal to $R^q f_* \underline{\mathbb{Q}}_X$, but rather to ${}^{\pi}H^q f_* \underline{\mathbb{Q}}_X$. This is not a single sheaf but a perverse complex of sheaves of which the hypercohomology groups can be considered. Using Lemma 14.6 we deduce that these have mixed Hodge structures. The above corollary 14.12 gives in fact:

Corollary 14.13. *The* perverse *Leray spectral sequence for the constant sheaf* $\underline{\mathbb{Q}}_X$

$$\mathbb{H}^p(Y, {}^{\pi}H^q[f_* \underline{\mathbb{Q}}_X]) \implies H^{p+q}(X, \mathbb{Q})$$

is a spectral sequence of mixed Hodge structures compatible with the (standard) mixed Hodge structure on $H^{p+q}(X; \mathbb{Q})$. If f is smooth and proper, it coincides with the usual Leray spectral sequence up to renumbering the indices. See (XIII–17).

Remark. For f smooth and proper we get back a special case of Theorem 6.5.

To obtain the non-perverse versions of the Leray spectral sequence, one has to reconsider the abelian category of mixed Hodge modules and its derived category. It has a natural t-structure and it is this t-structure which defines the usual canonical filtration on bounded complexes of mixed Hodge modules. But the same derived category has another t-structure which has been introduced in [Sa90, Remark 4.6(2)]. Let us call it the **anomalous t-structure**. As any t-structure, it comes with its own canonical filtration on complexes $M^\bullet$, the **anomalous canonical filtration** as well as its own, anomalous cohomology groups $\widetilde{H}^p(M^\bullet)$. These are mixed Hodge modules, but the rational component $\mathrm{rat}_X \widetilde{H}^p(M^\bullet)$ is the *ordinary* cohomology group of the rational component $\mathrm{rat}_X M^\bullet$, *not* the perverse cohomology. The preceding discussion, this time with the anomalous canonical filtration yields the ordinary Leray spectral sequence:

Corollary 14.14. *The* ordinary *Leray spectral sequence for the constant sheaf* $\underline{\mathbb{Q}}_X$

$$\mathbb{H}^p(Y, H^q[f_*\underline{\mathbb{Q}}_X]) \implies H^{p+q}(X, \mathbb{Q})$$

is a spectral sequence of mixed Hodge structures compatible with the (standard) mixed Hodge structure on $H^{p+q}(X; \mathbb{Q})$.

Remark. If in the above we use the functor $T = (a_Y)_!$ (global sections with compact support) instead of $T = (a_Y)_*$, we get the two versions of the Leray spectral sequence with compact supports.

14.1.4 Intersection Cohomology

Let $j : U \hookrightarrow X$ be the inclusion of a dense open smooth submanifold of X and let $\mathbb{V}$ be a local system on U. Recall that by Prop. 13.24 we have $\mathcal{IC}^\bullet_X(\mathbb{V}) = j_{!*}(\mathbb{V}[d_X])$. For $\mathbb{V} = \underline{\mathbb{Q}}_U$ we get a simple object in the category of perverse sheaves (Lemma 13.26) which by Prop. 13.18 has no sub-object or quotient object supported on $X - U$. The functor $j_{!*}$ can be defined as for complexes of sheaves (formula (XIII–18)), so replacing $\underline{\mathbb{Q}}_U$ by $\underline{\mathbb{Q}}_U^{\mathrm{Hdg}}$, we obtain

$$\mathcal{IC}^\bullet_X\, \mathbb{Q}^{\mathrm{Hdg}} := j_{!*}\underline{\mathbb{Q}}_U^{\mathrm{Hdg}}[d_X].$$

Because the functor rat is faithful this is the unique simple object in the category $\mathsf{MHM}(X)$ restricting to $\underline{\mathbb{Q}}_U[d_X]$ over U.

Lemma 14.15. *We have*

$$\mathrm{Gr}^W_n\, H^n \underline{\mathbb{Q}}_X^{\mathrm{Hdg}} = \mathcal{IC}^\bullet_X\, \mathbb{Q}^{\mathrm{Hdg}}, \quad n = d_X. \tag{XIV–8}$$

In particular, the intersection complex underlies a mixed Hodge module of pure weight.

Proof. This is clearly true over U and it suffices to show that the left hand side has no sub-objects or quotient objects over the complement $Z = X - U$. Since axiom C) guarantees that the pure mixed Hodge complex $\mathrm{Gr}_n^W H^n(\underline{\mathbb{Q}}_X^{\mathrm{Hdg}})$ is semi-simple, every quotient object is also a subobject and so it suffices to show it has no quotient objects $M \in \mathsf{MHM}(X)$ of pure weight n supported on Z. Note that a morphism of complexes $K^\bullet \to L$ in an Abelian category where L is in degree zero gives a morphism $H^0(K^\bullet) \to L$ and if K^0 is semi-simple the converse holds as well. Since the modules figuring in the complex $\mathrm{Gr}_n^W \underline{\mathbb{Q}}_X^{\mathrm{Hdg}}$ are all semi-simple this implies

$$\left.\begin{aligned}\mathrm{Hom}(H^n(\mathrm{Gr}_n^W \underline{\mathbb{Q}}_X^{\mathrm{Hdg}}), M) &= \mathrm{Hom}(H^0(\mathrm{Gr}_n^W \underline{\mathbb{Q}}_X^{\mathrm{Hdg}}[n]), M)\\ &= \mathrm{Hom}(\mathrm{Gr}_n^W \underline{\mathbb{Q}}_X^{\mathrm{Hdg}}[n], M)\\ &= \mathrm{Hom}(\underline{\mathbb{Q}}_X^{\mathrm{Hdg}}[n], M).\end{aligned}\right\} \qquad \text{(XIV–9)}$$

The last equation holds because of the functorial properties of weights and since M has pure weight n. Next, note that we can take "Hom" in the category MHM or in its derived category (see Remark A.30). In the derived category we can use adjunction for the pair (i_*, i^*) where $i : Z \hookrightarrow X$ is the inclusion. Indeed, $M = i_* i^* M$ and $i^* \underline{\mathbb{Q}}_X^{\mathrm{Hdg}} = \underline{\mathbb{Q}}_Z^{\mathrm{Hdg}}$ so that by adjunction

$$\mathrm{Hom}(\underline{\mathbb{Q}}_X^{\mathrm{Hdg}}[n]), i_* i^* M) = \mathrm{Hom}(\underline{\mathbb{Q}}_Z^{\mathrm{Hdg}}[n]), i_* M). \qquad \text{(XIV–10)}$$

Look at the rational components of the right hand side. The rational component of the first argument up to quasi-isomorphism is $\underline{\mathbb{Q}}_Z[n]$, a single sheaf in degree $-n$. Since i_* preserves perversity (see Example 13.18.1), the rational component of $i_* M$ is the perverse complex $i_*(\mathrm{rat}_X M)$. By definition since $\dim Z = n - 1$, this is a complex concentrated in degrees $[-n+1, 0]$. It follows that the left hand side of (XIV–10) vanishes. □

Note that $IH^k(X) = H^k((a_X)_* \mathcal{IC}^\bullet \underline{\mathbb{Q}}_X)$, and so, applying axiom H) to $(a_X)_*$ and then applying axioms B) and C) we deduce:

Corollary 14.16. *Let X be a compact complex algebraic variety. Then the intersection cohomology group $IH^k(X)$ has a pure Hodge structure of weight k.*

For later use we need:

Corollary 14.17. *We have a natural quotient morphism*

$$\underline{\mathbb{Q}}_X^{\mathrm{Hdg}} \twoheadrightarrow \mathcal{IC}_X^\bullet(\mathbb{Q}^{\mathrm{Hdg}})[-d_X], \qquad \text{(XIV–11)}$$

which is the identity on U.

Proof. Since perverse complexes are concentrated in degrees $[-n, 0]$ we have ${}^{\pi}H^k(\mathrm{Gr}_n^W \underline{\mathbb{Q}}_X[n]) = 0$ for $k > 0$. Put $M^\bullet = \underline{\mathbb{Q}}_X^{\mathrm{Hdg}}[n]$. Since the functor rat_X is faithful, we have

$$H^k(\mathrm{Gr}_n^W M^\bullet) = 0 \quad \text{for } k > 0.$$

Hence the natural (quotient) morphism $M^\bullet \to \tau_{\geq 0}M^\bullet$ in the derived category is equal to $M^\bullet \to H^0(M^\bullet)$ and we get a morphism of complexes of mixed Hodge modules

$$\underline{\mathbb{Q}}_X^{\mathrm{Hdg}} \to H^0(\underline{\mathbb{Q}}_X^{\mathrm{Hdg}}[n])[-n], \quad n = d_X.$$

Composing it with (XIV–8) we get the desired morphism. □

14.1.5 Refined Fundamental Classes

We have defined (XIV–7) the induced morphism $i^\sharp : \underline{\mathbb{Q}}_X^{\mathrm{Hdg}} \to i_*\underline{\mathbb{Q}}_Z^{\mathrm{Hdg}}$. Composing it with the morphism coming from the projection (XIV–11) we get a morphism $\gamma : \underline{\mathbb{Q}}_X^{\mathrm{Hdg}} \to i_*\mathcal{IC}_Z^\bullet\,\mathbb{Q}^{\mathrm{Hdg}}[-d_Z]$. Since the dual of $\mathcal{IC}_Z^\bullet\,\mathbb{Q}^{\mathrm{Hdg}}$ is the Hodge module $\mathcal{IC}_Z^\bullet\,\mathbb{Q}^{\mathrm{Hdg}}(d_Z)$ we can compose the morphism γ with its (shifted and twisted) dual $\mathcal{IC}_Z^\bullet\,\mathbb{Q}^{\mathrm{Hdg}} \to \mathbb{D}\underline{\mathbb{Q}}_X^{\mathrm{Hdg}}(-d_Z)[-2d_Z]$. So we get a homomorphism from the constant Hodge module on X to its Verdier dual up to a shift and twist depending on the codimension. By adjunction with respect to $(a_X)_*$, such a homomorphism is nothing but a homomorphism between the constant Hodge module $\mathbb{Q}^{\mathrm{Hdg}}$ on a point and the complex $(a_X)_*\mathbb{D}\underline{\mathbb{Q}}_X^{\mathrm{Hdg}}(-d_Z)[-2d_Z]$. The upshot is a **refined fundamental class**

$$\begin{aligned}\mathrm{cl}^{\mathrm{Hdg}}(Z) &\in \mathrm{Hom}\left(\mathbb{Q}^{\mathrm{Hdg}}, (a_X)_*\mathbb{D}\underline{\mathbb{Q}}_X^{\mathrm{Hdg}}(-d_Z)[-2d_Z]\right)\\ &= \mathrm{Ext}_{\mathrm{MHS}}^{-2d_Z}\left(\mathbb{Q}, (a_X)_*\mathbb{D}\underline{\mathbb{Q}}_X^{\mathrm{Hdg}}(-d_Z)\right)\end{aligned}$$

where for the last equality we work in the derived category of (polarizable) mixed Hodge structures. This makes sense also for singular X.

Using the constructions from § 3.5.2, especially formula (III–16), the latter Ext-group is the absolute Hodge cohomology in degree $-2d_Z$ of the complex

$$R\Gamma({}^{\mathrm{Ve}}\mathbb{D}\underline{\mathbb{Q}}_X(-d_Z)) = (a_X)_*[{}^{\mathrm{Ve}}\mathbb{D}\underline{\mathbb{Q}}_X(-d_Z)]$$

after marking. If X is smooth and projective, ${}^{\mathrm{Ve}}\mathbb{D}\underline{\mathbb{Q}}_X = \underline{\mathbb{Q}}_X(d_X)[2d_X]$ and so in this case, with $c = d_X - d_Z$, we find a refined class

$$\mathrm{cl}^{\mathrm{Hdg}}(Z) \in \mathrm{Ext}^{2c}_{\mathrm{MHS}}(\mathbb{Q}, R\Gamma(\underline{\mathbb{Q}}_X(c))).$$

By Theorem 7.15 the right hand side is exactly the Deligne group $H^{2c}_{\mathrm{Del}}(X, \mathbb{Q}(c))$. In fact, as Saito shows, if X is a projective manifold, the refined fundamental class as defined here coincides with the fundamental class in Deligne cohomology as defined in § 7.2.2.

14.2 The Kashiwara-Malgrange Filtration

14.2.1 Motivation

Let X be a complex manifold and let $\mathcal{K}^\bullet$ be a perverse complex on X. If X is the total space of a one-parameter degeneration (§ 11.2.2) $t : X \to \Delta$

we have seen that (Prop. 13.29) that the associated nearby and vanishing complexes $\psi_t(\mathcal{K}^\bullet)[-1]$ and $\phi_t(\mathcal{K}^\bullet)[-1]$ are perverse on X_0. The Riemann-Hilbert correspondence (Theorem 13.64) tells us that $\mathcal{K}^\bullet$ corresponds to a regular holonomic D-module, say $\mathcal{M}$. If X_0 is smooth this holds also for the associated nearby and vanishing complexes, but it fails when X_0 is singular. We shall explain that if we consider $\mathcal{M}$ as a D-module on $X \times \Delta$ supported on the graph of t, there is a filtration on $\mathcal{M}$, the V-filtration, such that the De Rham complex of each graded part gives a suitable eigenspace of the nearby and vanishing complex.

Notation. Let $\{V_\alpha\mathcal{M}\}_{\alpha\in\mathbb{Q}}$ be an increasing filtration (a **rational filtration**) on a $\mathcal{D}_X$-module $\mathcal{M}$. Put

$$V_{<\alpha}\mathcal{M} = \bigcup_{\beta<\alpha} V_\beta\mathcal{M}$$
$$\mathrm{Gr}^V_\alpha \mathcal{M} = V_\alpha\mathcal{M}/V_{<\alpha}\mathcal{M}.$$

The filtration $V_\bullet\mathcal{M}$ is called **discrete** if for any given interval I, $\mathrm{Gr}^V_\alpha \mathcal{M} \neq 0$ for finitely many $\alpha \in I$.

To motivate the Kashiwara-Malgrange filtration, which is such a rational filtration, we first look at the simplest non-trivial example.

Example 14.18. Consider the $\mathcal{D}$-module over the unit disk Δ coming from a vector bundle E of rank r equipped with a meromorphic connection ∇ which is holomorphic on Δ^*. The bundle E can be trivialized by a holomorphic frame $\{e_1,\ldots,e_r\}$ with meromorphic connection matrix $M(t)$ holomorphic on Δ^*. The covariant derivative in the direction of the vector field d/dt acts on the framed bundle E through the matrix $M(t)$ and this action defines the $\mathcal{D}_\Delta$-module structure on $E[t^{-1}]$. The connection has regular singularities if and only if with respect to a suitable frame

$$M(t) = t^{-1}R + M_0(t) \qquad \text{(XIV–12)}$$

with R a constant matrix, the **residue**, and $M_0(t)$ a matrix of holomorphic functions on the punctured disk, admitting a holomorphic extension to Δ. This is equivalent to saying that the **Fuchs field**

$$\partial_{\mathrm{Fcs}} = t\frac{d}{dt}$$

preserves the framed vector bundle $\mathcal{O}_\Delta(E)$. Let us introduce

$$A = \mathcal{O}_{\Delta,0}$$
$$K = \mathcal{O}_{\Delta,0}[t^{-1}].$$

A pair (E,∇) consisting of a holomorphic vector bundle E on the unit disk with a meromorphic connection ∇ then corresponds to the endomorphism

$M(t)$ of the K-vector space $V_K = K^r$, and ∇ is regular if V_K contains a lattice left stable by $tM(t)$. Conversely, every lattice L left stable by $t \cdot M(t)$ yields a rank r vector bundle on Δ with a regular meromorphic connection of order 1. Its **residue** $\mathrm{res}_0(\nabla)$ at 0 is the endomorphism on $\mathbb{C}^r = V_A/tV_A$ induced by the Fuchs field. It is represented by R in (XIV–12). The Fuchs field acts also as R on the complex vector space $B := \mathbb{C}^r \subset V$ generated by the standard basis. Since $\partial_{\mathrm{Fcs}} t^k = t^k(k + \partial_{\mathrm{Fcs}})$, the Fuchs field acts as $k + R$ on $t^k B$ which implies that if $\Sigma \subset \mathbb{C}$ is the spectrum of ∂_{Fcs} on B then $\Sigma + \mathbb{N}$ is the spectrum of the action on V_A. The monodromy T is related to R by $T = \exp(-2\pi \mathrm{i} R)$. Writing T_u, T_s respectively for the unipotent, respectively semi-simple part of T and $R = S + N$ for the additive Jordan decomposition in a semi-simple part S and a nilpotent part N, we have

$$T_s = \exp(-2\pi \mathrm{i} S) \qquad T_u = \exp(-2\pi \mathrm{i} N) \qquad N = \frac{-1}{2\pi \mathrm{i}} \log T_u.$$

The eigenvalues of R are rational numbers precisely when T is quasi-unipotent. Let us assume this from now on. Let $E_\beta(R) \subset V_A$ be the maximal sub-module on which $R - \beta\,\mathrm{id}$ acts nilpotently (the generalized "eigenspace" of R with eigenvalue β). We then introduce

$$V_\alpha = V^{-\alpha} := \sum_{\beta \geq -\alpha - 1} E_\beta(R) A \subset V_A. \qquad \text{(XIV–13)}$$

This defines an exhaustive filtration of V_A, since every $v \in V_A$ can be written uniquely as a powerseries $v = \sum v_\beta$ with $v_\beta \in E_\beta(R)$. Replacing the inequality by a strict inequality defines $V_{<\alpha}$ so that

$$\mathrm{Gr}^V_\alpha = V_\alpha / V_{<\alpha} = E_{-\alpha-1}(R).$$

It follows that $\partial_{\mathrm{Fcs}} + \alpha + 1$ acts on Gr^V_α as $R + \alpha + 1$ and thus acts simply as N and hence nilpotently on this space. Summarizing, this V-filtration is a discrete and rational filtration and has the following properties:

$$\left.\begin{array}{rl} t\partial_t V_\alpha \subset & V_\alpha \\ t^k V_\alpha = & V_{\alpha-k} \\ t\partial_t + \alpha + 1 & \text{acts nilpotently on } \mathrm{Gr}^\alpha_V\,. \end{array}\right\} \qquad \text{(XIV–14)}$$

In particular, R acting on V_{-1}/V_{-2} has eigenvalues in the interval $[0, 1)$. Since $V_{-2} = tV_{-1}$ the lattice V_{-1} defines a locally free rank r bundle $\tilde{E}$ extending E to the disk such that the connection ∇ has a logarithmic pole whose residue has eigenvalues in the interval $[0, 1)$. It is called the **canonical sublattice**. This is exactly the canonical extension of E (see Definition 11.4).

14.2.2 The Rational V-Filtration

We start by defining a V-filtration on $\mathcal{D}_X$ which originates from the way the differential operators along the codimension one smooth submanifold $Y \subset X$

are positioned with respect to $\mathcal{D}_X$. This V-filtration (indexed by $\mathbb{Z}$) is an increasing filtration in the usual sense. It can be introduced as follows. Let $\mathcal{I} \subset \mathcal{O}_X$ be the sheaf of ideals of Y and let $V_k\mathcal{O}_X = \mathcal{O}_k$ for $k \geq 0$ and $V_{-k}\mathcal{O}_X = \mathcal{I}^k$ for $k > 0$. It induces a filtration on the sheaf of $\underline{\mathbb{C}}_X$-endomorphisms by setting

$$V_k\mathcal{E}nd_{\mathbb{C}}\mathcal{O}_X := \{\varphi \mid \varphi V_j\mathcal{O}_X \subset V_{j+k}\mathcal{O}_X, \quad \forall j \in \mathbb{Z}\}.$$

This induces the V-filtration on $\mathcal{D}_X$. Let us describe this filtration in local coordinates $(t, x_1, \ldots, x_n)$ such that the hypersurface Y is given by a local equation $t = 0$. Write

$$\partial_t := \frac{\partial}{\partial t}.$$

The Fuchs field $t \cdot \partial_t$ together with the fields tangent to Y preserve the sheaf of ideals $\mathcal{I}$ of Y and together with $\mathcal{O}_X$ one gets the sheaf $V_0\mathcal{D}_X$. This implies that $V_k\mathcal{D}_X$ is locally generated as a (left or right) $V_0\mathcal{D}_X$-module by products $t^i\partial_t^j$ with $i - j \geq -k$. This V-filtration is clearly exhaustive and respects the multiplicative structure. Multiplication by t sends V_k onto V_{k-1} while ∂_t sends V_k to V_{k+1}. Also note that

$$\mathrm{Gr}_0^V \mathcal{D}_X = \mathcal{D}_Y[\partial_t].$$

Definition 14.19. Let X be a complex manifold, $Y \subset X$ a complex submanifold of codimension 1 and M a $\mathcal{D}_X$-module. A (rational) V**-filtration on** $\mathcal{M}$ **along** Y consists of a discrete increasing and exhaustive rational filtration by coherent $V_0\mathcal{D}_X$-modules $V_\alpha\mathcal{M}$ such that

1) The filtration is compatible with the V-filtration on $\mathcal{D}_X$ in the sense that $V_k\mathcal{D}_X V_\alpha\mathcal{M} \subset V_{\alpha+k}\mathcal{M}$. Furthermore, the inclusion $\mathcal{I}V_\alpha\mathcal{M} \subset V_{\alpha-1}\mathcal{M}$ should be an equality for $\alpha < 0$;
2) for all α the action of $t\partial_t + \alpha + 1$ is nilpotent on $\mathrm{Gr}_\alpha^V \mathcal{M}$.

If $\mathcal{M}$ admits such a filtration, we say that $\mathcal{M}$ is **specializable along** Y.

This notion can be generalized when Y is an *analytic subspace* of codimension 1. Then Y is locally given by one equation, so it is a Cartier divisor. The sheaf $\mathcal{O}_X(Y)$ of germs of meromorphic functions on X which have at most a pole of order 1 along Y is therefore locally free of rank 1. Let E be the associated line bundle and $\sigma : X \to E$ the section defined by the inclusion $\mathcal{O}_Y \hookrightarrow \mathcal{O}_Y(X)$. Then $Y = \sigma^{-1}E_0$, where E_0 is the zero section of E. It is an easy exercise to show that for Y smooth, a $\mathcal{D}_X$-module $\mathcal{M}$ is specializable along Y if and only if $\sigma_+\mathcal{M}$ is specializable along E_0. In view of this remark it makes sense to define:

Definition 14.20. Let X be a complex manifold and $Y \subset X$ a codimension 1 subvariety. A $\mathcal{D}_X$-module $\mathcal{M}$ is **specializable** along Y if $\sigma_+\mathcal{M}$ is specializable along E_0.

Examples 14.21. 1) The V-filtration construction in Example 14.18 satisfies these conditions in view of (XIV–14). So meromorphic connections with quasi-unipotent monodromy are specializable along the origin.
2) Suppose that $\mathcal{M}$ is supported on Y. Then by Theorem 13.47 we have

$$\mathcal{M} = \bigoplus_{i \geq 0} \mathcal{M}_i,$$

where $\mathcal{M}_0 = \operatorname{Ker}\{t : \mathcal{M} \to \mathcal{M}\}$, and $\mathcal{M}_i = \partial_t^i \mathcal{M}_0$. Moreover, $t^i : \mathcal{M}_i \xrightarrow{\sim} \mathcal{M}_0$ and $\partial_t^i : \mathcal{M}_0 \xrightarrow{\sim} \mathcal{M}_i$. Then, since t^{i+1} acts as the zero-morphism on $\mathcal{M}_i$, setting

$$V_\alpha \mathcal{M} := \operatorname{Ker}[t^{[\alpha+1]} : \mathcal{M} \to \mathcal{M}]$$

we have

$$V_\alpha \mathcal{M} = \bigoplus_{0 \leq i \leq [\alpha]} \mathcal{M}_i.$$

From these two descriptions of $V_\alpha \mathcal{M}$, the first condition for the V-filtration follows. Using the commutation relation $\partial_{\mathrm{Fcs}} = t\partial_t = \partial_t t - 1$ repeatedly we see that ∂_{Fcs} acts as multiplication by $-i-1$ on $\partial_t^i \mathcal{M}_0$. In other words, $\mathcal{M}_i \subset \operatorname{Ker}(\partial_{\mathrm{Fcs}} + i + 1)$ from which in fact equality follows. In other words, $\partial_{\mathrm{Fcs}} + [\alpha] + 1$ acts as the zero morphism on $\mathrm{Gr}_V^{[\alpha]}$. This shows that also the second condition for the V-filtration holds.

Remark. If $\mathcal{M}$ is specializable along Y, then the filtration V is uniquely defined [Sa88, Lemme 3.1.2]. Hence $\mathcal{M}$ is specializable along Y if and only if it is so locally. It is called the **Kashiwara-Malgrange filtration** or **the rational V-filtration**.

To explain for which D-modules this filtration exists, we first introduce:

Definition 14.22. Let t be a non-constant holomorphic function on a complex manifold X, $X_0 = t^{-1}(0)$ a (possibly singular) fibre. A holonomic $\mathcal{D}_X$-module $\mathcal{M}$ has **quasi-unipotent monodromy along** X_0 if the monodromy action on $\psi_t(\mathrm{DR}(\mathcal{M}))$ is quasi-unipotent.

Then we have:

Theorem 14.23 ([Malg79]). *Every regular holonomic $\mathcal{D}_X$-module with quasi-unipotent monodromy is specializable along X_0.*

In the setting where $Y = X_0 = t^{-1}0$ the reduction to the case of a smooth fibre X_0 can be made globally since $\mathcal{O}_Y(X) = \mathcal{O}_Y$: one replaces t by the composition of

$$i = i_t : X \to X \times \mathbb{C}, \quad x \mapsto (x, t(x)) \qquad \text{(XIV–15)}$$

followed by the projection $p : X \times \mathbb{C} \to \mathbb{C}$. If $\mathcal{M}$ is regular holonomic, by Prop. 13.63 also $i_+\mathcal{M}$ is regular holonomic. Since $\mathrm{DR}\, i_+ = i_* \mathrm{DR}$ we see that $\psi_p(\mathrm{DR}(i_+\mathcal{M})) = \psi_p(i_*(\mathrm{DR}\,\mathcal{M})) = i_*\psi_t(\mathrm{DR}(\mathcal{M}))$, and hence we deduce that if $\mathcal{M}$ has quasi-unipotent monodromy along X_0, then $i_+\mathcal{M}$ has quasi-unipotent monodromy along $X \times \{0\}$. In view of Thm. 14.23 it is thus specializable along $X \times \{0\}$ and admits its own rational V-filtration. As announced in 14.2.1 it can be linked to the monodromy action as made explicit in [Kash83] and [Malg83]. Saito summarizes it as follows.

Theorem 14.24 ([Sa88, 3.4.12]). *Let X be a complex manifold, let $t : X \to \mathbb{C}$ be a non-constant holomorphic function on X, and let $\mathcal{M}$ be a regular holonomic $\mathcal{D}_X$-module with quasi-unipotent monodromy along $X_0 = t^{-1}(0)$ (Def. 14.22).*

Introduce be the subcomplex $\psi_{t,\mathrm{e}(\alpha)}(\mathrm{DR}_X(\mathcal{M}))$ of $\psi_t(\mathrm{DR}_X(\mathcal{M}))$ corresponding to the eigenvalue $\mathrm{e}(\alpha) = \exp(2\pi \mathrm{i}\alpha)$ of the semi-simple part of the local monodromy operator, and similarly for $\phi_t(\mathcal{M})$. Then, setting $\widetilde{\mathcal{M}} = (i_t)_+\mathcal{M}$ (see (XIV–15)), there are canonical isomorphisms

$$\mathrm{DR}_{X\times\{0\}}(\mathrm{Gr}^V_\alpha \widetilde{\mathcal{M}}) \simeq \begin{cases} \psi_{t,\mathrm{e}(\alpha)}(\mathrm{DR}_X(\mathcal{M}))[-1] \text{ if } -1 \le \alpha < 0 \\ \phi_{t,\mathrm{e}(\alpha)}(\mathrm{DR}_X(\mathcal{M}))[-1] \text{ if } -1 < \alpha \le 0 \end{cases}$$

Moreover, under these isomorphisms

a) *$t\partial_t + \alpha + 1$ corresponds to* $\dfrac{-1}{2\pi \mathrm{i}} \log T_u = N$,
b) *∂_t corresponds to* can, *and*
c) *multiplication by t corresponds to* var.

Remark 14.25. Let $\mathcal{M}$ be a $\mathcal{D}_X$-module specializable along X_0. Then $\mathrm{Gr}^V_\alpha \widetilde{\mathcal{M}}$ is a coherent $\mathcal{D}_{X_0}$-module and it is (regular) holonomic if $\mathcal{M}$ is (regular) holonomic). Let $\mathcal{K}^\bullet$ be the perverse complex corresponding to $\mathcal{M}$ under the Riemann-Hilbert correspondence so that $\psi_t\mathcal{K}^\bullet[-1]$ is perverse on X_0. The above theorem states that

a) the $\mathcal{D}_{X\times\{0\}}$-module $\bigoplus_{-1\le\alpha<0} \mathrm{Gr}^V_\alpha \widetilde{\mathcal{M}}$, which is in fact supported on $X_0 \times \{0\}$ plays the role of $\mathrm{DR}^{-1}_{X_0}(\psi_t\mathcal{K}^\bullet[-1])$ with $\mathrm{Gr}^V_\alpha \widetilde{\mathcal{M}}$ corresponding to the eigenspace of the "logarithmic connection" $t\partial_t$ on $\widetilde{\mathcal{M}}$ for the eigenvalue $-\alpha$;
b) the $\mathcal{D}_{X\times\{0\}}$-module $\bigoplus_{-1<\alpha\le 0} \mathrm{Gr}^V_\alpha \widetilde{\mathcal{M}}$ plays the role of $\mathrm{DR}^{-1}_{X_0}(\phi_t\mathcal{K}^\bullet[-1])$. The component with $\alpha = 0$ corresponds to the eigenspace of the unipotent part of the local monodromy on $\phi_t\mathcal{K}^\bullet[-1]$ for the eigenvalue 1 .

Remark. In [Sa88] D-modules act from the *right*. So the left action of $t\partial_t$ would have to be replaced with the left action of $\partial_t t = t\partial_t + 1$. Working out what this means in Example 14.21 2) where $t\partial_t + \alpha + 1$ acts nilpotently on Gr^α_V we find that under this right action it is $t\partial_t - \alpha$ which acts nilpotently. This checks with the proof of [Sa88, Lemme 3.1.3]. It explains our convention of indexing the V-filtration as used in Def. 14.19. With these changes we can indeed use Saito's definitions for the nearby and vanishing D-modules to which Theorem 14.24 is a prelude. One further word of warning: Saito uses left D-modules and a *decreasing* V-filtration in the introduction to loc.cit.

Suppose now that $(\mathcal{M}, F)$ is a filtered $\mathcal{D}_X$-module such that $\mathcal{M}$ is specializable along X_0. By definition, it admits the rational V-filtration so that in particular $N_\alpha := t\partial_t + \alpha + 1$ acts nilpotently on the graded parts $\mathrm{Gr}_\alpha^V \mathcal{M}$. Let $W = W(N_\alpha)$ be the associated weight filtration centred at 0 (Def. 11.9). The following compatibility between F and the rational V-filtration is needed:

Definition 14.26. We say that $(\mathcal{M}, F)$ is **specializable along** X_0 if

1) $\mathcal{M}$ is specializable along X_0;
2) $t(F_p V_\alpha \mathcal{M}) \subset F_p V_{\alpha-1}\mathcal{M}$ with equality if $\alpha < 0$;
3) $\partial_t(F_p \mathrm{Gr}_\alpha^V) \subset F_{p+1} \mathrm{Gr}_{\alpha+1}^V$ with equality if $\alpha > -1$.
4) F induces good filtrations on the $\mathcal{D}_{X_0}$-modules $\mathrm{Gr}_i^W \mathrm{Gr}_\alpha^V$, i.e. (by Lemma 13.43) the $\mathcal{D}_{X_0}$-modules $\mathrm{Gr}^F \mathrm{Gr}_i^W \mathrm{Gr}_\alpha^V$ are coherent.

14.3 Polarizable Hodge Modules

14.3.1 Hodge Modules

We start with the definitions leading to the notion of a Hodge module on a *complex manifold.* The more general case of a complex analytic space can be handled by taking charts and a patching procedure as explained in [Sa90, §2.1]. The main difficulty is the definition of D-modules on a singular variety; local charts embedding X in a smooth variety U suggest to define a D-module on X as a $\mathcal{D}_U$-module with support X. One needs to check independence of charts.

Definition 14.27. Let X be a complex manifold. A **rational structure** on a holonomic $\mathcal{D}_X$-module $\mathcal{M}$ consists of a perverse complex $\mathcal{M}^\bullet_\mathbb{Q} \in \mathrm{Perv}(X;\mathbb{Q})$ together with a quasi-isomorphism in $\mathrm{Perv}(X;\mathbb{C})$

$$\alpha : \mathcal{M}^\bullet_\mathbb{Q} \otimes \mathbb{C} \xrightarrow{\sim} \mathrm{DR}_X(\mathcal{M}) \quad \textbf{(comparison isomorphism)}.$$

The triple $(\mathcal{M}, \alpha, \mathcal{M}^\bullet_\mathbb{Q})$ is called a **rational** $\mathcal{D}_X$-MODULE.

Let us next switch to *filtered* regular holonomic rational $\mathcal{D}_X$-modules. Forgetting the comparison isomorphism we denote the resulting rational $\mathcal{D}_X$-modules as triples $(\mathcal{M}, F, \mathcal{M}^\bullet_\mathbb{Q})$; the following standard operations can be defined:

1) Tate twists:

$$(\mathcal{M}, F, \mathcal{M}^\bullet_\mathbb{Q})(n) := (\mathcal{M}, F[n], \mathcal{M}^\bullet_\mathbb{Q}(n)). \tag{XIV–16}$$

A word of explanation. Since F is an *increasing* filtration we have $F[n]_p = F_{p-n}$. The notation $\mathcal{M}^\bullet_\mathbb{Q}(n)$ means that the same complex $\mathcal{M}_\mathbb{Q}$ is used, but the comparison morphism α gets replaced by $(2\pi \mathrm{i})^n \alpha$.

2) Proper direct images: let $f : X \to Y$ be a proper map between complex manifolds. Then, within the associated *derived category* we wish to define

$$f_*(\mathcal{M}, F, \mathcal{M}^\bullet_{\mathbb{Q}}) := (f_+(\mathcal{M}, F), Rf_*(\mathcal{M}^\bullet_{\mathbb{Q}})).$$

If we forget the filtration, using the Riemann-Hilbert correspondence (Theorem 13.64) we see that this makes sense. There is however a problem with F-filtration, since we want it to induce a filtration on the direct images. For this to be true one needs that the derivatives d preserve the filtration of $f_+(\mathcal{M}, F)$ strictly. This is automatic for a closed immersion $i : Z \hookrightarrow X$ where we have

$$i_*(\mathcal{M}, F, \mathcal{M}^\bullet_{\mathbb{Q}}) := (i_+(\mathcal{M}, F), i_*(\mathcal{M}^\bullet_{\mathbb{Q}}))$$

In general, if the strictness condition is verified we set

$$f_*(\mathcal{M}, F, \mathcal{M}^\bullet_{\mathbb{Q}}) := (Rf_+(\mathcal{M}, F), {}^{\pi}Rf_*(\mathcal{M}^\bullet_{\mathbb{Q}})), \tag{XIV–17}$$

where ${}^{\pi}Rf_*$ stands for the perverse direct image, i.e. the derived functor for f_* with respect to the t-structure given by the middle perversity (A–20).

3) Duality:

$$\mathbb{D}_X(\mathcal{M}, F, \mathcal{M}^\bullet_{\mathbb{Q}}) := (\mathbb{D}_X(\mathcal{M}, F), {}^{\mathrm{Ve}}\mathbb{D}_X \mathcal{M}^\bullet_{\mathbb{Q}}).$$

This makes sense, again because of the Riemann-Hilbert correspondence 13.64 i).

4) Nearby and vanishing cycles: let t be a non-constant holomorphic function on X. Suppose that $(\mathcal{M}, F)$ is specializable along $\{t = 0\}$ (Def. 14.26). Put as before $\widetilde{\mathcal{M}} = (i_t)_+\mathcal{M}$ (see (XIV–15)). We then have

$$\begin{array}{lcl} \mathrm{Gr}^{\alpha}_V(\widetilde{\mathcal{M}}, F[1]) & \xrightarrow{\partial_t} & \mathrm{Gr}^{\alpha+1}_V(\widetilde{\mathcal{M}}, F) \\ \mathrm{Gr}^{\alpha}_V(\widetilde{\mathcal{M}}, F) & \xrightarrow{t} & \mathrm{Gr}^{\alpha-1}_V(\widetilde{\mathcal{M}}, F). \end{array}$$

By Remark 14.25 taking on the left hand side the sum over $\alpha \in [-1, 0)$ we obtain a filtered $\mathcal{D}_{X_0}$-module which corresponds to $\psi_t(\mathrm{DR}_X(\mathcal{M}, F))$. So, following Saito [Sa88, (5.1.3.3)] we put

$$\left.\begin{array}{ll} \psi_t(\mathcal{M}, F, \mathcal{M}^\bullet_{\mathbb{Q}}) & := \bigoplus_{-1 \le \alpha < 0} \left(\mathrm{Gr}^{\alpha}_V(\widetilde{\mathcal{M}}, F[1]), \psi_{t,\mathrm{e}(\alpha)}(\mathcal{M}^\bullet_{\mathbb{Q}})[-1] \right) \\ \psi_{t,1}(\mathcal{M}, F, \mathcal{M}^\bullet_{\mathbb{Q}}) & := \left(\mathrm{Gr}^{-1}_V(\widetilde{\mathcal{M}}, F), \psi_{t,1}(\mathcal{M}^\bullet_{\mathbb{Q}})[-1] \right). \end{array}\right\} \tag{XIV–18}$$

Ideally we would like to put $\phi_t(\mathcal{M}, F) = \bigoplus_{0 \le \alpha < 1} \mathrm{Gr}^{\alpha}_V(\widetilde{\mathcal{M}}, F)$ since we then would have that $\partial_t : \psi_t(\mathcal{M}, F) \to \phi_t(\mathcal{M}, F)$ and $t : \phi_t(\mathcal{M}, F) \to \psi_t(\mathcal{M}, F[-1]) = \psi_t(\mathcal{M}, F)(-1)$. However, under the Riemann-Hilbert correspondence this module does *not* correspond to $\phi_t(\mathrm{DR}_X(\mathcal{M}, F))$. By Theorem 14.24 this is indeed $\bigoplus_{-1 < \alpha \le 0} \mathrm{Gr}^{\alpha}_V(\widetilde{\mathcal{M}}, F)$ which intersects the previous D-module only in $\mathrm{Gr}^0_V(\widetilde{\mathcal{M}}, F)$ which is the *unipotent part* of $\phi_t(\mathrm{DR}_X(\mathcal{M}, F))$. For this reason we put

$$\phi_{t,1}\left((\mathcal{M},F),\mathcal{M}^{\bullet}_{\mathbb{Q}}\right) := \left(\mathrm{Gr}^0_V(\widetilde{\mathcal{M}},F[1]),\phi_{t,1}(\mathcal{M}^{\bullet}_{\mathbb{Q}})[-1]\right) \qquad \text{(XIV–19)}$$

So now by Theorem 14.24 we have morphisms

$$\mathrm{can} : \psi_{t,1}(\mathcal{M},F,\mathcal{M}^{\bullet}_{\mathbb{Q}}) \to \phi_{t,1}(\mathcal{M},F,\mathcal{M}^{\bullet}_{\mathbb{Q}}) \qquad \text{(XIV–20)}$$
$$\mathrm{var} : \phi_{t,1}(\mathcal{M},F,\mathcal{M}^{\bullet}_{\mathbb{Q}}) \to \psi_{t,1}(\mathcal{M},F,\mathcal{M}^{\bullet}_{\mathbb{Q}})(-1). \qquad \text{(XIV–21)}$$

Before we can give a definition of a Hodge module we need two more concepts. One is the monodromy weight filtration.

Definition 14.28. A **filtration** on a rational $\mathcal{D}_X$-module $(\mathcal{M},\alpha,\mathcal{M}^{\bullet}_{\mathbb{Q}})$ consists of a pair of filtrations on $\mathcal{M}$ and on the complex $\mathcal{M}^{\bullet}_{\mathbb{Q}}$ that correspond under α.

The **monodromy weight filtration** on the complex $\psi_t(\mathcal{M}^{\bullet}_{\mathbb{Q}})$ can be defined as in the special case of the constant sheaf treated in § 11.2.6 and § 11.2.5; as in this special case there is a complex counterpart of this filtration on $\psi_t\mathcal{M}$ compatible with the rational filtration, and thus these define a monodromy weight filtration W on $\psi_t(\mathcal{M},F,\mathcal{M}^{\bullet}_{\mathbb{Q}})$ and similarly on $\phi_{t,1}(\mathcal{M},F,\mathcal{M}^{\bullet}_{\mathbb{Q}})$. In fact for all $i \geq 0$ we have isomorphisms

$$\left.\begin{array}{l} N^i : \mathrm{Gr}^W_{n-1+i}\,\psi_t(\mathcal{M},F,\mathcal{M}^{\bullet}_{\mathbb{Q}}) \xrightarrow{\sim} \mathrm{Gr}^W_{n-1-i}\,\psi_t(\mathcal{M},F,\mathcal{M}^{\bullet}_{\mathbb{Q}})(-i) \\ N^i :\ \mathrm{Gr}^W_{n+i}\,\phi_{t,1}(\mathcal{M},F,\mathcal{M}^{\bullet}_{\mathbb{Q}}) \xrightarrow{\sim} \mathrm{Gr}^W_{n-i}\,\phi_{t,1}(\mathcal{M},F,\mathcal{M}^{\bullet}_{\mathbb{Q}})(-i) \end{array}\right\} \qquad \text{(XIV–22)}$$

See [Sa88, § 5] for details.

We also need the concept of strict support.

Definition 14.29. A filtered D-module equipped with a rational structure has **strict support in a subvariety** $Z \subset X$, if first of all it has support on Z, and, in addition, no sub-object or quotient-object has strictly smaller support. A direct sum of such modules is said to **satisfy the strict support condition**.

Now we can give a recursive definition of the concept of a Hodge module:

Definition 14.30. Let X be a smooth complex algebraic variety. Consider the category of those rational filtered regular holonomic $\mathcal{D}_X$-modules which in addition satisfy the strict support condition. The category $\mathsf{MH}(X,n)$ of **Hodge modules on** X **of weight** n is the largest full subcategory of this category such that

1) If $\mathcal{M}$ has support $\{x\}$, then $\mathcal{M} = i_{x*}(H_{\mathbb{C}},F,H_{\mathbb{Q}})$, where $i_x : x \hookrightarrow X$ is the inclusion, and $(H_{\mathbb{C}},F,H_{\mathbb{Q}})$ is a rational Hodge structure of weight n.
2) For all Zariski open $U \subset X$, $t : U \to \mathbb{C}$ a non-constant holomorphic function, and $\mathcal{M}$ having support not contained in $t^{-1}(0)$, if we let W be the weight filtration centred at $(n-1)$, respectively n (see (XIV–22)) then for all $k \in \mathbb{Z}$ the graded modules $\mathrm{Gr}^W_k\,\psi_t(\mathcal{M},F,\mathcal{M}^{\bullet}_{\mathbb{Q}})$ and $\mathrm{Gr}^W_k\,\phi_{t,1}(\mathcal{M},F,\mathcal{M}^{\bullet}_{\mathbb{Q}})$ respectively, are Hodge modules of weight k with support on $t^{-1}(0)$.

For a Hodge module $(\mathcal{M}, F, \mathcal{M}^\bullet_{\mathbb{Q}})$ we call F the underlying **Hodge filtration** and $\mathcal{M}^\bullet_{\mathbb{Q}}$ its **rational component**. This association defines a functor

$$\mathrm{rat}_X : \mathsf{MH}(X, n) \to \mathrm{Perv}(X; \mathbb{Q}) \tag{XIV–23}$$

which is compatible with the direct image functors (when defined) and with Verdier duality.

The category $\mathsf{MH}_Z(X, n)$ is the full subcategory of $\mathsf{MH}(X, n)$ of those Hodge modules having strict support Z. The category of Hodge modules is artinian in a very strong sense: a Hodge module enjoys the strict support condition [Sa88, § 5.1.6]:

$$\mathsf{MH}(X, n) = \bigoplus_{Z \subset X} \mathsf{MH}_Z(X, n). \tag{XIV–24}$$

We have moreover:

Proposition 14.31 ([Sa88, Prop. 5.1.14]). *The categories $\mathsf{MH}(X, n)$ and $\mathsf{MH}_Z(X, n)$ are abelian and every morphism in these categories is strict with respect to the F-filtration. We have*

a) *Both categories are stable under direct summands.*
b) *Tate twisting k times maps $\mathsf{MH}(X, n)$ to $\mathsf{MH}(X, n - 2k)$.*
c) *Duality sends $\mathsf{MH}_Z(X, n)$ to $\mathsf{MH}_Z(X, -n)$.*

A basic example of a Hodge module on a complex algebraic manifold X is provided by a polarizable variation of (rational) Hodge structures on X as introduced in § 10.2.

Theorem 14.32 ([Sa88, Th. 5.4.3]). *Let X be a complex algebraic manifold and let $V = (\mathbb{V}, \mathcal{F}^\bullet)$ a polarizable variation of Hodge structures of weight n on X. Let ∇ be the canonical integrable connection on $\mathcal{V} := \mathbb{V} \otimes_{\mathbb{Q}} \mathcal{O}_X$. Then*

$$V^{\mathrm{Hdg}} := ((\mathcal{V}, \nabla), \mathcal{F}_{-\bullet}, \mathbb{V}[d_X])$$

is a Hodge module of weight $n + d_X$ with strict support X.

Proof (Indication). The proof is long and complicated. By way of explanation, the Hodge filtration, modified to make it an increasing filtration, is an F-filtration. Griffiths's transversality (X–4) translates into F being good (Example 13.42). The De Rham complex of the D-module $\mathcal{V}$ is quasi-isomorphic to $\mathbb{V} \otimes \mathbb{C}[d_X]$. This last assertion is not at all trivial and uses the polarization.
□

By the preceding Theorem 14.32, a polarized variation of Hodge structure V on a complex algebraic manifold X determines a unique Hodge module V^{Hdg} with strict support X, the **Hodge module defined by the variation of Hodge structure** V.

More generally, we can construct a Hodge module from a polarized variation of Hodge structure over a *dense open subset* of X:

Theorem 14.33 ([Sa90, 3.20, 3.21]). *Let U be the complement of a divisor $D \subset X$ having normal crossings. Let $\mathbb{V}$ be a local system of $\mathbb{Q}$-vector spaces over U such that the local monodromy operators of $\mathbb{V}$ around D are quasi-unipotent. If $\mathbb{V}$ underlies a polarized variation of Hodge structures of weight n, say V, then there is a unique Hodge module V_X^{Hdg} of weight $n + d_X$ having strict support X and which restricts over U to V^{Hdg}.*

Proof (Sketch). By the Riemann-Hilbert correspondence, there is a regular holonomic $\mathcal{D}_X$-module $\mathcal{M}$ whose De Rham complex equals the perverse extension of $\mathbb{V}$. Indeed, by Prop. 11.3 we have an extension of $\mathbb{V}$ to a vector bundle $\mathcal{V}$ on X with a meromorphic connection having poles of order at most 1 along D and with residues along each branch of D in the interval $[0, 1)$. Then the module ${}^{\pi}\mathcal{V}$ from Example 13.65 is the D-module $\mathcal{M}$ we are after. If now $\mathbb{V}$ underlies a polarizable variation of Hodge structures of weight n, there is an F-filtration on $j^*\mathcal{V}$ and hence on $\mathcal{M}$. By construction $\mathcal{M}$ is regular holonomic. What remains to be shown is that the corresponding triple $(\mathcal{M}, F, {}^{\pi}\mathbb{V})$ fulfills all the requirements for a Hodge module. □

Note that $\mathcal{IC}_X^{\bullet}\,\mathbb{V} = j_{!*}(\mathbb{V}[d])$ is the unique perverse extension of $\mathbb{V}[d]$ with strict support X and since rat_X is faithful, it follows that $V_X^{\mathrm{Hdg}} = j_{!*}V^{\mathrm{Hdg}}$. This inspires the following definition.

Definition 14.34. Let Z be an irreducible complex algebraic subvariety of an algebraic variety X and let V be a polarizable variation of weight n Hodge structures over $U \subset Z$, a dense open set in the regular part of Z. Consider the Hodge module V^{Hdg} of weight $n - d_Z$ defined by V. With $i : Z \hookrightarrow X$, $j : U \hookrightarrow Z$ the inclusions, the **Hodge module extension** is defined by

$$V_Z^{\mathrm{Hdg}} := i_*(j_{!*}V^{\mathrm{Hdg}}),$$

where the intermediate direct image functor $j_{!*}$ is defined as in the case of constructible complexes (XIII–18).

Note that the rational component of this Hodge module is precisely the perverse extension of the local system $\mathbb{V}$ underlying V in the sense of Def. 13.25, i.e. we have $\mathrm{rat}_X V_Z^{\mathrm{Hdg}} = {}^{\pi}\mathbb{V}_Z$.

14.3.2 Polarizations

In this subsection we characterize the Hodge modules which are Hodge module extensions of polarisable variations of Hodge structure.

Let $K^{\bullet}$, $L^{\bullet}$ be two bounded complexes of sheaves of $\mathbb{Q}$-vector spaces on a complex manifold X with constructible cohomology, i.e. belonging to $D_c^{\mathrm{b}}(X; \mathbb{Q})$. Their tensor product in the category $D_c^{\mathrm{b}}(X; \mathbb{Q})$ is defined replacing the complexes by any projective resolution (see Example A.29) and then taking the usual tensor product. As usual, the result will be denoted by $K^{\bullet} \overset{\mathrm{L}}{\otimes} L^{\bullet}$.

The dualizing complex as given in (XIII–3) for rational perverse sheaves gets an additional Tate twist:

$$ {}^{\mathrm{Ve}}\mathbb{D}_X := \underline{\mathbb{Q}}_X(d_X)[2d_X]. \tag{XIV–25} $$

The reason is that by Prop. 13.5 $H_c^{-q}(X, {}^{\mathrm{Ve}}\mathbb{D}_X) = H_q(X;\mathbb{Q})$ and hence, by (XIV–25):

$$ H_c^{2d_X-q}(X;\mathbb{Q})(d_X) \simeq H_q(X;\mathbb{Q}), $$

the Hodge theoretic form of Poincaré-duality (Def. 1.17).

Consider a morphism

$$ K^\bullet \overset{\mathrm{L}}{\otimes} L^\bullet \xrightarrow{S} {}^{\mathrm{Ve}}\mathbb{D}_X^\bullet(r), \quad r \in \mathbb{Z}. $$

By adjunction it induces

$$ K^\bullet \xrightarrow{S'} \mathrm{Hom}(L^\bullet, {}^{\mathrm{Ve}}\mathbb{D}_X^\bullet(r)) = {}^{\mathrm{Ve}}\mathbb{D}_X L^\bullet(r), $$

and we say that S is non-singular if S' is a quasi-isomorphism. In the special case where $K^\bullet = L^\bullet = \mathcal{M}_{\mathbb{Q}}^\bullet$ is the rational component of a Hodge module of weight n and where $r = -n$ we speak of a **non-singular paring** on $\mathcal{M}_{\mathbb{Q}}^\bullet$. In this setting, the adjunction map S' becomes a quasi-isomorphism

$$ \mathcal{M}_{\mathbb{Q}}^\bullet \to {}^{\mathrm{Ve}}\mathbb{D}_X \mathcal{M}_{\mathbb{Q}}^\bullet(-n). \tag{XIV–26} $$

To have a polarization on the entire Hodge module this quasi-isomorphism should first of all extend to the Hodge module level. Some extra ingredients are needed to make the theory of polarizable Hodge modules functorial. Saito's definition is very involved and we give a simplified treatment:

Definition 14.35. Suppose that $M = (\mathcal{M}, F, \mathcal{M}_{\mathbb{Q}}^\bullet)$ is a Hodge module of weight n with strict support in $Z \subset X$ which is of the form $M = V_Z^{\mathrm{Hdg}}$, the Hodge module extension of a polarized variation of Hodge structures V of weight $n - d_Z$ on a Zariski-open subset U of Z. A **polarization** on M is a non-singular pairing on the rational component $\mathcal{M}_{\mathbb{Q}}^\bullet$ such that

1) the quasi-isomorphism (XIV–26) extends to an isomorphism

$$ S' : (\mathcal{M}, F, \mathcal{M}_Q^\bullet) \to \mathbb{D}_X(\mathcal{M}, F, \mathcal{M}_{\mathbb{Q}}^\bullet)(-n) $$

of Hodge modules of weight n with strict support in Z;

2) S' induces a polarization of the variation V (defined on U), in the sense of Def. 10.8.

A Hodge module admitting a polarization is called **polarizable**.

Remark 14.36. 1. A posteriori Saito's definition definition turns out to be equivalent with the preceding definition. See [Sa88] and [Sa90].

2. Any polarizable Hodge module M with strict support X by assumption is of the form $M = V_X^{\mathrm{Hdg}}$ as in the preceding definition. In particular, the rational component of M is the perverse extension (Def. 13.25) ${}^{\pi}\mathbb{V}$ of the local system $\mathbb{V}$ underlying V.

Since, as we saw (Theorem 14.31), the category of Hodge modules is abelian, the strict support condition for Hodge modules then entails its semi-simplicity [Sa88, Cor. 5.2.13]. Summarizing, we have:

Theorem 14.37. *The category of weight n polarizable Hodge modules*

$$\mathsf{MH}^{\mathrm{pol}}(X, n)$$

on X is semi-simple with simple objects the polarizable Hodge modules of the form V_Z^{Hdg}, where V is a polarizable weight $n - d_Z$ variation of Hodge structure on a smooth dense Zariski open subset of an irreducible algebraic subvariety $Z \subset X$, and for which moreover the monodromy representation is irreducible.

From this theorem and Remark 14.36.2 we deduce:

Corollary 14.38. *The Grothendieck group $K_0\mathsf{MH}^{\mathrm{pol}}(X, n)$ is generated by classes $[V_Z^{\mathrm{Hdg}}]$, where V is a polarizable variation of Hodge structures of weight $n - d$ on a smooth Zariski open subset of a d-dimensional irreducible smooth subvarieties Z of X. The functor* rat_X (XIV–23) *associates to a polarizable variation of Hodge structure of weight $n - d$ supported on a d-dimensional subvariety of X the perverse extension to X of its underlying local system. This functor is faithful and induces a ring homomorphism $K_0(\mathsf{MH}^{\mathrm{pol}}(X, n)) \to K_0(\mathrm{Perv}(X; \mathbb{Q}))$.*

14.3.3 Lefschetz Operators and the Decomposition Theorem

We introduce the Lefschetz operator on the level of filtered $\mathcal{D}_X$-modules, where X is a projective manifold X. We first have to find a suitable incarnation of the first Chern class of a line bundle M in the filtered complex $(\mathcal{E}_X^\bullet(\mathbb{C}), F)$ where F is the "Hodge filtration" from § 2.3.1, i.e. $F_p\mathcal{E}_X^k(\mathbb{C}) = \bigoplus_{r \geq -p} \mathcal{E}_X^{r,k-r}$. The crucial remark is that given a hermitian metric h on M, which locally on a trivializing open cover $\{U_\alpha\}$ of X is given by functions h_α, the first Chern class is represented by the (global) Chern form

$$\gamma_h = \frac{\mathrm{i}}{2\pi}\partial\bar{\partial}h_\alpha \in F_{-1}\mathcal{E}_X^2(\mathbb{C}).$$

Wedging with this form thus induces a morphism

$$L_M := \wedge(\gamma_h) : (\mathcal{E}_X^\bullet(\mathbb{C}), F) \to (\mathcal{E}_X^\bullet(\mathbb{C}), F)(1)[2]. \tag{XIV–27}$$

Another choice h' of a metric leads to a form which differs from γ_h by a form of the type $\partial\bar{\partial}f$ where f is a global C^∞-function. The form $\sigma(h, h') = -\partial f \in$

$F_{-1}\mathcal{E}^1_X(\mathbb{C})$ thus has the property that $d\sigma(h,h') = \gamma_h - \gamma_{h'}$, i.e. it gives a homotopy between wedging with γ_h and wedging with γ'_h so that L_M is well defined up to homotopy. Moreover, since γ_h represents an integral class, this is compatible with the usual Lefschetz operator on rational cohomology if we take for M an ample line bundle.

To rephrase this, introduce the filtered left $\mathcal{D}_X$-module

$$(\mathcal{L}, F) := (\mathcal{D}_X, F^{\mathrm{ord}}) \otimes_{\mathcal{O}_X} (\mathcal{E}^\bullet_X(\mathbb{C}), F).$$

Then, by flatness of the sheaves $\mathcal{E}^{p,q}_X$ over $\mathcal{O}_X$, any filtered $\mathcal{D}_X$-module $(\mathcal{M}, F)$ is quasi-isomorphic to

$$(\mathcal{M}, F)' := (\mathcal{M}, F) \otimes_{\mathcal{O}_X} (\mathcal{L}, F)$$

and for an ample line bundle the operator (XIV–27) induces the the **Lefschetz operator**:

$$L := \mathrm{id} \otimes L_M : (\mathcal{M}, F)' \to (\mathcal{M}, F)'(1)[2]. \qquad \text{(XIV–28)}$$

which is well defined in the derived category. Moreover, for any rational filtered $\mathcal{D}_X$-module $((\mathcal{M}, F), \alpha, \mathcal{M}^\bullet_{\mathbb{Q}})$ this action of L is compatible under α with the usual action of L on $\mathcal{M}^\bullet_{\mathbb{Q}}$.

This operator plays an important role in the following theorem, where it is used in the relative situation of a projective morphism $f : X \to Y$, and on the level of cohomology sheaves. Before formulating the theorem we also need the concept of an induced pairing. Note that $f_! = Rf_*$ since f is proper, and the natural adjunction morphism (see (XIII–10)) now reads $Rf_* {}^{\mathrm{Ve}}\mathbb{D}\underline{\mathbb{Q}}_X \to {}^{\mathrm{Ve}}\mathbb{D}\underline{\mathbb{Q}}_Y$. Any morphism $S : K^\bullet \overset{\mathrm{L}}{\otimes} L^\bullet \to {}^{\mathrm{Ve}}\mathbb{D}\underline{\mathbb{Q}}_X(r)$ induces

$$Rf_*S : Rf_*K^\bullet \overset{\mathrm{L}}{\otimes} Rf_*L^\bullet \to {}^{\mathrm{Ve}}\mathbb{D}\underline{\mathbb{Q}}_Y(r).$$

Consider the special case where $K^\bullet = L^\bullet = \mathcal{M}^\bullet_{\mathbb{Q}}$, the rational component of a Hodge module M of weight n, and where in addition $r = -n$; take the perverse cohomology of $Rf_*\mathcal{M}^\bullet_{\mathbb{Q}}$ in degree $(-k)$ for the first argument in the tensor product and in degree k for the second argument. Finally, extend the resulting morphism to the Hodge modules of which these are the rational components. This yields the morphism

$$R^{-k}f_*S : R^{-k}f_*M \overset{\mathrm{L}}{\otimes} R^k f_*M \to {}^{\mathrm{Ve}}\mathbb{D}\underline{\mathbb{Q}}_Y(-n). \qquad \text{(XIV–29)}$$

We can now formulate Saito's result concerning proper direct images [Sa88, Thm. 5.3.1]:

Theorem 14.39. *Let $f : X \to Y$ be a projective morphism between smooth complex varieties and let L be the Lefschetz operator with respect to a relatively ample line bundle. Let $M = (\mathcal{M}, F, \mathcal{M}^\bullet_{\mathbb{Q}})$ be a Hodge module on X of weight n with strict support on an irreducible subvariety of X and polarized by S. Then*

i) *The filtration on the complex* $f_*(\mathcal{M}, F)$ *is strict so that* $R^k f_*(\mathcal{M}, F)$ *is a filtered* $\mathcal{D}_Y$*-module. In fact, for all* k*, the filtered rational* $\mathcal{D}_Y$*-module* $R^k f_*(\mathcal{M}, F, \mathcal{M}^\bullet_{\mathbb{Q}})$ *(see* (XIV–17)*) is a Hodge module on* Y *of weight* $n+k$*.*

ii) *On these modules the Lefschetz operators* L *are defined by the procedure* (XIV–28)*, and the hard Lefschetz theorem holds:*

$$L^k : R^{-k} f_* M \xrightarrow{\sim} R^k f_* M(k) \quad \text{for } k \geq 0.$$

iii) *Consider the morphism* (XIV–29)*. Define*

$$(R^{-k} f_* M)_{\text{prim}} := \operatorname{Ker}\left[L^{k+1} : R^{-k} f_* M \to R^{k+2} f_* M\right].$$

Then the form $(x, y) \mapsto (-1)^{\frac{k(k-1)}{2}} R^{-k} f_* S(x, L^k y)$ *is a polarization on the Hodge module* $(R^{-k} f_* \mathcal{M})_{\text{prim}}$*.*

Corollary 14.40. *Let* X *be a compact Kähler manifold and* $\mathbb{V}$ *a local system of* $\mathbb{Q}$*-vector spaces on a Zariski open subset of* X*. If* $\mathbb{V}$ *underlies a polarizable variation of Hodge structures of weight* n*, the intersection cohomology group (Definition 13.12)* $IH^k(X, \mathbb{V})$ *carries a polarizable Hodge structure of weight* $k + n$*.*

Proof. Let H be the polarizable Hodge module of weight n which by Prop. 14.33 corresponds to a polarizable variation on $\mathbb{V}$. Since X is compact $a_X : X \to$ point is proper and hence $(R^k a_X)_*(H)$ is a Hodge module of weight $n + k$. The rational component of this Hodge module is just $IH^k(X, \mathbb{V})$. □

Corollary 14.41. *Let* $K^\bullet$ *be a perverse complex of* $\mathbb{Q}$*-vector spaces on* X *which is the rational component of a polarizable Hodge module. Let* $f : X \to Y$ *be a projective morphism. Then the (perverse) Leray spectral sequence for* $K^\bullet$ *and* f *degenerates at the* E_2*-term; in fact the analogue of the decomposition theorem* (I–16) *holds:*

$$Rf_* K^\bullet \simeq \bigoplus_i {}^{\pi}H^i f_* K^\bullet[-i].$$

Proof. That the Leray spectral sequence degenerates follows from Theorem 14.39 in the same way as Prop. 1.38. Like in Remark 1.39 2), one can in fact show that the decomposition holds as stated. □

One can apply this result to perverse complexes of the form $\mathcal{IC}^\bullet_X \mathbb{V}$, with $\mathbb{V} = \underline{\mathbb{Q}}_X$, or more generally any local system $\mathbb{V}$ which is a direct factor of $R^k g_* \underline{\mathbb{Q}}_Y$, $g : Y \to X$ a smooth projective morphism. Such local systems are called **of geometric origin**. Recalling the notation (XIII–13) we have

Corollary 14.42 (Decomposition Theorem)**.** *Let* $f : X \to Y$ *be a projective morphism and let* $\mathbb{V}$ *be a local system on a smooth dense open subset of* X *which is of geometric origin. Then*

$$Rf_* \mathcal{IC}^\bullet_X \mathbb{V} \simeq \bigoplus_i {}^\pi R^i f_* \mathcal{IC}^\bullet_X \mathbb{V}[-i] \quad \text{in } D^{\mathrm{b}}_{\mathrm{cs}}(Y;\mathbb{Q})$$

$${}^\pi R^i f_* \mathcal{IC}^\bullet_X \mathbb{V} = \bigoplus_Z \mathcal{IC}^\bullet_Z \mathbb{V}^i_Z \quad \text{in } \mathrm{Perv}(Y;\mathbb{Q}).$$

Here Z runs over the irreducible subvarieties of Y and each $\mathbb{V}^i_Z$ is locally constant over a dense open smooth subset of Z.

In fact the last assertion follows since the category of Hodge modules satisfies the strict support condition of Def. 14.29.

Remark. For a proof using only (classical) mixed Hodge theory, see [dC-M].

If we apply the decomposition theorem to a resolution of singularities, we deduce:

Corollary 14.43. *Let $f : Y \to X$ be a projective resolution of singularities of a projective variety X. Then the complex $\mathcal{IC}^\bullet_X \underline{\mathbb{Q}}_X$ is a direct factor of $Rf_* \underline{\mathbb{Q}}_Y[d_X]$. In particular the (rational) intersection cohomology group $IH^k(X) \otimes \mathbb{Q}$ is a direct factor of the (rational) cohomology group $H^k(Y;\mathbb{Q})$ as weight k Hodge structures.*

14.4 Mixed Hodge Modules

14.4.1 Variations of Mixed Hodge Structure

Mixed Hodge structures can be viewed as successive extensions of pure Hodge structures of different weights, encoded in the weight filtration and its graded subquotients. Given a rational vector space V with filtrations $W_\bullet$ on V and $F^\bullet$ on $\mathbb{V}_\mathbb{C}$, we have a mixed Hodge structure if and only if F induces a Hodge structure of weight k on each $\mathrm{Gr}^W_k V$.

One of the subtleties of mixed Hodge modules is that arbitrary extensions of mixed Hodge modules do not result in mixed Hodge modules: some additional conditions on the extension data are required. In this subsection we describe these conditions in the case of mixed Hodge modules M on a smooth variety X such that $\mathrm{rat}_X M$ is a local system. These are called **smooth mixed Hodge modules**.

Definition 14.44. Let S be a complex manifold. A **variation of mixed Hodge structure** on S consists of the following data:

1) a local system $\mathbb{V}_\mathbb{Z}$ of finitely generated abelian groups on S;
2) a finite decreasing filtration $\{\mathcal{F}^p\}$ of the holomorphic vector bundle $\mathcal{V} := \mathbb{V}_\mathbb{Z} \otimes_\mathbb{Z} \mathcal{O}_S$ by holomorphic subbundles (the **Hodge filtration**).
3) a finite increasing filtration $\{W_m\}$ of the local system $\mathbb{V}_\mathbb{Q} := \mathbb{V}_\mathbb{Z} \otimes_\mathbb{Z} \mathbb{Q}$ by local subsystems (the **weight filtration**).

These data should satisfy the following conditions:

1) for each $s \in S$ the filtrations $\{\mathcal{F}^p(s)\}$ and $\{W_m\}$ of $\mathbb{V}(s) \simeq \mathbb{V}_{\mathbb{Z},s} \otimes_{\mathbb{Z}} \mathbb{C}$ define a mixed $\mathbb{Q}$-Hodge structure on the $\mathbb{Q}$-vectorspace $\mathbb{V}_{\mathbb{Q},s}$;
2) the connection $\nabla : \mathcal{V} \to \mathcal{V} \otimes_{\mathcal{O}_S} \Omega^1_X$ whose sheaf of horizontal sections is $\mathbb{V}_{\mathbb{C}}$ satisfies the **Griffiths' transversality condition**

$$\nabla(\mathcal{F}^p) \subset \mathcal{F}^{p-1} \otimes \Omega^1_S.$$

The notion of a **morphism of variations of mixed Hodge structure** is defined in the obvious way.

Definition 14.45. A variation of mixed Hodge structure will be called **graded-polarizable** if the induced variations of pure Hodge structure $\mathrm{Gr}^W_k \mathbb{V}$ are all polarizable.

Consider a graded polarizable variation of mixed Hodge structure $(\mathbb{V}, W, \mathcal{F})$ over the punctured unit disc Δ^* with parameter t. As W is a filtration of $\mathbb{V}_{\mathbb{Q}}$ by local subsystems, the monodromy operator preserves W; moreover the monodromy theorem 11.8 guarantees that the monodromy of each $\mathrm{Gr}^W \mathbb{V}_{\mathbb{Q}}$ is quasi-unipotent. Hence the monodromy of $\mathbb{V}$ is quasi-unipotent.

From now on we assume that the monodromy T of $\mathbb{V}$ is in fact unipotent. We let $V = (\psi_t \mathbb{V}_{\mathbb{Q}})_0$ and $N = \log T : V \to V$. We let W denote the induced filtration on V. Consider the weight filtration ${}^kM$ of the nilpotent endomorphism $\mathrm{Gr}_k(N)$ on $\mathrm{Gr}^W_k V$ centered at k (see Def. 11.9).

Definition 14.46. A **weight filtration** of N **relative to** W is a filtration M of V such that

i) $NM_i \subset M_{i-2}$
ii) M induces ${}^kM$ on $\mathrm{Gr}^W_k V$.

Proposition 14.47 ([Del80, (1.6.13)]). *There is at most one weight filtration M of N on V relative to W.*

Definition 14.48. If such M exists, it is called the **weight filtration of** N **relative to** W and we denote it by $M = M(N; W)$.

We let $\tilde{\mathcal{V}}$ denote the canonical extension of $\mathcal{V}$ to a holomorphic vector bundle on Δ such that the connection extends to one with a logarithmic pole at 0 with nilpotent residue. The filtration W extends to $\tilde{\mathcal{V}}$ and $\mathrm{Gr}^W_k \tilde{\mathcal{V}}$ is the canonical extension for $\mathrm{Gr}^W_k \mathcal{V}$. Let ${}^k\mathcal{F}$ denote the Hodge filtration on $\mathrm{Gr}^W_k \mathcal{V}$. It extends to a filtration ${}^k\tilde{\mathcal{F}}$ of $\mathrm{Gr}^W_k \tilde{\mathcal{V}}$.

Definition 14.49. 1. A variation of mixed Hodge structure $(\mathbb{V}, W, \mathcal{F})$ over the punctured unit disc Δ^* is called **admissible** if
   a) it is graded-polarizable;
   b) the monodromy T is unipotent and the weight filtration $M(N, W)$ of $N = \log T$ relative to W exists;
   c) the filtration $\mathcal{F}$ extends to a filtration $\tilde{\mathcal{F}}$ of $\tilde{\mathcal{V}}$ which induces ${}^k\tilde{\mathcal{F}}$ on $\mathrm{Gr}^W_k \tilde{\mathcal{V}}$ for each k.

2. Let X be a compact complex analytic space and $U \subset X$ a smooth Zariski-open subset. A graded polarizable variation of mixed Hodge structure $(\mathbb{V}, W, \mathcal{F})$ on U is called **admissible** (with respect to the embedding $U \subset X$) if for every holomorphic map $i : \Delta \to X$ which maps Δ^* to U and such that $i^*\mathbb{V}$ has unipotent monodromy, the variation $i^*(\mathbb{V}, W, \mathcal{F})$ on Δ^* is admissible.

Lemma 14.50. *If $(\mathbb{V}, W, \mathcal{F})$ is admissible over Δ^*, then for each k the triple $(\tilde{\mathcal{V}}(0), M, \tilde{\mathcal{F}}(0))$ is a mixed Hodge structure, and N is an endomorphism of type $(-1,-1)$ of it.*

Proof. See the appendix of [St-Z] or [Kash86, Prop. 5.2.1].

Suppose that $f : X \to S$ is a morphism of complex algebraic varieties and $k \in \mathbb{N}$. Then there exists a Zariski-open dense subset $U \subset S$ such that the restriction to U of $R^k f_* \underline{\mathbb{Z}}_X$ is a local system and underlies a variation of mixed Hodge structure. If f is quasi-projective, this variation of mixed Hodge structure is graded-polarizable. We refer to this situation as a **geometric variation of mixed Hodge structure**.

Theorem 14.51 ([St-Z, ElZ86, Kash86]). *Geometric variations of mixed Hodge structure are admissible.*

The fundamental result about admissible variations of Hodge structures is

Theorem 14.52 ([St-Z, ElZ03]). *Let $\mathbb{V}$ be an admissible variation of mixed Hodge structure on U. Then for each k the vector space $\mathbb{H}^k(U, \mathbb{V})$ carries a canonical mixed Hodge structure.*

Idea of the proof. To explain where the admissibility comes in, consider first the Hodge filtration. Assume that $D = S - U$ is a divisor with normal crossings, say $D = \sum D_k$. Let T_k be the local monodromy operator around D_k and let $N_k = \log(T_k)$. With $\tilde{\mathcal{V}}$, respectively $\tilde{\mathcal{F}}^p$ the canonical extension of $\mathcal{V} = \mathbb{V} \otimes \mathcal{O}_U$, respectively $\mathcal{F}^p$,we put

$$F^p(\Omega^\bullet_X(\log D) \otimes \tilde{\mathcal{V}}) = \left[\tilde{\mathcal{F}}^p \to \Omega^1_X(\log D) \otimes \tilde{\mathcal{F}}^{p-1} \to \cdots\right].$$

The naive weight filtration, obtained by taking the intersection complexes $\mathcal{IC}^\bullet_X(W_k)$ does not produce a mixed Hodge complex of sheaves on X. One needs instead the relative weight filtrations $M(N_i)$ with respect to the W_k in the sense of Def. 14.46. The precise descriptions and the proof that this gives a mixed Hodge complex of sheaves can be found in [St-Z] for the curve case, and in [ElZ03] for the general situation. □

Definition 14.53. Let $V = (\mathbb{V}_{\mathbb{Q}}, W, \mathcal{F})$ be an admissible variation of mixed Hodge structure on a smooth complex variety U. It gives rise to the holonomic $\mathcal{D}_U$-module $\mathcal{V} = \mathbb{V}_{\mathbb{Q}} \otimes_{\mathbb{Q}} \mathcal{O}_U$ which is filtered by

$$W_k\mathcal{V} := W_k\mathbb{V}_{\mathbb{Q}} \otimes_{\mathbb{Q}} \mathcal{O}_U$$
$$F_p\mathcal{V} := \mathcal{F}^{-p}\mathcal{V}.$$

These data, together with the comparison isomorphism

$$(\mathbb{V}_{\mathbb{Q}}[d_U], W) \otimes \mathbb{C} \xrightarrow{\sim} \mathrm{DR}(\mathcal{V}, W)$$

constitutes the **smooth mixed Hodge module** V^{Hdg}, or $\mathbb{V}_{\mathbb{Q}}^{\mathrm{Hdg}}[d_U]$ on U.

Examples 14.54. Let X be a smooth complex algebraic variety.

a) Polarizable variations of Hodge structure on X are smooth mixed Hodge modules.
b) The **constant mixed Hodge module** on X is the pure weight d_X Hodge module defined by the weight 0 constant variation of Hodge structure on $\underline{\mathbb{Q}}_X$. A concrete incarnation, used by Saito is as follows:

$$\underline{\mathbb{Q}}_X^{\mathrm{Hdg}}[d_X] := (\omega_X, F, \underline{\mathbb{Q}}_X[d_X], W), \quad \mathrm{Gr}^F_{-k} = 0 = \mathrm{Gr}^W_k \text{ if } k \neq d_X. \tag{XIV–30}$$

So the D-module component is the canonical bundle ω_X viewed as a right $\mathcal{D}_X$-module and its perverse component is the constant sheaf $\underline{\mathbb{Q}}_X$ placed in degree $-d_X$ (recall (XIII–24) that the De Rham complex of the associated left $\mathcal{D}_X$-module $\mathcal{O}_X$ is just the usual De Rham complex shifted d_X-places to the left, a complex quasi-isomorphic to $\underline{\mathbb{C}}_X[d_X]$).

14.4.2 Defining Mixed Hodge Modules

Just like Hodge structures, Hodge modules are not always stable under direct sums (e.g. if the weights are different). Mixed Hodge structures can be viewed as iterated extensions of pure Hodge structures of different weights, encoded in the weight filtration and their pure gradeds. The same holds for mixed Hodge modules on a complex algebraic variety X, except that the definition is much more involved. In this section we give an inductive definition for mixed Hodge modules on a smooth algebraic variety which is different from Saito's complicated definition, but is in fact equivalent to it.

We are going to extend Definition 14.53 to arbitrary mixed Hodge modules on a smooth algebraic X. The definition is local in the Zariski topology and the singular case can be treated by locally embedding X in a smooth variety, just as for Hodge modules.

First introduce the category of **bi-filtered rational D-modules** consisting of a filtered rational D-module $(\mathcal{M}, F, \mathcal{M}^\bullet_{\mathbb{Q}})$ together with a filtration W on $(\mathcal{M}, \mathcal{M}^\bullet_{\mathbb{Q}})$, the **weight filtration**. This is a *pair* of filtrations compatible with the comparison isomorphism. Note that $\mathrm{Gr}^W_i(\mathcal{M}, F, \mathcal{M}^\bullet_{\mathbb{Q}}, W)$ is a filtered rational D-module. We further demand that these be polarizable weight i Hodge modules on X. The full subcategory generated by such bi-filtered rational D-modules will be denoted $\mathsf{MHW}(X)$. This subcategory is not yet the category we are after, since it does not behave well with respect to the vanishing and nearby cycle functors. To remedy this we proceed as follows. Recalling the convention (XIV–18), (XIV–19) we put

$$\begin{aligned}\psi_{t,1}(\mathcal{M}, F, \mathcal{M}^\bullet_{\mathbb{Q}}, W) &:= (\psi_{t,1}(\mathcal{M}, F, \mathcal{M}^\bullet_{\mathbb{Q}}), W),\\ \phi_{t,1}(\mathcal{M}, F, \mathcal{M}^\bullet_{\mathbb{Q}}, W) &:= (\phi_{t,1}(\mathcal{M}, F, \mathcal{M}^\bullet_{\mathbb{Q}}), W).\end{aligned}$$

The formulas (XIV–20) and (XIV–21) define the two morphisms can and var making $(M|U, \psi_{t,1}M, \mathrm{can}, \mathrm{var})$ an example of a gluing datum:

Definition 14.55. Let $t : X \to \mathbb{C}$ be a holomorphic function, $Y = t^{-1}(0)$, $U = X - Y$. A **gluing datum in** MHW **for** (X, t) is a quadruple (M', M'', u, v) where $M' \in \mathsf{MHW}(U)$, $M'' \in \mathsf{MHW}(Y)$, and

$$\begin{array}{ccc}\phi_{t,1}M' & \xrightarrow{u} & M''\\ M'' & \xrightarrow{v} & \phi_{t,1}M'(1),\end{array}$$

are homomorphisms in the category $\mathsf{MHW}(Y)$ such that $v \circ u = N \otimes (2\pi \mathrm{i})^{-1}$.

Gluing data can be used to define mixed Hodge modules by induction on the dimension of the support as follows.

Definition 14.56. An object $M \in \mathsf{MHW}(X)$ is a **mixed Hodge module** if there is a Zariski-open cover $\{X_i\}$ of X and holomorphic functions $t_i : X_i \to \mathbb{C}$ such that, putting $Y_i = t_i^{-1}(0)$, $U_i = X_i - Y_i$, the restriction $M|Y_i$ is a mixed Hodge module on Y_i, and $M|U_i$ is a *smooth* mixed Hodge module on U_i (see Def 14.53). Finally, there should be gluing data $(M|U_i, M|Y_i, u_i, v_i)$ in MHW for the pairs (X_i, t_i).

If $(\mathcal{M}, F, \mathcal{M}^\bullet_{\mathbb{Q}}, W)$ is a mixed Hodge module, we say that the perverse complex $\mathcal{M}^\bullet_{\mathbb{Q}}$ is the **rational component** of the mixed Hodge module and that W, respectively F, are the **weight** and **Hodge** filtration respectively.

That this is indeed equivalent to Saito's original definition follows from [Sa90].

14.4.3 About the Axioms

The functor which assigns to a mixed Hodge module its rational component sends $\mathsf{MHM}(X)$ to $\mathrm{Perv}(X; \mathbb{Q})$. By [Sa88, 5.1.14] and by definition, the category of mixed Hodge modules on X is abelian and rat_X is faithful and exact. The functor then extends to give the functor $\mathrm{rat}_X : D^{\mathrm{b}}\mathsf{MHM}(X) \to D^{\mathrm{b}}_{\mathrm{cs}}(X; \mathbb{Q})$ from **axiom A**.

Since by definition mixed Hodge modules supported on a point are smooth they give mixed Hodge structures, this proves **Axiom B)**. See also [Sa90, Thm. 3.98].

Axiom C) being built into the definition of a mixed Hodge module does not present a difficulty. The asserted semi-simplicity is Prop. 14.37.

We next pass to **axioms D), E)**. Note that the direct image functors in general cannot be performed on the level of mixed Hodge modules. The reason is that they do not even preserve perverse complexes. As an example, let X be a smooth complex algebraic variety and let $a_X : X \to \mathrm{pt}$ be the constant map. Then $a_X^*\mathbb{Q} = \underline{\mathbb{Q}}_X$ is not perverse, but it becomes perverse when placed in degree $-d_X$, i.e. we need to work in the *derived category* $D^{\mathrm{b}}_{\mathrm{cs}}(X;\mathbb{Q})$. In particular, the constant mixed Hodge module $\underline{\mathbb{Q}}_X^{\mathrm{Hdg}}[d_X]$ defined by (XIV–30), when placed in degree $-d_X$ becomes a *complex*. Over a point it becomes $\mathbb{Q}^{\mathrm{Hdg}}$ and the functor a_X^* maps it to $\underline{\mathbb{Q}}_X^{\mathrm{Hdg}}$, since the functor rat_X is faithful (axiom A). The main result from [Sa90] confirms axioms D) and E):

Theorem 14.57. *Let $f : X \to Y$ be a morphism between complex algebraic varieties and let $M^\bullet$ and $N^\bullet$ be bounded complexes of mixed Hodge modules on X and Y respectively. Recalling the functors* rat_X *and* rat_Y (XIV–1), *we have:*

i) *there is a bounded complex of mixed Hodge modules $f_*M^\bullet$ on Y such that $f_*(\mathrm{rat}_X M^\bullet) = \mathrm{rat}_Y(f_*M^\bullet)$ in $D^{\mathrm{b}}_{\mathrm{cs}}(Y;\mathbb{Q})$;*
ii) *there is a bounded complex of mixed Hodge modules $f^!N^\bullet$ on X such that $f^!(\mathrm{rat}_Y N^\bullet) = \mathrm{rat}_X(f^!N^\bullet)$ in $D^{\mathrm{b}}_{\mathrm{cs}}(X;\mathbb{Q})$;*
iii) *the Verdier duality operator extends as an involution $\mathbb{D}_X$ on the derived category of bounded complexes Hodge modules on X, commuting with* rat_X*;*
iv) *there are operations f^*, $f_!$ in the appropriate derived categories of mixed Hodge modules such that $\mathbb{D}_Y f_* = f_!\mathbb{D}_X$ and $\mathbb{D}_X f^* = f^!\mathbb{D}_Y$ (see also Remark 13.66) .*

This is by no means a trivial result, since we have seen that for non-proper morphisms the direct image of a filtered holonomic D-module need not be holonomic, and the inverse image only behaves well for non-characteristic filtered modules.

Finally, for **axioms G), H)** we refer to [Sa90, Prop. 2.26]

14.4.4 Application: Vanishing Theorems

The techniques of mixed Hodge modules can be used to derive a vanishing result. The statement uses the filtered De Rham complex associated to a filtered $\mathcal{D}_X$-module $(\mathcal{M}, F)$ (XIII–41).

Theorem 14.58 (Saito's vanishing theorem[Sa90, Prop. 2.33])**.** *Let $Z \subset X = \mathbb{P}^N$ be an irreducible projective variety embedded by $\mathcal{O}_Z(1) = \mathcal{L}^m$. Let $\mathcal{E}^\bullet$ be a bounded complex of holomorphic vector bundles on Z. Suppose that there exists a filtered $\mathcal{D}_X$-module $(\mathcal{M}, F)$ supported on Z and an integer $p \in \mathbb{Z}$ such that the following equality holds in the derived category of bounded $\mathcal{O}_Z$-modules:*

$$\mathcal{E}^\bullet = \mathrm{Gr}_F^p(\mathrm{DR}_X(\mathcal{M}, F)) \otimes \mathcal{O}_Z.$$

Suppose moreover that $(\mathcal{M}, F)$ is the filtered D-module part of a mixed Hodge module on X supported on Z. Then the following vanishing results hold

$$\mathbb{H}^i(Z, \mathcal{E}^\bullet \otimes \mathcal{L}) = 0 \quad \textit{for } i > 0$$
$$\mathbb{H}^i(Z, \mathcal{E}^\bullet \otimes \mathcal{L}^{-1}) = 0 \quad \textit{for } i < 0.$$

In the applications below, the filtration on $\mathcal{M}$ is the trivial one-step filtration in degree p. In this case we record

$$\mathrm{Gr}^F_{p-j} \mathrm{DR}_X(\mathcal{M}, F) = \Omega^j_X(\mathcal{M})[-j + d_X]. \qquad \text{(XIV–31)}$$

Corollary 14.59 (Kodaira-Nakano vanishing theorem). *Let Z be a smooth projective variety and $\mathcal{L}$ an ample line bundle on Z. Then $H^i(Z, \Omega^j_Z \otimes \mathcal{L}) = 0$ for $i + j > d_Z$ and $H^i(Z, \Omega^j_Z \otimes \mathcal{L}^{-1}) = 0$ for $i + j < d_Z$.*

Proof. Take the mixed Hodge module $\underline{\mathbb{Q}}_Z[d_Z]$ whose filtered D-module part is ω_Z with the one-step filtration in degree $-d_Z$. Its De Rham complex is the usual (shifted) De Rham complex and so from (XIV–31) it follows that $\mathcal{E}^\bullet = \mathrm{Gr}^F_{-d_Z-j} \mathrm{DR}_Z(\omega_Z, F) = \Omega^j_Z[d_Z - j]$ and so $\mathbb{H}^i(Z, \mathcal{E}^\bullet \otimes \mathcal{L}) = H^{i-j+d_Z}(Z, \Omega^j_Z \otimes \mathcal{L}) = 0$ which proves the first vanishing result. The second is proven similarly.
□

Remark 14.60. It is not hard to derive the other vanishing results stated in § 7.3.2. Saito also shows how to derive the **Kollár-Ohsawa vanishing theorem** which states that $H^j(Z, R^k f_* \omega_Y \otimes \mathcal{L}) = 0$ for $k \geq 0$, $j > 0$, $\mathcal{L}$ ample and $f : Y \to Z$ a morphism between projective varieties with Y smooth.

14.4.5 The Motivic Hodge Character and Motivic Chern Classes

We have seen that over any algebraic variety X we have a canonically associated *complex* of mixed Hodge modules $\underline{\mathbb{Q}}^{\mathrm{Hdg}}_X$, in general only well-defined in the derived category. If X admits a morphism to S, say $f : X \to S$ we associate to X the complex $f_! \underline{\mathbb{Q}}^{\mathrm{Hdg}}_X \in D^{\mathrm{b}}\mathsf{MHM}(S)$. We want to show, following [B-S-Y] that this assignment is *motivic*: it respects the operation of cutting up X/S in locally closed subsets (relative to S).

We start by explaining the relative version of the Grothendieck group $K_0(\mathrm{Var})$ which came up in Remark 5.56. Let S be a complex algebraic variety and let $K_0(\mathrm{Var}/S)$ be the free abelian group on isomorphism classes of complex algebraic varieties over S (i.e. morphisms $X \to S$) modulo the **scissor relations** where we identify the class $[X]$ of X and $[X - Y] + [Y]$ whenever $Y \subset X$ is a closed subvariety. We identify $K_0(\mathrm{Var}/\mathrm{pt})$ with $K_0(\mathrm{Var})$.

The direct product between a variety over S and a variety over T gives a variety over $S \times T$. This is compatible with the scissor relations and defines an "exterior" product $K_0(\mathrm{Var}/S) \times K_0(\mathrm{Var}/T) \to K_0(\mathrm{Var}(S \times T))$. When $S = T$, taking instead the fibred product, defines a ring structure on $K_0(\mathrm{Var}/S)$ with unit the class $[S]$ of the identity morphism $S \xrightarrow{\mathrm{id}} S$. Taking $T = \mathrm{pt}$, the exterior product makes $K_0(\mathrm{Var}/S)$ into a $K_0(\mathrm{Var})$-module.

For a morphism $\varphi : S \to T$, composition defines a push forward morphism $\varphi_! : K_0(\mathrm{Var}/S) \to K_0(\mathrm{Var}/T)$ and the fibre product construction gives a pull back $\varphi^{-1} : K_0(\mathrm{Var}/T) \to K_0(\mathrm{Var}/S)$ which are $K_0(\mathrm{Var})$-linear.

Lemma 14.61. *There is a well-defined homomorphism, the* ***motivic Hodge-Grothendieck characteristic:***

$$\begin{aligned} \chi^c_{\mathrm{Hdg}}(S) : K_0(\mathrm{Var}/S) &\to K_0(\mathsf{MHM}(S)) \\ [f : X \to S] &\mapsto \chi^c_{\mathrm{Hdg}}[X/S] = [f_! \, \underline{\mathbb{Q}}^{\mathrm{Hdg}}_X]. \end{aligned}$$

It is compatible with $\varphi_!$ and ψ^{-1} for any morphism $\varphi : S \to T$, $\psi : T \to S$ respectively.

Proof. We only need to show that the above morphism is well defined. So let $i : Z \subset X$ be a closed subvariety and let $j : X - Z \hookrightarrow X$ be the inclusion of the complement. For any complex $M^\bullet \in D^{\mathrm{b}}\mathsf{MHM}(X)$ we have an adjunction triangle (XIII–15) for the underlying perverse complexes. By [Sa90, (4.4.1)] this triangle lifts to an adjunction triangle in the category $D^{\mathrm{b}}\mathsf{MHM}(X)$:

$$\begin{array}{ccc} j_! j^* M^\bullet & \longrightarrow & M^\bullet \\ & {\scriptstyle [1]}\nwarrow \quad \swarrow & \\ & i_* i^* M^\bullet & \end{array} \qquad \text{(XIV–32)}$$

The existence of this distinguished triangle immediately implies compatibility of the definition with the scissor-relations. □

Taking dimensions, we land into the ring of constructible functions on S. Let us recall the definition. For any ring R the ring $C_{\mathrm{cs}}(S; R)$ of R-constructible functions on S by definition is generated by the characteristic functions 1_Z of a subvariety $Z \subset S$. Given a morphism $\varphi : S \to T$, pulling back functions defines $\varphi^{-1} : C_{\mathrm{cs}}(T; R) \to C_{\mathrm{cs}}(S; R)$ and there is also a push forward $\varphi_! : C_{\mathrm{cs}}(S; R) \to C_{\mathrm{cs}}(T; R)$. This is completely analogous to the construction of the functor $R\varphi_!$ for bounded constructible complexes of R-modules.

Corollary 14.62. *The composition* $\dim \circ \chi^c_{\mathrm{Hdg}}(S) : K_0(\mathrm{Var}/S) \to C_{\mathrm{cs}}(S; \mathbb{Q})$ *is a $\mathbb{Q}$-linear map, compatible with $\varphi_!$ and ψ^{-1} for any morphism $\varphi : S \to T$, $\psi : T \to S$ respectively.*

Remark 14.63. Let us give an elementary construction for the Hodge-Grothendieck characteristic. First observe the following consequence of axiom C), D) and E):

Complement. *We have* $K_0(\mathsf{MHM}(X)) = K_0(\mathsf{MH}^{\mathrm{pol}}(X))$. *The functor* rat_X *is compatible with these identifications: it sends a mixed Hodge module to* *its underlying perverse complex. For any morphism* $f : X \to Y$ *the morphisms* $f_*, f_! : K_0(\mathsf{MHM}(X)) \to K_0(\mathsf{MHM}(Y))$ *and* $f^*, f^! : K_0(\mathsf{MHM}(Y)) \to K_0(\mathsf{MHM}(X))$ *under the functor* rat *correspond to* $Rf_*, Rf_! : K_0(\mathrm{Perv}(X;\mathbb{Q})) \to K_0(\mathrm{Perv}(Y;\mathbb{Q}))$, *and* $f^{-1}, f^! : K_0(\mathrm{Perv}(Y;\mathbb{Q})) \to K_0(\mathrm{Perv}(X;\mathbb{Q}))$ *respectively.*

So we have $K_0(D^{\mathrm{b}}\mathsf{MHM}(X)) \simeq K_0(\mathsf{MHM}^{\mathrm{pol}}(S;\mathbb{Q})$. We recall that the category $\mathsf{MHM}^{\mathrm{pol}}(S;\mathbb{Q})$ is the abelian category of variations of pure $\mathbb{Q}$-*polarisable* Hodge structures defined over a dense open smooth subset of a subvariety $Z \subset S$. It is not hard to see that $K_0(\mathrm{Var}/S)$ is generated by classes of morphisms $f : X \to S$ for which X is smooth and projective. So we assume that $f : X \to S$ is proper. Then there is a stratification of S so that over the open strata f restricts to a smooth morphism. By the scissor relations we may thus further assume that f itself is smooth with image a smooth subvariety $Z \subset S$. Then $R^j f_* \underline{\mathbb{Q}}_X$ is a polarizable variation of Hodge structures on Z and we have $\chi^c_{\mathrm{Hdg}}[X/S] = \sum(-1)^j[R^j f_*\underline{\mathbb{Q}}_X] \in K_0(\mathsf{MHM}^{\mathrm{pol}}(S;\mathbb{Q}))$. In particular we can use this formula to *define* the Euler-Hodge characteristic.

A mixed Hodge module on S contains as part of its data a filtered (holonomic) $\mathcal{D}_S$-module. We have seen that the *gradeds* of the De Rham complex of any filtered $\mathcal{D}_S$-module $(\mathcal{M}, F)$ are complexes of $\mathcal{O}_S$-modules (XIII–41). If $\mathcal{M}$ is coherent, these gradeds are complexes of coherent $\mathcal{O}_S$-modules. These are components of the De Rham characteristic (XIII–43) introduced in a previous chapter. However, there we only treated the case when S is *smooth*. The whole treatment for D-modules can also be adapted to the case of singular algebraic varieties, the De Rham functor can be defined and its gradeds give functors

$$\mathrm{Gr}^F_j \mathrm{DR}^\bullet_S : D^{\mathrm{b}}F\mathcal{D}_S \to D^{\mathrm{b}}_{\mathrm{coh}}(\mathcal{O}_S).$$

This is very well explained in [Sa00, § 1]. Using the definition of the Grothendieck group $K_0(S)$ of coherent sheaves of $\mathcal{O}_S$-modules as given in Lemma-Definition A.20, it follows that the De Rham characteristic can be defined in the singular case as well:

Lemma 14.64. *The De Rham functor induces the* ***De Rham characteristic***

$$\left.\begin{aligned} \chi_{\mathrm{DR}}(S) : K_0(D^{\mathrm{b}}F\mathcal{D}_S) &\to K_0(S)[u, u^{-1}] \\ [(\mathcal{V}, F)] &\mapsto \sum(-1)^j[\mathrm{Gr}^F_j \mathrm{DR}^\bullet_S(\mathcal{V}, F)]u^j. \end{aligned}\right\}$$

This is compatible with proper push forwards, in the sense that for $\varphi : S \to T$ *a proper map between algebraic manifolds, we have a commutative diagram*

$$\begin{array}{ccc} K_0(D^{\mathrm{b}}F\mathcal{D}_S) & \to & K_0(S)[u, u^{-1}] \\ \downarrow{\scriptstyle \varphi_+} & & \downarrow{\scriptstyle \varphi_*} \\ K_0(D^{\mathrm{b}}F\mathcal{D}_T) & \to & K_0(T)[u, u^{-1}]. \end{array}$$

The composition

$$\chi_{\mathrm{Mot}} = \chi_{\mathrm{DR}}(S) \circ \chi^{c}_{\mathrm{Hdg}}(S) : K_0(\mathrm{Var}/S) \to K_0(S)[u, u^{-1}]$$

is the **motivic Chern class transformation** mC_* from [B-S-Y]. If S is smooth and we consider S as a variety over S in a trivial way, setting $c_i := [\Lambda^i T_S] \in K_0(S)$, we have $\chi_{\mathrm{Mot}}(S/S) = c_0 - c_1 u + c_2 u^2 + \cdots$. For $u = -1$ this gives the total Chern class of S. So the motivic Chern class transformation can be viewed as generalization of the ordinary Chern character.

Historical Remarks. The notion of (mixed) Hodge module has been coined by Morihiko Saito [Sa88]. The axiomatic presentation given here is modelled on [Sa87]. The Kashiwara-Malgrange filtration has been introduced in [Kash83] and [Malg83].

Part V

Appendices

A

Homological Algebra

A.1 Additive and Abelian Categories

In this section we recall the definition of additive and abelian categories. We assume that the reader is familiar with the basic language of category theory. In particular we assume familiarity with the notion of a category itself (with its objects and morphisms) as well as with the notions of contra- and covariant functors between them. Standard references are [Iver, Ge-Ma] and [B-B-D, Ch. 1].

In abelian categories we have the usual notion of exact sequences and complexes. We can furthermore define homotopies between complexes. Long exact sequences are induced by triangles, a crucial notion for defining derived categories.

Many standard examples of categories are abelian, such as the category of modules over a fixed ring, but others, like the category of filtered modules over a fixed ring, are only additive. Sometimes when working with a category $\mathfrak{A}$ while using the same objects, we need to invert all arrows. The resulting category is denoted

$$\mathfrak{A}^\circ \quad \textbf{(the opposite category of } \mathfrak{A}). \tag{A–1}$$

In general there are no products in an abelian category and hence it does not make sense to speak of the tensor product of complexes. However, for the category of R-modules we do have a tensor product. Here R is a fixed commutative ring with unit 1. Complexes of *free* R-modules are also called **R-cochain complexes**.

The **tensor product** $C^\bullet \otimes_R D^\bullet$ of two R-cochain complexes $C^\bullet$ and $D^\bullet$ is defined as

$$\begin{aligned} (C \otimes_R D)^n &= \textstyle\bigoplus_{p+q=n} C^p \otimes_R D^q \\ d(x \otimes y) &= dx \otimes y + (-1)^p x \otimes dy, \quad x \in C^p, y \in D^q. \end{aligned} \tag{A–2}$$

A.1.1 Pre-Abelian Categories

Definition A.1. Let us fix a category $\mathfrak{A}$ with a an initial object $0 \in \mathfrak{A}$, i.e. an object admitting unique morphisms $0 \to A$ and $A \to 0$ for every object $A \in \mathfrak{A}$. We say that $f : A \to B$ is a **monomorphism**, respectively **epimorphism** if for all $h \in \mathrm{Hom}(C, A)$, C an object of $\mathfrak{A}$ the composition $f \circ h \in \mathrm{Hom}(C, B)$ is injective, surjective respectively.

1) The **zero morphism** $0 : A \to B$ is the morphism factoring as $A \to 0 \to B$;
2) A **kernel** $\mathrm{Ker}\, f$ for a morphism $f : A \to B$ in $\mathfrak{A}$ is a pair (K, i) consisting of a monomorphism $i : K \to A$ such that $f \circ i = 0$, which is universal with respect to this property: if $g : B \to A$ is such that $f \circ g = 0$, then g factors through i;
3) A **cokernel** $\mathrm{Coker}\, f$ is a pair (C, p) consisting of an epimorphism $p : B \to C$ with $p \circ f = 0$, and which is universal with respect to this property;
4) An **image** $\mathrm{Im}\, f$ is a kernel for a cokernel;
5) A **coimage** $\mathrm{Coim}\, f$ is a cokernel for a kernel;
6) A category is **pre-abelian** [1] if zero objects, kernels and cokernels exist such that for all morphisms f the canonical morphism $\mathrm{Coim}\, f \to \mathrm{Im}\, f$ is an isomorphism.

A 3-term sequence $A \xrightarrow{f} B \xrightarrow{g} C$ in $\mathfrak{A}$ is **exact** if $\mathrm{Ker}\, g = \mathrm{Im}\, f$; a sequence $\ldots A^{n-1} \to A^n \to A^{n+1} \to \ldots$ is exact if every 3-term sequence of consecutive terms is exact. In particular we have the **short exact sequences**

$$0 \to A \xrightarrow{f} B \xrightarrow{g} C \to 0,$$

i.e. f is a monomorphism, g an epimorphism, and $\mathrm{Ker}(g) = \mathrm{Im}(f)$.

A **complex** is a couple $(K^\bullet, d^\bullet)$, where $\{K^i\}_{i \in \mathbb{Z}}$ is an indexed set of objects in $\mathfrak{A}$ and $d^\bullet$, the **differential** of the complex, is a collection of morphisms $\{d^i : K^i \to K^{i+1}\}$ with the property that $d^{i+1} \circ d^i = 0$ for all $i \in \mathbb{Z}$. The **cohomology** of $K^\bullet$ is the complex $H^\bullet(K^\bullet)$ with trivial differential and where

$$H^q(K^\bullet) = \mathrm{Ker}(d^q) / \mathrm{Im}(d^{q-1}) = \mathrm{Ker}\left(\mathrm{Coker}(d^{q-1}) \to \mathrm{Coim}(d^q)\right). \quad \text{(A–3)}$$

We need the notion of a shifted complex:

Definition A.2. If $n > 0$ the complex $K^\bullet[n]$ is obtained from $K^\bullet$ by shifting it n places to the left and multiplying the differentials by $(-1)^n$; in other words, for any integer n the complex $K^\bullet[n]$ is obtained by placing K^{k+n} in degree k and taking $(-1)^n d^{k+n}$ for the k-th differential. In particular

$$H^i(K^\bullet[n]) = H^{i+n}(K^\bullet).$$

[1] In [Iver] this is called an *exact* category; however this is different from the more established notion due to Quillen [Quil72]

A complex $K^\bullet$ is **bounded below**, **bounded above**, respectively **bounded**, if $K^i = 0$ for i sufficiently small, i sufficiently large, respectively $|i|$ sufficiently large. A complex is **acyclic** if it has zero cohomology. A **morphism of complexes** $f : K^\bullet \to L^\bullet$ is a collection of morphisms $f^i : K^i \to L^i$ compatible with the differentials, i. e. $f^{i+1} \circ d^i = d^i \circ f^i$ for all i. Such a morphism induces a morphism $H^\bullet(f)$ in cohomology and we say that f is a **quasi-isomorphism** if $H^\bullet(f)$ is an isomorphism. We denote this by

$$f : K^\bullet \xrightarrow[\sim]{\text{qis}} L^\bullet.$$

If such an f exists, we say that $K^\bullet$ and $L^\bullet$ are quasi-isomorphic. For a morphism between complexes $f : K^\bullet \to L^\bullet$ the shifted morphism $f[n] : K^\bullet[n] \to L^\bullet[n]$ is defined by taking for $f[n]$ in degree p the morphism f in degree $p + n$.

A sequence of complexes

$$0 \to K^\bullet \xrightarrow{f} L^\bullet \xrightarrow{g} M^\bullet \to 0 \tag{A–4}$$

is called **short exact** if for every degree separately it is a short exact sequence. Such a short exact sequence can be shown [Iver, I.2] to induce a long exact sequence in cohomology

$$\cdots \to H^i(K^\bullet) \xrightarrow{H^i(f)} H^i(L^\bullet) \xrightarrow{H^i(g)} H^i(M^\bullet) \xrightarrow{\delta^i} H^{i+1}(K^\bullet) \to \cdots .$$

The map δ^i is called the **connecting homomorphism** and behaves functorially in the obvious way (loc. cit.).

Example A.3. Given a complex $K^\bullet$ in $\mathfrak{A}$ its two **truncations** are given by

$$\tau_{\le k} K^\bullet := \cdots K^{k-2} \to K^{k-1} \to \operatorname{Ker} d^k \to 0 \to \cdots \tag{A–5}$$
$$\tau_{\ge k} K^\bullet := \quad \cdots \quad \to 0 \to K^k/(\operatorname{Im} d^{k-1}) \to K^{k+1} \to K^{k+2} \to \cdots \tag{A–6}$$

These fit into a short exact sequence

$$0 \to \tau_{\le k} K^\bullet \to K^\bullet \to \tau_{\ge k+1}(K^\bullet) \to 0. \tag{A–7}$$

The associated long exact sequence shows that $K^\bullet$ has the same cohomology in degrees $\le k$ as its truncated complex $\tau_{\le k} K^\bullet$, and zero cohomology otherwise. A similar statement is true for $\tau_{\ge k} K^\bullet$.

A.1.2 Additive Categories

Basic for additive categories is the notion of **direct sum** for two objects $P, Q \in \mathfrak{A}$. This is a triple $(P \oplus Q, i, j)$ consisting of an object $P \oplus Q \in \mathfrak{A}$ and two morphisms $i : P \to P \oplus Q$ and $j : Q \to P \oplus Q$ enjoying a certain universality property: whenever there are morphisms $f : P \to X$ and $g : Q \to X$ to the same object X, there should be a unique morphism $P \oplus Q \to X$ making the following diagram commutative

$$\begin{array}{ccc} P & & \\ & \searrow^{i} \quad \searrow^{f} & \\ & P \oplus Q \longrightarrow & X. \\ & \nearrow^{j} \quad \nearrow_{g} & \\ Q & & \end{array}$$

In particular we have projectors $p : P \oplus Q \to P$ and $q : P \oplus Q$, i.e. $p \circ i = 1$, $q \circ j = 1$, $p \circ j = 0$, $q \circ i = 0$.

Definition A.4. 1) An **additive category** is category with a zero object, in which for all objects A and B the set $\mathrm{Hom}(A, B)$ has the structure of an abelian group such that compositions become bilinear, and in which every two objects have a direct sum.
2) An **abelian category** is a pre-abelian category which is moreover additive.
3) For an abelian category $\mathfrak{A}$ with the additional property that its isomorphism classes of objects are sets (such a category is called a *small category*) we define the associated **Grothendieck group** $K_0(\mathfrak{A})$ as the quotient of the free $\mathbb{Z}$-module on isomorphism classes $[V]$ of objects V in $\mathfrak{A}$ under the equivalence relation $[V] = [U] + [W]$ whenever there is an exact sequence $0 \to U \to V \to W \to 0$ in $\mathfrak{A}$. The group structure is the obvious one.

Examples A.5. 1) Modules over a fixed ring with 1, commutative or not, form an abelian category. The category of filtered modules over a fixed ring (see Sect. A.3.1) is only additive. The problem is that the natural map from the coimage to the image is not always an isomorphism. Indeed, tracing through the definitions, this holds precisely if the morphism is strict (see formula (A–26)). The category of filtered modules with strict morphisms is pre-abelian but not additive, because the sum of strict morphisms need not be strict.
2) The category of sheaves of abelian groups on a fixed topological space is abelian [Iver, II.2.5.].
3) Let $(X, \mathcal{O}_X)$ be a ringed space. The category of $\mathcal{O}_X$-modules is abelian [Gode, II.2].

In an additive category a sequence

$$C \xrightarrow{f} D \xrightarrow{g} E$$

is **split exact**, if for all objects X the sequence

$$0 \to \mathrm{Hom}(X, C) \to \mathrm{Hom}(X, D) \to \mathrm{Hom}(X, E) \to 0$$

is an exact sequence of abelian groups. It follows that g has a **section** $s : E \to D$, i.e. $g \circ s = \mathrm{id}_E$, and that f has a **retraction** $r : D \to C$, i.e. $r \circ f = \mathrm{id}_C$. Moreover, $g \circ f = 0$, $r \circ s = 0$. The sequence $P \xrightarrow{i} P \oplus Q \xrightarrow{q} Q$, the model split exact sequence, exemplifies this. We claim that a section s determines a unique retraction r for which $f \circ r + s \circ g = \mathrm{id}_D$ and similarly a given retraction r determines a section s with the stated property. To see this, note that $g \circ (\mathrm{id}_D - s \circ g) = 0$ and so by the splitting property there is a unique $r \in \mathrm{Hom}(D, C)$ for which $f \circ r = \mathrm{id}_D - s \circ g$. Since $f \circ r \circ f = f$ and hence $f \circ (r \circ f - \mathrm{id}_C) = 0$, by the splitting property one must have $rf = \mathrm{id}_C$, i.e. r is a retraction. The second statement can be seen in a similar way.

A split exact sequence of complexes in an additive category

$$K^\bullet \xrightarrow{f} L^\bullet \xrightarrow{g} M^\bullet \tag{A–8}$$

is a sequence of complexes such that $K^p \xrightarrow{f^p} L^p \xrightarrow{g^p} M^p$ is split exact for all p. Note that is not required to have a section $s : M^\bullet \to L^\bullet$ which is a morphisms of complexes. For this reason one writes $s : M \to L$ for the collection of sections $s^p : M^p \to L^p$. As noted above, one has a unique retraction r for which $\mathrm{id}_L = f \circ r + s \circ g$. We set $h = r \circ d \circ s$. Although r and s are not morphisms, h turns out to be a morphism of complexes

$$h : M^\bullet \to K^\bullet[1].$$

For reasons to become clear in § A.2.1 the map h will be called the **homotopy invariant** of the split exact sequence.

The notion of an exact sequence does not make sense in an additive category, so there are no long exact sequences associated to split exact sequences, but we can view h as a substitute for the connecting homomorphism. In fact, we have

Lemma A.6. *The homotopy invariant of a split exact sequence of complexes in an* abelian category *induces the connecting homomorphism in cohomology.*

A central construction in additive categories is that of the cone over a morphism:

Definition A.7. Let $K^\bullet, L^\bullet$ be two complexes in an additive category $\mathfrak{A}$ and let $f : K^\bullet \to L^\bullet$ be a morphism of complexes.

1) Let $f : K^\bullet \to L^\bullet$ be a morphism of complexes in $\mathfrak{A}$. The **(mapping) cylinder** $\mathrm{Cyl}^\bullet(f)$ **of** f is the complex

$$\begin{aligned} \mathrm{Cyl}^q(f) &= K^{q+1} \oplus K^q \oplus L^q \\ d(x, y, z) &= (-dx, -x + dy, f(x) + dz),\ x \in K^{q+1},\ y \in K^q,\ z \in L^q. \end{aligned}$$

2) The **cone** $\mathrm{Cone}^\bullet(f)$ **over** f is the complex

$$\begin{aligned} \mathrm{Cone}^q(f) &= K^{q+1} \oplus L^q \\ d(x, z) &= (-dx, -f(x) + dz),\ x \in K^{q+1},\ z \in L^q. \end{aligned}$$

There is a split exact sequence relating cylinder and cone

$$K^\bullet \xrightarrow{k} \mathrm{Cyl}^\bullet f \xrightarrow{\ell} \mathrm{Cone}^\bullet f. \tag{A–9}$$

The map k sends $x \in K^q$ to the second summand of $\mathrm{Cyl}^q f = K^{q+1} \oplus K^q \oplus L^q$ and ℓ sends $(x, y, z) \in K^{q+1} \oplus K^q \oplus L^q$ to $(-x, z) \in \mathrm{Cone}^q f = K^{q+1} \oplus L^q$. This map has a section given by $(y, z) \mapsto (-y, 0, z)$ showing that the sequence is split.

A.2 Derived Categories

A.2.1 The Homotopy Category

For the moment we work in a fixed additive category. Let us define a **homotopy** between two homomorphisms $f, g : K^\bullet \to L^\bullet$ to be a collection of morphisms $k^q : K^q \to L^{q-1}$ such that $f^q - g^q = d^{q-1} \circ k^q + k^{q+1} \circ d^q$:

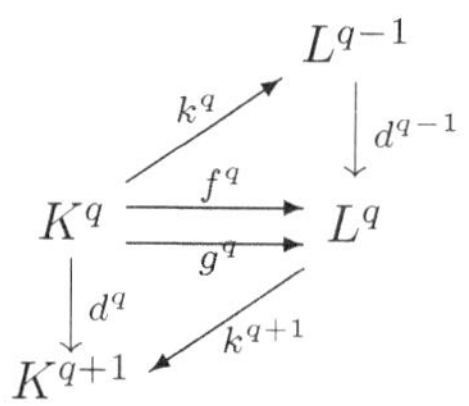

Homotopy is an equivalence relation compatible with composition. Introduce the additive group

$$[K^\bullet, L^\bullet] = \{\text{homotopy classes of morphisms } K^\bullet \to L^\bullet\}.$$

For any two complexes $K^\bullet$, $L^\bullet$ we introduce

$$\mathrm{Hom}^\bullet(K^\bullet, L^\bullet), \quad \mathrm{Hom}^n(K^\bullet, L^\bullet) = \prod_{i \in \mathbb{Z}} \mathrm{Hom}(K^i, L^{i+n}), \tag{A–10}$$

i.e. $f \in \mathrm{Hom}^n(K^\bullet, L^\bullet)$ can be written $f = (f^i)_{i \in \mathbb{Z}}$, $f^i \in \mathrm{Hom}(K^i, L^{i+n})$. To make this into a complex of abelian groups we define a boundary by $(df)^i = d^{i+n} f^i + (-1)^{n+1} f^{i+1} d^i$. This implies that a degree n homomorphism $f : K^\bullet \to L^\bullet$ satisfies $df = 0$ if and only if it gives a homomorphism of complexes $K^\bullet \to L^\bullet[n]$. Moreover, if $f - f' = dg$, then g defines a homotopy between the two resulting homomorphisms, and conversely. In other words, we have

$$[K^\bullet, L^\bullet[n]] = H^n \,\mathrm{Hom}^\bullet(K^\bullet, L^\bullet). \tag{A–11}$$

A homomorphism $f : K^\bullet \to L^\bullet$ admitting an inverse up to homotopy is called a **homotopy equivalence** and the two complexes $K^\bullet$ and $L^\bullet$ are said to be **homotopy equivalent**.

Definition A.8. The **homotopy category of complexes** in an additive category $\mathfrak{A}$ is the category $K(\mathfrak{A})$ of complexes in $\mathfrak{A}$ where the morphisms are the homotopy classes of morphisms between complexes. Similarly, using bounded, or bounded below, or bounded above complexes we get $K^{\mathrm{b}}(\mathfrak{A})$, $K^{+}(\mathfrak{A})$, $K^{-}(\mathfrak{A})$ respectively.

What makes these categories useful is the following two easily verified observations [Iver, I.4] and [Iver, I.5.3]:

Lemma A.9. *For any split exact sequence* (A–8) *with splitting* $s : M \to L$ *and related retraction* $r : L \to K$ *the homotopy invariant* $h = r \circ d \circ s$ *is a well defined invariant*

$$h \in [M^\bullet, K^\bullet[1]]$$

of the split sequence.

Lemma A.10. *In abelian categories homotopic maps induce the same maps in cohomology.*

Remark A.11. The homotopy invariant of the split exact sequence (A–9) is the projection of the cone onto its first factor. We similarly have the **split exact sequence of the cone**

$$L^\bullet \to \mathrm{Cone}^\bullet(f) \to K^\bullet[1]. \tag{A–12}$$

The homotopy invariant of this sequence turns out to be equal to $-f$.

Let us next make some remarks which show the flexibility of working in the homotopy category. First of all [Iver, p. 24]:

Lemma A.12. *The map* $(0, f, \mathrm{id})$ *induces a homotopy equivalence between the complexes* $\mathrm{Cyl}^\bullet(f)$ *and* $L^\bullet$ *so that within the homotopy category any morphism of complexes* $f : K^\bullet \to L^\bullet$ *is isomorphic to the injection* $K^\bullet \hookrightarrow \mathrm{Cyl}^\bullet(f)$ *figuring in the split exact sequence* (A–9). *In this sense* (A–12) *is a shifted version of* (A–9).

As a second observation, from the long exact sequence associated to (A–12) we deduce:

Observation A.13. *A homomorphism between complexes is a quasi-isomorphism if and only if its cone is acyclic.*

As an application, suppose that we have a commutative square of complexes

$$\begin{array}{ccc} A^\bullet & \xrightarrow{i} & B^\bullet \\ \downarrow{\scriptstyle \pi'} & & \downarrow{\scriptstyle \pi} \\ C^\bullet & \xrightarrow{j} & D^\bullet. \end{array}$$

Corollary A.14. *The cone over the morphism*

$$\mathrm{Cone}^\bullet(i) \xrightarrow{(\pi',\pi)} \mathrm{Cone}^\bullet(j)$$

is equal to the cone over

$$A^\bullet[1] \xrightarrow{(-i,\pi')} \mathrm{Cone}^\bullet\left(B^\bullet \oplus C^\bullet \xrightarrow{\pi+j} D^\bullet\right)$$

Hence (π',π) is a quasi-isomorphism if and only if $(-i,\pi')$ is a quasi-isomorphism.

We leave the (easy) proof to the reader.

A.2.2 The Derived Category

We refer [Iver, Chap. IX] and [B-B-D, Chap. 1] for the following discussion.

To define the derived category of a given category, we are going to invert the quasi-isomorphisms in the relevant homotopy category (Def. A.8). Formally this resembles the procedure of localizing a ring in a multiplicative set, except that here we localize in a family of morphisms and so we need to distinguish between left and right fractions as we now explain. Instead of fractions in the homotopy category, we start looking at any additive category $\mathfrak{H}$.

Definition A.15. A collection S of morphisms in an additive category $\mathfrak{H}$ is a **multiplicative system** if

i) The composition of two (composable) morphisms in S belongs again to S; the identity of every object belongs to S,
ii) Any diagram in $\mathfrak{H}$

$$\begin{array}{ccc} & & \bullet \\ & & \downarrow{\scriptstyle s} \\ \bullet & \rightarrow & \bullet \end{array}$$

with $s \in S$ can be completed to a commutative diagram

$$\begin{array}{ccc} \bullet & \rightarrow & \bullet \\ \downarrow{\scriptstyle t} & & \downarrow{\scriptstyle s} \\ \bullet & \rightarrow & \bullet \end{array}$$

with $t \in S$. Similarly with the arrows reversed.
iii) For $f, g : K \to L$ in $\mathfrak{H}$ the existence of $s : K' \to K$ in S with $fs = gs$ is equivalent to the existence of $t : L \to L'$ in S such that $tf = tg$.

Definition A.16. Let $\mathfrak{H}$ be any additive category and let S be a multiplicative system. The objects of the category $S^{-1}\mathfrak{H}$ are the objects of $\mathfrak{H}$. The morphisms are the equivalence classes of right fractions, where a right fraction a/s from K to L is a diagram

$$a/s : K \xleftarrow{s} \bullet \xrightarrow{a} L, \quad s \in S,$$

and where the central dot is any object in $\mathfrak{H}$. Any two right fractions a/s and b/t from K to L are defined to be equivalent if there is a fraction c/u fitting into a commutative diagram

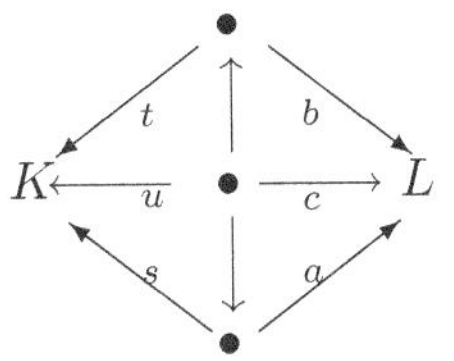

(A–13)

We note that various verifications are in order here, for instance one should show that this indeed defines an equivalence relation and that replacing the morphisms by right (or left) fractions indeed defines a category.

Definition A.17. Let $\mathfrak{A}$ be an additive category. The set S of homotopy classes of quasi-isomorphisms in the associated homotopy category $K(\mathfrak{A})$ is a multiplicative system. The **derived category** $D(\mathfrak{A})$ is the category $S^{-1}K(\mathfrak{A})$ having the same objects as $K(\mathfrak{A})$, i.e. the complexes in $\mathfrak{A}$; however the morphisms are equivalence classes of right fractions

$$a/s : K^\bullet \underset{s}{\overset{\text{qis}}{\underset{\sim}{\longleftarrow}}} \bullet \xrightarrow{a} L^\bullet$$

with a a homotopy class of a morphism between complexes and s a homotopy class of a quasi-isomorphism. The equivalence relation is generated by diagrams (A–13). Analogously, using bounded, or bounded below, or bounded above complexes we obtain the derived categories $D^{\mathrm{b}}(\mathfrak{A})$, $D^{+}(\mathfrak{A})$, and $D^{-}(\mathfrak{A})$ respectively.

Derived categories are in general *not* abelian since short exact sequences are not preserved under quasi-isomorphisms. In the derived category one should instead work with triangles:

Definition A.18. A **triangle** in $K(\mathfrak{A})$, respectively $D(\mathfrak{A})$ is a diagram

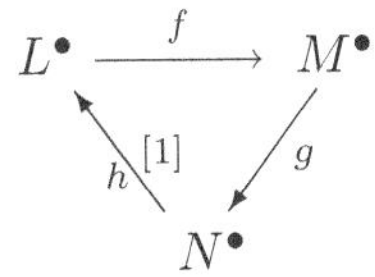

(A–14)

whose morphisms are in the homotopy category, respectively the derived category for $\mathfrak{A}$; the notation [1] means that the left hand morphism in fact has the shifted complex for its target: $h : N^\bullet \to L^\bullet[1]$. It will be clear what a morphism or an isomorphism between triangles should be.

The standard example of a triangle is given by the triangle defined by any split exact sequence of complexes (A–4), with h the homotopy invariant. Triangles isomorphic to such triangles are called **distinguished triangles**. In fact, in the derived category, all exact sequences are isomorphic to split exact sequences and hence give rise to distinguished triangles:

Lemma A.19. *Within the derived category an exact sequence of complexes* $0 \to K^\bullet \xrightarrow{f} L^\bullet \xrightarrow{g} M^\bullet \to 0$ *in* $\mathfrak{A}$ *(split or not) is isomorphic to the split exact sequence* (A–9) *and so defines a distinguished triangle*

$$\begin{array}{ccc} K^\bullet & \xrightarrow{\quad f \quad} & L^\bullet \\ {\scriptstyle [1]}\nwarrow & & \swarrow \\ & \mathrm{Cone}^\bullet(f) & \end{array} \qquad \text{(A–15)}$$

Sketch of the proof: By [Iver, XI,3] the morphism $(0, g) : \mathrm{Cone}^\bullet(f) \to M^\bullet$ is a quasi-isomorphism. So $M^\bullet$ can be replaced by the cone over f. We may also (Lemma A.12) replace $L^\bullet$ by the cylinder on f, thereby obtaining an exact commutative diagram

$$\begin{array}{ccccccc} 0 \to K^\bullet & \xrightarrow{k} & \mathrm{Cyl}^\bullet(f) & \xrightarrow{\ell} & \mathrm{Cone}^\bullet(f) & \to 0 \\ \Vert & & {\scriptstyle (0,f,1)}\downarrow\wr & & {\scriptstyle (0,g)}\downarrow\wr & \\ 0 \to K^\bullet & \xrightarrow{f} & L^\bullet & \xrightarrow{g} & M^\bullet & \to 0, \end{array} \qquad \text{(A–16)}$$

where the last two vertical arrows are quasi-isomorphisms. This completes the proof. □

Any triangle induces a long sequence of homomorphisms in cohomology

$$\cdots H^i(K^\bullet) \xrightarrow{H^i(f)} H^i(L^\bullet) \xrightarrow{H^i(g)} H^i(M^\bullet) \xrightarrow{H^i(h)} H^{i+1}(K^\bullet) \to \cdots$$

and it is exact when the triangle is distinguished. Isomorphic triangles give the same exact sequences so that in the derived category isomorphism classes of distinguished triangles play the role of short exact sequences. This remark leads us to the following considerations for the Grothendieck group.

Lemma A.20. *Let* $K^\bullet$ *be a bounded complex in* $\mathfrak{A}$. *Define*

$$[K^\bullet] := \sum_{i \in \mathbb{Z}} (-1)^i [K^i] \in K_0(\mathfrak{A}). \qquad \text{(A–17)}$$

Then

i) $[K^\bullet] = \sum_{i\in\mathbb{Z}} (-1)^i [H^i(K^\bullet)]$ *and hence only depends on the homotopy class of* $K^\bullet$.
ii) *For every distinguished triangle* (A–14) *we have* $[M^\bullet] = [L^\bullet] + [N^\bullet]$.

Proof. i) $\sum_{i\in\mathbb{Z}}(-1)^i[K^i] = \sum_{i\in\mathbb{Z}}(-1)^i[H^i(K^\bullet)]$ is standard and can easily be verified by induction.
ii) This is because distinguished triangles induce long exact sequences in cohomology. □

Remark. These relations imply that there is no need to define a Grothendieck-group $K_0(D^{\mathrm{b}}\mathfrak{A})$: by (A–17) complexes $K^\bullet$ in $\mathfrak{A}$ define an element $[K^\bullet]$ in $K_0(\mathfrak{A})$.

We have seen that the sequence relating cylinder and cone (A–9) can be shifted to give the sequence of the cone (A–12). The triangle for this sequence is turned with respect to the triangle for the first sequence. Since all distinguished triangles are isomorphic to such triangles this shows that distinguished triangles may always be turned to give new distinguished triangles:

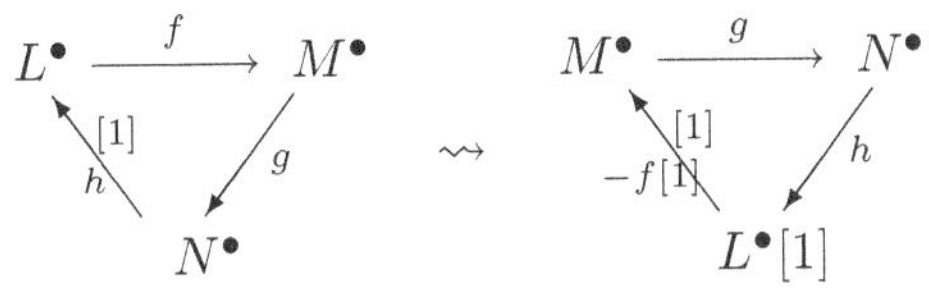

See also [Iver, I.4.16]. So, in the derived category the three terms of an exact sequence can be cyclically interchanged up to signs.

Suppose that instead of $D(\mathfrak{A})$ we start off with any additive category $\mathfrak{D}$ equipped with a shift $X \mapsto X[1]$. Then the concept of a triangle like (A–14) makes sense. A collection of triangles verifying the axioms "TRI, TRII, TRIII, TRIV" from [Verd96, Chap. II] is called a set of **distinguished** triangles. If $\mathfrak{D}$ is equipped with such a set we say that it is a **triangulated category**. As expected, a typical example is the derived category of an abelian category together with its set of distinguished triangles. Indeed, the axiom TRI is lemma A.12 and TRII says that we may turn triangles. The axiom TRIII tells us that if you have two distinguished triangles such that two corresponding sides of the triangles are related by morphisms forming a commutative diagram, this holds for the other sides as well in an obvious compatible way. This is clear in the case of the derived category, since the triangles are built from one side using the cone. The last axiom is called the octahedral diagram and is too involved to state here. See loc. cit. Verdier's axioms will not be explicitly used in this this book.

The derived category of $\mathfrak{A}$ contains the category $\mathfrak{A}$ as the full subcategory of complexes $K^\bullet$ having only cohomology in degree 0, its **core**. Let $D^{\leq k}\mathfrak{A}$ respectively $D^{\geq k}\mathfrak{A}$ be the full subcategory of complexes having only cohomology in degree $\leq k$, respectively $\geq k$. For any complex $K^\bullet$ in $\mathfrak{A}$ its truncated complexes $\tau_{\leq k}K^\bullet$, respectively $\tau_{\geq k}K^\bullet$ define objects in the category $D^{\leq k}\mathfrak{A}$, $D^{\geq k}\mathfrak{A}$ respectively and the exact sequence (A–7) transforms into a distinguished triangle. Note also that in $D(\mathfrak{A})$ we have

$$\tau_{\leq 0}\tau_{\geq 0}[K^\bullet[k]] = H^k(K^\bullet).$$

These truncated complexes give the proto-type of a so-called t-structure on a triangulated category.

Definition A.21. Let $\mathfrak{D}$ be a triangulated category with a shift $X \mapsto X[1]$. For any object X of $\mathfrak{D}$, let $X[k]$ be the k-th iterate of the shift applied to X. A **t-structure** on $\mathfrak{D}$ consists two full subcategories $\mathfrak{D}^{\leq 0}$ and $\mathfrak{D}^{\geq 0}$ such that, setting

$$\begin{aligned}\mathfrak{D}^{\leq k} &:= \mathfrak{D}^{\leq 0}[-k] = \{X[-k] \mid X \in \mathfrak{D}^{\leq 0}\}\\ \mathfrak{D}^{\geq k} &:= \mathfrak{D}^{\geq 0}[-k] = \{X[-k] \mid X \in \mathfrak{D}^{\geq 0}\},\end{aligned}$$

the following conditions are satisfied

i) $\mathfrak{D}^{\leq -1} \subset \mathfrak{D}^{\leq 0}$ and $\mathfrak{D}^{\geq 1} \subset \mathfrak{D}^{\geq 0}$;
ii) There are no non-trivial morphisms from objects in $\mathfrak{D}^{\leq 0}$ to objects in $\mathfrak{D}^{\geq 1}$;
iii) For any object X in $\mathfrak{D}$, there is an associated object $X_{\leq 0}$ in $\mathfrak{D}^{\leq 0}$, respectively $X_{\geq 1}$ in $\mathfrak{D}^{\geq 1}$ which together fit into a distinguished triangle

$$\begin{array}{ccc} X_{\leq 0} & \longrightarrow & X \\ & {\scriptstyle [1]}\nwarrow \quad \swarrow & \\ & X_{\geq 1} & \end{array}$$

We have a full subcategory

$$C(\mathfrak{D}, t) := \mathfrak{D}^{\leq 0} \cap \mathfrak{D}^{\geq 0} \quad (\textbf{core} \text{ of the } t\text{-structure})$$

which turns out to be an *abelian* subcategory of $\mathfrak{D}$. The functor

$$ {}^{t}H^{k}(X) := [\tau_{\geq 0}\tau_{\leq 0}(X[k]) \in C(\mathfrak{D}, t). \tag{A–18}$$

is a cohomological functor [B-B-D, Thm. 1.3.6], i.e. for any distinguished triangle (A–14) the short sequence $0 \to {}^{t}H^{0}(L) \to {}^{t}H^{0}(M) \to {}^{t}H^{0}(N)$ is exact, and can be prolonged to a long exact sequence in the obvious way.

A.2.3 Injective and Projective Resolutions

An **injective object** of an abelian category $\mathfrak{A}$ is an object I of $\mathfrak{A}$ such that any morphism from a subobject K of L (in $\mathfrak{A}$) to I extends to L. In other words, given i and f in the following diagram, $\bar{f}$ exists making it commutative.

$$\begin{array}{ccccc} 0 & \longrightarrow & K & \xrightarrow{\ i\ } & L \\ & & \big\downarrow f & \swarrow \bar{f} & \\ & & I & & \end{array}$$

Similarly, reversing the arrows, we define a **projective object**: every morphism to a quotient object K of L (i.e. K is image of a surjection $L \twoheadrightarrow K$) factors over L.

If every object in $\mathfrak{A}$ is a subobject of an injective object, one says that $\mathfrak{A}$ **has enough injectives.** If every object is a quotient object of a projective objects, one says that $\mathfrak{A}$ **has enough projectives.**

Examples A.22. 1) The category of (left or right) modules over a not necessarily commutative ring with unit is an abelian category with enough injectives since every module is a submodule of an injective module. See [Gode, §I.1.4]. It also has enough projectives since any module is a quotient of a free module and free modules are projective.
2) The category of sheaves of abelian groups on a fixed topological space has enough injectives [Gode, II.3.1], but not in general enough projectives.
3) Let $(X, \mathcal{O}_X)$ be a ringed space. The category of sheaves of $\mathcal{O}_X$-Modules is an abelian category with enough injectives. See [Gode, II, 7.1.1].

Definition A.23. A (right) **resolution** of an object A in an abelian category consists of a complex $I^\bullet$ which fits into an exact sequence

$$0 \to A \xrightarrow{\epsilon} I^0 \xrightarrow{d^0} I^1 \xrightarrow{d^1} I^2 \cdots .$$

Equivalently, this is a quasi-isomorphism $A \overset{\text{qis}}{\xrightarrow{\sim}} I^\bullet$ where A is considered as a complex concentrated in degree 0. We say that the resolution is **injective** if the I^q are all injective.

Similarly, by reversing arrows we get a left resolution, and by replacing "injective" by "projective", we arrive at the notion of a **projective resolution.**

One can easily prove that any two injective resolutions of A are related by a quasi-isomorphism unique up to homotopy. Moreover, if $I^\bullet$ is an injective resolution of A, $J^\bullet$ an injective resolution of B, any given homomorphism $f : A \to B$ can be extended to a homomorphism of complexes

$$\{0 \to A \to I^\bullet\} \xrightarrow{f, f^\bullet} \{0 \to B \to J^\bullet\}$$

and any other extension of f is homotopic to it. More generally, one has [Iver, Theorem I.6.1, I.6.2]:

Theorem A.24. *Suppose $\mathfrak{A}$ has enough injectives. Then*

1) *Every bounded below complex admits an injective resolution, i.e. a quasi-isomorphism to a bounded below injective complex.*
2) *Let $f : K^\bullet \to L^\bullet$ be a quasi-isomorphism and $I^\bullet$ a bounded below complex of injective objects. The morphism*

$$\begin{aligned} [L^\bullet, I^\bullet] &\to [K^\bullet, I^\bullet] \\ g &\mapsto g \circ f \end{aligned}$$

is an isomorphism.

This theorem implies that any bounded below complex $K^\bullet$ maps quasi-isomorphically to a complex $I^\bullet$ of injectives, unique up to homotopy. We call any such complex $I^\bullet$ an **injective resolution of** $K^\bullet$ and denote it by $I(K^\bullet)$. Using the existence and uniqueness up to homotopy of such injective resolutions, one can show [Iver, Chapt. XI] that the derived category of bounded below complexes can be described concretely as follows.

Lemma A.25. *Let $\mathfrak{A}$ be an abelian category with enough injectives. The derived category $D^+(\mathfrak{A})$ is the category whose objects are the bounded below complexes in $\mathfrak{A}$, and whose morphisms are the homotopy classes of maps between injective resolutions of the respective complexes.*

Needless to say that similar results hold for projective resolutions if these exist.

A.2.4 Derived Functors

We start with two abelian categories $\mathfrak{A}$ and $\mathfrak{B}$ having enough injectives and a **left exact** additive functor $T : \mathfrak{A} \to \mathfrak{B}$. This means that

$$0 \to T(K) \xrightarrow{T(f)} T(L) \xrightarrow{T(g)} T(M)$$

is exact whenever $0 \to K \xrightarrow{f} L \xrightarrow{g} M$ is an exact sequence in $\mathfrak{A}$.

For any bounded below complex $K^\bullet$ in $\mathfrak{A}$ we choose an injective resolution $I(K^\bullet)$ and for any map $f : K^\bullet \to L^\bullet$ we choose an extension $I(f) : I(K^\bullet) \to I(L^\bullet)$ of f as before. We define the **right derived functor**

$$RT : D^+(\mathfrak{A}) \to D^+(\mathfrak{B})$$

upon setting $RT(K^\bullet) = T(I(K^\bullet))$ and $RT(f) = T(I(f))$. Theorem A.24 implies that this is well defined up to a quasi-isomorphism. The **i-th derived functor of T**

$$R^iT(K^\bullet) := H^i(RT(K^\bullet))$$

is then independent of the chosen injective resolution. In particular, for any object K of $\mathfrak{A}$, the derived functors $R^iT(K)$ are defined as

$$H^i\left(T(IK^0) \to T(IK^1) \to \cdots\right)$$

where $IK^\bullet$ is an injective resolution of K.

Definition A.26. We say that any object K of $\mathfrak{A}$ is **T-acyclic** if $R^iTK{=}0$ for all $i > 0$ and a complex $K^\bullet$ is T-acyclic if all its components K^p are T-acyclic.

There are long exact sequences for the derived functor associated to short exact sequences:

Lemma A.27. *Let $\mathfrak{A}$ and $\mathfrak{B}$ two abelian categories having enough injectives and let $T : \mathfrak{A} \to \mathfrak{B}$ be a left exact functor. For any short exact sequence of bounded below complexes in $\mathfrak{A}$*

$$0 \to K^\bullet \xrightarrow{f} L^\bullet \xrightarrow{g} M^\bullet \to 0$$

connecting homomorphisms $d : T(M^i) \to T(K^{i+1})$ exist inducing a long exact sequence

$$\cdots \to R^iT(K^\bullet) \xrightarrow{R^if} R^iT(L^\bullet) \xrightarrow{R^ig} R^iT(M^\bullet) \xrightarrow{d} R^{i+1}T(K^\bullet) \to \cdots$$

Examples A.28. 1) The function "taking global sections" defines a functor

$$\Gamma : \{\text{Sheaves on } X\} \longrightarrow \{\text{Abelian groups}\}$$

whose derived functor $R\Gamma$ by definition computes the **hypercohomology groups**

$$\mathbb{H}^i(X, \mathcal{F}^\bullet) := R^i\Gamma(\mathcal{F}^\bullet). \tag{A–19}$$

We can also use this for a single sheaf $\mathcal{F}$ viewed as a complex in degree 0 and write $H^i(X, \mathcal{F})$ instead of $\mathbb{H}^i(X, \mathcal{F})$. Later, in § B.2.4 we discuss another definition of sheaf cohomology and compare the two.

2) Fix any abelian category $\mathfrak{A}$, a bounded *above* complex $N^\bullet$, and consider

$$\mathrm{Hom}^\bullet(N^\bullet, -) : \mathfrak{A} \longrightarrow \{\text{Complexes of abelian groups}\}$$

which is a left exact additive functor on bounded *below* complexes. Its derived functor is the Ext-functor:

$$\mathrm{Ext}^i(N^\bullet, K^\bullet) = R^i\,\mathrm{Hom}^\bullet(N^\bullet, K^\bullet), \quad N^\bullet \in D^-(\mathfrak{A}),\ K^\bullet \in D^+(\mathfrak{A}).$$

3) Fix a topological space X and a commutative ring R with unit. For two sheaves $\mathcal{F}$, $\mathcal{G}$ of R-modules on a fixed topological space, the presheaf

$$U \mapsto \mathrm{Hom}_R(\mathcal{F}|U, \mathcal{G}|U)$$

defines the sheaf $\mathcal{H}om(\mathcal{F}, \mathcal{G})$. This gives a bi-functor

$$\mathcal{H}om : \{R\text{-sheaves on } X\} \times \{R\text{-sheaves on } X\} \longrightarrow \{R\text{-sheaves on } X\}.$$

If instead we have two complexes of R-sheaves, say $\mathcal{F}^\bullet$ and $\mathcal{G}^\bullet$, it will be clear how to define the Hom-complex $\mathcal{H}om^\bullet(\mathcal{F}^\bullet, \mathcal{G}^\bullet)$ of R-sheaves. The procedure from the previous example can then be used to define the Ext-sheaves

$$\begin{aligned}&\mathcal{E}xt^i(\mathcal{F}^\bullet, \mathcal{G}^\bullet) = R^i\,\mathcal{H}om(\mathcal{F}^\bullet, \mathcal{G}^\bullet),\\ &\quad \mathcal{F}^\bullet \in D^-(R\text{-sheaves on } X)^\circ, \quad \mathcal{G}^\bullet \in D^+(R\text{-sheaves on } X),\end{aligned}$$

where the index $^\circ$ means that we work in the opposite category (A–1).

4) Let $(X, \mathcal{O}_X)$ be a ringed space. Working with complexes of $\mathcal{O}_X$-modules $\mathcal{F}^\bullet$ (bounded above), $\mathcal{G}^\bullet$ (bounded below), and replacing R by $\mathcal{O}_X$ in the above example, we get the Hom-complex of $\mathcal{O}_X$-modules $\mathcal{H}om^\bullet_{\mathcal{O}_X}(\mathcal{F}^\bullet, \mathcal{G}^\bullet)$ leading to the Ext-sheaves

$$\begin{aligned}&\mathcal{E}xt^i_{\mathcal{O}_X}(\mathcal{F}^\bullet, \mathcal{G}^\bullet) = R^i\,\mathcal{H}om_{\mathcal{O}_X}(\mathcal{F}^\bullet, \mathcal{G}^\bullet),\\ &\quad \mathcal{F}^\bullet \in D^-(\mathcal{O}_X\text{-modules on } X)^\circ, \quad \mathcal{G}^\bullet \in D^+(\mathcal{O}_X\text{-modules on } X).\end{aligned}$$

Finally a few words about **right exact** functors $T : \mathfrak{A} \to \mathfrak{B}$. To define the left derived functor LT, we assume that $\mathfrak{A}$ has enough projectives and we use projective resolutions in $\mathfrak{A}$ instead of injective resolutions. So if $K \in \mathfrak{A}$, choose a projective resolution $P_\bullet$ and define

$$LTK := T(P_\bullet), \quad L_iTK = H^{-i}(T(P_{-\bullet}).$$

The minus sign is needed, because we work with cohomology of complexes.

Example A.29. Fix a commutative ring R with unit and consider the category of R-modules. The tensor product is right exact and for two R-modules M and N we define

$$M \overset{\mathrm{L}}{\otimes}_R N := P_\bullet \otimes_R N, \quad \mathrm{Tor}_i(M, N) := H^{-i}(M \overset{\mathrm{L}}{\otimes}_R N).$$

In fact, instead of projective resolutions, one may use resolutions by **flat** R-**modules**, i.e. modules having the property that tensoring by them preserve exactness.

Another version of this example is the tensor product of R-sheaves on a topological space.

We turn now to functors between *triangulated categories.* We say that $T : \mathfrak{D} \to \mathfrak{D}'$ is an **additive functor** if T intertwines the shifts and maps distinguished triangles into distinguished triangles. If both categories have a t-structure, denoting $\epsilon : C(\mathfrak{D}) \to \mathfrak{D}$ the natural inclusion of the core, we can introduce the associated functor between the cores:

$$^tT := {}^tH^0 \circ T \circ \epsilon : C(\mathfrak{D}, t) \to C(\mathfrak{D}', t). \tag{A–20}$$

We say that T is **left exact**, respectively **right exact** with respect to the t-structure if T maps $\mathfrak{D}^{\leq 0}$ to $\mathfrak{D}'^{\leq 0}$, respectively $T(\mathfrak{D}^{\geq 0}) \subset (\mathfrak{D}'^{\geq 0})$. If T is both left and right exact it is called **exact**. Note that this terminology is compatible with the usual notions; if for instance F is a left exact functor between abelian categories, the derived functor RT is left exact with respect to the natural t-structure on the derived category.

Remark. For functors $T : \mathfrak{A} \to \mathfrak{B}$ between abelian categories $\mathfrak{A}, \mathfrak{B}$ which fail to have enough projectives (or injectives) there is an abstract approach giving conditions on T which guarantee the existence of RT (or LT) enjoying certain universal properties. For such T the derived functor RT (or LT) having these properties then still exists. See for instance [Hart69, Verd77]. We shall not use this theory except for the Ext-functor where we follow Yoneda's version of Verdier's approach. See § A.2.6.

A.2.5 Properties of the Ext-functor

Let $N^\bullet$, respectively $K^\bullet$ be a bounded, a bounded below complex in $\mathfrak{A}$ respectively. Let $I(K^\bullet)$ be an injective resolution of $K^\bullet$. Using (A–11) and Theorem A.24 we find

$$\left.\begin{aligned}\mathrm{Ext}^i(N^\bullet, K^\bullet) :=& [N^\bullet, I(K^\bullet))[i]]\\ =& H^i(\mathrm{Hom}^\bullet(N^\bullet, I(K^\bullet)) = \mathrm{Hom}_{D^+(\mathfrak{A})}(N^\bullet, K^\bullet[i]).\end{aligned}\right\} \quad \text{(A–21)}$$

As an example, consider an exact sequence

$$0 \to K^\bullet \xrightarrow{f} L^\bullet \to M^\bullet \to 0.$$

of bounded complexes in $\mathfrak{A}$. We have seen (A–16) that in $D^{\mathrm{b}}(\mathfrak{A})$ such a sequence can be represented by the split exact sequence relating the cylinder and the cone over f. The **extension class** of the exact sequence is the class in $\mathrm{Ext}^1(M^\bullet, K^\bullet)$ which corresponds to the homotopy invariant $h \in [M^\bullet, K^\bullet[1]]$. It appears in the long exact sequences for the Ext-groups by way of cup product. To explain this notion, observe that the composition of morphisms induces the **composition product**

$$\left.\begin{array}{lcl}\mathrm{Ext}^i(M^\bullet, P^\bullet) \times \mathrm{Ext}^j(P^\bullet, N^\bullet) & \longrightarrow & \mathrm{Ext}^{i+j}(M^\bullet, N^\bullet)\\ (e, f) & \mapsto & f[i] \circ e := f \cup e\end{array}\right\} \quad \text{(A–22)}$$

and this defines cup product. The first long exact sequence for the Ext-groups for the above short exact sequence in the derived category $D^{\mathrm{b}}(\mathfrak{A})$ reads:

$$\begin{array}{ccccccccc}\cdots \to & [N^\bullet, K^\bullet[i]] & \to & [N^\bullet, L^\bullet[i]] & \to & [N^\bullet, M^\bullet[i]] & \to & [N^\bullet, K^\bullet[i+1]] & \to \cdots\\ & \| & & \| & & \| & & \| & \\ \cdots \to & \mathrm{Ext}^i(N^\bullet, K^\bullet) & \to & \mathrm{Ext}^i(N^\bullet, L^\bullet) & \to & \mathrm{Ext}^i(N^\bullet, M^\bullet) & \to & \mathrm{Ext}^{i+1}(N^\bullet, K^\bullet) & \to \cdots,\end{array}$$

where the connecting homorphism, say d^i is given by $[1, h[i]]$ [Iver, I,4.8] i. e $d^i a = h \cup a$

There is a second long exact sequence

$$\begin{array}{ccccccccc}\cdots \to & [M^\bullet, N^\bullet[i]] & \to & [L^\bullet, N^\bullet[i]] & \to & [K^\bullet, N^\bullet[i]] & \to & [M^\bullet, N^\bullet[i+1]] & \to \cdots\\ & \| & & \| & & \| & & \| & \\ \cdots \to & \mathrm{Ext}^i(M^\bullet, N^\bullet) & \to & \mathrm{Ext}^i(L^\bullet, N^\bullet) & \to & \mathrm{Ext}^i(K^\bullet, N^\bullet) & \to & \mathrm{Ext}^{i+1}(M^\bullet, N^\bullet) & \to \cdots,\end{array}$$

where the connecting homorphism d^i is given by $(-1)^{i+1}[h, 1]$ [Iver, I,4.9] i. e $d^i a = (1)^{i+1} a \cup h$.

A.2.6 Yoneda Extensions

In this section we work with any abelian category $\mathfrak{A}$, with or without enough injectives. The formula (A–21) gives a definition for the Ext-groups in general:

$$\mathrm{Ext}^n(A, B) := \mathrm{Hom}_{D\mathfrak{A}}(A, B[n]) \quad \text{(A–23)}$$

Remark A.30. There is one thing to check here: the natural map $\mathrm{Hom}_{\mathfrak{A}}(A,B) \to \mathrm{Hom}_{D\mathfrak{A}}(A,B)$ should be a bijection. This can be seen as follows. Any homomorphism in the derived category is represented by a fraction f/s with $f : L^\bullet \to B$ a morphism of complexes and $s : L^\bullet \to A$ a quasi-isomorphism. Then f induces a morphism $g : \mathrm{Ker}\, d_L^0 / \mathrm{Im}\, d_L^{-1}$ and s gives an isomorphism $t : H^0(L) = \mathrm{Ker}\, d_L^0 / \mathrm{Im}\, d_L^{-1} \to A$. Hence f/s yields a true morphism $g \circ t^{-1}$ in $\mathfrak{A}$ which in the derived category equals f/s. This proves surjectivity. As to injectivity: if $f : A \to B$ is a morphism in $\mathfrak{A}$, it is zero in the derived category, if for some quasi-isomorphism $s : L^\bullet \to A$ and some map $g : L^\bullet \to B$ homotopic to zero we have $f \circ s = g$. Again, replacing L by $H^0(L)$, we may assume that L is a single object and hence $g = 0$ and hence also $f = 0$.

We give Yoneda's alternative path to extensions as explained for instance in [Iver, XI. 4]. Let us start with a fraction f/s with $f : L^\bullet \to B[n]$ a morphism and $s : L^\bullet \to A$ a quasi-isomorphism. The fact that s is a quasi-isomorphism means that $H^k(L^\bullet) = 0$ for $k \neq 0$ and that $H^0(s) : H^0(L^\bullet) \xrightarrow{\sim} A$. Replacing $L^\bullet$ by the doubly truncated complex $K^\bullet = \tau_{\leq 0}\tau_{\geq -n} L^\bullet$ (see Example A.3 for the notation) we get a bounded complex concentrated in degrees $[-n, 0]$ with the same cohomology in degrees $-n, \ldots, 0$. Now extend $K^\bullet$ to degree 1 by placing in this degree the object $K^1 := \mathrm{Ker}\, d_L^0 / \mathrm{Im}(d_L^{-1}) \simeq A$ and by taking for the boundary $K^0 \to K^1$ the natural map

$$K^0 = \mathrm{Ker}\, d_L^0 \to K^1 := \mathrm{Ker}\, d_L^0 / \mathrm{Im}(d_L^{-1}) \simeq A \to 0.$$

The resulting complex $K^\bullet$ is then exact. The morphism of complexes $f : L^\bullet \to B[n]$ induces $g : K^\bullet \to B[n]$ whence an inclusion $h : B[n] \to \mathrm{Cone}^\bullet(g) = C^\bullet$. Adding h on the left, we obtain an *exact* sequence of the form

$$E : \left[0 \to B \to C^{-n} = B \oplus K^{-n+1} \to \cdots \to C^{-1} = K^0 \xrightarrow{i^0} A \to 0 \right]. \quad \text{(A-24)}$$

This is an n**-fold extension** of A by B defined by f/s. Since fractions like f/s can be represented in various ways the extension is not unique; we need an equivalence relation on extensions which takes care of the ambiguity. It turns out that the equivalence relation generated by congruence is the right one, where we say that E and

$$E' = \left[0 \to B \to K'^{-n+1} \cdots \to K'^0 \xrightarrow{i'^0} A \to 0 \right] \quad \text{(A-25)}$$

are **congruent** if there is a commutative diagram

$$\begin{array}{ccccccccccc} 0 & \to & B & \to & K^{-n+1} & \to \cdots \to & K^0 & \to & A & \to & 0 \\ & & \| & & \downarrow & & \downarrow & & \| & & \\ 0 & \to & B & \to & K'^{-n+1} & \to \cdots \to & K'^0 & \to & A & \to & 0. \end{array}$$

The set $\mathrm{Ex}^n(A,B)$ of n-fold extension of A by B is made into a bi-functor as for the usual Ext-groups: given $u : A' \to A$ and $v : B \to B'$ there are induced maps $\mathrm{Ex}^n(u) : \mathrm{Ex}^n(A', B) \to \mathrm{Ex}^n(A, B)$, the **pull-back along** u and $\mathrm{Ex}^n(v) : \mathrm{Ex}^n(A, B) \to \mathrm{Ex}^n(A, B')$, the **push-out along** v. The pull-back is defined as follows. Replace A by A', K^0 by

$$K'^0 = \mathrm{Ker}[(-i^0, u) : K^{n-1} \oplus A' \to A].$$

and i^0, i^{-1} by the obvious maps. For the push-out, replace B by B', K^{-n+1} by

$$K'^{-n+1} = \mathrm{Coker}[(-i^0, v) : B \to K^{-n+1} \oplus B'].$$

and i^{-n+1}, i^{-n+2} by the obvious maps. Let $\Delta : A \to A \oplus A$ be the diagonal embedding and let $s : B \oplus B \to B$ be the codiagonal (or addition map). Taking termwise the direct sum of two n fold extensions of A by B followed by $\mathrm{Ex}^n(\Delta)\mathrm{Ex}^n(s)$ defines the **Baer sum** and makes $\mathrm{Ex}^n(A, B)$ into a group.

Conversely, start with an n-fold extension of A by B, say

$$E : [0 \to B = K^{-n} \to K^{-n+1} \cdots \to K^0 \xrightarrow{s} A \to 0].$$

It can be written in the form of a quasi-isomorphism $s : K^\bullet \xrightarrow[\sim]{\text{qis}} A$. Together with the morphism $f : K^\bullet \to B[n]$ of complexes induced by the identity map on B this produces the fraction f/s, a morphism form A to $B[n]$ in the derived category of $\mathfrak{A}$. One can furthermore show that congruent extensions define the same fraction. This motivates the following definition.

Definition A.31. Given an n-fold extension of A by B as above, the class of $(-1)^{n(n+1)/2} f/s$ in $\mathrm{Hom}_{D\mathfrak{A}}(A, B[n])$ is called the **Yoneda class** of the extension.

The preceding discussion can be summarized as follows:

Lemma A.32. *Taking the Yoneda class establishes a 1-1 correspondence between* $\mathrm{Ex}^n(A, B)$ *and* $\mathrm{Ext}^n(A, B) = \mathrm{Hom}_{D\mathfrak{A}}(A, B[n])$. *We shall therefore identify the two groups.*

Let us discuss the composition product of two extensions $h \in \mathrm{Hom}_{D\mathfrak{A}}(L^\bullet, M^\bullet[m])$ and $k \in \mathrm{Hom}_{D\mathfrak{A}}(K^\bullet, L^\bullet[n])$. Recall (A–22) that this is the extension $h \cup k \in \mathrm{Hom}_{D\mathfrak{A}}(K^\bullet, M^\bullet[n + m])$ defined by $h[m] \circ k$. In the language of Yoneda-extensions, given an extension of B by C

$$G = [0 \to C \to G^{-m-1} \to G^{-m-2} \cdots \to G^0 \xrightarrow{s} B \to 0]$$

and an extension of A by B

$$H = [0 \to B \xrightarrow{s'} H^{-n-1} \to H^{-n-2} \cdots \to H^0 \to A \to 0],$$

one defines the **spliced extension** by

$$G \cup H = [0 \to C \to G^{-m-1} \cdots \to G^0 \xrightarrow{s \circ s'} H^{-n-1} \cdots \to H^0 \to A \to 0].$$

By [Iver, XI.4.6] the Yoneda-class of $G \cup H$ is the composition product of the Yoneda-classes of G and H. This explains the somewhat strange sign which is needed since the cup product is graded commutative and not commutative.

As a consequence, we have

Lemma A.33. *If* $\operatorname{Ext}^k(A, -)$ *is right exact for all* A *in the category, then* $\operatorname{Ext}^n(A, B) = 0$ *for* $n \geq k + 1$ *and all* A *and* B *in the category.*

Proof. To calculate the Yoneda-class of an n-fold extension E of A by B we may splice E from an k-fold extension of X by B and an $n - k$-fold extension of A by X. It suffices therefore to prove that $\operatorname{Ext}^{k+1}(A, B) = 0$. Now we view a $(k + 1)$-fold extension of A by B as spliced from a simple extension

$$0 \to B \to H \to C \to 0$$

and a k-fold extension from A to C. We consider the connecting homomorphism $\operatorname{Ext}^k(A, C) \to \operatorname{Ext}^{k+1}(A, B)$ from the long exact sequence for $\operatorname{Hom}(A, -)$ with respect to the preceding short exact sequence. Since $\operatorname{Ext}^k(A, -)$ is right exact, this connecting homomorphism is zero. Now we apply this to the Yoneda class $f \in \operatorname{Ext}^k(A, C)$ of the second extension. If the Yoneda-class of the short exact sequence is e, the connecting homomorphism is given by taking the composition product with e (see loc. cit. I.8.8 in conjunction with XI.4.5]). But this gives the Yoneda class $e \cup f$ of the extension we started with. This class is therefore zero. □

A.3 Spectral Sequences and Filtrations

A.3.1 Filtrations

Let there be given an object A in some abelian category $\mathfrak{A}$. A *decreasing*, respectively *increasing*, filtration $F^\bullet A$, $F_\bullet A$, is a family of subobjects of A with $\cdots \subset F^{p+1}A \subset F^pA \subset F^{p-1}A \subset \cdots$, respectively $\cdots \subset F_{p-1}A \subset F_pA \subset F_{p+1}A \subset \cdots$. The **associated graded** (of a decreasing filtration) is defined by

$$\operatorname{Gr}_F A = \bigoplus_p \operatorname{Gr}_F^p A, \quad \operatorname{Gr}_F^p A = F^pA/F^{p+1}A.$$

In case of an *increasing* filtration W we write

$$\operatorname{Gr}^W A = \bigoplus_p \operatorname{Gr}_p^W A, \quad \operatorname{Gr}_p^W A = W_pA/W_{p-1}A.$$

We say that a morphism $f : (A, F^\bullet) \to (B, F^\bullet)$ between two filtered objects in $\mathfrak{A}$ is a a **filtered morphism** if for all $p \in \mathbb{Z}$ we have $f(F^pA) \subset F^pB$. It induces morphism $\operatorname{Gr}_F^p(f) : \operatorname{Gr}_F^p A \to \operatorname{Gr}_F^p B$ between the gradeds. Such a morphism f is **strict** if the natural map $\operatorname{Coim}(f) \to \operatorname{Im}(f)$ is an isomorphism of filtered complexes.

Example. If $\mathfrak{A}$ is a category of modules, strictness means:

$$F^p A \cap \operatorname{Im}(f) = f(F^p A) \subset F^p B. \tag{A–26}$$

Let $K^\bullet$ be a complex in $\mathfrak{A}$, A decreasing filtration of $K^\bullet$ is a family of subcomplexes $F^p K^\bullet$ such that $\cdots \subset F^{p+1}K^\bullet \subset F^p K^\bullet \subset F^{p-1}K^\bullet \subset \cdots$. Likewise for an increasing filtration. Such a filtration is called a **biregular filtration** if it is finite on every K^m.

Examples A.34. 1) The **trivial filtration** σ is the decreasing filtration, obtained by placing 0 in degrees $< p$ while keeping the K^m in all other degrees:

$$\sigma^{\geq p} = \{0 \to 0 \cdots \to 0 \to K^p \to K^{p+1} \to \cdots\}$$

The p-graded part is the complex with K^p in degree p and zero elsewhere:

$$\operatorname{Gr}^p_\sigma K^\bullet = K^p[-p].$$

2) The **canonical filtration** τ is the increasing filtration obtained from the truncation (A–5):

$$\tau_{\leq p} K^\bullet = \{\cdots \to K^{p-1} \to \operatorname{Ker}(d) \to 0 \to 0 \to \cdots\}.$$

Its p-graded part is the complex $0 \to K^{p-1}/\operatorname{Ker} d \to \operatorname{Ker}(d) \to 0$, quasi-isomorphic to the complex $H^p K^\bullet$ concentrated in degree p, i.e.

$$\operatorname{Gr}^\tau_p K^\bullet \xrightarrow[\sim]{\text{qis}} H^p(K^\bullet)[-p]. \tag{A–27}$$

We say that $f : (K^\bullet, F^\bullet) \to (L^\bullet, F^\bullet)$ is a **morphism of filtered complexes**, if it is a morphism of complexes preserving the filtration, i.e. $f(F^p K^\bullet) \subset F^p L^\bullet$. It induces a *morphism of complexes*

$$\operatorname{Gr}_F f : \operatorname{Gr}_F K^\bullet \to \operatorname{Gr}_F L^\bullet$$

between the associated gradeds. We speak of a **filtered quasi-isomorphism** if this is a quasi-isomorphism. Note that a filtered quasi-isomorphism in general need *not* be a quasi-isomorphism, but for a biregular filtration this is always the case. Let us now consider the cohomology $H^*(K^\bullet)$. If (A, F) is any filtered object in $\mathfrak{A}$, then by [Del71, Lemme 1.1.9] there is a canonical way to put a filtration on any subquotient of A. Since cohomology groups are subquotients (see (A–3)), given a filtered complex $(K^\bullet, F^\bullet)$, the filtration induced by F on K^p yields a canonical filtration on $H^p(K^\bullet)$. It is called the filtration on $H^*(K^\bullet)$ **induced by** $F^\bullet$. We want to compare it with the cohomology $H^*(\operatorname{Gr}^p_F K^\bullet)$ of the associated gradeds. One easily shows:

Lemma-Definition A.35 ([Del71, Lemme 1.1.11]). *If the differentials of a filtered complex* $(K^\bullet, F^\bullet)$ *are strict, then*

$$H^p(\operatorname{Gr}^q_F K^\bullet) \simeq \operatorname{Gr}^q_F H^p(K^\bullet).$$

We say that $(K^\bullet, F^\bullet)$ *is strict, or that the F-**filtration is strict**.*

If we have bi-filtered complexes equipped with a decreasing filtration $F^\bullet$ and an increasing filtration $W_\bullet$, the filtration induced by $F^\bullet$ on the W_k induces one on Gr^W and one can form the induced bigraded $\mathrm{Gr}_F \mathrm{Gr}^W$. A morphism respecting the two filtrations is called a **bi-filtered quasi-isomorphism** if the induced bi-graded is a quasi-isomorphism.

We recall (Example A.5.1) that the (bi-)filtered objects of a given abelian category $\mathfrak{A}$ in general only form the objects of an additive category $F\mathfrak{A}$. In this category the filtered quasi-isomorphisms form a multiplicative system [Hart69, Prop. 4.2 p.35]. This makes it possible to define the derived filtered category as follows.

Definition A.36. Let $\mathfrak{A}$ be an abelian category.

1) The **derived filtered category** $D^+F\mathfrak{A}$ has as its objects the bounded below filtered complexes in $\mathfrak{A}$ and its morphisms are the right fractions

$$(K^\bullet, F^\bullet) \overset{\text{qis}}{\underset{s}{\xleftarrow{\sim}}} \bullet \xrightarrow{a} (L^\bullet, F^\bullet)$$

where s is a filtered quasi-isomorphism and a a (homotopy class of) a filtered morphism. Two such right fractions are equivalent if a diagram as (A–13) exists where all the arrows respect the filtration.

2) Let $F^\bullet$, resp. $W_\bullet$ be a decreasing resp. increasing filtration. The **derived bi-filtered category** $D^+FW\mathfrak{A}$ has as its objects the bounded below bi-filtered complexes in $\mathfrak{A}$ and its morphisms are the right fractions

$$(K^\bullet, F^\bullet, W_\bullet) \overset{\text{qis}}{\underset{s}{\xleftarrow{\sim}}} \bullet \xrightarrow{a} (L^\bullet, F^\bullet, W_\bullet)$$

where s is a bi-filtered quasi-isomorphism and a is a (homotopy class of) a bi-filtered morphism. Two such right fractions are equal if a diagram as (A–13) exists where all the arrows respect both filtrations.

Next we treat the derived functors in the derived (bi)-filtered category. Start with an *object* A of $\mathfrak{A}$ equipped with a *finite* filtration F and $T : \mathfrak{A} \to \mathfrak{B}$ a left-exact functor. Since T is left-exact, the objects TF^pA are subobjects of TA and so we have a filtered object (TA, TF) in $\mathfrak{B}$. If $\mathrm{Gr}_F A$ is T-acyclic, also the F^pA are T-acyclic since they are successive extensions of T-acyclic objects. Similarly, if we have two filtrations F and W such that $\mathrm{Gr}_F \mathrm{Gr}^W A$ are T-acyclic, the objects F^pW_qA are T-acyclic. The same considerations hold for *complexes* of objects in $\mathfrak{A}$ equipped with a biregular filtration.

Definition A.37. A **filtered T-acyclic resolution** of a biregularly filtered complex $(K^\bullet, F)$ consists of a biregularly filtered complex $(L^\bullet, F)$ and a filtered quasi-isomorphism $i : (K^\bullet, F) \to (L^\bullet, F)$ such that the $\mathrm{Gr}_F^p L^\bullet$ are all T-acyclic.

For such a resolution, the filtered complex $(TL^\bullet, TF)$ has the property that the $\mathrm{Gr}_F^p L^q$ are also T-acyclic, yielding a T-acyclic resolution of the complex

$\mathrm{Gr}_F^p K^\bullet$. Also, since the $F^p L^q$ are T-acyclic, the following definition makes sense:

$$RT(K^\bullet, F) := (TL^\bullet, TFL^\bullet).$$

In case of biregularly bi-filtered complexes $(K^\bullet, F, W)$, one has to choose a bi-filtered quasi-isomorphism of the given complex onto a biregularly bi-filtered complex $(L^\bullet, F, W)$ such that all the $\mathrm{Gr}_F^p \mathrm{Gr}_m^W L^n$ are T-acyclic and one puts

$$RT(K^\bullet, F, W) := (TL^\bullet, TFL^\bullet, TWL^\bullet).$$

A.3.2 Spectral Sequences and Exact Couples

Fix an abelian category $\mathfrak{A}$.

Definition A.38. A **spectral sequence** starting at E_a consists of the following data.

1) A family $\{E_r^{p,q}\}, p, q \in \mathbb{Z}, r \in \mathbb{Z}, r \geq a$ of objects in $\mathfrak{A}$;
2) morphisms, the **differentials**

$$d_r : E_r^{p,q} \to E_r^{p+r,q-r+1}$$

such that $d_r \circ d_r = 0$;
3) an isomorphism

$$E_{r+1}^{p,q} \simeq H(E_r^{p-r,q+r-1} \xrightarrow{d_r} E_r^{p,q} \xrightarrow{d_r} E_r^{p+r,q-r+1}).$$

It should be clear what is meant by a **morphism of spectral complexes**.

An **exact couple** is a pair (D, E) of objects in $\mathfrak{A}$ together with an exact triangle

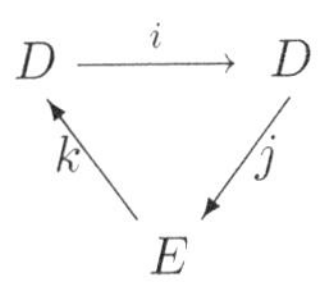

(A–28)

of morphisms such that each each vertex of the triangle exactness holds. The map $d = j \circ k : E \to E$ satisfies $d \circ d = 0$. Replacing D by $D_1 = i(D)$, E by $E_1 = H(d : E \to E)$ we get a new exact couple; i, j and k get replaced by the maps i_1, j_1 respectively k_1 defined by $i_1(i(x)) = i(x)$, $j_1(i(x)) = j(x)$, respectively $k_1[y] = k(y)$, $[y] \in E_1$ represented by $y \in \mathrm{Ker}(d)$. Putting $d_1 = j_1 \circ k_1 : E_1 \to E_1$, we continue in this way and we get the exact couple D_r, E_r with maps i_r, j_r, k_r.

Suppose next that D and E are bigraded, that i has degree $(-1, 1)$, j degree $(0, 0)$ and k degree $(1, 0)$, then one finds that i_r, k_r each have the same bidegrees for all r, while j_r has bidegree $(r, -r)$, forcing $d_r = j_{r-1} \circ k_{r-1}$, $r \geq 1$ to have bidegree $(r, 1-r)$. Instead of starting at $E = E_0$, one can likewise start at any degree $a \geq 0$, provided j has bidegree $(a, -a)$. We have [Weib, 5.9]:

Lemma A.39. *An exact couple (D, E) of bigraded objects in $\mathfrak{A}$ (A–28) where i, j, k have bidegrees $(-1, 1)$, $(a, -a)$, respectively $(1, 0)$ defines a spectral sequence starting with $E = E_{a+1}$. Morphisms between such bigraded exact couples induce morphisms of spectral sequences.*

A.3.3 Filtrations Induce Spectral Sequences

Let us start with a complex $K^\bullet$ in an abelian category equipped with a biregular decreasing filtration. The spectral sequence associated to such a filtration is defined by

$$\begin{aligned} Z_r^{p,q} &= \operatorname{Ker}\left(d : F^p K^{p+q} \to K^{p+q+1}/(F^{p+r} K^{p+q+1})\right) \\ B_r^{p,q} &= F^{p+1} K^{p+q} + d(F^{p-r+1} K^{p+q-1}) \\ E_r^{p,q} &= Z_r^{p,q}/(B_r^{p,q} \cap Z_r^{p,q}). \end{aligned}$$

This makes also sense for $r = \infty$. The fact that the filtration is biregular implies that, for p and q fixed, from a certain index r on we have

$$\begin{aligned} Z_r^{p,q} &= Z_{r+1}^{p,q} = \cdots = Z_\infty^{p,q} := \operatorname{Ker}(d : F^p K^{p+q} \to K^{p+q+1}) \\ B_r^{p,q} &= B_{r+1}^{p,q} = \cdots = B_\infty^{p,q} := F^{p+1} K^{p+q} + dK^{p+q-1} \end{aligned}$$

and so the $E_r^{p,q} = E_\infty^{p,q}$ from a certain index r on. For the first terms of the spectral sequence we have

$$\left.\begin{aligned} E_0^{p,q} &= \operatorname{Gr}_F^p(K^{p+q}) \\ E_1^{p,q} &= H^{p+q}(\operatorname{Gr}_F^p(K^\bullet)) \end{aligned}\right\} \qquad \text{(A–29)}$$

An easy calculation shows:

Lemma A.40. *The differential in $K^\bullet$ induces a homomorphism $d_1 : E_1^{p,q} \to E_1^{p+1,q}$ which can be identified with the connecting homomorphism*

$$H^{p+q}(\operatorname{Gr}_F^p K^\bullet) \to H^{p+q+1}(\operatorname{Gr}_F^{p+1} K^\bullet)$$

of the long exact sequence associated to

$$0 \to \operatorname{Gr}_F^{p+1} \to F^p/F^{p+2} \to \operatorname{Gr}_F^p \to 0.$$

More generally, the differential of the complex $K^\bullet$ induces homomorphisms

$$d_r : E_r^{p,q} \to E_r^{p+r,q-r+1}$$

which are the differentials of the spectral sequence. A short calculation indeed shows that E_{r+1} is the cohomology of the resulting complex so that the E_r form a spectral sequence. In fact it comes from an exact couple as follows.

Proposition A.41. *Consider the long exact sequence in cohomology for the sequence*

$$0 \to F^{p+1} K^\bullet \to F^p K^\bullet \to \operatorname{Gr}_F^p K^\bullet \to 0$$

in which we set $D^{p,q} = H^{p+q}(F^pK^\bullet)$ *and* $E^{p,q} = H^{p+q}(\mathrm{Gr}^p_F K^\bullet)$*:*

$$\cdots \to D^{p+1,q-1} \xrightarrow{i} D^{p,q} \xrightarrow{j} E^{p,q} \xrightarrow{k} D_1^{p+1,q} \to \cdots.$$

Then (D, E) *is a bigraded exact couple with* $d_1 = j \circ k$ *of bidegree* $(1, 0)$ *and the associated spectral sequence is the same as the spectral sequence for the filtration* F *on* $K^\bullet$*.*

Next, a computation shows that

$$E_\infty^{p,q} = \mathrm{Gr}^p_F H^{p+q}(K^\bullet),$$

where F is the filtration induced on cohomology:

$$F^p(H^n(K^\bullet)) = \mathrm{Im}(H^n(F^p(K^\bullet)) \xrightarrow{H^n(i)} H^n(K^\bullet)),$$
$$\text{with } i : F^p(K^\bullet) \hookrightarrow K^\bullet \text{ the inclusion.}$$

One summarizes this by saying that **the spectral sequence converges to the filtered cohomology of the complex** or that **the spectral sequence abuts to** $H^{p+q}(K^\bullet)$ or that $H^{p+q}(K^\bullet)$ is the **abutment** of the spectral sequence, and one writes

$$E_r^{p,q} \Longrightarrow H^{p+q}(K^\bullet).$$

If for all (p, q) one has $E_r^{p,q} = E_\infty^{p,q}$, we say that **the spectral sequence degenerates at** r. Moreover, the spectral sequence degenerates at E_r if and only if the differentials d_s vanish for all $s \geq r$.

Suppose that we have a finite first quadrant spectral sequence in the sense that the only non-zero terms $E_1^{p,q}$ occur when $0 \leq p \leq N$ and $0 \leq q \leq M$. We want to explain how certain **edge-homomorphisms** can be defined.

At the left edge $(0, q)$ there are no in-coming arrows and so

$$E_2^{0,q} = \mathrm{Ker}\{d_1 : E_1^{0,q} \to E_1^{1,q}\} \subset E_1^{0,q}$$

and similarly for the higher groups $E_k^{0,q}$. So, since in this case $E_\infty^{0,q}$ is a graded quotient of all of $H^q(K^\bullet)$ we get the first edge-homomorphism

$$e^q : H^q(K^\bullet) \twoheadrightarrow E_\infty^{0,q} \subset E_1^{0,q} = H^q(\mathrm{Gr}^0_F K^\bullet). \tag{A–30}$$

Similarly, looking at the right hand edge (N, q), we obtain a surjection $E_1^{N,q} \twoheadrightarrow E_\infty^{N,q}$, and since the latter is the smallest graded quotient, it is in fact a subspace of $H^N(K^\bullet)$. This yields the second edge-homomorphism:

$$f^q : H^q(\mathrm{Gr}^N_F K^\bullet) = E_1^{N,q} \twoheadrightarrow E_\infty^{N,q} \subset H^q(K^\bullet). \tag{A–31}$$

Lemma A.42. *Let* $(L^\bullet, F^\bullet)$ *be a biregularly filtered complex. Then the following are equivalent.*

1) *The F-filtration is strict (Lemma-Definition A.35).*
2) *For all m, k, the sequence*

$$0 \to H^k(F^m L^\bullet) \to H^k(L^\bullet) \to H^k(L^\bullet/F^m L^\bullet) \to 0$$

is exact.
3) *The spectral sequence for $(L^\bullet, F^\bullet)$ degenerates at E_1.*

Proof. Strictness is equivalent to the injectivity of the maps $H^k(F^m L^\bullet) \to H^k(L^\bullet)$, which is equivalent to 2). in view of the long exact sequence in cohomology associated to

$$0 \to F^m L^\bullet \to L^\bullet \to L^\bullet/F^m L^\bullet \to 0.$$

On the other hand, strictness of d implies that the natural morphisms

$$\begin{array}{ccc} E_\infty^{p,q} & & \\ \| & & \\ \dfrac{\{x \in F^p K^{p+q} \mid dx = 0\}}{[F^{p+1} + dK^{p+q-1}] \cap F^p K^{p+q}} & \hookrightarrow & \dfrac{\{x \in F^p K^{p+q} \mid dx \in F^{p+1} K^{p+q+1}\}}{[F^{p+1} + dK^{p+q-1}] \cap F^p K^{p+q}} \\ & & \| \\ \dfrac{\{x \in F^p K^{p+q} \mid dx \in F^{p+1} K^{p+q+1}\}}{[F^{p+1} K^{p+q} + d(F^p K^{p+q-1})] \cap F^p K^{p+q}} & \to & \dfrac{\{x \in F^p K^{p+q} \mid dx \in F^{p+1} K^{p+q+1}\}}{[F^{p+1} + dK^{p+q-1}] \cap F^p K^{p+q}} \\ \| & & \\ E_1^{p,q} & & \end{array}$$

are isomorphisms and so the spectral sequence degenerates at E_1. Conversely, if the spectral sequence degenerates at E_1, the maps d can be shown to be strict (see [Del71, Proposition 1.3.2]). □

The following result summarizes the effect of morphisms between complexes on spectral sequences. We omit the easy proofs which make use of the recursive nature of a spectral sequence.

Lemma A.43. *If $f : K^\bullet \to L^\bullet$ is a filtered homomorphism between complexes equipped with biregular filtrations, there are induced homomorphisms*

$$E_r^{p,q}(f) : E_r^{p,q}(K^\bullet) \to E_r^{p,q}(L^\bullet)$$

and if for some s the map $E_s^{p,q}(f)$ is an isomorphism for all p and q, then $E_r^{p,q}(f)$ is an isomorphism for $r \geq s$ and all p and q as well, and the spectral sequences have the same abutment.

If f is a filtered quasi-isomorphism, for $r \geq 1$ the maps $E_r^{p,q}(f)$ are isomorphisms.

Finally we want to introduce the two **spectral sequences of a double complex**

Definition A.44. Let $\{K^{\bullet\bullet}, d', d''\}$ be a double complex, i.e.

- $d' : K^{p,q} \to K^{p+1,q}$, $d' \circ d' = 0$,
- $d'' : K^{p,q} \to K^{p,q+1}$, $d'' \circ d'' = 0$,
- $d' \circ d'' + d'' \circ d' = 0$.

The associated **simple complex** is given by

$$s(K)^n = \bigoplus_{p+q=n} K^{p,q}$$
$$d = d' + d''.$$

Remark A.45. One sometimes uses the convention $d' \circ d'' = d'' \circ d'$. This calls for replacing d'' by $(-1)^p d''$ on $K^{p,q}$ so that the total derivative becomes $d = d' + (-1)^p d''$ on $K^{p,q}$. The alternation of signs is necessary in order to have $d \circ d = 0$. An instance where this happens is the Godement resolution (§ B.2.1).

A special case of a biregularly filtered complex arises for a first quadrant double complex, i. e. $K^{p,q} = 0$ whenever $p < 0$ or $q < 0$. In fact, in this setting the following two filtrations on $s(K)$ are biregular

$$'F^p = \bigoplus_{r \geq p} K^{r,s} \quad \textbf{(horizontal filtration)}$$
$$''F^q = \bigoplus_{s \geq q} F^{r,s}. \quad \textbf{(vertical filtration)}$$

and these induce then filtrations $'F$ and $''F$ on the associated simple complex. The associated spectral sequences are denoted by $'E_r^{p,q}$ and $''E_r^{p,q}$. Writing out the first two terms using (A–29) gives:

$$\left.\begin{aligned} &'E_1^{p,q} = H^q(K^{p,\bullet}, d'') \quad ''E_1^{p,q} = H^q(K^{\bullet,p}, d') \\ &'E_2^{p,q} = H^p(H^q(K^{\bullet\bullet}, d''), d') \quad ''E_2^{p,q} = H^p(H^q(K^{\bullet,\bullet}, d'), d'') \end{aligned}\right\} \tag{A–32}$$

A.3.4 Derived Functors and Spectral Sequences

We start with two abelian categories $\mathfrak{A}$ and $\mathfrak{B}$ having enough injectives and a left exact functor $T : \mathfrak{A} \to \mathfrak{B}$. A crucial rôle is played by the two spectral sequences for derived functors:

Lemma-Definition A.46. There are **two spectral sequences for derived functors** given by

$$'E_1^{pq} = R^q T(K^p) \Longrightarrow R^{p+q} T(K^\bullet)$$
$$''E_2^{pq} = R^p T(H^q(K^\bullet)) \Longrightarrow R^{p+q} T(K^\bullet)$$

Proof. Choose an injective resolution for each K^p, say $I^{p,0} \to I^{p,1} \to \cdots$. By the remarks made in Example A.23 the derivatives $d^p : K^p \to K^{p+1}$ can be extended to a map between complexes $d^p : I^{p,\bullet} \to I^{p+1,\bullet}$ (unique up to homotopy) and we can use the homotopy to arrange for $d^{p+1} \circ d^p = 0$ so that we get a double complex $I^{p,q}$. A calculation then shows that the associated simple complex is quasi-isomorphic to $K^\bullet$. Now apply T and consider two spectral sequences for the double complex $T(I^{\bullet\bullet})$ from § A.3.3. Writing this out in this case gives precisely the two spectral sequences above. □

Corollary A.47. *Let $K^\bullet$ be a resolution of K. Then there is a canonical identification*

$$R^pT(K) = R^pT(K^\bullet).$$

If moreover $K^\bullet$ is a T-acyclic resolution of K then is a canonical identification

$$H^p(TK^\bullet) = R^pT(K^\bullet).$$

Proof. If $K^\bullet$ is any resolution of K, the second spectral sequence degenerates at E_2 since the E_2-term consists only of the terms $''E_2^{n,0} = R^nT(H^0(K^\bullet)) = R^nT(K)$. If moreover K^p is T-acyclic for all p, the first spectral sequence degenerates at $'E_2^{p,0} = H^p(TK^\bullet)$ and hence there is a canonical identification $H^p(TK^\bullet) = R^pT(K^\bullet)$.

Example A.48. Recall (A–19) that we defined hypercohomology as the derived global section functor. The two spectral sequences for this functor read

$$E_1^{p,q} = H^p(X, \mathcal{K}^q) \implies \mathbb{H}^{p+q}(X, \mathcal{K}^\bullet) \tag{A–33}$$

$$E_2^{p,q} = \mathbb{H}^p(X, H^q(\mathcal{K}^\bullet)) \implies \mathbb{H}^{p+q}(X, \mathcal{K}^\bullet). \tag{A–34}$$

Spectral sequences for the derived functors can be obtained directly by using the canonical and trivial filtrations on $K^\bullet$:

Lemma A.49. 1) *The first spectral sequence for the derived functor of T comes from the trivial filtration, i.e*

$$'E_r(TK^\bullet) = E_r(T(K^\bullet, \sigma)).$$

2) *The second spectral sequence comes from the canonical filtration up to a shift:*

$$''E_{r+1}^{2p+q,-p}(TK^\bullet) = E_r^{p,q}(T(K^\bullet, \tau)).$$

Proof. For the trivial filtration we have

$$E_1^{p,q}(T(K^\bullet, \sigma)) = R^qT(K^p) \Longrightarrow R^{p+q}T(K^\bullet),$$

which is just the first spectral sequence in hypercohomology.

The canonical filtration is an increasing filtration and so we have to change p by $-p$ to obtain a decreasing filtration. When we do this, the p-th graded complex is a complex quasi-isomorphic to a complex having zeroes everywhere except in degree $-p$, where the $-p$-th cohomology $H^{-p}(K^\bullet)$ is located. Since the derivatives d_r are induced by d, it follows that

$$E_1^{p,q}T(K^\bullet,\tau)) = R^{2p+q}T(H^{-p}(K^\bullet)) \Longrightarrow R^{p+q}T(K^\bullet),$$

which is indeed the second spectral sequence up to renumbering $E_r^{p,q} \mapsto E_{r+1}^{2p+q,-p}$. $\square$

Remark A.50. Deligne [Del71, 1.3] explains the renumbering from the point of view of filtrations as follows. Starting from a given filtered complex $(K^\bullet, F^\bullet)$, one introduces the **backshifted filtered complex** $(K^\bullet, (\mathrm{Dec}\, F)^\bullet)$ by

$$(\mathrm{Dec}\, F)^p K^n = \{x \in F^{p+n}K^n \mid dx \in F^{p+n+1}K^{n+1}\}.$$

The right hand side is $Z_1^{p+n,-p}$ and by d it is mapped to $Z_1^{p+n+1,-p}$ and so one obtains a filtered complex.

As an example, if S is the simplest non-trivial filtration with $\mathrm{Gr}_S^i K = 0$ for $i \neq 0$, its backshifted filtered complex is the canonical decreasing filtration $\mathrm{Dec}(S)^p K = \tau_{\leq -p}K$.

To compare the associated spectral sequence with the spectral sequence of the original filtered complex, observe that the group $Z_1^{p+1+n,-p-1}$ is contained in $F^{p+1+n}K^n \subset B_1^{p+n,-p} \subset Z_1^{p+n,-p}$ and so there are natural morphisms

$$E_0^{p,n-p}(\mathrm{Dec}(F)) = Z_1^{p+n,-p}/Z_1^{p+1+n,-p-1} \to Z_1^{p+n,-p}/B_1^{p+n,-p} = E_1^{p+n,-p}(F).$$

One verifies that these form morphisms of graded complexes that are isomorphisms in cohomology and then inductively give isomorphisms

$$E_r^{p,n-p}(K, \mathrm{Dec}(F)) \xrightarrow{\sim} E_{r+1}^{p+n,-p}(K, F).$$

So, if $I^{p,\bullet}$ is an injective resolution of K^p, we have $''E_{r+1}^{2p+q,-p}(TK^\bullet) = E_r^{pq}(s(I^{\bullet,\bullet}), \mathrm{Dec}(''F))$.

B

Algebraic and Differential Topology

B.1 Singular (Co)homology and Borel-Moore Homology

We review the basic definitions and properties of singular (co)-homology referring to [Greenb], [Span] and [Hatch] for the details.

B.1.1 Basic Definitions and Tools

Let R be a ring and X a topological space. A **singular q-simplex** is a continuous map $\Delta_q \to X$. The free R-module $S_q(X;R)$ generated by the singular q-simplices fits in a chain-complex $S_\bullet(X;R)$ whose q-th homology group $H_q(X;R)$ by definition is the q-th **singular homology group**. The module $S^q(X;R) := \mathrm{Hom}_R(S_q(X;R), R)$ fits dually in a cochain complex whose q-th cohomology $H^q(X;R)$ is the q-th **singular cohomology group**. The homology functors are covariant on the category of topological spaces with continuous maps, while cohomology behaves contravariantly. For any continuous map $f : X \to Y$, the induced maps are denoted $f_* : H_q(X;R) \to H_q(Y;R)$ and $f^* : H^q(Y;R) \to H^q(X;R)$. It follows that the (co)homology modules are topological invariants. Even more is true: homotopic maps induce homotopic maps in the associated singular (co)chain complexes and hence induce the same maps in (co)homology. It follows that spaces with the same homotopy type have the same (co)homology.

Let us next introduce the relative (co)homology groups for the pair (X, A) where $A \subset X$. Note that $S_\bullet(A;R)$ is a subcomplex of $S_\bullet(X;R)$ and dually $S^\bullet(A;R)$ is a quotient of $S^\bullet(X;R)$. Set

$$S_\bullet(X, A; R) := S_\bullet(X;R)/S_\bullet(A;R)$$
$$S^\bullet(X, A; R) := \mathrm{Ker}(S^\bullet(X;R) \to S^\bullet(A;R)).$$

Then $H_q(X, A; R)$ is the q-the homology of the former chain complex and $H^q(X, A; R)$ is the cohomology of the latter cochain complex. There are the usual long exact sequences associated to the two defining sequences. For instance, we have the **exact sequence in cohomology** for the pair (X, A):

$$\cdots H^q(X, A; R) \to H^q(X; R) \to H^q(A; R) \xrightarrow{\delta^q} H^{q+1}(X, A; R) \cdots . \quad \text{(B–1)}$$

To introduce compactly supported cohomology we need to explain the phrase $f \in S^q(X; R)$ *has support in* A, where A is closed in X. It means that f vanishes on the chains from the submodule $S_q(X - A; R) \subset S(X; R)$. Those cochains form the R-module $S^q(X, X - A; R)$. Varying A over all compact sets, we get a direct system and the limit is the module of compactly supported q-cochains

$$S^q_c(X; R) := \varinjlim S^q(X, X - K), \quad K \text{ compact}$$

with cohomology groups $H^\bullet_c(X; R)$. Dually we have

$$S^{\text{BM}}_q(X; R) := \varprojlim S_q(X, X - K), \quad K \text{ compact},$$

the **Borel-Moore** q-chains with homology groups $H^{\text{BM}}_\bullet(X; R)$. For a pair (X, A) one has the notion of compactly supported relative q-cochains yielding the group $H^q_c(X, A; R)$, and, dually $H^{\text{BM}}_q(X, A; R)$. Cohomology with compact support and Borel-Moore homology are topological invariants, but not homotopy invariants.

Remark B.1. 1) One has $H^q_c(X; R) = \varinjlim H^q(X, X - K; R)$ and dually $H^{\text{BM}}_q(X; R) = \varprojlim H_q(X, X - K; R)$.

2) Borel-Moore homology was originally defined using sheaves [Bor-M], and only for locally compact spaces of finite dimension (see § 13.1.1), as follows:

$$H^{\text{BM}}_q(X; R) = \mathbb{H}^{-q}(X, {}^{\text{Ve}}\mathbb{D}\underline{R}_X).$$

3) Recall that for the one-point compactification $X^* = X \cup \{\infty\}$ the open neighbourhoods of ∞ are the complements of the compact sets in X to which the extra point ∞ has been added. For locally compact Hausdorff X one may define Borel-Moore homology as

$$H^{\text{BM}}_q(X; R) = H_q(X^*, \infty; R). \quad \text{(B–2)}$$

4) If X is a paracompact space whose one-point compactification X^* is a finite CW-complex one can show that the three definitions are the same. For details see [Fult, Example 19.1.1]. This applies to complex algebraic varieties. In the more general situation where X is embeddable as a *closed* subset of $\mathbb{R}^N$ the preceding definitions can be replaced by one in terms of singular *cohomology*:

$$H^{\text{BM}}_q(X; R) = H^{N-q}(\mathbb{R}^N, \mathbb{R}^N - X; R). \quad \text{(B–3)}$$

This applies to complex analytic spaces, even if they are not algebraic or compatifyiable in the sense that they embed in a compact analytic space with complement a finite number of irreducible proper analytic subvarieties.

By definition, for proper maps $f : X \to Y$ the inverse image of a compact subset L of Y is compact. By remark B.1.1) it then follows that there are induced maps

$$f_* : H_q^{\mathrm{BM}}(X;R) \to H_q^{\mathrm{BM}}(Y;R)$$

so that Borel-Moore homology is a covariant functor from locally compact spaces equipped with proper maps to abelian groups.

By the excision theorem stated below (Theorem B.2), if $U \subset X$ is open and $K \subset U$ compact, the inclusion $(U, U-K) \hookrightarrow (X, X-K)$ induces isomorphisms in homology. The canonical projection maps

$$\lim_{\leftarrow} H_q(X, X-K : R) \to H_q(X, X-K; R) \quad \text{(limit over all compacts } K \subset X\text{)}$$

followed by the inverses $H_q(X, X-K; R) \to H_q(U, U-K; R)$ of the excision isomorphisms induce contravariant restrictions in the limit, this time taken over compacts $K \subset U$:

$$j^* : H_q^{\mathrm{BM}}(X;R) \to H_q^{\mathrm{BM}}(U;R).$$

Under suitable hypotheses on X, for instance if X is embeddable as a closed subset of $\mathbb{R}^N$, using definition (B–3), one shows [Fult, Example 19.1.1] that for a closed immersion $i : Z \hookrightarrow X$ and $j : U = X - Z \hookrightarrow X$ the complementary inclusion one has an exact sequence

$$\begin{aligned} \cdots \to H_{q+1}^{\mathrm{BM}}(U;R) \to H_q^{\mathrm{BM}}(Z;R) &\xrightarrow{j_*} H_q^{\mathrm{BM}}(X;R) \\ &\xrightarrow{i^*} H_q^{\mathrm{BM}}(U;R) \to \cdots . \end{aligned} \tag{B–4}$$

For a point P, the only homology is in degree 0, generated by the singular 0-cycle $[P] := \{\Delta_0 \to P\}$. Likewise for the cohomology of a point. The homotopy property then implies that for any path connected space X, $H_0(X;R)$ is generated by the class of $[P]$, where $P \in X$ is any point, and similarly for $H^0(X;R)$. Introducing the constant map

$$a_X : X \to P$$

this motivates the introduction of **reduced (co)homology**

$$\begin{aligned} \tilde{H}_q(X;R) &= \mathrm{Ker}((a_X)_* : H_q(X;R) \to H_q(P;R)) \\ \tilde{H}^q(X;R) &= \mathrm{Coker}((a_X)^* : H^q(P;R) \to H^q(X;R)). \end{aligned}$$

It is customary to drop the coefficient ring R from the notation if $R = \mathbb{Z}$ and so

$$\begin{aligned} H_q(X,A) &= H_q(X,A;\mathbb{Z}) & H^q(X,A) &= H^q(X,A;\mathbb{Z}), \\ H_q^{\mathrm{BM}}(X) &= H_q^{\mathrm{BM}}(X;\mathbb{Z}) & H_c^q(X) &= H_c^q(X;\mathbb{Z}). \end{aligned}$$

The universal coefficient theorem [Greenb, 29.12] gives a recipe to determine (co)-homology with coefficients in any commutative ring R (with unit) from the knowledge of the groups $H^q(X)$, $q = 0, 1, \ldots$. We shall only need the following special case.

Theorem (UNIVERSAL COEFFICIENT THEOREM). *When R is a field, there is a natural isomorphism*

$$H_q(X) \otimes R \xrightarrow{\sim} H_q(X; R). \tag{B–5}$$

To calculate relative cohomology one frequently uses the following property.

Theorem B.2 (EXCISION THEOREM). *Let (X, A) be a pair and let $Z \subset X$ such that its closure is contained in the interior of A, then the excision map $(X - Z, A - Z) \subset (X, Z)$ induces isomorphisms in (co)homology.*

We also need the variant given in [Span, Thm. 4.8.9]:

Theorem B.3 (STRONG EXCISION THEOREM). *Let (X, Z) be a compact Hausdorff pair, Y Hausdorff and T closed in Y. Suppose that Z is a strong deformation retract of a closed neighbourhood G of Z in X (i.e. with $j : Z \hookrightarrow X$ the inclusion, there is a retraction $r : G \to Z$ such that $j \circ r$ is homotopic to the identity by a homotopy that point-wise fixes Z). If, moreover there exists a continuous map $f : (X, Z) \to (Y, T)$ restricting to a homeomorphism from $X - Z$ to $Y - T$, then $f^* : H^q(Y, T; R) \xrightarrow{\sim} H^q(X, Z; R)$ (and similarly in homology).*

For calculations the Mayer-Vietoris sequence is useful. First we need a definition.

Definition B.4. Two subsets X_1, X_2 of a topological space form an **excisive couple** if excision induces an isomorphism $H^*(X_1 \cup X_2, X_2; R) \xrightarrow{\sim} H^*(X_1, X_1 \cap X_2; R)$ (or, equivalently $H^*(X_1 \cup X_2, X_1; R) \xrightarrow{\sim} H^*(X_2, X_1 \cap X_2; R)$).

Examples B.5. 1) If X is the union of the interiors of X_1 and X_2, the couple (X_1, X_2) is excisive.

2) Two subcomplexes of a CW-complex form an excisive couple. We apply this to subvarieties X_1, X_2 inside a complex variety X. This follows from the (non-trivial) fact that X can be triangulated in such a way that X_1 and X_2 are subcomplexes. See e.g. [Hir74] and the references given therein.

Theorem B.6 (MAYER-VIETORIS SEQUENCE). *For an excisive couple X_1, X_2 let $i_k : X_k \to X_1 \cup X_2$ and $j_k : X_1 \cap X_2 \to X_k$ be the inclusion maps, $k = 1, 2$. Then there is a long exact sequence*

$$\cdots \to H^q(X_1 \cup X_2; R) \xrightarrow{(i_1^*, i_2^*)} H^q(X_1; R) \oplus H^q(X_2; R) \xrightarrow{j_1^* - j_2^*} H^q(X_1 \cap X_2; R)$$
$$\to H^{q+1}(X_1 \cup X_2; R) \to \cdots.$$

B.1.2 Pairings and Products

The tautological pairing

$$S^q(X, A; R) \times S_q(X, A; R) \to R$$

is compatible with boundary and coboundary and defines the **Kronecker pairing**

$$\begin{aligned} H^q(X, A; R) \times H_q(X, A; R) \to R \\ H_c^q(X, A; R) \times H_q^{\mathrm{BM}}(X, A; R) \to R \end{aligned} \tag{B–6}$$

This pairing will be denoted by $\langle\, , \,\rangle$, so that

$$\langle [f], [c] \rangle = f(c) \quad f \text{ a } q\text{-cocycle, } c \text{ a } q\text{-cycle,}$$

and the square brackets denote the corresponding classes in (co)-homology. It induces the **Kronecker homomorphisms**

$$\begin{aligned} H^q(X, A; R) &\to \mathrm{Hom}_R(H_q(X, A; R), \\ H_c^q(X, A; R) &\to \mathrm{Hom}_R(H_q^{\mathrm{BM}}(X, A); R) \end{aligned}$$

which are isomorphisms if R is a field.

Recalling the definition (A–2) of the tensor product of complexes, we introduce the **Alexander-Whitney homomorphism**

$$\left.\begin{aligned} S_n(X \times Y; R) &\xrightarrow{A} [S(X; R) \otimes S(Y; R)]_n \\ (\sigma, \tau) &\mapsto (\text{front } p\text{-face of } \sigma) \times (\text{back } (n-p)\text{-face of } \tau) \end{aligned}\right\} \tag{B–7}$$

which can be used to define a product on $S^\bullet(X; R)$ as follows. Let $\Delta : X \hookrightarrow X \times X$ be the diagonal. The **cup-product** is defined as the composition

$$\bigcup : [S^\bullet(X; R) \otimes S^\bullet(X; R)] \xrightarrow{{}^tA} S^\bullet(X \times X; R) \xrightarrow{S^\bullet(\Delta)} S^\bullet(X; R).$$

Obviously, we have the relation

$$\partial(f \cup g) = \partial f \cup g + (-1)^p f \cup \partial g$$

which shows that cup-product induces a ring structure on the cohomology

$$H^*(X; R) = \bigoplus_q H^q(X; R)$$

with unit $1 \in H^0(X; R)$ given by the constant cochain $x \mapsto 1$, $x \in X$. The cup-product can be shown to be (graded) commutative:

$$a \cup b = (-1)^{pq} b \cup a, \quad a \in H^p(X; R),\ b \in H^q(X; R).$$

If $f : X \to Y$ is continuous, the induced homomorphism $f^* : H^*(Y; R) \to H^*(X; R)$ preserves cup-products.

If we evaluate cup-product on $\sigma \otimes \tau$, where σ is any cochain, but τ a cochain with compact support, the result is a cochain with compact support so that we also get induced cup-products

$$H^p(X;R) \otimes H^q_c(X;R) \to H^{p+q}_c(X;R). \tag{B–8}$$

Dual to cup-products, we have cap-products

$$\begin{aligned} &S_{p+q}(X;R) \times S^p(X;R) \xrightarrow{\cap} S_q(X;R) \\ &g(c \cap f) = (f \cup g)(c), \quad f \in S^p(X;R), g \in S^q(X;R), c \in S_{p+q}(X;R) \end{aligned}$$

inducing

$$\bigcap : H_{p+q}(X;R) \otimes H^p(X;R) \to H_q(X;R) \tag{B–9}$$

$$\bigcap : H^{\mathrm{BM}}_{p+q}(X;R) \otimes H^p(X;R) \to H^{\mathrm{BM}}_q(X;R). \tag{B–10}$$

We close this subsection with a recipe to calculate the cohomology of a product:

Theorem B.7 (Künneth formula). *For R a field there is a natural isomorphism*

$$H^n(X \times Y;R) \cong \bigoplus_{p=0}^{n} H^p(X;R) \otimes H^{n-p}(Y;R).$$

B.2 Sheaf Cohomology

We refer to [Gode] and [Iver] for more details for proofs of the results in this section.

B.2.1 The Godement Resolution and Cohomology

Let $\mathcal{F}$ be a sheaf of abelian groups on a topological space X. A "discontinuous section" over an open subset U of X consists of a collection of germs $\{a_x \in \mathcal{F}_x \mid x \in U\}$. The set of all such sections is denoted $\mathcal{C}^0_{\mathrm{Gdm}}\mathcal{F}(U)$. Varying U we obtain a presheaf $\mathcal{C}^0_{\mathrm{Gdm}}\mathcal{F}$ which is in fact a sheaf. By definition it comes equipped with an injective homomorphism $\mathcal{F} \hookrightarrow \mathcal{C}^0_{\mathrm{Gdm}}\mathcal{F}$. Following [Gode, II.4.3] one inductively defines

$$\begin{aligned} \mathcal{Z}^0\mathcal{F} &= \mathcal{F} \\ \mathcal{Z}^p\mathcal{F} &:= \mathcal{C}^{p-1}_{\mathrm{Gdm}}\mathcal{F}/\mathcal{Z}^{p-1}\mathcal{F} \\ \mathcal{C}^p_{\mathrm{Gdm}}\mathcal{F} &:= \mathcal{C}^0_{\mathrm{Gdm}}\left(\mathcal{C}^{p-1}_{\mathrm{Gdm}}\mathcal{F}/\mathcal{Z}^{p-1}\mathcal{F}\right). \end{aligned}$$

The sheaves $\mathcal{C}^p_{\mathrm{Gdm}}\mathcal{F}$ are **flabby**, i.e. any section over an open set extends to the entire space X. The natural maps $d : \mathcal{C}^p_{\mathrm{Gdm}}\mathcal{F} \to \mathcal{Z}^{p+1}\mathcal{F} \hookrightarrow \mathcal{C}^{p+1}_{\mathrm{Gdm}}\mathcal{F}$ are the coboundary maps of a resolution

$$\mathcal{F} \to \mathcal{C}^0_{\mathrm{Gdm}}(\mathcal{F}) \xrightarrow{d} \mathcal{C}^1_{\mathrm{Gdm}}(\mathcal{F}) \xrightarrow{d} \cdots$$

of the sheaf $\mathcal{F}$, by definition the **Godement resolution** of $\mathcal{F}$. We define:

$$H^p(X, \mathcal{F}) := H^p(\Gamma(X, \mathcal{C}^\bullet_{\mathrm{Gdm}}\mathcal{F})).$$

From the definition of the Godement resolution it follows that any morphism of sheaves $f : \mathcal{F} \to \mathcal{G}$ induces a morphism of complexes $\mathcal{C}^\bullet_{\mathrm{Gdm}}(f)$ between the respective Godement resolutions. Moreover, for two such morphisms f and g, we have:

$$\mathcal{C}^\bullet_{\mathrm{Gdm}}(f \circ g) = \mathcal{C}^\bullet_{\mathrm{Gdm}}(f) \circ \mathcal{C}^\bullet_{\mathrm{Gdm}}(g).$$

If $f : \mathcal{F} \to \mathcal{G}$ is a morphism of sheaves on X, the induced morphism $\mathcal{C}^\bullet_{\mathrm{Gdm}}(f)$ induce maps $H^q(f) : H^q(X, \mathcal{F}) \to H^q(X, \mathcal{G})$; these behave functorially, i.e. sheaf cohomology is a contra-variant functor on sheaves of abelian groups on X.

Secondly, an exact sequence of sheaves of R-modules $0 \to \mathcal{F} \xrightarrow{i} \mathcal{G} \xrightarrow{j} \mathcal{H} \to 0$ induces a short exact sequence on the level of their Godement resolutions and hence a long exact sequence

$$\ldots H^q(X, \mathcal{F}) \xrightarrow{H^q(i)} H^q(X, \mathcal{G}) \xrightarrow{H^q(j)} H^q(X, \mathcal{H}) \longrightarrow H^{q+1}(X, \mathcal{F}) \ldots \qquad \text{(B–11)}$$

Next, we pass to a *complex of sheaves* $\mathcal{F}^\bullet$ on X which is bounded below. For every $\mathcal{F}^p$ take its Godement resolution $\mathcal{C}^\bullet_{\mathrm{Gdm}}\mathcal{F}^p$. The differentials $d^p : \mathcal{F}^p \to \mathcal{F}^{p+1}$ induce maps of complexes $d^p : \mathcal{C}^\bullet_{\mathrm{Gdm}}\mathcal{F}^p \to \mathcal{C}^\bullet_{\mathrm{Gdm}}\mathcal{F}^{p+1}$ and by functoriality, $d^{p+1} \circ d^p = 0$ so that we have a double complex $\mathcal{C}^\bullet_{\mathrm{Gdm}}\mathcal{F}^\bullet$. Since the Godement sheaves are flabby, the associated simple complex $s(\mathcal{C}^\bullet_{\mathrm{Gdm}}\mathcal{F}^\bullet)$ (a finite sum since $\mathcal{F}^q = 0$ for $q \ll 0$) gives a flabby resolution of $\mathcal{F}^\bullet$. Its complex of global sections is called the **derived complex**

$$R\Gamma(X, \mathcal{F}^\bullet) := \Gamma(X, s[\mathcal{C}^\bullet_{\mathrm{Gdm}}\mathcal{F}^\bullet]). \qquad \text{(B–12)}$$

This complex computes the hypercohomology (see e.g. [Gode, Example II, 7.2.1] where the case of a single sheaf is explained):

Lemma B.8. *Let X be a topological space and let $\mathcal{F}^\bullet$ be a bounded below complex of sheaves on X. There is a canonical identification*

$$\mathbb{H}^p(X, \mathcal{F}^\bullet) := R^p\Gamma(\mathcal{F}^\bullet) = H^p(R\Gamma(\mathcal{F}^\bullet)) = H^p(\Gamma(X, s[\mathcal{C}^\bullet_{\mathrm{Gdm}}\mathcal{F}^\bullet])).$$

We finish this section with the relation of the Godement resolution to tensor products. Start with two sheaves $\mathcal{F}$ and $\mathcal{G}$ on X. There are natural morphisms $\mathcal{Z}^p(\mathcal{F}) \otimes \mathcal{Z}^q(\mathcal{G}) \to \mathcal{Z}^{p+q}(\mathcal{F} \otimes \mathcal{G})$ and $\mathcal{C}^p_{\mathrm{Gdm}}(\mathcal{F}) \otimes \mathcal{C}^q_{\mathrm{Gdm}}(\mathcal{G}) \to \mathcal{C}^{p+q}_{\mathrm{Gdm}}(\mathcal{F} \otimes \mathcal{G})$ which can be constructed inductively with respect to the total degree $p + q$. The Godement resolutions are bounded below complexes and one can form the tensor product of two such complexes. The morphisms we just constructed are compatible with differentials in an obvious way and hence we get a natural map of *complexes*

$$\mathcal{C}^\bullet_{\mathrm{Gdm}}(\mathcal{F}) \otimes \mathcal{C}^\bullet_{\mathrm{Gdm}}(\mathcal{G}) \to \mathcal{C}^\bullet_{\mathrm{Gdm}}(\mathcal{F} \otimes \mathcal{G}).$$

This also works for complexes, whence a natural map of complexes

$$R\Gamma(\mathcal{F}^\bullet) \otimes R\Gamma(\mathcal{G}^\bullet) \to R\Gamma(\mathcal{F}^\bullet \otimes \mathcal{G}^\bullet)]. \tag{B–13}$$

B.2.2 Cohomology and Supports

The **support** of a section s of a sheaf $\mathcal{F}$ of abelian groups on any topological space X is the (closed) set of points $x \in X$ such that the germ s_x of the section at x does not vanish.

Definition B.9. A **family of supports** is a collection Φ of closed subsets of X such that

a) whenever two members belong to Φ, their union belongs to Φ;
b) any closed subset of a member of Φ belongs to Φ.

If moreover all members of Φ are paracompact, and we have

c) any member of Φ admits a (paracompact) neighbourhood belonging to Φ,

we say that Φ is **paracompactifying**. The group of (global) sections of a sheaf $\mathcal{F}$ of abelian groups, having support in a family of supports Φ is denoted $\Gamma_\Phi(X, \mathcal{F})$.

We recall that X is **paracompact** if every open covering of X admits a locally finite refinement.

For a family of supports (not necessarily paracompactifying) one can introduce cohomology with supports generalizing what we did before without supports (Prop. B.8):

$$\mathbb{H}^p_\Phi(X, \mathcal{F}^\bullet) = R^p\Gamma_\Phi(\mathcal{F}^\bullet) = H^p(\Gamma_\Phi(X, s[\mathcal{C}^\bullet_{\mathrm{Gdm}}(\mathcal{F}^\bullet)])).$$

Examples B.10. 1) If we take the family of all closed subsets of X we get back the usual cohomology groups;
2) If X is Hausdorff, compact subsets of X are closed and give a family of supports. So for X Hausdorff we may define

$$\mathbb{H}^p_c(X, \mathcal{F}^\bullet) := H^p(\Gamma_c(X, s[\mathcal{C}^\bullet_{\mathrm{Gdm}}(\mathcal{F}^\bullet)])).$$

In general, the compacts do not form a paracompactifying family, but they do if in addition X is paracompact and locally compact.

An exact sequence of complexes of sheaves $0 \to \mathcal{F}^\bullet \xrightarrow{i} \mathcal{G}^\bullet \xrightarrow{j} \mathcal{H}^\bullet \to 0$ gives rise to a corresponding long exact sequence

$$\cdots \mathbb{H}^q_\Phi(X, \mathcal{F}^\bullet) \xrightarrow{\mathbb{H}^q(i)} \mathbb{H}^q_\Phi(X, \mathcal{G}^\bullet) \xrightarrow{\mathbb{H}^q(j)} \mathbb{H}^q_\Phi(X, \mathcal{H}^\bullet) \to \mathbb{H}^{q+1}_\Phi(X, \mathcal{F}^\bullet) \cdots \tag{B–14}$$

provided Φ is *paracompactifying*. It generalizes (B–11).

We apply this to one important example. Let $Z \subset X$ be a closed subset of a Hausdorff space, and let $U = X - Z$ the open complement and $\mathcal{F}$ any sheaf on X. Then the restriction $\mathcal{F}|U$ extended by zero over Z is a subsheaf of $\mathcal{F}$ with quotient sheaf $\mathcal{F}|Z$. Taking for Φ the set of compact subsets of X, we remarked that it is paracompactifying if X is paracompact and locally compact. The sequence (B–14) in this case gives:

$$\cdots H^q_c(U, \mathcal{F}|U) \to H^q_c(X, \mathcal{F}) \to H^q_c(Z, \mathcal{F}|Z) \to H^{q+1}_c(U, \mathcal{F}|U) \to \cdots . \quad \text{(B–15)}$$

Singular cohomology with R-coefficients may well be different from the corresponding cohomology in the constant sheaf. We need to make certain assumptions on the topology of X.

Proposition B.11. *Let X be a Hausdorff space such that every open subset is paracompact. Assume moreover that X is locally contractible. Then there are natural isomorphisms*

$$H^q(X; R) \xrightarrow{\sim} H^q(X, \underline{R}_X), \quad q = 0, 1, \ldots .$$

If, moreover X is locally compact, there are natural isomorphisms

$$H^q_c(X; R) \xrightarrow{\sim} H^q_c(X, \underline{R}_X).$$

Sketch of Proof. We first sketch the argument for ordinary cohomology. By [Gode, Example II.3.91] paracompactness for X and $U \subset X$ implies that the sheaf $\widehat{S^q}$ associated to the presheaf S^q defined by $U \mapsto S^q(U; R)$ is flabby. Moreover, since X is locally contractible, $\widehat{S^q}$ is a resolution of the constant sheaf $\underline{R}_X$. Hence, by [Gode, Exemple II, 7.2.1] the complex $\widehat{S^\bullet}(X)$ of its global sections computes the cohomology of $\underline{R}_X$.

Since the above presheaf is not a sheaf, in general $\widehat{S^q}(X) \neq S^q(X; R)$. Instead we have $\widehat{S^q}(X) = S^q(X; R)/S^q_0(X; R)$ where $S^q_0(X; R)$ are those cochains f which are zero on all elements of a suitable open cover of X (which may depend on f). Then $S^\bullet_0(X; R)$ is a subcomplex of the singular complex and it suffices to show that this complex is acyclic. By definition we have an open cover $\mathfrak{U} = \{U_j\}$ of X such that $f|U_j = 0$. The R-submodule $S^{\mathfrak{U}}_q(X; R) \subset S_q(X; R)$ consisting of chains on singular simplices with image in one of the U_j consists of so-called $\mathfrak{U}$-*small* q-*chains*; dually we have $S^q_{\mathfrak{U}}(X; R) = \mathrm{Hom}_R(S^{\mathfrak{U}}_q(X; R), R)$, the $\mathfrak{U}$-small q-cochains, a natural quotient of $S^q(X; R)$ whose kernel K^q contains our f. Note that the K^q form a cochain complex $K^\bullet$ fitting in a short exact sequence

$$0 \to K^\bullet \to S^\bullet(X; R) \xrightarrow{\pi} S^\bullet_{\mathfrak{U}}(X; R) \to 0.$$

A classical argument [Warn, § 5.32] shows that π induces an isomorphism in cohomology and hence $H^q(K^\bullet) = 0$. So taking a cocycle $f \in S^q_0(X; R)$ it becomes a coboundary in $K^\bullet$ and so a fortiori a coboundary in $S^\bullet_0(X; R)$.

If X is Hausdorff, by Example B.10.2 compactly supported cohomology of $\underline{R}_X$ can be computed as the hypercohomology of the functor Γ_c. In particular, we may use the same flabby resolution $\widehat{S^q}$ of $\underline{R}_X$ as before. Since X is moreover locally compact and paracompact, the compacts form a paracompactifying family and we have long exact sequences associated to cohomology with compact support. The rest of the argument is essentially as before, using instead compactly supported cohomology. □

The class of topological spaces we are working with in this book have all of the above properties and a few more which will be needed later on. We collect the desired properties in the following provisional

Definition B.12. A topological space X is called **perfect** if it has the following properties:

a) X is Hausdorff;
b) X is locally compact;
c) X has a countable basis for the topology;
d) X is locally contractible.

Examples include topological manifolds, but also complex analytic spaces as we shall see later (Prop. C.9).

It is classical [Warn, Lemma 1.9] that a perfect space is in particular paracompact. Moreover, every open subset is paracompact. So, for a perfect space we can apply Prop. B.11:

Corollary B.13. *For a perfect space X (e.g. a manifold or an analytic space) singular and compactly supported singular cohomology (with values in R) coincides with cohomology of the constant sheaf $\underline{R}_X$, respectively compactly supported cohomology of $\underline{R}_X$.*

If X is Hausdorff and locally compact, its one-point compactification $X^* = X \cup \{\infty\}$ is compact and Hausdorff. For such a space we have $H^k_c(X;R) = H^k(X^*,\infty;R) = \tilde{H}^k(X^*)$, which gives an alternative definition of compactly supported cohomology. In fact, we need a bit more. Replace X^* by any compact perfect space X and replace ∞ by any closed subset Z. Compare now the long exact sequence for the pair (X, Z) (B–1) with the long exact sequence (B–15) for the constant sheaf $\underline{R}_X$. We find, making use of Cor. B.13:

Corollary B.14. *Let X be a perfect compact topological space, and let $Z \subset X$ be closed. Put $U = X - Z$. Then the natural restriction map induces isomorphisms $H^k(X, Z; R) \xrightarrow{\sim} H^k_c(U; R)$. This applies in particular to manifolds and to complex analytic spaces.*

B.2.3 Čech Cohomology

Let $\mathfrak{U} = \{U_i \mid i \in I\}$ be an open covering of a topological space X. The nerve $N(\mathfrak{U})$ of the covering is the set of the non-empty intersections in the covering.

Inclusions induce maps between the elements of the nerve. Its q-simplices correspond to non-empty intersections of exactly $(q+1)$ sets U_{i_j}, $i_j \in I$, $j = 0, \ldots, q$ of the covering. We denote this simplex by $\sigma = \{i_0, \ldots, i_q\}$, and its support $|\sigma|$ is by definition $U_{i_0} \cap \cdots \cap U_{i_q}$. A q-Čech-cochain is a function f which associates to each q-simplex σ an element $f(\sigma)$ of $\mathcal{F}(|\sigma|)$. These form the group

$$C^q(\mathfrak{U}, \mathcal{F}) := \prod_{\sigma \text{ a } q\text{-simplex of } \mathfrak{U}} \mathcal{F}(|\sigma|).$$

To make a complex out of it, we consider the j-th face of σ, i. e. the $(q-1)$-simplex $\sigma^j = \{i_0, \ldots, i_{j-1}, i_{j+1}, \ldots, i_q\}$, and we take the alternating sum of the various restriction maps $\mathcal{F}(|\sigma|) \xrightarrow{\rho^i} \mathcal{F}(|\sigma^i|)$:

$$d : C^q(\mathfrak{U}, \mathcal{F}) \to C^{q+1}(\mathfrak{U}, \mathcal{F})$$

$$df(\sigma) = \sum_{i=0}^{q+1} (-1)^i \rho^i \left(f(\sigma^i) \right).$$

The resulting complex is the **Čech complex**. The open coverings of X form a directed set under refinement and the direct limit of its cohomology groups is the **Čech cohomology**:

$$\check{H}^q(X, \mathcal{F}) := \varinjlim H^q(\mathfrak{U}, \mathcal{F}),$$

where the direct limit is taken over the set of coverings, partially ordered under the refinement relation. This cohomology computes $H^q(X, \mathcal{F})$ on paracompact spaces:

Theorem B.15. *Suppose that X is paracompact. Then the canonical homomorphisms*

$$\check{H}^q(X, \mathcal{F}) \to H^q(X, \mathcal{F})$$

are isomorphisms.

Sometimes there is no need to pass to the limit, making calculations simpler:

Theorem B.16 (Leray's theorem). *Let $\mathfrak{U} = \{U_i\}_{i \in I}$ be an open cover of X which is* ***$\mathcal{F}$-acyclic*** *in the sense that for any non-empty intersection $U = U_{i_1} \cap U_{i_2} \cdots \cap U_{i_k}$ of the cover we have*

$$H^p(U, \mathcal{F}) = 0 \quad \text{for } p \geq 1. \tag{B–16}$$

Then the natural homomorphisms

$$H^p(\mathfrak{U}, \mathcal{F}) \to \check{H}^p(X, \mathcal{F})$$

are isomorphisms.

Examples B.17. 1) For a complex manifold X one can find a cover simultaneously acyclic for all $\mathcal{O}_X$-modules $\mathcal{F}$. Indeed, any open set which is biholomorphic to a ball and intersections of such sets are examples of **Stein manifolds** and for these Cartan's famous "Théorème B" [Cart] (or [Gr-R77] for a modern proof) says that the cohomology groups $H^k(U, \mathcal{F})$ vanish for $k \geq 1$.

2) Any differentiable manifold can be covered by geodesic balls. Intersections of such balls are geodesically convex sets and hence contractible. Such covers are therefore acyclic with respect to locally constant sheaves.

B.2.4 De Rham Theorems

The two spectral sequences for the global section functor (A–33) and (A–34) yield:

Theorem B.18 (Abstract De Rham theorem). *Let X be a topological space and let $\mathcal{F}$ be a sheaf with a Γ-acyclic resolution $\mathcal{F}^\bullet = \{\mathcal{F}^0 \to \mathcal{F}^1 \to \cdots\}$, i.e. $H^k(X, \mathcal{F}^j) = 0$ for $k \geq 1$, $j \geq 0$. We have canonical identifications*

$$H^k(X, \mathcal{F}) = \mathbb{H}^k(X, \mathcal{F}^\bullet) = H^k\left[\Gamma(X, \mathcal{F}^\bullet)\right].$$

This formal result enters indeed crucially in the proof of the classical De Rham Theorem which states that for a differentiable manifold X integration induces a canonical isomorphism

$$H^p_{\mathrm{DR}}(X) = H^p_{\mathrm{DR}}(X, \mathcal{E}^\bullet_X) \xrightarrow{\cong} H^p(X; \mathbb{R}).$$

That there exists an isomorphism can be seen as follows. For a differentiable manifold, the Poincaré lemma states that the De Rham complex

$$0 \to \underline{\mathbb{R}}_X \xrightarrow{i} \mathcal{E}^0_X \xrightarrow{d^0} \mathcal{E}^1_X \xrightarrow{d^1} \cdots$$

is exact. In other words, viewing $\underline{\mathbb{R}}_X$ as a complex totally concentrated in degree 0, the map $i : \underline{\mathbb{R}}_X \to \mathcal{E}^\bullet_X$ is a quasi-isomorphism. The sheaves $\mathcal{E}^k_X$ are not injective in general but a partition of unity argument implies that these sheaves are still Γ-acyclic. In fact, each of these sheaves is a so called **fine sheaf** which means that the identity isomorphism is a sum of homomorphisms each of which is zero outside the open sets of a given cover of X. Then by Theorem B.18 the cohomology of the complex of global sections computes $H^k(X; \mathbb{R})$ and hence a functorial isomorphism $H^p_{\mathrm{DR}}(X) \xrightarrow{\cong} H^p(X; \mathbb{R})$. On the other hand, integration over smooth singular simplices defines a homomorphism $\omega \mapsto \left\{\sigma \mapsto \int_\sigma \omega\right\}$. By Stokes' theorem this homomorphism is a homomorphism of complexes. Since singular cohomology on smooth manifolds is known to be computable with cochains on smooth simplices, we have therefor a well defined homomorphism $H^p_{\mathrm{DR}}(X) \to H^p(X; \mathbb{R})$. That this is exactly the preceding canonical isomorphism lies deeper. See [Warn] where one finds a proof of the following version of

De Rham's Theorem. *Wedge-product induces a graded-commutative algebra structure on* $H^*_{\rm DR}(X) = \bigoplus_p H^p_{\rm DR}(X)$ *and with the algebra structure on* $H^*(X;\mathbb{R})$ *induced by cup-product, the integration map induces a graded algebra isomorphism*

$$H^*_{\rm DR}(X) \xrightarrow{\cong} H^*(X;\mathbb{R}). \tag{B–17}$$

Next, for a complex manifold the Dolbeault complex

$$0 \to \Omega^p_X \to \mathcal{E}^{p,0}_X \xrightarrow{\bar\partial} \mathcal{E}^{p,1}_X \xrightarrow{\bar\partial} \mathcal{E}^{p,2}_X \xrightarrow{\bar\partial} \cdots,$$

is a resolution of Ω^p_X (by the Dolbeault lemma) by fine sheaves $\mathcal{E}^{p,q}_X$ and so by Theorem B.18, for $H^{p,q}_{\bar\partial}(X) := H^q_{\rm DR}(\mathcal{E}^{p,\bullet})$ we have a canonical isomorphism

$$H^{p,q}_{\bar\partial}(X) \xrightarrow{\sim} H^q(X, \Omega^p_X) \qquad \textbf{(Dolbault's theorem)}. \tag{B–18}$$

Finally, as a last application we have the **holomorphic De Rham complex**

$$0 \to \underline{\mathbb{C}}_X \to \mathcal{O}_X \xrightarrow{\partial} \Omega^1_X \xrightarrow{\partial} \cdots$$

and the holomorphic Poincaré lemma says that this is an exact complex. Hence the inclusions $\underline{\mathbb{C}}_X \to \Omega^\bullet_X$ and $\Omega^\bullet_X \to \mathcal{E}^\bullet_X(\mathbb{C})$ are quasi-isomorphisms. Again, by Theorem B.18 we get an isomorphism

$$\mathbb{H}^n(X, \Omega^\bullet_X) \cong H^n(X, \mathbb{C}).$$

Remark B.19. We can also approach hypercohomology in the Čech setting. So let X be a paracompact space and $\mathcal{F}^\bullet$ a complex of sheaves on X. Let $\mathfrak{U}$ an open covering of X. Then the Čech groups $C^p(\mathfrak{U}, \mathcal{F}^q)$ form a double complex whose associated simple complex is denoted $sC^\bullet(\mathfrak{U}, \mathcal{F}^\bullet)$ and we have

$$\mathbb{H}^p(X, \mathcal{F}^\bullet) = \varinjlim H^p(sC^\bullet(\mathfrak{U}, \mathcal{F}^\bullet)),$$

where the limit is over all open coverings $\mathfrak{U}$.

B.2.5 Direct and Inverse Images

Let $f : X \to Y$ be a continuous map, $\mathcal{F}$ a sheaf on X. The sheaf associated to the presheaf

$$U \mapsto \mathcal{F}(f^{-1}U)$$

defines the **direct image sheaf** $f_*\mathcal{F}$ on Y. The resulting functor f_* is left exact. It sends flabby sheaves to flabby sheaves and extends to complexes of sheaves. Its right derived functors, the higher direct images are denoted

$$Rf_* : D^+(\text{sheaves of } R\text{-modules on } X) \to D^+(\text{sheaves of } R\text{-modules on } Y),$$
$$R^q f_* : D^+(\text{sheaves of } R\text{-modules on } X) \to \{\text{sheaves of } R\text{-modules on } Y\}.$$

Since direct images of flabby sheaves are flabby (this follows immediately from the definitions, see also [Gode, Théorème 3.1.1]) the Godement resolution is also f_*-acyclic and we have:

$$Rf_*(\mathcal{F}^\bullet) = f_* s[\mathcal{C}^\bullet_{\text{Gdm}}\mathcal{F}^\bullet] \tag{B–19}$$
$$R^q f_*(\mathcal{F}^\bullet) = H^q\left(f_* s[\mathcal{C}^\bullet_{\text{Gdm}}\mathcal{F}^\bullet]\right). \tag{B–20}$$

From this it follows that $R^q f_* \mathcal{F}^\bullet$ is the sheaf associated to the presheaf

$$U \mapsto \mathbb{H}^q(f^{-1}U, \mathcal{F}^\bullet).$$

As to functoriality, we have

$$R(f \circ g)_* \mathcal{F}^\bullet \xrightarrow{\overset{\text{qis}}{\sim}} Rf_*(Rg_* \mathcal{F}^\bullet).$$

If $\mathcal{F}^\bullet$ carries a biregular filtration F, the Godement resolution $s[\mathcal{C}^\bullet_{\text{Gdm}}\mathcal{F}^\bullet]$ is filtered by the subcomplexes $s[\mathcal{C}^\bullet_{\text{Gdm}}F^p\mathcal{F}^\bullet]$, which are f_*-acyclic. So the filtered Godement resolution can be used to define the higher direct images in the filtered setting

$$Rf_*(\mathcal{F}^\bullet, F\mathcal{F}^\bullet) = (f_* s[\mathcal{C}^\bullet_{\text{Gdm}}\mathcal{F}^\bullet], f_* s[\mathcal{C}^\bullet_{\text{Gdm}}F\mathcal{F}^\bullet])$$

and likewise in the bi-filtered setting.

Using the Godement resolution, one sees that there are natural identifications

$$\mathbb{H}^p(X, \mathcal{F}^\bullet) = \mathbb{H}^p(Y, Rf_* \mathcal{F}^\bullet). \tag{B–21}$$

Using Remark A.50 one concludes:

Lemma-Definition B.20. The spectral sequence for the double complex $\Gamma(Y, s[f_* \mathcal{C}^\bullet_{\text{Gdm}}\mathcal{F}^\bullet])$ with the vertical filtration is the Leray spectral sequence $E_r^{p,q}(f, \mathcal{F}^\bullet)$ satisfying

$$E_2^{p,q}(f, \mathcal{F}^\bullet) = \mathbb{H}^p(X, R^q f_* \mathcal{F}^\bullet) \Longrightarrow \mathbb{H}^{p+q}(X, \mathcal{F}^\bullet).$$

If $\mathcal{G}$ is a sheaf on Y one can define the **inverse image sheaf** $f^{-1}\mathcal{G}$ on X as follows. View a sheaf over a topological space X as a topological covering space over X. The covering space defining $f^{-1}\mathcal{G}$ is the fibre product $X \times_Y \mathcal{G}$. Alternatively,

$$f^{-1}\mathcal{F}(U) = \varinjlim \mathcal{F}(U), \quad U \text{ open in } Y, U \supset f(V). \tag{B–22}$$

This defines an exact functor f^{-1} on sheaves over Y. Suppose that $\mathcal{F}$ is any sheaf on X. Then, one directly verifies that $f_* \mathcal{H}om(f^{-1}\mathcal{G}, \mathcal{F}) = \mathcal{H}om(\mathcal{G}, f_*\mathcal{F})$. We say that f^{-1} is **left adjoint** to f_* (or f_* is **right adjoint** to f^{-1}). Applying the preceding formula to $\mathcal{F} = f^{-1}\mathcal{G}$ and the identity morphism, we obtain the **adjunction morphism** [Iver, II.4]:

$$f^\sharp : \mathcal{G} \mapsto f_* f^{-1}\mathcal{G}. \tag{B–23}$$

For complexes of sheaves $\mathcal{F}^\bullet$ and $\mathcal{G}^\bullet$ on X, respectively Y, we have to replace the functors by their derived versions so that we get $Rf_* R\mathcal{H}om(f^{-1}\mathcal{G}^\bullet, \mathcal{F}^\bullet) = R\mathcal{H}om(\mathcal{G}^\bullet, Rf_*\mathcal{F}^\bullet)$. The adjunction morphism then becomes

$$f^\sharp : \mathcal{G}^\bullet \to Rf_* f^{-1}\mathcal{G}^\bullet \tag{B–24}$$

which induces the natural maps

$$f^* : \mathbb{H}^p(Y, \mathcal{G}^\bullet) \longrightarrow \mathbb{H}^p(Y, Rf_* f^{-1}\mathcal{G}^\bullet) = \mathbb{H}^p(X, f^{-1}\mathcal{G}^\bullet) \tag{B–25}$$

In particular, since for the constant sheaf $\underline{R}_Y$ we have $f^{-1}\underline{R}_Y = \underline{R}_X$, we get back the induced maps for cohomology

$$f^* : H^p(Y; R) \longrightarrow H^p(X; R). \tag{B–26}$$

For cohomology with supports, functoriality is more complicated except for maps that are universally closed. For locally compact spaces universally closed maps are exactly the proper maps (i.e. the inverse image of a compact set is compact). So, for $f : X \to Y$ is a proper map between locally compact Hausdorff spaces, with Φ the family of compact subsets of Y, the set $f^{-1}\Phi$ consists of compact subsets of X and the above discussion thus extends to compactly supported cohomology without any change. In particular, in this situation there is an identification as in (B–21) and a Leray spectral sequence.

In general, if f is not necessarily proper, (but still within the category of Hausdorff locally compact spaces) we should replace the functor f_* by another functor, as we now explain. Let $V \subset Y$ open and take for Φ the family of closed sets Z in $f^{-1}V$ such $f|Z \to V$ is proper. Again, in view of the assumptions, this defines a family of supports. The sheaf $f_!\mathcal{F}$ is associated to the presheaf

$$V \mapsto \Gamma_\Phi(f^{-1}(V), \mathcal{F}).$$

There results a left exact functor, the **proper direct image functor** $f_!$ going from sheaves on X to sheaves on Y. There is a canonical injective map

$$0 \to f_!\mathcal{F} \to f_*\mathcal{F}$$

which is an isomorphism if f is proper. On the other end of the spectrum, when $f = a_X$ is the constant map to a point, we have

$$(a_X)_!\,\mathcal{F} = \Gamma_c(X, \mathcal{F}). \tag{B–27}$$

Parallel to (B–21) valid for ordinary cohomology, for any continuous map $f : X \to Y$ between locally compact spaces, we have a canonical identification

$$\mathbb{H}^q_c(X, \mathcal{F}^\bullet) = \mathbb{H}^q_c(Y, Rf_!\mathcal{F}^\bullet) \tag{B–28}$$

B.2.6 Sheaf Cohomology and Closed Subspaces

Let $i : Z \hookrightarrow X$ be a closed subset of a topological space X, and let $j : U = X - Z \hookrightarrow X$ be the inclusion of its complement. For any sheaf $\mathcal{F}$ on X the sheaf $j_! j^{-1}\mathcal{F}$ is just the restriction $\mathcal{F}|U$ extended by zero. It is a subsheaf of $\mathcal{F}$ with quotient $i_* i^{-1}\mathcal{F}$, i.e. we have a short exact sequence

$$0 \to j_! j^{-1}\mathcal{F} \to \mathcal{F} \to i_* i^{-1}\mathcal{F} \to 0. \tag{B–29}$$

This generalizes to a *complex* of sheaves $\mathcal{F}^\bullet$ on X provided we replace i_* by the derived functor Ri_*. So, if we define

$$\mathbb{H}^p(X, Z; \mathcal{F}^\bullet) := \mathbb{H}^p(X, j_! j^{-1}\mathcal{F}^\bullet), \tag{B–30}$$

the long exact sequence for ordinary hypercohomology gives

$$\cdots \to \mathbb{H}^p(X, Z; \mathcal{F}^\bullet) \to \mathbb{H}^p(X, \mathcal{F}^\bullet) \to \mathbb{H}^p(Z, \mathcal{F}^\bullet) \to \cdots . \tag{B–31}$$

Comparing this sequence with the sequence (B–1) (and using Prop. B.13) shows that definition (B–30) is compatible with the definition for relative cohomology for the constant sheaf $\underline{\mathbb{Z}}_X$.

If, instead, we use hypercohomology with compact supports, for a perfect space X we get the hypercohomology version of (B–15):

$$\cdots \to \mathbb{H}^p_c(U, \mathcal{F}^\bullet) \to \mathbb{H}^p_c(X, \mathcal{F}^\bullet) \to \mathbb{H}^p_c(Z, \mathcal{F}^\bullet) \to \mathbb{H}^{p+1}_c(U, \mathcal{F}^\bullet) \to \cdots . \tag{B–32}$$

Using Cor. B.14 we thus get:

Lemma B.21. *Let X be perfect and compact, and let $Z \subset X$ be closed. Then there is a natural isomorphism*

$$\mathbb{H}^k_c(X - Z, \mathcal{F}^\bullet) \xrightarrow{\sim} \mathbb{H}^k(X, Z; \mathcal{F}^\bullet).$$

We next introduce **local (hyper) cohomology** groups or cohomology groups with support in Z by defining

$$\mathbb{H}^p_Z(X, \mathcal{F}^\bullet) := H^p(\Gamma_Z(X, s[\mathcal{C}^\bullet_{\mathrm{Gdm}}(\mathcal{F}^\bullet)])).$$

Here Γ_Z is the functor of taking global sections with support in Z. Local cohomology can also be described as ordinary cohomology of a complex of sheaves on Z as follows. Consider the functor γ_Z defined for *sheaves* on X defined by

$$\gamma_Z(\mathcal{F})(U) := \mathrm{Ker}\,(\mathcal{F}(U) \to \mathcal{F}(U - U \cap Z)$$

and introduce

$$i^!(\mathcal{F}^\bullet) = i^{-1} R\gamma_Z(\mathcal{F}^\bullet) = i^{-1}\gamma_Z s[\mathcal{C}^\bullet_{\mathrm{Gdm}}(\mathcal{F}^\bullet)]. \tag{B–33}$$

Since $\Gamma \circ \gamma_Z = \Gamma_Z$, writing out the definitions we get

$$\mathbb{H}^p_Z(X,\mathcal{F}^\bullet) = \mathbb{H}^p(Z, i^!\mathcal{F}^\bullet) = \mathbb{H}^p(X, Ri_*i^!\mathcal{F}^\bullet), \tag{B–34}$$

where the last equation follows from (B–21). The complex $Ri_*i^!\mathcal{F}^\bullet$ figures in a distinguished triangle

$$\begin{array}{ccc} Ri_*i^!\mathcal{F}^\bullet & \longrightarrow & \mathcal{F}^\bullet \\ & {\scriptstyle [1]}\nwarrow \quad \swarrow {\scriptstyle \alpha_{Z,X}(\mathcal{F}^\bullet)} & \\ & Rj_*j^{-1}\mathcal{F}^\bullet & \end{array} \tag{B–35}$$

In this **adjunction triangle** the **attachment homomorphism**:

$$\alpha_{Z,X}(\mathcal{F}^\bullet) : \mathcal{F}^\bullet \longrightarrow Rj_*j^{-1}\mathcal{F}^\bullet$$

is defined using (B–24). The associated long exact sequence

$$\cdots \mathbb{H}^p_Z(X,\mathcal{F}^\bullet) \to \mathbb{H}^p(X,\mathcal{F}^\bullet) \to \mathbb{H}^p(X-Z,\mathcal{F}^\bullet) \to \mathbb{H}^{p+1}_Z(X,\mathcal{F}^\bullet)\cdots$$

is functorial in the sense that if $f : X \to Y$ is continuous and $T \subset Y$ is a closed subset such that $f(X-Z) \subset Y-T$, for any complex of sheaves $\mathcal{G}^\bullet$ on Y one has a commutative diagram of exact sequences

$$\begin{array}{ccccccc} \cdots\mathbb{H}^p_T(Y,\mathcal{G}^\bullet) & \to & \mathbb{H}^p(Y,\mathcal{G}^\bullet) & \to & \mathbb{H}^p(Y-T,\mathcal{G}^\bullet) & \to & \mathbb{H}^{p+1}_T(Y,\mathcal{G}^\bullet)\cdots \\ \downarrow & & \downarrow & & \downarrow & & \downarrow \\ \cdots\mathbb{H}^p_Z(X,f^*\mathcal{G}^\bullet) & \to & \mathbb{H}^p(X,f^*\mathcal{G}^\bullet) & \to & \mathbb{H}^p(X-Z,f^*\mathcal{G}^\bullet) & \to & \mathbb{H}^{p+1}_Z(X,f^*\mathcal{G}^\bullet)\cdots \end{array}$$

Suppose now that in addition to a closed set $Z \subset X$, we have an open set $V \subset X$ as well. The **excision exact sequence** is the sequence induced by restrictions

$$\cdots \to \mathbb{H}^p_{Z-Z\cap V}(X,\mathcal{F}^\bullet) \to \mathbb{H}^p_Z(X,\mathcal{F}^\bullet) \to \mathbb{H}^p_{Z\cap V}(V,\mathcal{F}^\bullet) \to \cdots \tag{B–36}$$

In case $Z \subset V$ this gives the **excision isomorphism**

$$\mathbb{H}^p_Z(X,\mathcal{F}^\bullet) \xrightarrow{\sim} \mathbb{H}^p_Z(V,\mathcal{F}^\bullet).$$

B.2.7 Mapping Cones and Cylinders

Let us recall a few notions from topology. Details can be found in [Hu, Chapter II]. Given a topological space X, the **cylinder** over X is the topological space $\mathrm{Cyl}(X) = X \times I$, where I is the unit interval. The **cone** $\mathrm{Cone}(X)$ is obtained from the cylinder by identifying $X \times \{0\}$ to a single point v, the vertex. Equivalence classes of points $(x,t) \in X \times I$ are denoted $[(x,t)]$. Cones are contractible onto the vertex. There is a natural inclusion $i : X \to \mathrm{Cyl}(X)$ of X as the "top" of the cylinder (i.e. $i(x) = [(x,0)]$ and an inclusion $j : X \to \mathrm{Cone}(X)$ of X as the "bottom" of the cone ($j(x) = [(x,1)]$). Next, if $f : X \to Y$ is a continuous map, the **cylinder** $\mathrm{Cyl}(f)$ **over** f is obtained by gluing the cylinder over X to

Y upon identifying a bottom point $(x, 1)$ of $\mathrm{Cyl}(X)$ with $f(x) \in Y$. The map $i : x \mapsto (x, 0)$ identifies X as a subspace of $\mathrm{Cyl}(f)$. The inclusion $Y \to \mathrm{Cyl}(f)$ of Y as the bottom of this cylinder is a homotopy equivalence since the cylinder retracts onto Y. Under this retraction the inclusion $X \to \mathrm{Cyl}(f)$ as the top deforms into $f : X \to Y$. Similarly one defines the **mapping cone over** f, $\mathrm{Cone}(f)$ by collapsing the top of the mapping cylinder to a single point v. The quotient space $\mathrm{Cyl}(f)/X$ is canonically homeomorphic to $\mathrm{Cone}(f)$. Since for any pair (X, A) with A closed, we have $H^*(X, A) = \tilde{H}^*(X/A)$ the long exact sequence associated to the pair $(\mathrm{Cyl}(f), X)$ therefore gives rise to the exact sequence

$$\to \tilde{H}^{q-1}(X) \to \tilde{H}^q(\mathrm{Cone}(f)) \xrightarrow{j^*} \tilde{H}^q(Y) \xrightarrow{f^*} \tilde{H}^q(X) \to . \tag{B–37}$$

As an example, if f is the inclusion $A \hookrightarrow X$, we find $\tilde{H}^q(\mathrm{Cone}(f)) = H^q(X, A)$.

Let us now make the link with the algebraic cone-construction from § A.1:

Theorem B.22. *Let $f : X \to Y$ be a continuous map between two topological spaces and let $\mathcal{L}^\bullet$ be a resolution of $\underline{\mathbb{Z}}_Y$. Consider the adjunction map* (B–24)

$$f^\sharp : \mathcal{L}^\bullet \to Rf_* f^{-1}\mathcal{L}^\bullet.$$

There is a natural isomorphism

$$\mathbb{H}^q(Y, \mathrm{Cone}^\bullet(f^\sharp)) = \tilde{H}^{q+1}(\mathrm{Cone}(f))$$

Proof. The exact sequence of the cone (A–12) yields

$$\cdots \to \mathbb{H}^q(Y, \mathcal{L}^\bullet) \to \mathbb{H}^q(X, f^{-1}\mathcal{L}^\bullet) \to \mathbb{H}^q(\mathrm{Cone}^\bullet(f^\sharp)) \to \mathbb{H}^{q+1}(Y, \mathcal{L}^\bullet) \to \cdots.$$

Since $\mathcal{L}^\bullet$ is a resolution of $\underline{\mathbb{Z}}_Y$, the complex $f^{-1}\mathcal{L}^\bullet$ resolves $f^{-1}\underline{\mathbb{Z}}_Y = \underline{\mathbb{Z}}_X$. The above sequence becomes

$$\cdots \to \mathbb{H}^q(Y) \to \mathbb{H}^q(X) \to \mathbb{H}^q(\mathrm{Cone}^\bullet(f^\sharp)) \to H^{q+1}(Y) \to \cdots.$$

Comparison with the exact sequence (B–37), shows that indeed

$$\mathbb{H}^q(Y, \mathrm{Cone}^\bullet(f^\sharp)) \simeq \tilde{H}^{q+1}(\mathrm{Cone}(f)) \qquad \square$$

Remark B.23. Compare this with [Weib, 1.5]. Note however that there the convention for indexing the cone-complex is different: Weibel's convention is better adapted to the topological situation because the cohomology of his cone-complex on singular chains gives the cohomology of the topological cone without shifts. In this book we follow the convention of [Iver].

B.2.8 Duality Theorems on Manifolds

Suppose that X is an n-dimensional oriented manifold. It has an **orientation class** $[X] \in H_n^{\mathrm{BM}}(X)$. Taking cap-product (B–10) with this class yields an isomorphism

$$D_X^{\mathrm{BM}} : H^{n-p}(X) \xrightarrow{\sim} H_p^{\mathrm{BM}}(X)$$

which is **Poincaré-duality** for Borel-Moore homology. The classical form of Poincaré duality is not stated using cap products involving Borel-Moore homology. Instead, one takes a limit over cap product with classes $[X]_K \in H_n(X, X-K) \xrightarrow{\cong} \mathbb{Z}$ induced by the orientation, where $K \subset X$ is a compact set. However, capping with the fundamental homology class class $[X]$ in Borel-Moore homology as in (B–9) directly gives the duality isomorphism of [Greenb, Theorem 26.6] $D_X : H_c^q(X) \to H_{n-q}(X)$. Summarizing, we have:

Theorem B.24 (Poincaré duality theorem). *Let X be a connected oriented n-dimensional topological manifold. The orientation induces a fundamental homology class $[X] \in H_n^{\mathrm{BM}}(X)$ and isomorphisms*

$$\begin{aligned} D_X : H_c^q(X) &\xrightarrow{\cap[X]} H_{n-q}(X) \\ D_X^{\mathrm{BM}} : H^q(X) &\xrightarrow{\cap[X]} H_{n-q}^{\mathrm{BM}}(X). \end{aligned}$$

In particular we find that orientation gives a generator for $H_c^n(X)$, defining the **trace map**

$$\mathrm{tr}_X : H_c^n(X) \xrightarrow{\sim} \mathbb{Z}. \tag{B–38}$$

It is compatible with cup-product (B–8) and Kronecker pairings (B–6) in the sense that

$$\mathrm{tr}(x \cup y) = \langle y, D_X x \rangle, \quad x \in H_c^q(X),\ y \in H^{n-q}(X). \tag{B–39}$$

One deduces

Corollary B.25. *Let X be a connected n-dimensional manifold. Cup-product pairing*

$$\begin{array}{ccc} H_c^q(X) \times H^{n-q}(X) & \longrightarrow & \mathbb{Z} \\ (x, y) & \longmapsto & \mathrm{tr}(x \cup y) \end{array}$$

is perfect in the sense that if $\mathrm{tr}(x \cup y) = 0$ for all $x \in H_c^q(X)$ then y is torsion and similarly if $\mathrm{tr}(x \cup y) = 0$ for all $y \in H^{n-q}(X)$ then x is torsion.

Remark. For *differentiable* manifolds, using the De Rham isomorphism (B–17) translates (B–39) into

$$\langle a, D_X b \rangle = \int_X b \wedge a, \quad a \in H_{\mathrm{DR}}^p(X),\ b \in H_{\mathrm{DR}}^{n-p}(X). \tag{B–40}$$

Using the duality isomorphisms, for any continuous map between oriented manifolds $X \to Y$, say $\dim X = n$. $\dim Y = m$, we can replace the induced homomorphism in homology $f_* : H_k(X) \to H_k(Y)$ by the associated **Gysin homomorphism**

$$f_! = D_Y^{-1} \circ f_* \circ D_X : H_c^k(Y) \to H_c^{k+m-n}(X) \tag{B–41}$$

and similarly for cohomology we put $f^! = D_X \circ f^* \circ D_Y^{-1}$. This is indeed compatible with the notation used elsewhere in this book, since first of all formulas (B–26) and (XIII–11) show that f^*, respectively f_* correspond to f^{-1} and Rf_* respectively, while the commutative diagram (XIII–9) then show that $f^!$ and $f_!$ correspond to $f^!$ and $Rf_!$ respectively.

Using this compatibility, dualizing the projection formula as stated on [Iver, p.390]) we get:

Lemma B.26 (Projection formula). *Let $f : X \to Y$ be a continuous map between oriented manifolds. We have*

$$f_!(x \cup f^* y) = f_!(x) \cup y, \quad x, y \in H^*(X).$$

Observing that the Poincaré-dual over any field is completely determined by the formula (B–39) one now shows:

Proposition B.27. *Let $f : X \to Y$ be a continuous map between two compact connected oriented manifolds of the same dimension n. Let R be a field. Then $f_! f^* : H^n(Y; R) \to H^n(Y; R)$ is multiplication with $\deg(f)$; if this degree does not vanish, for all q the homomorphism $f^* : H^q(Y; R) \to H^q(X, R)$ is injective.*

Proof. First observe that for all $a \in H^q(Y; R)$ and $b \in H^{n-q}(Y; R)$ we have

$$\begin{aligned} \mathrm{tr}_Y(f_! f^*(a \cup b)) &= \langle b, D_Y f_! f^* a\rangle = \langle b, f_* D_X f^* a\rangle \\ &= \langle f^* b, D_X f^* a\rangle = \mathrm{tr}_X(f^* a \cup f^* b) \\ &= \mathrm{tr}_X(f^*(a \cup b)) = \deg(f)\, \mathrm{tr}_Y(a \cup b). \end{aligned}$$

The last equality holds by the definition of degree. Next, applying this for $a \cup b$ mapping onto $1 \in \mathbb{Z}$ under the trace map, the first assertion follows. For the second assertion, assuming $f^* a = 0$ all terms in the above formula vanish so that $\mathrm{tr}_Y(a \cup b) = 0$ for all $b \in H^{n-q}(X)$ and hence $a = 0$ by Cor. B.25. □

We also have a more general version of duality, Poincaré-Lefschetz duality, classically stated in singular cohomology. The following is a more general variant in Borel-Moore homology (see [Fult, 19.1]):

Theorem B.28 (Poincaré-Lefschetz duality). *Let X be a connected oriented manifold and Z a closed subset. Then cap product with the orientation class induces an isomorphism*

$$D^{\mathrm{BM}}_{(X,Z)} : H^q_Z(X) = H^q(X, X - Z) \xrightarrow{\cong} H^{\mathrm{BM}}_{n-q}(Z).$$

B.2.9 Orientations and Fundamental Classes

Let X be any n-dimensional differentiable manifold and let $Y \subset X$ be an m-dimensional connected oriented submanifold. Using Poincaré-Lefschetz duality, let us introduce:

$$\tau(Y) = (D_X^{\mathrm{BM}})^{-1}[Y] \in H_Y^{n-m}(X) \quad (\textbf{Thom class} \text{ of } Y).$$

Its image under restriction $r : H_Y^{n-m}(X) \to H^{n-m}(X)$ is denoted

$$\mathrm{cl}(Y) = r\tau(Y) \in H^{n-m}(X) \quad (\text{the } \textbf{fundamental cohomology class} \text{ of } Y \text{ in } X).$$

With $i : Y \hookrightarrow X$ the inclusion inducing the lower horizontal arrow, there is a commutative diagram

$$\begin{array}{ccc} H_Y^{n-m}(X) & \xrightarrow{r} & H^{n-m}(X) \\ \downarrow D_{(X,Y)}^{\mathrm{BM}} & & \downarrow D_X \\ H_m^{\mathrm{BM}}(Y) & \xrightarrow{i_*} & H_m^{\mathrm{BM}}(X) \end{array}$$

which shows that $\mathrm{cl}(Y)$ is Poincaré-dual to $i_*[Y] \in H_m^{\mathrm{BM}}(X)$. We use the following incarnation of the Thom isomorphism:

Theorem B.29 (Thom's isomorphism theorem). *Taking cup-product with the Thom class induces an isomorphism for all q*

$$H^q(Y) \xrightarrow{\sim} H_Y^{q+c}(X), \quad c = n - m$$

Proof. The usual Thom isomorphism ([Span, Chapt 5 §7 Theorem 10]) for the total space N of the normal bundle of Y in X states that cup-product with the Euler class for the normal bundle of Y in X induces isomorphisms

$$H^q(Y) \xrightarrow{\sim} H^{q+c}(N, N - \{\text{zero section}\}).$$

The right hand side is isomorphic to $H^{q+c}(X, X - Y)$ by the tubular neighbourhood Theorem (there is a neighbourhood $T \subset X$ of Y such that (T, Y) is diffeomorphic to $(N, \{\text{zero section}\})$, see [Lang, Ch. 4.5]) and the excision isomorphism applied to the inclusion $(T, T - Y) \hookrightarrow (X, X - Y)$. Secondly, one checks easily that the Thom class corresponds to the Euler class of the normal bundle under the isomorphism $H_Y^c(X) = H^c(X, X - Y) \xrightarrow{\sim} H^c(N, N - \{\text{zero section}\})$. □

Next, suppose that X (and hence Y) are compact. Using the Gysin morphisms

$$i_! : H^q(Y) \to H^{q+c}(X), \quad c = \dim X - \dim Y$$

one has

$$\mathrm{cl}(Y) = i_!(1) \in H^c(X)$$

since $D_Y(1) = [Y]$.

Proposition B.30. 1) *For all $\alpha \in H^q(Y)$ one has the relation*

$$i^* i_!(\alpha) = i^* \,\mathrm{cl}(Y) \cup \alpha.$$

2) *For all* $\beta \in H^p(X)$ *one has the relation*

$$i_! i^*(\beta) = \mathrm{cl}(Y) \cup \beta.$$

Proof. 1) One proves this formula after applying the Thom isomorphism and making use of the formula $a \cup i^* r(b) = r(a) \cup b$ valid for $a \in H^p(X)$ and $b \in H^q_Z(X)$. For details see [Iver, proof of Formula 2.7 on p. 339].
2) This is an easy exercise in the definitions. Alternatively, one may apply the projection formula (Prop. B.26) to the first formula. □

Remark B.31. There is still another way to introduce fundamental classes by means of currents. This method has the advantage that it treats homology and cohomology on a uniform basis. A **current** of degree q on a differentiable manifold X is a q-form with distributional coefficients. These form a topological vector space $\mathcal{D}^q(X)$. Alternatively, we can consider currents of degree q as elements in the dual of the vector space of compactly supported $(n-q)$-forms on X, the duality being given by integration. The transpose of the d-operator on forms thus defines a d-operator on currents. The sheaf associated to the presheaf

$$U \mapsto \mathcal{D}^q(U) = \Gamma_c(U, \mathcal{E}^{n-q})^\vee$$

is the sheaf $\mathcal{D}^q_X$ of degree q-currents. These sheaves are fine. The d-operators define a complex $\mathcal{D}^\bullet_X$ which is a cohomological resolution of the constant sheaf $\underline{\mathbb{R}}_X$. So the cohomology of X can also be computed by means of currents, and we have canonical isomorphisms

$$H^q_{DR}(X) = H^q(\Gamma(\mathcal{D}^\bullet_X)) \cong H^q(X, \mathbb{R}).$$

Let $i : Y \hookrightarrow X$ be the embedding of a compact m-dimensional oriented submanifold $Y \subset X$ into X (possibly with boundary ∂Y). The **integration current** $[Y]$ is the degree m-current defined by its values on m-forms α on X:

$$\langle [Y], \alpha \rangle = \int_Y \alpha.$$

Stokes' formula shows that $d[Y] = (-1)^{m-1}[\partial Y]$ and so this defines a closed current if Y has no boundary. We denote the resulting cohomology class by $\mathrm{cl}(Y)$. This is indeed the Poincaré dual of the orientation class in homology, since

$$\int_Y \alpha = \langle \alpha, i_*[Y] \rangle = \int_X \alpha \wedge \mathrm{cl}(Y) \quad \alpha \in H^{n-m}(X, \mathbb{R}).$$

This approach can be used to define integration currents for compact subvarieties Y of a complex manifold X. We just integrate compactly supported forms over the smooth part of Y. This is possible, since the volume of Y in any coordinate polydisc is bounded: if $\dim Y = d$ choose local coordinates such that projection onto any set I of d coordinates is finite of degree n_I onto the image Y_I; the volume of Y is then bounded by $\sum_I n_I \mathrm{Vol} Y_I$. What is less trivial is the fact that this current is *closed*. See for this [Lel]. It should be clear that the resulting cohomology class is dual to the orientation class in Borel-Moore homology introduced before.

We shall indicate how to construct the fundamental class for an *irreducible analytic space* X. The complement U in X of the singular locus X_{sing} is an $2n$-dimensional connected manifold with a natural orientation coming from the complex structure and hence one has an orientation class $[U]$ which generates $H_{2n}^{\text{BM}}(U)$. The long exact sequence

$$\cdots \to H_{2n}^{\text{BM}} X_{\text{sing}} \to H_{2n}^{\text{BM}} X \to H_{2n}^{\text{BM}} U \to H_{2n-1}^{\text{BM}} X_{\text{sing}} \cdots$$

together with induction on the dimension, shows that $H_{2n}^{\text{BM}}(X)$ is cyclic and generated by a unique class $[X]$ which maps to $[U]$.

More generally, if Y is an irreducible m-dimensional subvariety of X, one defines a fundamental class $\text{cl}_X(Y)$ by

$$\text{cl}_X(Y) = i_*[Y] \in H_{2m}^{\text{BM}}(X),$$

where $i : Y \hookrightarrow X$ is the inclusion. In addition, if X is smooth, Poincaré duality for Borel-Moore homology provides fundamental cohomology classes

$$\tau(Y) \in H_Y^{2c}(X), \quad c = n - m = \text{codim}(Y).$$

One can construct this class also directly from the corresponding class $\tau(Y_{\text{reg}}) \in H_{Y_{\text{reg}}}^{2m}(X)$ with support on the regular locus Y_{reg} of Y as follows. One first notes that for any subvariety Z of codimension c in X the groups $H_Z^r(X)$ vanish for $r < 2c$, This can be shown inductively using the excision sequence for cohomology with support, starting with Z smooth and the Thom isomorphism. Then the excision sequence

$$H_{Y_{\text{sing}}}^{2c}(X) \to H_Y^{2c}(X) \to H_{Y_{\text{reg}}}^{2c}(X) \to H_{Y_{\text{sing}}}^{2c+1}(X)$$

shows that in fact restriction induces an isomorphism between $H_Y^{2c}(X)$ and $H_{Y_{\text{reg}}}^{2c}(X)$.

B.3 Local Systems and Their Cohomology

Local systems are defined on any topological space. One can define (co)homology with values in a local system. This generalizes the concept of (co)homology with constant coefficients and is a special case of sheaf cohomology. On manifolds there is also a version of Poincaré duality for cohomology with values in a locally constant system.

B.3.1 Local Systems and Locally Constant Sheaves

Let R be some commutative ring with a unit.

Definition B.32. Let X be a topological space. A **local system of R-modules** on X consists of a collection V_x of R-modules, one for each point $x \in X$, together with a collection of isomorphisms $\rho([\gamma]) : V_x \xrightarrow{\sim} V_y$, one for each homotopy class $[\gamma]$ of paths from x to y. Furthermore, one requires that this assignment is functorial in the sense that $\rho_{[e_x]} = \mathrm{id}_{V_x}$ for the class of the constant path e_x at x and that $\rho_{[\gamma * \gamma']} = \rho_{[\gamma']} \circ \rho_{[\gamma]}$ for two classes of composable paths. Here the product of two composable paths γ and γ' is denoted $\gamma * \gamma'$, which means *first* traverse γ and *then* γ', both with double speed.

Usually we denote a local system with fibres V_x by $\mathbb{V}$. The **constant system** with fibre V on X is denoted $\underline{V}_X$. If (X, o) is a pointed path-connected topological space, the collection $\{[\rho([\gamma]) \mid \gamma \text{ a loop at } o\}$ defines the associated **monodromy representation**

$$\rho : \pi_1(X, o) \to \mathrm{GL}(V_o).$$

Suppose now that X is locally 1-connected so that it admits a covering $\{U_i\}_{i \in I}$ by 1-connected open subsets. Then for any two points $x, y \in U_i$, there is a unique isomorphism $f_{x,y} : V_x \xrightarrow{\sim} V_y$ defined by any path connecting x and y within U_i so that there is a canonical trivialization of the local system $\mathbb{V}$ above U_i, say

$$\phi_i : \mathbb{V}|U_i \xrightarrow{\sim} \underline{V}_{U_i}.$$

Comparing the two trivialisations in the overlaps we see that the resulting fibre bundle has constant transition functions with values in $\mathrm{Aut}(V)$. Alternatively, one may view the trivialisations as a way to describe $\mathbb{V}$ as a locally constant sheaf, the sheaf of local sections of the associated fibre bundle. Here we recall

Definition B.33. A **locally constant sheaf** $\mathcal{F}$ on X is a sheaf with the property that for some open cover $\{U_i\}_{i \in I}$ of X, the restrictions $\rho_{U_i,x} : \mathcal{F}(U_i) \to \mathcal{F}_x$, $x \in U_i$ are isomorphisms.

Conversely, let $\mathcal{F}$ be a locally constant sheaf of R-modules on a path connected and locally 1-connected space X. There is a locally constant system $\mathbb{V}$ associated to it as follows. For V_x one takes the stalk of $\mathcal{F}$ at x and for any path $\gamma : I \to X$, one defines $\rho([\gamma])$ by subdividing first I so that each segment $[a, b]$ maps to one of the U_i and then one takes the composition of the isomorphisms $\rho_{U_i,b} \circ \rho_{U_i,a}^{-1}$, where $\rho_{U_i,x}$ are the restrictions. This is independent of the subdivision and depends only on the homotopy class of the path. This gives indeed a bijection:

Lemma B.34. *On a path connected and locally 1-connected space X there is a one to one correspondence between locally constant sheaves of R-modules and local systems of R-modules on X.*

As a consequence, a locally constant sheaf $\mathbb{V}$ becomes constant on any simply connected space. In particular, it pulls back to a constant system on the universal cover $(\tilde{X}, \tilde{o})$ of a pointed connected and locally path connected space (X, o). The fundamental group acts on the universal cover from the left by translation and the monodromy representation ρ gives back the local system on X by taking the quotient under the product action

$$\begin{aligned} \mathbb{V} \cong V_\rho &= (\tilde{X} \times V)/\pi_1(X, o) \\ g(\tilde{x}, v) &= (\tilde{x}g^{-1}, \rho(g)v), \quad \forall g \in \pi_1(X, o). \end{aligned}$$

B.3.2 Homology and Cohomology

To define homology and cohomology of a local system $\mathbb{V}$ in the singular framework, we slightly modify the definition of singular chains and their (co)-boundary maps as follows. We consider finite formal sums $\sum v_\sigma \sigma$ over singular p-simplices σ and $v \in V_{\sigma(e_0)}$. Associate to σ the path γ_σ from the zeroth vertex $\sigma(e_0)$ to the first vertex $\sigma(e_1)$. The boundary then is defined by

$$\delta_p(\sum v\sigma) = \sum \rho(\gamma_\sigma)(v) + \sum_{q=1}^{p} (-1)^q v\sigma^q.$$

The coboundary is defined similarly. Taking homology, respectively cohomology of the resulting complexes defines $H_p(X, \mathbb{V})$, respectively $H^p(X, \mathbb{V})$. One can likewise define (co)-homology of a pair and cohomology with compact support. See [Span, Exercise I, J of Chap. 5] for details.

Note that a perfect space (Def. B.12) is in particular locally 1-connected so local systems and locally constant sheaves are the same on a connected such space. One can modify the proof of Prop. B.11 to obtain the following analogue of Cor. B.13.

Lemma B.35. *For a connected perfect space X the group $H^p(X, \mathbb{V})$ coincides with the p-th cohomology of the (locally constant) sheaf defined by $\mathbb{V}$. A similar assertion holds for cohomology with compact support.*

If $\mathbb{V}$ and $\mathbb{W}$ are two local systems, their tensor product is likewise a local system and the classical definition of cup-product can be extended to obtain a product

$$H^p(X, \mathbb{V}) \otimes H^q_c(X, \mathbb{W}) \to H^{p+q}_c(X, \mathbb{V} \otimes \mathbb{W}).$$

In particular, when $\mathbb{W} = \mathbb{V}^\vee$ is the dual local system, evaluation maps the right hand side to $H^{p+q}(X)$ and we get

$$\bigcup : H^p(X, \mathbb{V}) \otimes H^q_c(X, \mathbb{V}^\vee) \to H^{p+q}_c(X).$$

There is also a weak generalization of Poincaré duality in the sense that Cor. B.25 generalizes:

Theorem B.36 (Poincaré duality for local systems). *For an oriented connected manifold X, the pairing*

$$H_c^p(X, \mathbb{V}) \otimes H^{n-p}(X, \mathbb{V}^\vee) \xrightarrow{\cup} H_c^n(X) \xrightarrow{\text{tr}} \mathbb{Z}$$

is non-degenerate.

This can either be proved directly by imitating the classical proof (see e.g. ([Grif-Ha]), or it can be viewed as a special case of Verdier-duality (13.7).

B.3.3 Local Systems and Flat Connections

Let E be any (real or complex) vector bundle on a smooth manifold M. Let $\mathcal{E}_X^p(E)$ be the bundle of smooth E-valued p-forms on X.

Definition B.37. 1. A **connection** on E is an operator

$$\nabla : \mathcal{E}_X^0(E) \to \mathcal{E}_X^1(E)$$

which satisfies the Leibniz rule: $\nabla(f \cdot s) = df \cdot s + f\nabla s$ for any smooth function f on X and any section s of E.

2. For any vector field v on X the **covariant derivative** ∇_v **in the direction of** v is the section of E defined by the rule

$$\nabla_v(s) := v(\nabla s).$$

Consider first a trivial vector bundle $U \times \mathbb{R}^m$ on an open set U of $\mathbb{R}^n$. A differentiable section is nothing but a row vector $\mathbf{x}$ of functions. Any connection ∇ on this bundle can be given by

$$\nabla \mathbf{x} = d\mathbf{x} - \mathbf{x}\omega, \quad \omega \in \mathcal{E}^1\left(\mathfrak{gl}(n, \mathbb{R})\right). \tag{B–42}$$

This formula is valid in over any trivializing coordinate chart U for E, i.e. after the choice of an m-frame for $E|U$. The resulting matrix-valued 1-form ω in (B–42) is then called the **connection matrix** in U of the connection.

Example B.38. If the transitions functions are locally constant, one can take $\nabla = d$, i.e., all connection matrices can be taken to be zero.

Using the Leibniz rule $\nabla(\alpha \cdot e) = d\alpha \cdot e + (-1)^{\deg \alpha} \alpha \wedge \nabla e$, we can extend the connection to an operator

$$\nabla : \mathcal{E}_X^p(E) \to \mathcal{E}_X^{p+1}(E)$$

on p-forms with values in E.

Lemma-Definition B.39. – The **curvature** operator of the connection ∇ is defined by

$$F_\nabla = \nabla \circ \nabla : \mathcal{E}_X^0(E) \to \mathcal{E}_X^2(E).$$

This operator can be checked to be linear over the differentiable functions and thus can be viewed as an $\mathrm{End}(E)$-valued 2-form on X.

– The connection is **flat** if its curvature is zero.

Example B.40. Any locally constant vector bundle, or equivalently, any local system (see Sect. B.3) admits a flat connection, since, as we saw before, in a trivializing coordinate chart d defines a global connection for which trivially $F = d \circ d = 0$.

Let us now consider parallel transport along a smooth path $\gamma : I \to X$, say with $\gamma(0) = p$ and $\gamma(1) = q$. Assume that the path is contained in a trivializing chart and that we have chosen a frame. The pull back along γ of a section of E can then be given by a vector valued function $\mathbf{x}(t)$ on I. The connection form pulls back to a form Adt, with $A \in \mathfrak{gl}(m)$. The section is parallel along γ if $d\mathbf{x} = \mathbf{x}\gamma^*\omega$ which translates into the differential equation

$$\frac{d\mathbf{x}}{dt} = \mathbf{x}A.$$

By the theory of ordinary differential equations this equation always has a unique solution if the initial value $\mathbf{x}(0) = \mathbf{v}$ is fixed. An explicit solution $\mathbf{x}(1) = \mathbf{w}$ using Picard iteration can be found by converting the preceding equation into an integral equation:

$$\mathbf{x}(s) = \mathbf{x}(0) + \int_0^s \mathbf{x}(t)A(t)dt.$$

With as a zeroth approximate solution the initial value $\mathbf{x}_0(t) = \mathbf{v}$, we define successively

$$\mathbf{x}_n(s) = \mathbf{v} + \int_0^s \mathbf{x}_{n-1}(t)A(t)dt.$$

Thus, we find

$$\begin{aligned} \mathbf{x}_1(s) &= \mathbf{v} + \textstyle\int_0^s \mathbf{v}A(t_1)dt_1 \\ \mathbf{x}_2(s) &= \mathbf{v} + \textstyle\int_0^s \mathbf{x}_1(t_2)A(t_2)dt_2 \\ &= \mathbf{v} + \textstyle\int_0^s (\mathbf{v} + \int_0^{t_2} \mathbf{v}A(t_1)dt_1)A(t_2)dt_2 \\ &= \mathbf{v}\left(1 + \textstyle\int_{0\le t_1\le s} A(t_1)dt_1 + \int_{0\le t_1\le t_2\le s} A(t_1)A(t_2)dt_1dt_2\right) \end{aligned}$$

The limit of this process yields the true solution, namely,

$$\mathbf{x}_\infty(s) = \mathbf{v}\left(1 + \int_{\Delta_1(s)} A(t_1)dt_1 + \int_{\Delta_2(s)} A(t_1)A(t_2)dt_1dt_2 + \cdots\right),$$

where

$$\Delta_n(s) = \{0 \le t_1 \le \cdots \le t_n \le s\}$$

The expression on the right is the value of an iterated integral:

Lemma-Definition B.41. Let X be a differentiable manifold.

- Given an ordered set of r matrix valued one-forms $\omega_1, \omega_2, \ldots, \omega_r$, the associated **iterated integral** is the function on paths $\gamma : I \to X$ given by the formula

$$\int_\gamma \omega_1\omega_2\cdots\omega_r = \int\cdots\int_{0\le t_1\le\cdots\le t_r} f_1(t_1)f_2(t_2)\cdots f_r(t_r)dt_1\cdots dt_r,$$

where we have written

$$\gamma^*\omega_i = f_i(t)dt.$$

- Let E be a trivial rank m vector bundle with connection ∇ and connection matrix ω. Parallel transport along a piecewise differentiable path γ is given by the **transport map**

$$\mathbf{w} = \mathbf{v}\,T(\gamma),$$

where $T(\gamma)$ is the following (convergent) series of iterated integrals evaluated on γ

$$T(\gamma) = 1 + \int_\gamma \omega + \int_\gamma \omega\omega + \int_\gamma \omega\omega\omega + \cdots$$

If we want that the result of parallel transport from p to q does not change under homotopies fixing the end points, an integrability condition is needed which follows from the curvature being zero.

Lemma B.42. *For a flat connection the result of parallel transport between two points p and q depends only on the homotopy class of the path (relative to p and q).*

For a proof see [Don-Kr, Th. 2.2.1].

Corollary B.43. *Given a flat connection in a differentiable vector bundle E over a differentiable manifold X, there is a local trivialization of E by a parallel frame. Hence, the existence of a flat connection in E implies that E is a locally constant vector bundle.*

C

Stratified Spaces and Singularities

C.1 Stratified Spaces

Large classes of analytic spaces, including those that underlie algebraic varieties, admit a so-called Whitney-stratification. This is a decomposition into smooth submanifolds of smaller and smaller dimension having the property that the topology in normal directions to strata is locally constant. As such the stratification gives the space the structure of a pseudomanifold, a very special topological space of finite dimension. The existence of Whitney-stratifications has a great many important topological consequences which are reviewed below.

C.1.1 Pseudomanifolds

We give the following *inductive definition* of a topological stratified space.

Definition C.1. Let X be a paracompact Hausdorff space. A filtration

$$X = X_n \supset X_{n-1} \supset \cdots \supset X_1 \supset X_0$$

by closed subspaces is called an **n-dimensional topological stratification** if

1) The $X_k - X_{k-1}$ are k-dimensional manifolds (or empty); its connected components are called the (open) **strata**;
2) local normal triviality: for each point x of a connected component S of $X_k - X_{k-1}$ there exists an $(n-k-1)$-dimensional topological stratified compact Hausdorff space $L_x = L_{n-k-1} \supset L_{n-k-2} \supset \cdots \supset L_0$, the **link** of x in X_k, and an an open neighbourhood $U \subset X$ together with a homeomorphism

$$h : U \xrightarrow{\sim} \{\text{open ball } B_k \text{ centered at } x \text{ in } S\} \times \text{open cone on } L_x,$$

which preserves stratifications in the sense that h maps $U \cap X_{k+j}$ homeomorphically to $B \times$ open cone on L_j, $j = 0, \dots, n-k-1$.

Any topological space admitting a topological stratification with the extra conditions that $X_{n-1} = X_{n-2}$ and $X - X_{n-2}$ dense in X is called an n-dimensional **pseudomanifold** and a pseudomanifold endowed with a stratification is called a **stratified** pseudomanifold. An **orientable** pseudomanifold by definition admits an **orientation**, i.e. an orientation for the n-dimensional manifold $X - X_{n-2}$.

A pseudomanifold is **topological normal** if every point has a fundamental system of neighbourhoods which do not get disconnected by leaving out the non-manifold locus.

Remark C.2. 1) A **stratification** of a topological space is a locally finite partition into locally closed non-empty subsets, the strata; each stratum having the property that its closure is a union of strata. So the open strata given by the filtration occurring in the definition of a topological stratification do give a stratification in this sense.
2) Any complex analytic space of pure dimension n is an orientable $(2n)$-dimensional pseudomanifold. In fact, these have special stratifications called **Whitney stratifications**. We treat these in detail in § C.1.2. The strata are analytic subspaces and hence are even-dimensional. If X is normal, X is also topological normal.

C.1.2 Whitney Stratifications

A good reference for this subsection is the book [G-M88] where further background and proofs can be found

Definition C.3. Let X be a complex space of pure dimension n. A **complex analytic stratification** consists of a filtration

$$X = X_n \supset X_{n-1} \supset \cdots \supset X_1 \supset X_0 \supset X_{-1} = \varnothing$$

by closed *analytic* subspaces such that the **strata**, i. e. the connected components of $X_k - X_{k-1}$ are k-dimensional complex manifolds.

The Whitney regularity conditions control the way two strata meet. So let $x \in X$ and view the germ (X, x) as being embedded in $(\mathbb{C}^N, 0)$ so that the tangent spaces along strata near x can be viewed as points in appropriate Grassmannians, and similarly for the linear joins of a finite set of points. When speaking of convergence we always mean convergence inside these Grassmannians. We can now explain the regularity conditions:

Definition C.4. 1) A stratification of X with strata $S, S', \ldots$ satisfies the **first regularity condition** at $x \in S \subset \overline{S'}$ if for all sequences $y_n \to x$, $y_n \in S'$ for which the corresponding sequence of tangent spaces converges to T, we have $T_x S \subset T$. The **second regularity condition** is satisfied if in addition for all $x_n \in S$ converging to x for which the linear joins $\langle x_n, y_n \rangle$ converge to a line, say $\lim_{n\to\infty} \langle x_n, y_n \rangle = L$, we have $L \subset T$. If

both conditions are satisfied for all points in the intersection of a stratum and the closure of another stratum, the stratification is called a **Whitney stratification**.
2) We say that a morphism $f : X \to Y$ between two complex spaces is **stratified** if X and Y are Whitney stratified and every stratum of X is mapped submersively and surjectively onto a stratum of Y.

For simplicity, we assume now that $X \subset U$, where $U \subset \mathbb{C}^N$ is a small open ball. We use the standard metric d on $\mathbb{C}^N$. For any $x \in X$ we let

$$B(x,r) := \{y \in \mathbb{C}^N \mid d(x,y) \leq r\}$$

be the closed ball of radius r and centre x and

$$S(x,r) := \partial B(x,r) = \{y \in \mathbb{C}^N \mid d(x,y) = r\}$$

the boundary. Let $S \subset X$ be a stratum of a Whitney stratification, $x \in S$. Choose a complex submanifold $N' \subset U$ passing through x, transverse to all strata of X, meeting S in x and such that $\dim S + \dim N' = N$. Choose r so small that $S(x,\rho)$ meets each stratum of X transversely for $0 < \rho \leq r$. Then

$$N(S,x) = N' \cap X \cap B(x,r)$$

is a Whitney stratified space, the **normal slice** through (S,x). The **link** $L(S,x)$ of the stratum is the *compact* Whitney stratified space

$$L(S,x) = N' \cap X \cap S(x,r).$$

For r small enough the topological type of the pair $(N(S,x), L(S,x))$ does not depend on r, the embedding $X \hookrightarrow U$, and it stays constant. Because of this we also use the notation $(N(S), L(S))$. Moreover, the normal slice has a cone-like structure:

Proposition C.5. *There is a stratified homeomorphism*

$$N(S,x) \xrightarrow{\sim} \operatorname{Cone} L(S,x)$$

which takes x to the vertex of the cone.

It follows that a Whitney stratified space is stratified in the topological sense:

Corollary C.6. *Let X_{2k} be the union of the strata of a Whitney-stratification of the n-dimensional complex space X having complex dimension $\leq k$. Then $X = X_{2n} \supset X_{2n-2} \supset \cdots \supset X_0$ is a topological $(2n)$-dimensional stratification giving X the structure of an oriented $(2n)$-dimensional pseudomanifold.*

The basic properties concerning Whitney stratifications are:

Properties C.7. 1) Closed subvarieties of complex manifolds admit a Whitney stratification. In particular, any analytic space can locally be Whitney stratified.

2) If X can be Whitney stratified, any given analytic stratification can be refined to a Whitney-stratification.

3) Proper morphism between complex spaces can be stratified. In particular, compact analytic spaces can be Whitney stratified.

4) Transverse intersections of two Whitney stratified spaces become Whitney stratified upon taking the intersection of strata.

5) Whitney stratified spaces are locally topologically trivial along the strata: there is a neighbourhood $T(S)$ of a stratum S in X such that the projection $T(S) \to S$ is locally trivial and each fibre is homeomorphic to the cone over $L(S)$. In other words, if $x \in S$ a stratum of complex dimension d, then x has a neighbourhood U homeomorphic to $B^{2d} \times \operatorname{Cone} L(S)$. In particular $H^k(U,\mathbb{Q}) = 0$ for $k \neq 0$.

We need also a result about cohomology with compact support. Consider the pair $(U, U \cap S) \cong (B^{2d} \times \operatorname{Cone}(L(S)), B^{2d}) = B^{2d} \times (\operatorname{Cone}(L(S), x)$. Since $U - U \cap S \cong B^{2d+1} \times L(S)$, by [Bor84, Lemma V.3.8] one has

$$H^k_c(U - U \cap S; \mathbb{Q}) = H^k_c(B^{2d+1} \times L(S); \mathbb{Q}) = H^{k-2d-1}(L(S); \mathbb{Q}).$$

The exact sequence of cohomology with compact supports for the pair $(U, U \cap S)$ then shows

$$H^k_c(U, \mathbb{Q}) = \begin{cases} 0 & \text{if } k \leq 2d+1 \\ H^{k-2d-1}(L(S); \mathbb{Q}) & \text{if } k > 2d+1. \end{cases}$$

We treat the special case where X is a local complete intersection.

Lemma C.8. *Let X be a local complete intersection of dimension n and let $x \in S$, a stratum of dimension d. Then for all sufficiently small neighbourhoods U of x, $H^k(U;\mathbb{Q}) = 0$ unless $k = 0$ and $H^k_c(U;\mathbb{Q}) = 0$ unless $k = n+d, n+d+1, 2n$.*

Proof. This follows from the previous example and the fact (see [Hamm71]) that $L(S)$, the link of a local complete intersection singularity of dimension $n-d$, is $(n-d-2)$-connected, and hence it has no cohomology in positive dimensions up to dimension $n-d-2$, and by duality neither in all remaining dimensions except perhaps in dimensions $n-d-1, n-d$ and $2n-2d-1$. □

The existence of local Whitney stratifications implies:

Proposition C.9. *A complex space X is perfect so that (see Corr. B.13 and Theorem B.15) all cohomology theories give the same for X, all good equivalences for local systems (Lemma B.34) apply to X. Moreover X has also a universal cover space.*

C.2 Fibrations, and the Topology of Singularities

A **fibration** is a proper surjective holomorphic map $f : X \to S$ between connected complex manifolds with connected fibres. There is a dense open subset $U \subset S$ over which f is a submersion. The **discriminant** of f is the smallest closed subvariety $\Delta(f)$ of S such that f is a submersion over the complement. Smooth fibrations are topological fibrations as a consequence of:

Theorem C.10 (EHRESMANN'S THEOREM). *A smooth fibration is locally differentiably trivial.*

For an easy proof see [Mor-Ko, p. 19]). A smooth fibration over a manifold S of dimension m is also called an **smooth m-parameter family**.

A fibration is locally differentiably trivial over the complement of the discriminant locus. We shall now consider in some more detail what happens over a 1-dimension base near a critical value , i.e. a point in the discriminant locus. We speak of a **one-parameter degeneration**.

C.2.1 The Milnor Fibration

The Milnor fibration arose classically [Mil68] as follows. Let $f : (\mathbb{C}^{n+1}, 0) \to (\Delta, 0)$ be a holomorphic map which has rank 1 everywhere, except possibly at points over the origin. Let $\Delta(\rho) = \{t \in \mathbb{C} \mid |t| < \rho\}$, the disk of radius ρ centred at 0 and put $\Delta^*(\rho) = \Delta(\rho) - \{0\}$. Milnor has shown [Mil68] that for $0 < \rho \ll r$ small enough

$$f : B(r) \cap f^{-1}\Delta^*(\rho) \to \Delta^*(\rho). \tag{C–1}$$

is a smooth locally trivial fibration. The fibre of this fibration, the **Milnor fibre** $\mathrm{Mil}_{f,x}$, has the homotopy type of a CW-complex of dimension $\leq n$. Moreover, if 0 is an isolated singularity, it has the homotopy type of a wedge of μ_x n-spheres. The number μ_x is called the **Milnor number** of the singularity (X, x). An n-cycle supported in the Milnor fibre of course is a boundary of a chain in $B(r)$. Therefore any such cycle is called a **vanishing cycle**. A collection of μ_x cycles which give a basis for $H_n(\mathrm{Mil}_{f,x})$ is called a **basis of vanishing cycles**.

Let $f : X \to S$ be a one-parameter fibration. For $x \in X$, t a regular value of f near $f(x)$, by means of the inclusion $\mathrm{Mil}_{f,x} \hookrightarrow X_t$ any vanishing cycle can be viewed as an n-cycle in X_t. If the induced map

$$H_n(\mathrm{Mil}_{f,x}) \to H_n(X_t) \tag{C–2}$$

is injective, we say that **vanishing cycles survive globally**.

With appropriate changes some of this discussion remains valid if $(\mathbb{C}^{n+1}, 0)$ gets replaced by the germ of an analytic subvariety (X, x) inside a ball $(B(r), 0) \subset (\mathbb{C}^N, 0)$. Indeed, Lê has shown [Le78] that in this set-up again for $0 < \rho \ll r$ small enough (C–1) is topologically locally trivial with fibre given by

$$\mathrm{Mil}_{f,x} := B(x, r) \cap f^{-1}(t), \quad 0 < |t| \ll r \text{ small enough.} \tag{C–3}$$

C.2.2 Topology of One-parameter Degenerations

We now turn to one-parameter degenerations $f : X \to \Delta$, i.e. f is proper holomorphic map from a complex manifold X onto some small disk Δ and having 0 as its only critical value. Over $\Delta^* = \Delta - \{0\}$ the fibration is a fibre bundle and hence completely described by its monodromy when we move along a circle $S^1 \subset \Delta$. Let us now assume that $x \in f^{-1}(0)$ is the only critical point of f so that we can speak of the Milnor fibre $\mathrm{Mil}_{f,x}$. We consider at the same time the local and global situation and put

$$F = \begin{cases} X_t, t \in S^1 & \text{if } f \text{ is proper,} \\ \mathrm{Mil}_{f,x} & \text{if } f \text{ is a local Milnor fibration.} \end{cases} \tag{C–4}$$

Let θ be the angle on S^1. Any lifting to $f^{-1}S^1$ of the associated vector field $\partial/\partial\theta$ on S^1 integrates to a flow and induces

$$h : F \xrightarrow{\sim} F \quad \text{(the \textbf{geometric monodromy} of the fibration).}$$

which is well defined up to isotopy and it induces

$$T := (h^*)^{-1} : H^q(F) \to H^q(F) \quad \text{(the \textbf{local monodromy}).} \tag{C–5}$$

In the local situation, since h can be realized by a homeomorphism which is the identity near the boundary of the Milnor fibre, $h - \mathrm{id}$ maps Borel-Moore cycles on the Milnor fibre to compactly supported cycles. In cohomology it induces

$$\mathrm{var} := T - I : H^q(\mathrm{Mil}_{f,x}) \to H^q_c(\mathrm{Mil}_{f,x}) \quad \text{(the \textbf{variation})} \tag{C–6}$$

The topological description of the degeneration is now completed by means of the following basic result:

Proposition C.11. *Let X be a manifold and let $f : X \to \Delta$ be a proper map which is smooth over Δ^*. Put $X_0 = f^{-1}(0)$. There is a fibrewise retraction $r : X \to X_0$. In particular, the homotopy type of X is that of the central fibre. Let $i_t : X_t \hookrightarrow X$ be the inclusion. If x is an isolated singularity for a suitable choice of a retraction r putting $r_t = r \circ i_t$, we have an inclusion $r_t^{-1} x \hookrightarrow \mathrm{Mil}_{f,x}$ which is a homotopy equivalence.*

The map $r_t : X_t \to X_0$ is called the **specialization map**. Topologically it can be considered as a map $r : F \to X_0$ compatible with the monodromy action $h : F \to F$ and X is obtained in two steps. First let X' be the quotient of $F \times \mathbb{R}$ by the relation $(x,t) \sim (h(x), t+1)$; then X is homeomorphic to the mapping cone of $r : X' \to X_0$.

Remark C.12. i) In the above setting, several constructions for a retraction have been proposed, see e.g. [Clem69, Clem77], [A'Cam, p. 238].

ii) The above proposition remains true when X is any complex variety mapping properly to the disk Δ. Since f is proper, we may view X as embedded in some complex manifold M. We put $X_t = f^{-1}(t)$. Following [G-M88, Part II, §6.13], we may then assume that f is stratified (Prop. C.7,3) and we can assume that the strata of Δ consist of Δ^* and 0. Then f is a topological fibration over Δ^* and there is a fibrewise retraction $r : X \to X_0$ of the total space onto X_0, but $r|X_0$ might not be the identity.

Introduce the complex of sheaves

$$\psi_f \underline{\mathbb{Z}}_X := R r_{t*} i_t^* \underline{\mathbb{Z}}_X,$$

the **complex of nearby cocycles**. Prop. C.11 can be used to describe its cohomology sheaves $H^q(\psi_f \underline{\mathbb{Z}}_X)$. The stalk at x is the limit over open neighbourhoods U_x of x of the (ordinary integral) cohomology group $H^q(r_t^{-1}(U_x))$. Since for all small enough U_x the inverse image $r_t^{-1}(U_x)$ is homotopy equivalent to $\mathrm{Mil}_{f,x}$, we find

$$[H^q(\psi_f \underline{\mathbb{Z}}_X)]_x = (R^q r_{t*} \underline{\mathbb{Z}}_X)_x = H^q(\mathrm{Mil}_{f,x}) \qquad \text{(C–7)}$$

The specialization map induces a natural map of complexes of sheaves:

$$\mathrm{sp} : \underline{\mathbb{Z}}_{X_0} \to R r_{t*} \underline{\mathbb{Z}}_X.$$

On the level of cohomology sheaves at x, this induces the same map as the one induced by the "specialization" map $\mathrm{sp} : r_t^{-1}(U_x) \to U_x$. Its cone is the suspension over the Milnor fibre so that $H^q(\mathrm{Cone}^\bullet(\mathrm{sp})) = \tilde{H}^{q+1}(\mathrm{Mil}_f)$. Recall (Theorem B.22) that there is a shift of 1 when we calculate this instead with cones of maps between complexes. So, if we define the **complex of vanishing cocycles** by

$$\phi_f \underline{\mathbb{Z}}_X := \mathrm{Cone}^\bullet(\mathrm{sp} : \underline{\mathbb{Z}}_{X_0} \to R r_{t*} \underline{\mathbb{Z}}_X),$$

we have

$$\tilde{H}^q(\mathrm{Mil}_{f,x}) \simeq [\phi_f \underline{\mathbb{Z}}_X]_x. \qquad \text{(C–8)}$$

We want to investigate under which condition the direct images $R^k f_* \underline{\mathbb{Z}}_X$ are locally constant sheaves. Two properties play a role. The first has to do with the monodromy action. We explain this first. Since there is a retraction of X onto the singular fibre X_0, the inclusion $X_t \hookrightarrow X$ induces a homomorphism

$$H^k(X_0) \cong H^k(X) \to H^k(X_t)_{\mathrm{inv}}$$

to the submodule of classes invariant under the monodromy-operator T. We say that the **local invariant cycle property holds** if this map is a surjection. Equivalently, the adjunction homomorphism

$$a_k : R^k f_* \underline{\mathbb{Z}}_X \to j_* j^* R^k f_* \underline{\mathbb{Z}}_X$$

is a surjection. Indeed, this is clear on Δ^* while for any local system $\mathbb{L}$ on Δ, the stalk of the sheaf $j_*\mathbb{L}$ at 0 consists of global sections of $\mathbb{L}|\Delta^*$, and these can be identified with the T-invariants in a stalk $\mathbb{L}_x$ over a point $x \in \Delta^*$.

The second property concerns the behaviour of vanishing cycles. We reformulate for cohomology the property that the map (C–2) is into. Consider F, the Milnor fibre as sitting in a general (smooth) fibre X_*. We let

$$H^n(X_*) \xrightarrow{r^*} H^n(F) \tag{C–9}$$

be the restriction map. Since it is dual to (C–2), vanishing cycles survive globally precisely when r^* is onto.

Using the above two properties we formulate our criterion:

Lemma C.13. *Let $f : X \to S$ be a one-parameter family of n-dimensional varieties with critical locus $\Delta(f)$. Assume that f has isolated singularities. Then $R^k f_*\mathbb{Z}_X$ is locally constant in the following cases:*

1) *If $k \neq n, n+1$;*
2) *if $k = n+1$ and the vanishing cycles survive globally, i.e if the restriction map* (C–9) *is surjective near each critical point.*

If $n = k$ and the local invariant cycle property holds near all critical values, then the adjunction morphism

$$R^n f_*\mathbb{Z}_X \to j_*j^*R^n f_*\mathbb{Z}_X$$

is an isomorphism. In this case $R^n f_*\mathbb{Z}_X$ is not locally constant, but it is completely determined by the local system $j^*R^n f_*\mathbb{Z}_X$.

Proof. The result is clear away from the critical locus and so the assertion is local with respect to the base. We may thus assume that $S = \Delta$ and that f is smooth over Δ^*. For simplicity we assume that X_0 has only one singular point. We use the shorthand notation $B \cap (X - X_0) \to \Delta^*$ for its Milnor fibration (C–1). Let X_* be the general (smooth) fibre of f. Excision (Theorem B.2) shows that

$$\tilde{H}^{k+1}(X, X_*) \cong \tilde{H}^{k+1}(B, F) \cong \tilde{H}^k(F)$$

and since F has the homotopy type of a bouquet of n-spheres this is non-zero only for $k \neq n-1$. The long exact sequence for the pair (X, X_*) then shows

$$H^k(X) \xrightarrow{\sim} H^k(X_*), \quad k \neq n, n+1.$$

Since X retracts onto X_0 this proves 1) and also yields an exact sequence

$$0 \to \tilde{H}^n(X_0) \to \tilde{H}^n(X_*) \xrightarrow{r} \tilde{H}^n(F) \to \tilde{H}^{n+1}(X_*) \to \tilde{H}^{n+1}(X_*) \to 0 \tag{C–10}$$

which proves 2): the direct image $R^{n+1}f_*\mathbb{Z}$ is locally constant if r is surjective.

The last assertion for $n = k$ is clear. □

C.2.3 An Example: Lefschetz Pencils

A classical method to describe the cohomology of a projective variety inductively is by means of its hyperplane sections. This method was developed by Lefschetz. The idea is to vary a hyperplane section in a pencil and build the cohomology from that of a hyperplane section and using the monodromy. As a first step, there is Lefschetz hyperplane theorem which states roughly that new cohomology appears only in the middle dimension. Modern proofs use the fact that the complement of a hyperplane section being affine has no cohomology in ranks beyond its complex dimension:

Theorem C.14. *An affine variety of pure dimension n has the homotopy type of a CW complex of dimension $\leq n$.*

For a proof, see [G-M88, Part II, 5.1*]. As announced, we will deduce:

Theorem C.15 (Lefschetz' hyperplane theorem). *Let X be an $n+1$-dimensional projective variety and $i : Y \hookrightarrow X$ a hyperplane section containing all the singularities of X. Then*

$$i^* : H^k(X) \to H^k(Y) \quad \begin{cases} \text{is an isomorphism} & \text{if } k \leq n-1 \\ \text{injective} & \text{for } k = n. \end{cases}$$

Dually

$$i_* : H_k(Y) \to H_k(X) \quad \begin{cases} \text{is an isomorphism} & \text{if } k \leq n-1 \\ \text{surjective} & \text{for } k = n. \end{cases}$$

Proof. By the long exact sequence in cohomology for the pair (X,Y) it suffices to show that $H^k(X,Y) = 0$ for $k \leq n$. Now By Lemma B.21 and the Poincaré Duality Theorem B.24 (note that here it is essential that $X - Y$ is smooth) we have $H^k(X,Y) = H^k_c(X-Y) = H_{2n+2-k}(X-Y)$. By Theorem C.14 this group indeed vanishes in the range we are interested in. The dual statement follows from the fact that the surjective Kronecker homomorphism $0 = H^k(X,Y) \to \mathrm{Hom}(H_k(X,Y),\mathbb{Z})$ is an isomorphism for all $k \leq n$, recalling that its kernel is the torsion submodule of $H_{k-1}(X,Y)$ which is also the torsion submodule of $H^k(X,Y)$. □

Remark C.16. 1) The homotopy type of a smooth quasi-projective variety X is that of a *general* hyperplane section Y modulo adjoining cells of dimension $n+1$. In particular, as before $H^k(X,Y) = 0$ for $k \leq n$. For a proof see [Hamm83, Theorem 5].
2) In the statement and proof of Theorem C.15 we may replace the constant coefficient $\mathbb{Z}$ by a local system. This follows from the fact that the exact sequences that we use remain valid when we use local systems as coefficients.

This result motivates the following definition.

$$H^n_{\text{fixed}}(Y;\mathbb{Q}) := \operatorname{Im}(i^* : H^n(X;\mathbb{Q}) \to H^n(Y;\mathbb{Q})), \tag{C–11}$$
$$H^n_{\text{var}}(Y;\mathbb{Q}) := \operatorname{Ker}(i_! : H^n(Y;\mathbb{Q}) \to H^{n+2}(X;\mathbb{Q})). \tag{C–12}$$

It is not hard to see that we have a direct sum decomposition, orthogonal with respect to cup product

$$H^n(Y;\mathbb{Q}) = H^n_{\text{fixed}}(Y;\mathbb{Q}) \oplus H^n_{\text{var}}(Y;\mathbb{Q}).$$

The theory of Lefschetz pencils describes the second summand in terms of the monodromy of the tautological family of all smooth hyperplane sections of X.

Definition C.17. A **Lefschetz pencil** on X is a pencil of hyperplane sections $\{X_u\}_{u\in\mathbb{P}^1}$, which has at most one ordinary double point.

It is well known that Lefschetz pencils abound [Katz73, Proposition 3.3].

Definition C.18. The **Lefschetz fibration** associated to the Lefschetz pencil is defined as follows. Let $B \subset X$ be the base locus of the pencil, the codimension two submanifold cut out by the linear space common to all members of the pencil and introduce

$$\tilde{X} := \operatorname{Bl}_B X \to X$$

the blow up of X in this base locus. The Lefschetz fibration is the natural fibration

$$f : \tilde{X} = \{(m,u) \in X \times \mathbb{P}^1 \mid m \in X_u\} \longrightarrow \mathbb{P}^1 .$$

The Zariski-Van Kampen theorem [Kamp] states:

Proposition C.19. *Let $X \subset \mathbb{P}^N$ be a projective manifold of dimension $(n+1)$ and let $X_U \to U$ be the tautological family of its smooth hypersurface sections and let $X_\ell \to \ell$ be a Lefschetz pencil. The natural map $\pi_1(U \cap \ell) \to \pi_1(U)$ is a surjection.*

As a consequence, the monodromy actions of the two agree on $H^k(Y)$, where Y is any smooth fibre of f.

Let $\Delta(f) \subset \mathbb{P}^1$ be the discriminant locus of f. Each singular fibre X_σ $\sigma \in \Delta(f)$ has exactly one ordinary double point x_σ, so there is exactly one vanishing cycle. Take a generator for $H_n(F_\sigma)$, where F_σ is the Milnor fibre and let $\delta_\sigma \in H_n(X_t)$ be the class of its image in a nearby fibre X_t. A central result is the following theorem, a proof of which can be found in [Lamot].

Theorem C.20. *The self-intersection of the image of a vanishing cycle δ_σ is given by*

$$(\delta_\sigma, \delta_\sigma) = 0 \qquad \textit{for } n \textit{ odd},$$
$$(\delta_\sigma, \delta_\sigma) = (-1)^{n/2}2 \quad \textit{for } n \textit{ even}.$$

The action of the local monodromy T_σ is trivial, except for rank n, where it is described by the ***Picard-Lefschetz formulas***

$$T_\sigma(\alpha) = \alpha + (-1)^{\frac{1}{2}(n+1)(n+2)} \langle \alpha, \delta_\sigma \rangle \delta_\sigma^\vee, \quad \alpha \in H^n(X_t; \mathbb{Q}),$$

where $\delta_\sigma^\vee$ *is Kronecker dual to* δ_σ *and* $\langle -, - \rangle$ *denotes the Kronecker duality pairing* (B–6).

In particular α is invariant under local monodromy if and only if α annihilates the space $\mathbb{Q}[\delta_\sigma]$ generated by the vanishing cycle. Let Δ_σ be a sufficiently small disk centred at σ. The sequence dual to (C–10) shows that

$$\begin{aligned}\mathbb{Q}[\delta_\sigma]^\perp &= \operatorname{Ker}\left[H_n(X_t;\mathbb{Q}) \to H_n(f^{-1}\Delta_\sigma;\mathbb{Q})\right]^\perp \\ &= \operatorname{Im}\left[H^n(f^{-1}\Delta_\sigma;\mathbb{Q}) \to H^n(X_t;\mathbb{Q})\right].\end{aligned}$$

We deduce:

Corollary C.21. *For a Lefschetz fibration, the local invariant cycle property holds.*

The restriction map r in (C–10) is surjective, exactly when $\delta_\sigma \neq 0$ (in rational homology) which is always the case when n is even since then $(\delta_\sigma, \delta_\sigma) \neq 0$. However for n odd δ_σ can be a torsion class. Indeed: for any Lefschetz fibration of odd-dimensional quadric hypersurfaces in projective space, the fibres have no middle cohomology and so $\delta_\sigma = 0$. However, it can be shown that this anomaly does not occur if the degree is sufficiently large. The next theorem C.23 shows that there is no rational vanishing cohomology if $H^n(X) \to H^n(Y)$ is an isomorphism. It also states that all vanishing co-cycles are conjugate under monodromy, hence if the class in rational cohomology is zero for one of them this must be the case for all of them. So the anomalous case occurs precisely when the restriction $H^n(X) \to H^n(Y)$ is an isomorphism.

Corollary C.22. *Let* $f : \tilde{X} \to \mathbb{P}^1$ *be a Lefschetz fibration associated to* X *with general (smooth) fibre* Y. *Let* $\Delta(f)$ *be the critical locus and* $j : \mathbb{P}^1 - \Delta(f) \hookrightarrow \mathbb{P}^1$ *the inclusion. Assume that the restriction* $H^n(X) \to H^n(Y)$ *is not an isomorphism (this is always the case when n is odd and generically when n is even). Then* $R^k f_* \mathbb{Q}$ *is locally constant for* $k \neq n$, *while* $R^n f_* \mathbb{Q} = j_* j^* R^n f_* \mathbb{Q}$.

As to the global situation, we have the following classical result [Lamot], [Katz73b]:

Theorem C.23. 1) *The variable cohomology* $H^n_{\text{var}}(Y;\mathbb{Q})$, Y *a smooth fibre of the Lefschetz fibration* $f : \tilde{X} \to \mathbb{P}^1$, *coincides with the* ***vanishing cohomology****, i.e. it is spanned by the classes* $\delta_\sigma^\vee$ *dual to the vanishing cycles. These classes are conjugate under the global monodromy of the fibration.*
2) *The fixed cohomology* $H^n_{\text{fixed}}(Y;\mathbb{Q})$ *coincides with the image under restriction*

$$H^n(\tilde{X};\mathbb{Q}) \to H^n(Y;\mathbb{Q}).$$

Corollary C.24. 1) *The fixed cohomology equals the subgroup of* $H^n(Y;\mathbb{Q})$ *left invariant by the global monodromy group of the Lefschetz fibration.*

2) *Introducing the constant sheaf* $\mathbb{I}$ *of* ***invariant cohomology*** *with fibre*

$$\mathbb{I}_t := H^n_{\text{fixed}}(X_t; \mathbb{Q}), \quad t \in \mathbb{P}^1 - \Delta(f)$$

and the locally constant sheaf $\mathbb{V}$ *of* ***vanishing cohomology*** *with fibre*

$$\mathbb{V}_t := H^n_{\text{var}}(X_t; \mathbb{Q}), \quad t \in \mathbb{P}^1 - \Delta(f)$$

there is an orthogonal direct sum decomposition

$$j^* R^n f_* \mathbb{Q} = \mathbb{I} \oplus \mathbb{V}$$

and $\mathbb{V}$ *is absolutely irreducible.*

Historical Remarks. Stratified spaces have been introduced by Whitney and Thom. Unfortunately much of this is folklore. The relevant literature can be collected from [G-M88].

[Mil68] remains the standard reference for Milnor fibres. The localized version, as developed in § C.2.2 which is well-adapted to both mixed Hodge theory and D-modules is probably due to Deligne. The terminology "the local invariant cycle theorem holds" as well as "vanishing cycles survive globally" is ours.

Lefschetz pencils made their appearance for the first time in Lefschetz' fundamental treatise [Lef] in which an inductive approach to the study of algebraic cycles has been proposed. In § C.2.3 we follow closely the presentation from [Katz73] and [Katz73b]. Several attempts have been made to make Lefschetz' arguments meet modern standards of rigor [Lamot].

References

[A-dJ] Abramovich, D. and A. J. de Jong: Smoothness, semistability and toroidal geometry, J. Algebr. Geom. **6** 789–801 (1997).

[A'Cam] A'Campo, N.: La fonction zêta d'une monodromie, Comm. Math. Helv. **50**, 233–248 (1975)

[Adams] Adams, J. F.: On the cobar construction, Colloque de Topologie Algébrique (Louvain 1956), George Thone, Paris, 81–87 (1957)

[A-N] Akizuki, Y. and S. Nakano: Note on the Kodaira-Spencer's proof of Lefschetz' theorems, Proc. Japan Acad. **30**, 166–172 (1954)

[Ara] Arapura, D: The Leray spectral sequence is motivic, Invent. Math. **160**, 567–589 (2005)

[At-Hir] Atiyah, M. F. and F. Hirzebruch: Analytic cycles on complex manifolds, Topology **1**, 25–45 (1962)

[BaV02] Barbieri-Viale, L.: On algebraic 1-motives related to Hodge cycles, in *Algebraic Geometry (A Volume in Memory of P. Francia)*, Walter de Gruyter, Berlin/New York, 25–60 (2002)

[BaV07] Barbieri-Viale, L.: On the theory of 1-motives, in *Algebraic Cycles and Motives, August 30– September 3, 2004, Leiden*, Vol. **1**, ed. J. Nagel and C. Peters, Lect. Notes of the London Math. Soc., Cambridge Univ. Press, 2007.

[Barl] Barlet, D.: Familles analytiques de cycles et classes fondamentales relatives, in *Fonctions de plusieurs variables complexes IV*, Springer Lect. Notes in Math. **807**, 1–24 (1980)

[Barth] Barth, W.: Transplanting cohomology classes in complex projective space, Amer. J. Math. **92**, 951–967 (1970)

[B-B-D] Beilinson, A., J. Bernstein and P. Deligne: Faisceaux pervers, in *Analyse et topologie sur les espaces singuliers I*, Astérisque **100**, (1982)

[Beil85] Beilinson, A.: Higher regulators and the values of L-functions, J. Soviet Math **30**, 2036–2070 (1985)

[Beil86] Beilinson, A.: Notes on absolute Hodge cohomology, in *Applications of K-theory to Algebraic Geometry and Number Theory, I*, Contemp. Math. A. M. S., Providence, R.I. **55**, 35–68 (1986)

[Bi04] Bittner, F.: The universal Euler characteristic for varieties of characteristic zero, Comp. Math. **140**, 1011-1032 (2004)

[Bi05] Bittner, F.: On motivic zeta functions and the motivic nearby fibre, Math. Z. **249**, 63–83 (2005)

[B-H-P-V] Barth, W., K. Hulek, C. Peters and A. Van de Ven: *Compact Complex Surfaces, Second Enlarged Edition*, Ergebnisse der Math., **3**, Springer Verlag (2003)

[Bj] Björk, J-E.: *Analytic $\mathcal{D}$-Modules and Applications*, in Mathematics and Its Applications **247**, Kluwer Academic Publishers, Dordrecht etc. (1993).

[B-M] Bloch, S. and J. P. Murre: On the Chow groups of certain types of Fano threefolds, Comp. Math. **39**, 47–105 (1974)

[Bor84] Borel, A. et al.: *Intersection cohomology*, Progress in Math. **50**, Birkhäuser Verlag (1984)

[Bor87] Borel, A. et al.: *Algebraic D-modules*, Perspectives in Math. **2**, Academic Press, Boston, etc (1987)

[Bor-M] Borel, A. and J. Moore: Homology for locally compact spaces, Mich. Math. J. **4**, 137–159 (1960)

[Bott] Bott, R.: Homogeneous vector bundles, Ann. Math. **6**, 203–248 (1957)

[B-P] Bogomolov, F. A. and T. G. Pantev: Weak Hironaka theorem, Math. Res. Lett. **3**, 299–307 (1996)

[Bre] Brélivet, Th.: The Hertling conjecture in dimension 2, preprint math.AG/0405489

[Br-H] Brélivet, Th. and C. Hertling: Bernoulli moments of spectral numbers and Hodge numbers, preprint math.AG/0405501

[B-S-Y] Brasselet, J.-P., J. Schuermann and S. Yokura: Hirzebruch classes and motivic Chern classes for singular spaces, preprint math.AG/0503492

[Bry] Brylinski, J.-L.: Transformations canoniques, dualité projective, théorie de Lefschetz, transformation de Fourier et sommes trigonométriques, in *Géométrie et Analyse Microlocales*, Astérisque **140–141**, Soc. Math. France, 3–134 (1986)

[Car79] Carlson, J.: Extensions of mixed Hodge structures, in *Journées de Géométrie Algébriques d'Angers 1979*, Sijthoff, Noordhoff, Alphen a/d Rijn, 107–127 (1979)

[Car85a] Carlson, J.: Polyhedral resolutions of algebraic varieties, Tr. A.M.S. **292**, 595–612 (1985)

[Car85b] Carlson, J.: The one-motif of an algebraic surface, Comp. Math. **56**, 271–314 (1985)

[Car87] Carlson, J.: The geometry of the extension class of a mixed Hodge structure, in *Algebraic Geometry, Bowdoin 1985* Proc. Symp. Pure Math. A.M.S. **46-2**, 199–222 (1987)

[Cart] Cartan, H.: Faisceaux analytiques sur les variétés de Stein (Exposé de H. Cartan, 19-5-52), in *Séminaire H. Cartan, 1951-52* See also: `http://archive.numdam.org/article/SHC_1951-1952__4__A18_0.djvu`

[Cart57] Cartan, H.: Quotients d'un espace analytique par un groupe d'automorphismes, in *Algebraic Geometry and Topology*, Princeton Univ. Press, Princeton, NJ (1957), 165–180

[C-C-M] Carlson, J., H. Clemens and J. Morgan: On the mixed Hodge structure associated to π_3 of a simply connnected projective manifold, Ann. Sci. E. N. S. **14**, 323–338 (1981).

[C-K-S86] Cattani, E., A. Kaplan and W. Schmid: Degeneration of Hodge structures, Ann. Math. **123**, 457–535 (1986)

[C-K-S87] Cattani, E., A. Kaplan and W. Schmid: L^2 and intersection cohomologies for a polarizable variation of Hodge structures, Inv. Math. **87**, 217–252 (1987)

[C-D-K] Cattani, E., P. Deligne and A. Kaplan: On the locus of Hodge classes, J. Amer. Math. Soc. **8**, 483–506 (1995)

[Ch] Chern, S.-S.: On a generalization of Kähler geometry, Lefschetz jubilee volume. Princeton Univ. Press (1957) 103–121.

[C-G] Clemens, C. H. and P. Griffiths: The intermediate jacobian of the cubic threefold, Ann. Math., **95**, 281–356 (1972)

[Clem69] Clemens, C. H.: Picard Lefschetz theorem for families of non singular algebraic varieties acquiring ordinary singularities, Trans. A.M.S. **136** 93–108 (1969)

[Clem77] Clemens, C. H.: Degeneration of Kähler manifolds, Duke Math. J. **44** 215–290 (1977)

[C-K-M] Clemens, H., J. Kollár, J. and S. Mori: *Higher dimensional complex geometry*, Astérisque **166** (1988)

[Chen76] Chen, K.-T.: Reduced bar construction on de Rham complexes, in *Algebra, Topology and Category Theory*, ed. Heller, A, & M. Tierney, Academic Press, New-York, 19–32 (1976)

[Chen77] Chen, K.-T.: Iterated path integrals, Bull. A.M.S., **83**, 831–879 (1977)

[Chen79] Chen, K.-T.: Extension of C^∞ function algebras and Malcev completion of π_1, Adv. Math. **23**, 181–210 (1979)

[Coll] Collino, A.: The Abel-Jacobi map is an isomorphism for cubic fivefolds, Pacif. J. Math. **122**, 43–56 (1986)

[Cor] Corlette, K.: Flat G-bundles with canonical metrics, J. Diff. Geom. **28** 361–382 (1988).

[dC-M] de Cataldo, M. and L. Migliorini: The Hodge Theory of Algebraic maps, Ann. Sci. École Norm. Sup. **38** 693–750 (2005)

[Del68] Deligne, P.: Théorème de Lefschetz et critère de dégénérescence de suites spectrales, Publ. Math. IHÉS **35**, 107–126 (1968)

[Del70] Deligne, P.: *Équations différentielles à points singuliers réguliers*, Springer Lecture Notes in Math., **163**, (1970)

[Del71] Deligne, P.: Théorie de Hodge II, Publ. Math. I.H.E.S **40**, 5–58 (1971)

[Del72] Deligne, P: La conjecture de Weil pour les surfaces K3, Invent. Math. **15**, 206–226 (1972)

[Del73] Deligne, P.: Intersections complètes, SGA 7, Exposé XI, Springer Lecture Notes in Math. **340**, 39–61 (1973)

[Del74] Deligne, P.: Théorie de Hodge III, Publ. Math., I. H. E. S **44**, 5-77 (1974)

[Del80] Deligne, P.: La conjecture de Weil II, Publ. Math. I.H.E.S. **52**, 137–252 (1980)

[DMOS] Deligne, P. (Notes by J. Milne): Hodge cycles on abelian varieties, in *Hodge cycles, motives and Shimura varieties*, Springer Lecture Notes in Math. **900**, 9-100 (1982)

[Del-Ill] Deligne, P. and L. Illusie: Relèvements modulo p^2 et décomposition du complexe de De Rham, Inv. Math. **89**, 247–270 (1987)

[Del-G-M-S] Deligne, P. P.A. Griffiths, J.W. Morgan and D. Sullivan: Real homotopy theory of Kähler manifolds, Invent. Math. **29**, 245–274 (1975)

[Dem] Demailly, J.-P.: Théorie de Hodge L^2 et Théorèmes d'annulation, in *Panoramas et Synthèse* **3**, Soc. Math. France, 3–111 (1997)

[Dim] Dimca, A.: *Sheaves in Topology*, Universitext , Springer Verlag, (2004)

[DL99] Denef, J. and F. Loeser: Motivic exponential integrals and a motivic Thom-Sebastiani Theorem, Duke Math. J. **99**, 285–309 (1999)

[D-K] Danilov, V. and A. Khovanski: Newton polyhedra and an algorithm for computing Hodge-Deligne numbers, Math. U.S.S.R. Izvestia **29**, 274–298 (1987)

[DL] Denef, J., Lœser, F.: Motivic exponential integrals and a motivic Thom-Sebastiani theorem, Duke Math. J. **99**, 285–309 (1999)

[Don-Kr] Donaldson, S.K. and P.B. Kronheimer: *The Geometry of Four-Manifolds*, Oxford Univ. Press, Oxford (1990)

[D-S03] Douai, A and C. Sabbah: Brieskorn lattices and Frobenius structures (I), Ann. Institut Fourier (Grenoble) **53** 1055–1116 (2003)

[D-S04] Douai, A and C. Sabbah: Gauss-Manin systems, Brieskorn lattices and Frobenius structures (II), in *Frobenius manifolds (Quantum cohomology and singularities)*, C. Hertling and M. Marcolli eds, Aspects of Mathematics **E36**, Vieweg,1–18 (2004)

[Du83] Durfee, A.: Neighborhoods of algebraic sets, Transactions A.M.S, **276**, 517–530 (1983)

[Du83b] Durfee, A.: Mixed Hodge structures on punctured neighborhoods, Duke Math. J. **50**, 1017–1040 (1983)

[Du87] Durfee, A.: Algebraic varieties which are a disjoint union of subvarieties, in *Geometry and Topology, Manifolds, varieties, and knots*, Clint McCrory, Theodore Shifrin eds. Marcel Dekker, Inc., New York and Basel, 99–102 (1987)

[DuB] Du Bois, Ph.: Complexe de De Rham filtré d'une variété singulière Bull. Soc. Math. Fr. **109**, 41–81 (1981)

[Du-H] Durfee, A. and R. M. Hain: Mixed Hodge structures on the homotopy of links, Math. Ann. **280**, 69–83 (1988)

[ES98] Ebeling, W. and J. Steenbrink: Spectral pairs for isolated complete intersection singularities, J. Alg. Geom. **7**, 55–76 (1998)

[EGA] Grothendieck, A. and J. Dieudonné: *Éléments de Géométrie Algébrique, I–IV*, Publ.Mat. I.H.E.S. **4** (1960) **6** (1961) **11** (1961), **17** (1963) **20** (1964) **24** (1966) **32** (1967)

[ElZ] El Zein, F.: Complexe dualisant et applications à la classe fondamentale, Bull. Soc. Math. Fr., Mém., **58**, 5–66 (1978)

[ElZ86] El Zein, F.: Théorie de Hodge des cycles évanescents, Ann. scient. É.N.S. **19**, 107–184 (1986)

[ElZ03] El Zein, F.: Hodge-De Rham theory with degenerating coefficients, Rapport de Recherche 03/10-1, Univ. de Nantes, Lab. de Mathématiques Jean Leray, UMR 6629.

[ElZ-Z] El Zein, F. and S. Zucker: Extendability fo normal functions associated to algebraic cycles in *Topics in transcendental algebraic geometry*, P. Griffiths ed., Princeton Un. Press **106**, 269–288 (1986)

[Es-V88] Esnault, H. and E. Viehweg: Deligne-Beilinson cohomology, in *Perspectives in Mathematics* **4**, Academic Press, Inc. 43–92 (1988)

[Es-V92] Esnault, H. and E. Viehweg: *Lectures on Vanishing Cohomology*, DMV Seminar **20**, Birkh. Verlag, Basel etc. (1992)

[Falt] Faltings, G.: *p*-adic Hodge theory, Journ. A.M.S. **1**, 255-299 (1988)

[F-M] Friedlander, E., B. Mazur: Filtration on the homology of algebraic varieties, Memoirs of the A.M.S. **529** (1994)

[Fuj] Fujiki, A.: Duality of Mixed Hodge Structures of Algebraic Varieties, Publ. RIMS, Kyoto Univ. **16**, 635–667 (1980)

[Fult] Fulton, W.: *Intersection Theory*, Erg. Math. 3. Folge Bnd. **2**, Springer Verlag, Berlin Heidelberg etc. (1984)

[Ge-Ma] Gelfand, S.I., and Yu. I. Manin: Homological algebra, in *Algebra V*, A. I. Kostrikin, I.R. Shafarevich (Eds), Springer Verlag, Berlin Heidelberg etc. (1994)

[G-M82] Goresky, M. and R. MacPherson: On the topology of complex algebraic maps, in *Algebraic Geometry, La Rábida 1982*, Springer Lecture Notes in Math. **961**, 119–129 (1982)

[G-M83] Goresky, M. and R. MacPherson: Analyse et topolgie sur les espaces singuliers (II–III), Astérisque **101–102**, 135–192 (1983)

[G-M88] Goresky, M. and R. MacPherson: *Stratified Morse Theory*, Erg. Math. 3. Folge Bnd. **14**, Springer Verlag, Berlin etc. (1988)

[G-N] Guillén, F. and V. Navarro Aznar: Sur le théorème local des cycles invariants, Duke Math. J. **61**, 133–155 (1990)

[G-N-P-P] Guillén, F., V. Navarro Aznar, P. Pascual-Gainza and F. Puerta: *Hyperrésolutions cubiques et descente cohomologique*, Springer Lecture Notes in Math. **1335**, (1988)

[Gode] Godement, R.: *Théorie des faisceaux*, Hermann, Paris (1964)

[Gr60] Grauert, H.: Ein Theorem der analytischen Garbentheorie und die Modulraüme komplexer Strukturen, Publ. Math. IHES **5**, 460–472 (1960)

[Gr] Grauert, H.: Über Modifikationen und exzeptionelle analytische Mengen, Math. Annalen **146**, 331–368 (1962)

[Greenb] Greenberg, M: *Lectures on algebraic topology*, W. A. Benjamin, Reading Mass (1967)

[Grif68] Griffiths, P.: Periods of rational integrals on algebraic manifolds, I, resp. II, Amer. J. Math. **90** 568–626, resp. 805–865 (1968)

[Grif69] Griffiths, P.: On the periods of certain rational integrals I, Ann. Math. **90**, 460–495 (1969)

[Grif70] Griffiths, P.: Periods of integrals on algebraic manifolds: Summary of main results and discussion of open problems, Bull. Amer. Math. Soc. **76**, 228–296 (1970)

[Grif-Ha] Griffiths, P. and J. Harris: *Principles of algebraic geometry*, John Wiley & Sons, New York etc. (1978)

[Grif-Mo] Griffiths, P. and J. W. Morgan: *Rational Homotopy Theory and Differential Forms*, Birkhäuser, Boston etc, Progress in Mathematics **16** (1981)

[Groth67] Grothendieck, A.: *Local Cohomology*, Lect. Notes in Math **41**, Springer Verlag, Berlin etc (1967)

[Groth69] Grothendieck, A.: Hodge's general conjecture is false for trivial reasons, Topology,**8**, 299–303 (1969)

[Gr-R71] Grauert, H. and R. Remmert: *Analytischen Stellenalgebren*, Grundl. Math. **176**, Springer Verlag, Berlin Heidelberg etc. (1971)

[Gr-R77] Grauert, H. and R. Remmert: *Theorie der Steinschen Räume*, Grundl. Math. **227**, Springer Verlag, Berlin Heidelberg etc. (1977)

[Gr-Rie] Grauert, H. and O. Riemenschneider: Verschwindungssätze für analytische Kohomologiegruppen auf komplexen Räumen, Invent. Math. **11**, 263–292 (1970)

[Gu-Ro] Gunning, R. C. and H. Rossi: *Analytic functions of several complex variables*, Prentice-Hall, Englewood Cliffs, N.J. 1965

[Hain86] Hain, R.: On the indecomposables of the bar construction, Proc. A.M.S. **98**, 312–316 (1986)

[Hain87] Hain, R.: The de Rham homotopy of complex algebraic varieties I, II, K-theory **1**, 271–324, resp. 481–497 (1987)

[Hain87b] Hain, R.: The geometry of the mixed Hodge structure on the fundamental group, Proc. Symp. Pure Math. A.M.S. **6-2**, 247–282 (1987)

[Halp] Halperin, S.: *Lectures on minimal models*, Mem. Soc. Math. Fr. **9/10** (1983)

[Hamm71] Hamm, H.: Lokale topologische Eigenschaften komplexer Räume, Math. Ann. **191**, 235–252 (1971)

[Hamm83] Hamm, H.: Lefschetz theorems for singular varieties, Proc. Symp. Pure Math. A.M.S. **40**, 547–557 (1983)

[Hart69] Hartshorne, R.: *Residues and Duality*, Lect. Notes in Math., Springer Verlag **20** (1966)

[Hart70] Hartshorne, R.: *Ample Subvarieties of Algebraic varieties*, Lect. Notes in Math., Springer Verlag **156** (1970)

[Hart75] Hartshorne, R.: Equivalence relations on algebraic cycles and Subvarieties of small codimension, Proc. Symp. Pure Math. A.M.S. **29**, 129–164 (1975)

[Hart77] Hartshorne, R.: *Algebraic Geometry*, Springer Verlag, Berlin Heidelberg etc., Graduate Texts in Mathematics **52**, (1977)

[Hatch] Hatcher, A.: *Algebraic Topology*, Cambr. Univ. Press (2002)

[Helg] Helgason, S.: *Differential Geometry and Symmetric Spaces*, Pure and Applied Math. **12**, Academic Press, Inc. New-York, London (1962)

[Hert01] Hertling, C.: Frobenius manifolds and variance of the spectral numbers, in: *New Developments in Singularity Theory*, D. Siersma et al. eds. Kluwer Academic Publishers 2001

[Hert03] Hertling, C.: *tt*∗- geometry, Frobenius manifolds, their connections, and the construction for singularities, J. reine angew. Math **555**, 7–161 (2003)

[Hir64] Hironaka, H.: Resolution of singularities of an algebraic variety of characteristic zero, Ann. Math. **79**, 109–326 (1964)

[Hir74] Hironaka, H.: Triangulation of algebraic sets, in *Algebraic Geometry Arcata 1974*, Proc. Symp. Pure Math. **29**, A.M.S. Providence R.I, 165–186 (1975)

[Ho47] Hodge, W.: *The Theory and Applications of Harmonic Integrals*, Cambridge University Press, Cambridge (1947).

[Ho50] Hodge, W.: The topological invariants of algebraic varieties, Proc. Int. Congr. Math., 182–192 (1950)

[Hu] Hu, S-T: *Homology theory. A first course in algebraic topology*, Holden-Day, Calif.-London-Amsterdam (1966)

[Ill71] Illusie, L.: Complexe cotangent et déformations I, Lecture Notes in Math. **239**, Berlin, Heidelberg, New York: Springer Verlag 1971

[Ill94] Illusie, L.: Logarithmic spaces (according to K. Kato), in *Barsotti Symposium in Algebraic Geometry* (Ed. V. Christante, W. Messing), Perspect. Math. **15**, Academic Press, 183–203 (1994)

[Is85] Ishii, S.: On isolated Gorenstein singularities, Math. Ann. **270**, 541–554 (1985)

[Is86] Ishii, S.: Small deformations of normal singularities, Math. Ann. **275**, 139–148 (1986)

[Is87] Ishii, S.: Du Bois singularities on a normal surface, in *Complex analytic singularities, Proc. Semin., Ibaraki/Jap. 1984*, Adv. Stud. Pure Math. **8**, 153–163 (1987)

[Iver] Iversen, B.: *Cohomology of Sheaves*, Springer Verlag, Berlin Heidelberg etc., Universitext (1986)

[Jann] Jannsen, U: *Mixed motives and algebraic K-theory*, Lect. Notes in Math. **1400**, Springer Verlag (1990)

[Jong] Jong, A. J., de: Smoothness, semi-stability and alterations, Publ. Math., Inst. Hautes Étud. Sci. **83** 51–93 (1996)

[Jo] Jouanolou, J.-P.: Une suite exacte de Mayer-Vietoris en K-théorie algébrique, in *Algebraic K-theory*, Lect. Notes Math. **341**, Springer Verlag, Berlin Heidelberg etc. (1973)

[Kar] Karras, U.: Local cohomology along exceptional sets, Math. Ann. **275**, 673–682 (1986)

[Kamp] Kampen, E.R. van: On the fundamental group of an algebraic curve, Am. J. Math. **55**, 255–260 (1933)

[Kash74] Kashiwara, M.: On the maximally overdetermined system of linear differential equations I. Publ. Res. Inst. Math. Kyoto Univ. **10** 563–579 (1974/75).

[Kash80] Kashiwara, M.: *Systems of microdifferential equations*, Progress in Math. **34** Birkhäuser Verlag, Boston etc. (1983)

[Kash83] Kashiwara, M.: Vanishing cycle sheaves and holonomic systems of differential equations, in *Algebraic geometry (Tokyo/Kyoto, 1982)*, Lect. Notes in Math. **1016**, Springer Verlag (1983)

[Kash86] Kashiwara, M.: A study of variation of mixed Hodge structure. Publ. Res. Inst. Math. Kyoto Univ. **2** 991–1024 (1986).

[Kash-Ka86] Kashiwara, M. and T. Kawai: Hodge Structure and Holonomic Systems, Proc. Japan Acad. **62**, Ser.A, No. 1, 1–4 (1986)

[Kash-Ka87a] Kashiwara, M. and T. Kawai: *Algebraic analysis, Papers dedicated to Professor Mikio Sato on the occasion of his sixtieth birthday, Volumes I and II*, Academic Press Inc. Boston, MA (1987)

[Kash-Ka87b] Kashiwara, M. and T. Kawai: The Poincaré lemma for variations of polarized Hodge structure, Publ. Res. Inst. Math. Sci. **23** 345–407 (1987)

[Kash-S] Kashiwara, M. and P. Schapira: *Sheaves on manifolds*, Grundl. math. Wissensch. **292**, Springer Verlag, Berlin Heidelberg etc. (1990)

[Kato88] Kato, K: Logarithmic structures of Fontaine-Illusie, in *Algebraic Analysis, Geometry and Number Theory*, J.-I. Igusa ed., Johns Hopkins Univ., 191–224 (1988)

[Katz73] Katz, N.: Pinceaux de Lefschetz: théorème d'existence SGA 7, Exposé XVII, Springer Lecture Notes in Math. **340**, 212–253 (1973)

[Katz73b] Katz, N.: Étude cohomologique des pinceaux de Lefschetz SGA 7, Exposé XVIII, Springer Lecture Notes in Math. **340**, 254–326 (1973)

[Katz-Oda] Katz, N., Oda: On the differentiation of the De Rham cohomology classes with respect to parameters, J. Math. Kyoto Univ. **1**, 199–213 (1968)

[King] King, J.: The currents defined by analytic varieties, Acta Math. **127**, 185–220 (1973)

[Kir] Kirwan, F.: *An introduction to intersection homology*, Pitman Research Notes in Mathematics **187**, Longman Scientific & Technical, Harlow (1988)

[K-K-M-S] Kempf, G., F. Knudsen, D. Mumford and B. Saint-Donat: *Toroidal embeddings I*, Springer Lect. Notes **336**, 1973)

[Kod53] Kodaira, K.: On a differential geometric method in the theory of analytic stacks, Proc. Natl. Acad. Sci. USA **39**, 1268–1273 (1953)

[Kod54] Kodaira, K.: On Kähler varieties of restricted type (an intrinsic characterization of algebraic varieties), Ann.Math. **60** 28–48, (1954)

[Kov] Kovács, S.: Rational, Log Canonical, Du Bois Singularities: On the Conjectures of Kollár and Steenbrink, Comp. Math. **188**, 123–133 (1999)

[Ku] Kulikov, Va.: *Mixed Hodge structures and singularities*, Cambridge tracts in Math **132**, Cambridge University Press, Cambridge (1998)

[Lamot] Lamotke, K.: The topology of complex projective varieties after S. Lefschetz, Topology **20**, 15–52 (1981)

[Lang] Lang, S.: *Differential Manifolds*, Springer Verlag, Berlin etc.(1985)

[Lef] Lefschetz, S.: *L'Analyse Situs et la Géométrie Algébrique*, Gauthier-Villars (1924)

[Lel] Lelong, P.: Intégation sur un ensemble analytique complexe, Bull. Soc. Math. France **85**, 239–262 (1957)

[Leray] Leray, J.: Le calcul différentiel et intégral sur une variété analytique complexe (Problème de Cauchy III), Bull. Soc. Math. France **87**, 81–180 (1959)

[Le78] Lê Dung Tráng: The geometry of the monodromy theorem, in *C.P. Ramanujam. - A tribute. Collect. Publ. of C.P. Ramanujam and Pap. in his Mem.*, Tata Inst. fundam. Res., Stud. Math. **8**, 157–173 (1978)

[Le79] Lê Dung Tráng: Sur les cycles évanouissants des espaces analytiques, C.R. Acad. Sc. Paris, Ser A. 283–285 (1979)

[Le88] Lê Dung Tráng: Singularités isolées des intersections complètes, in *Introduction la théorie des singularités, I*, Travaux en Cours **36**, Hermann, Paris,1–48 (1988)

[Let] Letitia, M.: The Abel-Jacobi mapping for the quartic threefold, Invent. Math. **75**, 477–492 (1984)

[Lewis] Lewis, J.D.: *A survey of the Hodge conjecture with an appendix by B. Brent Gordon* CRM monograph series **10**, A.M.S. Providence, (1999)

[Lo] Looijenga, E.: Motivic measures. Séminaire Bourbaki, 52ème année, 1999-2000, no. 874

[Ma] Manin, Y.: Moduli Fuchsiani, Ann. Sc. Norm. Sup. Pisa Ser III **19**, 113–126 (1965)

[Malg74] Malgrange, B.: Intégrales asymptotiques et monodromie, Ann. Sc. ENS **7**, 405–430 (1974)

[Malg79] Malgrange, B.: L'involutivité des caractéristiques des systèmes différentiels et microdifférentiels. Sém. Bourbaki **522** (1977–1978), Lect. Notes in Math. **710**, 277-289, Springer Verlag, Berlin etc. (1979)

[Malg83] Malgrange, B.: Polynômes de Bernstein-Sato et cohomologie évanescente, in *Analysis and topology on singular spaces, II, III (Luminy, 1981)*, Astérisque **101**, 243–267, Soc. Math. France, (1983)

[Malg94] Malgrange, B.: Connexions méromorphes, in *Congrès Singularités, Lille 1991*, Cambridge Univ. Press, Cambridge etc., 251–261 (1994)

[Ma-Sa] Maisonobe, P. and C. Sabbah: *Aspects of the theory of $\mathcal{D}_X$-modules*, Lect. Notes Kaiserslautern 2002, `http://math.polytechnique.fr/cmat/sabbah/sabbah.html`

[MC] McCrory, C.: Massey products in singularity links, Duke Math. J. **51**, 691–697 (1984)

[Mil63] Milnor, J.: *Morse theory*, Princeton Univ. Press (1963)

[Mil68] Milnor, J.: *Singular points of complex hypersurfaces*, Princeton Univ. Press (1968)

[Mil-Mo] Milnor, J., J. Moore: On the structure of Hopf algebras. Ann. Math. **81**, 137–204 (1965)

[Mor-Ko] Morrow, J. and K. Kodaira: *Complex manifolds*, Holt-Rinehart & Winston, New-York (1971)

[Mor] Morgan, J.W.: The algebraic topology of smooth algebraic varieties, Publ. Math. I.H.E.S. **48**, 137–204 (1978) Correction, Publ. Math. I.H.E.S.,**64**, 185 (1986)

[Munk] Munkres, E. : *Elementary differential topology*, Princeton University Press, Princeton Ann. Math. Studies **54** (1963)

[Naga] Nagata, M.: Imbedding of an abstract variety in a complete variety, J. Math. Kyoto Univ. **2**, 1–10 (1962)

[Nar] Narasimhan, R.: *Analysis on real and complex manifolds*, North Holland, Amsterdam (1973)

[Nav] Navarro Aznar, V.: Sur la théorie de Hodge des variétés algébriques à singularités isolées, in: *Systèmes différentiels et singularités*, Astérisque **130**, 272–307 (1985)

[Nav87] Navarro Aznar, V.: Sur la théorie de Hodge-Deligne, Inv. Math. **90** 11–76 (1987)

[Nori] Nori, M.: Constructible sheaves, in *Proc. Int. Conf. on Algebra*, TIFR, Bombay (2002)

[Oda] Oda, T.: *Convex bodies in algebraic geometry*, Springer Verlag, Berlin, Heidelberg etc. (1988)

[Pham] Pham, F.: *Singularités des systèmes différentiels de Gauss-Manin*, Progr. in Math. **2**, Birkhäuser Verlag (1979)

[Pul] Pulte, M.: The fundamental group of a Riemann surface: Mixed Hodge structures and algebraic cycles, Duke Math. J. **57**, 721–760 (1988)

[Quil86] Quillen, D.: Rational homotopy theory, Ann. Math. **90**, 205–295 (1969)

[Quil72] Quillen, D.: *Higher Algebraic K-theory. I*, Lecture Notes in Mathematics **341**, 85–147 (1972)

[Ra] Ran, Z.: Cycles on Fermat hypersurfaces, Comp.Math., **42** 121-142 (1980)

[R-S-S] Rapoport, M., N. Schapacher and P. Schneider: *Beilinson's Conjectures and Special Values of L-functions*, Perspectives in Math. **4** Academic Press, Boston etc. (1988)

[R-R-V] Ramis, J.-P., Ruget, G. and J.-L. Verdier: Dualité relative en géométrie analytique complexe, Invent. Math. **13**, 261–283 (1971)

[Reid] Reid, M.: Young person's guide to canonical singularities, in *Algebraic Geometry (Bowdoin 1985)*, Proc. Symp. Pure Math. **46**, A.M.S. Providence, 345–416 (1987)

[Sa83] Saito, M.: On the exponents and the geometric genus of an isolated hypersurface singularity, Proc. Symp. Pure Math. **40** Part 2, A.M.S. Providence, 465–472 (1983)

[Sa85] Saito, M.: Hodge structure via filtered **D**-modules, in: Astérisque **130** 342–351 (1985)

[Sa87] Saito, M.: Introduction to Mixed Hodge modules, Actes du Colloque de Théorie de Hodge (Luminy, 1987), Astérisque **179-180** 145–162 (1989)

[Sa88] Saito, M.: Modules de Hodge polarisables, Publ. RIMS. Kyoto Univ. **24** 849–995 (1988)

[Sa89] Saito, M.: Mixed Hodge modules and admissible variations, C. R. Acad. Sci. Paris Sér. I Math. **309** 351–356 (1989)

[Sa89b] Saito, M.: Duality for vanishing cycle functors, Publ. Res. Inst. Math. Sci. **25** 889–921 (1989)

[Sa89c] Saito, M.: Induced $\mathcal{D}$-modules and differential complexes, Bull. Soc. Math. France **117** 361–387 (1989)

[Sa90] Saito, M.: Mixed Hodge Modules, Publ. Res. Inst. Math. Sci. **26** 221–333 (1990)

[Sa00] Saito, M.: Mixed Hodge complexes on algebraic varieties, Math. Ann. **316**, 283–331 (2000)

[Sata56] Satake, I.: On a generalization of the notion of manifold, Proc. Natl. Acad. Sci. USA **42**, 359–363 (1956)

[Sata80] Satake, I.: *Algebraic Structures of Symmetric Domains*, Princeton University Press, Princeton (1980)

[Sch73] Schmid, W.: Variation of Hodge structure: the singularities of the period mapping, Invent. Math. **22**, 211–319 (1973)

[SchS85] Scherk, J. and J.H.M. Steenbrink: On the Mixed Hodge Structure on the Cohomology of the Milnor Fibre, Mathematische Annalen **271**, 641–665 (1985)

[Se65] Serre, J.-P.: *Lie Algebras and Lie Groups*, Springer Lect. Notes in Math. **1500** (1965)

[Shio79] Shioda, T.: The Hodge conjecture for Fermat varieties, Math. Ann. **245**, 175–184 (1979)

[Shio81] Shioda, T.: Algebraic cycles on abelian varieties of Fermat type, Math. Ann. **258**, 65–80 (1981)

[Shio83] Shioda, T.: What is known about the Hodge conjecture? In: *Advanced studies in Pure Math.* **1**, Kinokuniya Comp., Tokyo and North Holland Publ. Comp. Amsterdam, 55–68 (1983)

[Shi-Ka] Shioda, T. and T. Katsura: On Fermat varieties, Tohoku Math. J. **31**, 97–115 (1979)

[Si92] Simpson, C.: Higgs bundles and local systems, Publ. Math. IHES **75**, 5–95 (1992)

[Si94] Simpson, C.: Moduli of representations of the fundamental group of a smooth variety, Publ. Math. IHES **79** 47–129 (1994)

[Si95] Simpson, C.: Moduli of representations of the fundamental group of a smooth projective variety, II, Publ. Math. IHES **80**, 5–79 (1995)

[Span] Spanier, E.: *Algebraic Topology*, Springer Verlag Berlin, Heidelberg, etc. (1966)

[S-T] Siu, Y.-T. and G. Trautmann: *Gap-sheaves and extensions of coherent analytic sheaves*, Lect. Notes **172** (Springer Verlag Berlin, Heidelberg, etc. (1971)

[Stal] Stallings, J. : Quotients of the powers of the augmentation ideal in a group ring, in *Knots, Groups and 3-Manifolds, Papers Dedicated to the Memory of R. H. Fox*, ed. L. Neuwirth, Princeton Univ. Press (1975)

[Ste76] Steenbrink, J.: Limits of Hodge structures, Inv. Math. **31**, 229–257 (1976)

[Ste77a] Steenbrink, J.: Mixed Hodge structures on the vanishing cohomology, in *Real and Complex Singularities, Oslo, 1976*, Sijthoff-Noordhoff, Alphen a/d Rijn, 525–563 (1977)

[Ste77b] Steenbrink, J.: Intersection form for quasi homogeneous singularities, Comp. Math. **34**, 211–223 (1977)

[Ste81] Steenbrink, J.: Cohomologically insignificant degenerations, Comp. Math. **42**, 315–320 (1981)

[Ste83] Steenbrink, J.: Mixed Hodge structures associated with isolated singularities. Singularities, in *Singularities, Part 2, Arcata 1981*, Proc. Symp. Pure Math. **40**, A.M.S. Providence R.ZI, 513–536 (1983)

[Ste85] Steenbrink, J.: Semicontinuity of the singularity spectrum, Inventiones math. **79**, 557–566 (1985)

[Ste85a] Steenbrink, J.: Vanishing theorems on singular spaces, in *Systèmes différentiels et singularités, Colloq. Luminy/France 1983*, Astérisque **130**,330–341 (1985)

[Ste87] Steenbrink, J.: Some remarks about the Hodge conjecture, in *Hodge Theory. Proceedings of the U.S.-Spain Workshop, Sant Cugat 1985*, Springer Lecture Notes in Math. **1246**, 165–175 (1987)

[Ste95bis] Steenbrink, J.: Monodromy and weight filtration for smoothings of isolated singularities, Compos. Math. **97**, 285–293 (1995)

[Ste95] Steenbrink, J: Logarithmic embeddings of varieties with normal crossings and mixed Hodge structures, Math. Ann. **301**, 105–118 (1985)

[Ste-St] Steenbrink, J.H.M. and J. Stevens: Topological invariance of the weight filtration, Indag. Math. **87**, 63–76 (1984)

[Str] Straten, D. van: *Weakly normal surface singularities and their improvement*, Ph. D. Thesis, Leiden (1987)

[St-Z] Steenbrink, J. and S. Zucker: Variation of mixed Hodge structure I, Invent. Math. **80**, 489–542 (1985)

[Sull] Sullivan, D.: Infinitesimal computations in topology, Publ. Math. IHES **47**, 269–331 (1977)

[Tj72] Tjurin, A.N.: Five lectures on three dimensional varieties, Russian Math. Surveys **27**, 1–53 (1972)

[Tj75] Tjurin, A.N.: On intersections of quadrics, Russian Math. Surveys **30**, 51–105 (1975)

[Var80] Varchenko, A.N.: Asymptotics of holomorphic forms define mixed Hodge structure, Dokl. Akad. Nauk SSSR **22**, 1035–1038 (1980).

[Var81] Varchenko, A.N.: Asymptotic mixed Hodge structure in the vanishing cohomology. Izv. Akad. Nauk SSSR, Ser. Mat. **45**, 540–591 (1981) (in Russian). [English transl.: Math. USSR Izvestija **18:3**, 469–512 (1982)]

[Var83] Varchenko, A.N.: On semicontinuity of the spectrum and an upper bound for the number of singular points of propjective hypersurfaces. Dokl. Akad. Nauk. **270** 1294–1297 (1983) (in Russian) [English transl.: Sov. Math. Dokl. **27**, 735–739 (1983)]

[Verd77] Verdier, J.-L.: Catégories Dérivées (Etat 0), in: SGA $4\frac{1}{2}$, Springer Lecture Notes in Math. **569**, 262–311 (1977)

[Verd96] Verdier, J.-L.: *Des catégories dérivées des catégories abéliennes*, Soc. Math. de France, Astérisque **239** (1996)

[Vois02] Voisin, C.: A counterexample to the Hodge conjecture extended to Kähler varieties, Int. Math. Res. Notes **20**, 1057–1075 (2002)

[Vois07] Voisin, C.: Hodge loci and absolute Hodge classes, Preprint Inst. Math. de Jussieu, 2007.

[Wang] Wang, H.C.: The homology groups of the fibre bundles over a sphere, Duke math. J. **16**, 33–38 (1949)

[Warn] Warner, F.W.: *Foundations of Differentiable Manifolds and Lie groups*, Springer Verlag Berlin, Heidelberg, etc. (1983)

[Weib] Weibel. C.: *An introduction to homological algebra*, Cambridge Univ. Press (1994)

[Weil] Weil, A.: *Variétés kählériennes*, Hermann, Paris (1971)

[Wells] Wells, R.: *Differential Analysis on Complex Manifolds*, Springer Verlag Berlin, Heidelberg, etc. (1980)

[Wh] Whitehead, G. W. : *Elements of homotopy theory*, Graduate Texts in Mathematics **61**, Springer-Verlag , New York, Heidelberg, Berlin (1978)

[Yone] Yoneda, N.: On Ext and exact sequences, J. Fac. Sci. Univ. Tokyo. Sec. I. **8**, 507–576 (1960)

[Zuc76] Zucker, S.: On the degeneration of the Leray spectral sequence in algebraic geometry, preprint (1976).

[Zuc77] Zucker, S.: The Hodge conjecture for cubic fourfolds, Comp. Math. **34**, 199–209 (1977)

[Zuc79] Zucker, S.: Hodge theory with degenerating coefficients: L_2-cohomology in the Poincaré-metric, Ann. Math. **109**, 415–476 (1979)

[Zuc85] Zucker, S.: Variation of mixed Hodge structure II, Inv. Math. **80**, 543–565 (1985)

Index of Notations

Index

SPRINGER NATURE

GPSR Compliance

The European Union's (EU) General Product Safety Regulation (GPSR) is a set of rules that requires consumer products to be safe and our obligations to ensure this.

If you have any concerns about our products, you can contact us on ProductSafety@springernature.com

In case Publisher is established outside the EU, the EU authorized representative is:

Springer Nature Customer Service Center GmbH
Europaplatz 3
69115 Heidelberg, Germany

Kolibri 360 GmbH
Ferdinand-Jühlke-Straße 7
99095 Erfurt, Deutschland
produktsicherheit@kolibri360.de